Soft Chemistry Routes to New Materials

Soft Chemistry Routes to New Materials

- *Chimie Douce* -

Proceedings of the International Symposium
held in Nantes, France, September 6-10, 1993

Edited by

J. Rouxel, M. Tournoux and R. Brec

TRANS TECH PUBLICATIONS
Switzerland - Germany - UK - USA

Seplae
Chem

Front cover illustration (book edition):

Soft Chemistry deals with intercalation-deintercalation reactions (lower left), grafting-pillaring (lower right), polycondensation of solid units or molecules through acid-base reactions and sol-gel processes (lower middle). The latter can end up with mixed inorganic organic constructions (upper part).

TP155
.7
S64
1994
CHEM

Copyright © 1994 Trans Tech Publications Ltd, Switzerland

ISBN 0-87849-677-7

Volumes 152-153 of
Materials Science Forum
ISSN 0255-5476

Distributed *by*

Trans Tech Publications Ltd
Hardstrasse 13
CH-4714 Aedermannsdorf
Switzerland
Fax: (++41) 62 74 10 53

CHIMIE DOUCE
SOFT CHEMISTRY ROUTES TO NEW MATERIALS

ORGANISATEURS
(CHAIRMEN OF THE ORGANIZATION)

Jean Rouxel ⎫

Michel Tournoux ⎬ Institut des Matériaux de Nantes

Raymond Brec ⎭

COMITE SCIENTIFIQUE INTERNATIONAL
(INTERNATIONAL ADVISORY COMMITTEE)

M. Armand	Ionique et Electrochimie du Solide - Grenoble	France
A. Clearfield	Texas A and M University - College Station	U.S.A.
F. Di Salvo	Baker Chemistry Laboratory - Cornell	U.S.A.
M. Figlarz	Réactivité et Chimie des Solides - Amiens	France
P. Hagenmuller	Chimie du Solide - Bordeaux	France
F. Kanamaru	ISIR - Osaka	Japan
J. Livage	Chimie de la Matière Condensée - Paris	France
P. Mac Millan	Arizona state University - Tempe	U.S.A.
D. Murphy	AT & T - Murray Hill	U.S.A.
P. Pinnavaia	Department of Chemistry - East Lansing	U.S.A.
R. Schöllhorn	Anorganische und analytical Chemie - Berlin	Germany
J. Thomas	The Royal Institution of G.B. - London	U.K.
A.A.G. Tomlinson	ITSE - CNR - Roma	Italy
H. Van Damme	CRSOCI - Orléans	France
S. Whittingham	Material Research Center - Binghamton	U.S.A.

COMITE LOCAL D'ORGANISATION
(LOCAL ORGANIZING COMMITTEE)

Luc Brohan,	Victoria Cajipe,	Michel Danot,
Philippe Deniard,	Michel Evain,	Marcel Ganne,
Marie-Pierre Guilbaud,	Dominique Guyomard,	Stéphane Jobic,
Alain Meerschaut,	Guy Ouvrard,	Yves Piffard,
Eric Prouzet,	Armelle Radigois,	Jean-Charles Ricquier

SOUTIEN FINANCIER

(Sponsors)

C.N.R.S. (Centre National de la Recherche Scientifique)
D.R.E.T. (Ministère de la Défense)
Ministère de l'Enseignement Supérieur et de la Recherche
Région Pays de la Loire
Département de Loire-Atlantique
Mairie de Nantes
Atlantech, Nantes
Atlanpole, Nantes
La Communauté Européenne et COMETT-OUEST
Rhône-Poulenc, France
Saint-Gobain, France
Saft-Leclanché, France

PREFACE

Within the last few decades, solid-state chemistry has often aimed to establish the relationships between the structures of compounds that are prepared and their physical properties. From the point of view of preparative chemistry, various classical methods have long been used, involving heating, grinding, or powder-pressing, in order to obtain compounds formed at thermodynamic equilibrium. At the same time, crystal growth techniques have been well-developed. Thus, the conditions have been ideal for unravelling the structures of solid-state compounds, measuring their physical properties, and applying suitable theoretical models. These advances have resulted in substantial progress towards a better understanding of the solid state, and have even led to the development of some predictive capabilities in crystal chemistry.

Nonetheless, in more recent years, significant changes have occurred. Growing emphasis has been placed on finding novel preparative routes. Truly, a profound renaissance has taken place in preparative methods of solid-state compounds. Most frequently, these procedures involve low or medium range temperatures. Sometimes the motivation to find new synthetic methods has been driven by industrial aims, such as the preparation of finely dispersed powders. Often, however, entirely new ways of thinking have appeared. A good case in point is desintercalation chemistry, the converse proposition to concepts developed in intercalation chemistry since 1960. The acido-basic chemistry of the condensation of structural blocks through local protonation, which yields for example $TiO_2(B)$, can be regarded as a transposition to the solid state of the processes that lead to polycation formation in solution. Other procedures represent, still more closely, updated or improved versions of older techniques. Sometimes there have also been a few lucky cases, subsequently examined in closer detail, that have led to new synthetic routes. The term "*chimie douce*", or soft chemistry, is often applied in a general manner to refer to these new routes, these new ways of thinking.

In any case, much work has now been carried out in different directions, and, for many of us, it seemed to be an appropriate moment to pause and take stock of all that has been done, and try to formulate some concepts or, at least, some general guidelines in "*chimie douce*". This was the purpose of this meeting.

The field of *chimie douce* is not restricted to a well-defined area of science. While sol-gel processes represent one essential component of soft chemistry, other reactions dealing with intercalation-deintercalation, pillaring-grafting, exchange-condensation -- are no less important. The unifying theme behind these various processes is a common scientific approach, similar methods of characterization, and analogous problematics related to metastability and phase transitions. Perhaps the term "soft chemistry" should be regarded more as a way of thinking, a new paradigm in solid state chemistry.

In all cases, it is the precursor compound that holds considerable importance. Chemists working on sol-gel processes thus devote a major portion of their effort to designing specific new precursors, or at least modifying existing ones. In particular, they may vary the nature and the configuration of ligands in organometallic compounds. Doing this, a reaction is accelerated or driven in a particular direction. Alternatively, organic "arms" may be generated that are capable of linking inorganic aggregates together ; this mixed organic-inorganic solid state chemistry is currently in rapid development and has already led to the preparation of ormosils, ormoglasses, and other ormocers. When a solid precursor is used in soft chemistry, the nature of this precursor will also influence the rate of reaction and indeed whether the reaction will take place at all. Moreover, the soft chemistry here is largely topotactic : the final product retains the memory of the precursor structure. For example, in the acido-basic processes of exchange-reconstruction, the condensation of blocks takes place at the protonated sites that are the most basic anionic sites. But this bacisity depends on the geometry of the sites, the distance to the neighboring cations, and the nature of these cations. Thus the modifying chemistry, performed at a preliminary stage, is crucial in controlling the reactivity of the precursor.

In the examples above, the condensation processes lead to an increase in dimensionality of the resulting product. A three-dimensional structure, $TiO_2(B)$, can be obtained from a layered titanate. This is also the case in sol-gel processes that involve multiple condensation steps, either simultaneous or successive. Likewise, the dimensionality increases in pillaring-grafting reactions, which represent another possible way of carrying out organic-inorganic chemistry by fixing, for example, the organic pillars in an inorganic matrix. In contrast, deintercalation chemistry leads to a decrease in dimensionality, or at least to an extrication of the tunnels or cages in a three-dimensional framework ; e.g. layered VS_2 can be prepared from $LiVS_2$.

Chimie Douce is certainly not easy, especially when it implies a redox process. This point is well illustrated by deintercalation reactions. Consider the framework A_xTY_n (A = alkali metal or Cu ; T = transition metal ; Y = O, S, Se, Te) : if A is removed by chemical or electrochemical routes, it may turn out that the desired framework TY_n is unstable even at room temperature, at least without some modification of the initial framework, as often occurs when the framework bond ionicity is above a critical value. The process also involves an oxidation of the framework, and, in this case, for elements T situated at the right-hand part of the periodic table, one may suspect that oxidation of T (whose d-levels are low) is in competition with the formation of anion pairs, at least when Y = S, Se, Te (whose sp levels are relatively high). As well, this process may entail some structural modifications that we have perhaps not clearly recognized until now.

Because it is a chemistry that takes place at low or medium range temperatures, *Chimie Douce* often leads to phases that are amorphous or poorly-ordered. However, the study of these phases has been made possible through the development and proliferation of analytical techniques that probe at a more local level. These techniques, such as EXAFS, XANES, and NMR, allow rational steps to be made in soft chemistry and aid the researcher in reaching a fuller understanding thereof. It can be said that without them *Chimie Douce* will not have developed so rapidly and so efficiently.

In this book, the introductory remarks made above will be illustred through numerous examples. As well, it will be seen that the field of soft chemistry is still much broader. Most importantly, it should be noted that there has truly been a profound renewal in the methods of synthesis in solid state chemistry and as such, there has also been a real change in the scientific thought process itself.

———————————

TABLE OF CONTENTS

Preface   ix

Colloidal Dispersions of Compounds with Layer and Chain Structures
A.J. Jacobson   1

Ambient Temperature Solid State Reactions in Battery Electrodes
J.O. Besenhard   13

Reactive Flux Syntheses at Low and High Temperatures
J.A. Cody, M.F. Mansuetto, S. Chien and J.A. Ibers   35

The Sol-Gel Route to Advanced Materials
J. Livage   43

Soft Chemistry: Thermodynamic and Structural Aspects
M. Figlarz   55

New Functional Solids derived from Layer Structured Crystals by Pillaring and Grafting
S. Yamanaka   69

Application of Host-Guest Reactions to Synthesis of New Materials: The Case of S.T.A. Project in Japan
M. Watanabe   81

Layered and Pillared Zirconium Phosphonates with α- and γ-Structures
G. Alberti, S. Murcia Mascarós and R. Vivani   87

Hydrothermal Synthesis of New Oxide Materials using the Tetramethyl Ammonium Ion
M.S. Whittingham, J. Li, J.D. Guo and P. Zavalij   99

A Pillared Rectorite Clay With Highly Stable Supergalleries
J. Guan and T.J. Pinnavaia   109

Synthesis of Novel Metal Phosphonate Complex Structures through Soft Chemistry
   E.W. Stein, Sr., C. Bhardwaj, C.Y. Ortiz-Avila, A. Clearfield and
   M.A. Subramanian   115

Structural Consequences of Synthesis Parameters for Oxyfluorinated Microporous Compounds
   G. Férey, T. Loiseau and D. Riou   125

The NiO$_2$ Slab: A Very Convenient Structural Unit for Chimie Douce Reactions
   C. Delmas   131

Metal Chalcogenide Clusters, M$_x$E$_y$: Generation and Structure
   I. Dance and K. Fisher   137

Some Chalcogenides Syntheses via Soft Chemistry
   G. Ouvrard, E. Prouzet, R. Brec and J. Rouxel   143

Nanophase Ceramics by the Sol-Gel Process
   C. Guizard, C. Mouchet, J.C. Achddou, S. Durand, J. Rouvière
   and L. Cot   149

Physico-Chemical Properties of Solid One-Dimensional Inclusion Compounds
   K.D.M. Harris   155

Soft Chemistry Routes to Oxide Catalysts
   K.R. Poeppelmeier and D.C. Tomczak   163

Acid-Base Routes in Soft Chemistry Involving Prefabricated Oxides
   M. Tournoux   169

Synthesis of Novel Metal Oxides by Soft-Chemistry Routes
   J. Gopalakrishnan, S. Uma, K. Kasthuri Rangan and N.S.P. Bhuvanesh   175

Ternary Nitride Synthesis: Ammonolysis of Ternary Oxide Precursors
   D.S. Bem, J.D. Houmes and H.-C. zur Loye   183

Room Temperature Tailoring of Electrical Properties of Semi- and Superconductors via Controlled Ion Migration
   D. Cahen, L. Chernyak, K. Gartsman, I. Lyubomirsky, Y. Scolnik,
   O. Stafsudd and R. Triboulet   187

Electrochemical Oxidation of La$_2$CuO$_4$. Phase Equilibria Among Superconductors
    S. Ondoño-Castillo, P. Gomez-Romero, A. Fuertes and
    N. Casañ-Pastor   193

Obtention of Superconducting YBa$_2$Cu$_3$O$_{7-\delta}$ Deposits by Electrodeposition and Electrophoresis
    S. Ondoño-Castillo, J. Bassas, P. Gomez-Romero, A. Fuertes and
    N. Casañ-Pastor   197

Interstratification in the Substituted Nickel Hydroxides
    Y. Borthomieu, L. Demourgues-Guerlou and C. Delmas   201

Low Temperature Synthesis of PbMo$_6$S$_8$ Superconducting Material
    S. Even-Boudjada, L. Burel, V. Bouquet, R. Chevrel, M. Sergent,
    J.C. Jegaden, J. Cors and M. Decroux   205

**Tert-Butyllithium, a Selective and Convenient Reagent for the Reductive Intercalation of Interstitial Solids:
A Novel Room Temperature Synthesis of (LiMo$_3$Se$_3$)$_n$, a Soluble Polymeric Chevrel Cluster Compound**
    J.H. Golden, F.J. DiSalvo and J.M.J. Fréchet   209

Lithium Deintercalation in the Spinel LiMn$_2$O$_4$
    F. Coowar, J.M. Tarascon, W.R. McKinnon and D. Guyomard   213

Mechanisms of the Reversible Electrochemical Insertion of Lithium Occuring with NCIM$_s$ (Nano-Crystallite-Insertion-Materials)
    S.D. Han, N. Treuil, G. Campet, J. Portier, C. Delmas, J.C. Lassègues
    and A. Pierre   217

Elaboration by Reaction in Molten Nitrates and Characterization of Pure Titanium Oxide with Large Surface Area
    V. Harlé, J.P. Deloume, L. Mosoni, B. Durand, M. Vrinat and
    M. Breysse   221

Binary and Ternary Telluride Solids from the Oxidation of Zintl Anions in Solution
    C.J. O'Connor, B. Wu and Y.S. Lee   237

Synthesis, Characterization and Properties of new Silicoantimonic Acids: H$_4$Sb$_4$O$_8$(Si$_4$O$_{12}$)·xH$_2$O and H$_3$Sb$_3$O$_6$(Si$_2$O$_7$)·xH$_2$O
    C. Pagnoux, A. Verbaere, F. Taulelle, M. Suchaud, Y. Piffard and
    M. Tournoux   241

Investigations on Layered Perovskites: $Na_2Nd_2Ti_3O_{10}$, $H_2Nd_2Ti_3O_{10}$ and $Nd_2Ti_3O_9\square$

M. Richard, L.Brohan and M. Tournoux   245

A Protonated Form of the Titanate with the Lepidocrocite-Related Layer Structure

T. Sasaki, M. Watanabe, Y. Fujiki and S. Takenouchi   251

The Layered Phosphatoniobic Acid $HNbP_2O_7 \cdot xH_2O$: Synthesis, Structure, Thermal Behavior and Ion Exchange Properties

J.J. Zah Letho, P. Houenou, A. Verbaere, Y. Piffard and
M. Tournoux   255

New Phases Obtained by Acid Delithiation of Layered $LiMO_2$ (M=Co,Ni)

E. Zhecheva and R. Stoyanova   259

Titanium-Carbon Coatings Prepared by Chemical Method at Mild Conditions

M. Wysiecki, A. Biedunkiewicz, W. Jasinski, S. Lenart and
A.W. Morawski   263

Synthesis and Characterization of Silica Gels Obtained in Lamellar Media

T. Dabadie, A. Ayral, C. Guizard, L. Cot, C. Lurin, W. Nie and
D. Rioult   267

Is Soft Chemistry Always so Soft ?

N. Allali, J.F. Favard, M. Rambaud, A. Goloub and M. Danot   271

Synthesis, Structure and Reactions of Oxides with the Hexagonal MoO_3 Structure

Y. Hu and P.K. Davies   277

Exfoliation of Graphite Intercalation Compounds: Classification and Discussion of the Processes from New Experimental Data Relative to Graphite-Acid Compounds

A. Hérold, D. Petitjean, G. Furdin and M. Klatt   281

Novel Properties of $(LiMo_3Se_3)_n$, a Polymeric Chevrel Cluster Compound: Thin Films of Macromolecular Wires Cast from Solution

J.H. Golden, F.J. DiSalvo and J.M.J. Fréchet   289

Reactions of Synthesis and Reduction of Mixed Cu-Ti Perovskites: Structure and Order-Disorder Characteristics

M.R. Palacín, A. Fuertes, N. Casañ-Pastor and P. Gómez-Romero   293

Structural Study of a New Iron Vanadium Oxide $Fe_{0.12}V_2O_{5.15}$ Synthesized via a Sol-Gel Process

S. Maingot, Ph. Deniard, N. Baffier, J.P. Pereira-Ramos, A. Kahn-Harari and R. Brec   297

Synthesis and Acid-Base Properties of "ALPON" Nitrided Aluminophosphates

R. Conanec, R. Marchand, Y. Laurent, Ph. Bastians and P. Grange   305

Control of Porosity and Surface Area in Sol-Gel Prepared TiO_2-Al_2O_3 Mixed Oxides by Means of Organic Solvents

T.E. Klimova and J.R. Solis   309

Chemistry of Hybrid Organic-Inorganic Materials Synthesized via Sol-Gel

C. Sanchez, F. Babonneau, F. Banse, S. Doeuff-Barboux, M. In and F. Ribot   313

A New Soft Chemistry Synthesized Vanadium Oxysulfide

G. Tchangbédji, E. Prouzet and G. Ouvrard   319

Lithium Doping of Cobalt-Nickel Spinel Oxides at Low Temperature

E. Zhecheva and R. Stoyanova   323

Preparation of Solids Constituted of Nanocrystallites by Reactions in Molten Salts

B. Durand, J.-P. Deloume and M. Vrinat   327

Solution Synthesis of Vanadium and Titanium Phosphates

A. Ennaciri and P. Barboux   331

Synthesis and Structure of Calcium Aluminate Hydrates Intercalated by Aromatic Sulfonates

V. Fernon, A. Vichot, P. Colombet, H. Van Damme and F. Béguin   335

Preparation and Morphological Characterization of Fine, Spherical, Monodisperse Particles of ZnO

D. Jezequel, J. Guenot, N. Jouini and F. Fievet   339

Intercalation of Organic Pillars in [Zn-Al] and [Zn-Cr] Layered Double Hydroxides
M. Guenane, C. Forano and J.P. Besse   343

A New Variety of Micronic Graphite and the Reduction of Its Intercalation Compounds
C. Hérold, J.F. Marêché, A. Mabchour and G. Furdin   347

Novel Preparation of CdS Nanocrystals in a Sodium Borosilicate Glassy Matrix
W. Granier, L. Boudes, A. Pradel, M. Ribes, J. Allegre, G. Arnaud,
P. Lefebvre and H. Mathieu   351

Partial Charges Distributions in Crystalline Materials through Electronegativity Equalization
M. Henry   355

Ruthenium Species in Layered Compounds
Part 1. The Direct Self-Catalysed Intercalation of Cationic Ruthenium(II) Ammine Complexes from Aqueous Media into Alpha-Tin(IV) Hydrogen Phosphate
M.J. Hudson and A.D. Workman   359

Synthesis, Structure and Magnetic Properties of some New Metal(II) Phosphonates: Layered $Fe(C_2H_5PO_3) \cdot H_2O$ and α-$Cu(C_2H_5PO_3)$, Tubular β-$Cu(CH_3PO_3)$
B. Bujoli, J. Le Bideau, C. Payen, P. Palvadeau, and J. Rouxel   365

Hard-Soft Chemistry in New Materials Synthesis
P.F. McMillan, W.T. Petuskey, C.A. Angell, J.R. Holloway,
G.H. Wolf, M. O'Keeffe and O. Sankey   371

Vanadate-Pillared Hydrotalcite Containing Transition Metal Cations
F. Kooli, V. Rives and M.A. Ulibarri   375

Intercalation of Thionine in Colloidal α-Zirconium Phosphate
E. Rodríguez-Castellón, A. Jiménez-López, P. Olivera-Pastor,
J.M. Mérida-Robles, F.J. Pérez-Reina, M. Alcántara-Rodríguez,
F.A. Souto-Bachiller, L. de los A. Rodríguez-Rodríguez and
G.G. Siegel   379

$TaNi_{2.05}Te_3$ and $Ta_2Ni_3Se_5$, New Metal-Rich Ternary Tantalum Chalcogenides
J. Neuhausen, W. Tremel and R.K. Kremer   383

Non-Stoichiometry and Solid State in Spinel-Type Catalysts
    F. Trifirò and A. Vaccari   387

Synthesis and Properties of High Surface Area Ni/Mg/Al Mixed Oxides via Anionic Clay Precursors
    D. Matteuzzi, F. Trifirò, A. Vaccari, M. Gazzano and O. Clause   391

Author Index   395

Subject Index   399

Materials Science Forum Vols. 152 - 153 (1994) pp. 1-12
© 1994 Trans Tech Publications, Switzerland

COLLOIDAL DISPERSIONS OF COMPOUNDS WITH LAYER AND CHAIN STRUCTURES

A.J. Jacobson

University of Houston, Department of Chemistry, Houston, Texas 77204-5641, USA

Keywords: Colloidal Dispersions, Exfoliation, Layered and Chain Structures, Transition Metal Chalcogenides, Intercalation

ABSTRACT:
Stable colloidal dispersions of several classes of compounds with layered or chain structures can be prepared by suitable manipulation of the interlayer or interchain interactions. The dispersion phenomenon is best known for the smectite clays which spontaneously exfoliate in water to form sols or gels depending on the concentration of colloidal particles. Such dispersions have been shown to contain significant concentrations of single layers or isolated chains and can be used in synthesis of new intercalation compounds and other materials. In this paper, general methods for the preparation of dispersions and aspects of their physical and chemical behavior will be reviewed with reference to specific examples.

INTRODUCTION
The anisotropy in chemical bonding in low dimensional solids with layer or chain structures is reflected in their chemical reactivity. The separation between individual layers or chains can be increased by the intercalation or exchange of molecules or ions. The limiting case of such lattice expansion is when the individual structural elements completely disperse to form colloidal solutions containing highly anisotropic particles and significant numbers of individual layers or chains. As with other colloidal systems, these dispersions are thermodynamically unstable with respect to flocculation, but can be given useful stability by the appropriate choice of cation/solvent combination or by the addition of surfactants. Dispersions can be used for the formation of new intercalation compounds, particularly with very large guest molecules; for the formation of high surface area materials useful in catalysis; and to obtain films and organic-inorganic composites.

The scope of this review paper is to discuss the synthesis, characterization, and reactions of transition metal dichalcogenide dispersions prepared by starting from a preformed layer or chain structure compound. The alternative approach of assembly of small particles from solution has been used, for example, to form V_2O_5 by the polymerization of decavanadic acid (1), to synthesize colloidal $VOPO_4 \cdot 2H_2O$ (2), and for the formation of dichalcogenides by reaction of metal halides with lithium sulfide (3). The assembly approach has the advantage of producing very uniform dispersions, but will not be considered further in this review.

A wide range of layered compounds have been studied and examples are tabulated in order of increasing charge per unit area. The layer charge correlates with the ease of synthesis; the higher layer charges generally requiring additional layer modification to achieve stable dispersions. The table also contains examples of clay minerals as reference points and two examples of chain structures that will be discussed separately.

TABLE 1

Layered and Chain Compounds forming Colloidal Dispersions

	Compound	Area $Å^2$/Charge		Ref.
LAYERS				
	kaolinite	‸neutral		4
	MoS_2	‸neutral		5
	Smectites	40-120		4
	M_xMoO_3	117	(x=1/4)	6
	M_xFeOCl	100	(x=1/4)	7
	M_xMS_2 (x=1/3)	60	M=H, Li, Na	8,9,10,11
	$Li_{2x}Mn_{1-x}PS_3$	64	(x =0.25)	12
	Vermiculites	27-40		4
	$Zr(HPO_4)_2 \cdot H_2O$	34	Require surface	13
	$HCa_2Nb_3O_{10}$	28	modification by	14
	$HTiNbO_5$	25	preintercalation	15
	Mica	24		4
CHAINS				
	$MFeS_2$	40-60		16
	$M_2Mo_6Se_6$	40		17

SYNTHESIS

Three general techniques have been used to obtain dispersions of layered compounds: (i) chemical exfoliation; (ii) preintercalation or layer surface modification; and (iii) mechanical routes.

(i) chemical exfoliation:

At low layer charges, with the appropriate interlayer cation and solvent, exfoliation can occur spontaneously or by the application of weak shear forces either by mechanically stirring or sonication. This phenomenon has been discussed extensively for smectite clays such as montmorillonite which spontaneously exfoliate in water to give colloidal solutions containing largely single layers (4). Two expansion regimes have been noted. One corresponds to solvation of the interlayer cations and the other to repulsion of the double layers arising from the layer charge and an equivalent amount of ionic charge accumulated in the liquid phase near the particle surface. The double layer repulsion is related to its thickness that is determined by the electrolyte concentration, the charge on the interlayer cation, and the dielectric constant. The thickness of the double layer is largest and layers have the greatest tendency to disperse when the layer charge is low, the interlayer cations are monovalent and the dielectric constant of the solvent is large. Conversely exfoliation is strongly inhibited by the addition of electrolytes or by the presence of impurities in solution. Some systems, for example clays in water exhibit a smooth transition from one regime to the next and the interlayer separation is dependent on the electrolyte concentration. In other cases solvation in water produces discrete hydrated phases with no tendency to further disperse. The hydrated alkali metal intercalation compounds of the dichalcogenides are examples of this behavior forming discrete monolayer or bilayer hydrates (9). In general, chemical exfoliation requires a low layer charge obtained by chemical reduction, solvation of a small monovalent cation to begin the exfoliation process and a high solvent dielectric constant to expand into the double layer repulsion regime.

Early examples of this synthesis approach were the work of Lerf and Schöllhorn on hydrated alkali metal dichalcogenides (9) and of Murphy and Hull (8) on hydrogen intercalation in 2H-TaS$_2$. Lerf and Schöllhorn reported that in solvents with high dielectric constants (e.g. formamide and N-methyl formamide), solvated alkali metal dichalcogenides $A^+_x(solv)[MS_2]^{x-}$ exfoliated to form intensely colored homogeneous colloidal solutions under the application of weak shear forces. Studies of $Na^+_x(NMF)_y[TiS_2]^{x-}$ showed that these dispersions contained negatively charged dichalcogenide layers that could be flocculated by the addition of trivalent cations (e.g. La^{3+})

Murphy and Hull investigated the dispersion behavior of H_xTaS_2 in water prepared by electrolysis of 2H-TaS$_2$ in 1M H$_2$SO$_4$. Removal of the solid from the electrolyte followed by the addition of deionized water produced an enormous swelling due to intercalation of water that is reversible as a function of pH. The stability of the dispersions is enhanced by the addition of neutral molecules such as glycerol and polyether surfactants. Similar results were subsequently obtained for 2H-NbS$_2$ and TaS$_2$ (11).

A modification of Lerf and Schöllhorns' approach was used to prepare very stable and well behaved dispersions of 2H-TaS$_2$ (18). TaS$_2$ was reduced in a 50/50 mixture of N-methyl formamide and water with sodium dithionite to give an intermediate phase $Na_{1/3}TaS_2(solv)_y$. Immediate swelling of the solid is observed when the crystals, after washing with distilled water, were placed in NMF and H$_2$O added (or vice versa). Stable dispersions with concentrations in the range of 0.01to 0.5 wt% were readily prepared by sonication of the mixtures. The degree of dispersion depends somewhat on the solvent composition, the best results in terms of stability and reactivity being obtained at a 1/1 ratio of NMF/H$_2$O (18)

Analogous to the dichalcogenides, the lithium intercalation compound of molybdenum trioxide $Li_{0.25}MoO_3(H_2O)_y$ also exfoliates in water to give very stable dispersions of reduced MoO$_3$ layers (6). The corresponding sodium compound swells in water, but does not completely disperse. Similar results have been obtained for $Li_{2x}Mn_{1-x}PS_3$ (12) and by aqueous reduction of FeOCl (7), though the formation of stable dispersions is complicated in these cases by dissolution and hydrolysis, respectively.

(ii) preintercalation or layer surface modification:

The layered oxides $Zr(HPO_4)_2 \cdot H_2O$ (19), $H_mM_nO_{2n+1}$, M=Ti, Nb (20) and $VO(HPO_4)$ $1/2H_2O$ (21), have strong interlayer hydrogen bonds and consequently, do not spontaneously exfoliate in water. These layered compounds behave as solid Brønsted acids and undergo acid-base intercalation reactions with, for example, ammonia and primary amines. Intercalation compounds obtained by complete exchange also do not readily exfoliate because the strong hydrogen bonds are replaced by van der Waals interactions between the alkyl chains. An intermediate situation was discovered by Alberti and coworkers (13) in measurements of the pH dependence of the reaction of α-Zr(HPO$_4$)$_2$·H$_2$O with n-propylamine. At a point where one half of the protons are titrated, the αZrP layers spontaneously exfoliate. Addition of more propylamine causes the layers to reflocculate. The exfoliation presumably results from intercalation of sufficient propylamine to disrupt the hydrogen bonding, but not enough to provide significant interchain interactions in the interlayer.

The presence of a polar solvent is necessary for exzfoliation to occur. Similar experiments in methanol permit the isolation of the intermediate phase α-[Zr(PO$_4$)(HPO$_4$)]·C$_3$H$_7$NH$_3$·H$_2$O which though expanded does not exfoliate. Addition of the solid phase to distilled water causes spontaneous exfoliation with formation of opalescent dispersions or gels depending on the concentration of the solid. Addition of acid displaces propylamine and causes reflocculation, though the reorganization is imperfect and the surface area of the product is high. Similar behavior was observed in formation of C$_2$H$_5$NH$_3$TiNbO$_5$ by reaction of HTiNbO$_5$ with aqueous ethylamine (15) and in the reaction of VO(HPO$_4$)·1/2H$_2$O with propylamine (22).

A different approach to the exfoliation of the layered perovskites K[Ca$_2$Na$_{n-3}$Nb$_n$O$_{10}$], n=3, 4, 5 was suggested by previous work on the stabilization of H$_x$TaS$_2$ dispersions using polyether surfactants and by exfoliation of zirconium organophosphonates containing polyether groups (23). After ion exchange in acid to obtain the proton exchange forms, monoamine surfactant molecules were intercalated. The specific surfactants used had average compositions n-C$_4$H$_9$(OCH$_2$CH$_2$)$_4$ (OCH$_2$CH(CH$_3$))$_2$NH$_2$ (M360) and CH$_3$OC$_2$H$_4$(CH$_3$(CH(CH$_3$)O)$_8$CH$_2$CHCH$_3$NH$_2$ (M600)(24). Reaction of excess surfactant with the layered solid produced intercalation compounds with very large c axis expansions (e.g. Δc=48.9Å, M360, n=3; Δc=25.7Å, n=3, M600). Exfoliation of the intercalation compounds was achieved by suspending the solid (.1 to .5 wt%) in water for the M360 phases or acetone for the M600 compounds. The suspensions were then placed in an ultrasonic cleaning bath for 15 minutes to induce the maximum amount of exfoliation. After the large particles were removed by sedimentation, the final dispersions were stable for several days (14).

(iii) mechanical routes:

One approach to the synthesis of dispersions has been to use mechanical methods such as the application of high intensity ultrasound. High intensity ultrasound is known to increase the rate of intercalation reactions (25), but this was attributed to particle size reduction effects. However, in recent studies of the exfoliation of the misfit layer compounds PbNb$_2$S$_5$ and SmNb$_2$S$_5$, some degree of dispersion was obtained in specific solvents such as ethanol and isopropanol (26). The solvents were not intercalated, but acted to stabilize the dispersions. The dispersions contained a substantial fraction of particles with less than 10 layers, but only a small number of single layers and only a small fraction of the total material exfoliated.

A second approach which also has a mechanical component has been used with success both for the misfit layer compounds (26) and for MoS$_2$ (5). In this method, the layered compounds is first intercalated with an alkali metal cation from non aqueous solution. The alkali metal intercalation compound is then exposed to water where reaction occurs with the liberation of hydrogen. The rapid reaction accompanied by the evolution of large amounts of hydrogen gas causes the layers to be blown apart. The approach has been studied for MoS$_2$. 2H-MoS$_2$ was soaked in 1.6M n-butyl lithium in hexane for 48h to produce Li$_x$MoS$_2$ (x=1). The intercalated MoS$_2$ was then washed to remove excess n-butyl lithium. On exposure to water, copious hydrogen gas evolution occurs with the formation of a highly opaque suspension in water. Some ultrasonication was used to assist the exfoliation and addition of a surfactant enhances the stability. After washing in water to remove LiOH exfoliated MoS_2 is stable in water at neutral to basic pH but reduction of the pH to less than 2 by addition of acid causes flocculation.

Dispersions of the layered compound $PbNb_2S_5$ have been prepared similarly (26). Treatment of $PbNb_2S_5$ with a solution of sodium naphthalide in THF was used to prepare the intercalation compound $NaPbNb_2S_5$ as an amorphous black product. Reaction of this phase with 0.1M HCL caused oxidation of the solid, liberation of hydrogen and the formation of suspensions of well dispersed layers, which in the acidic solution reflocculate over a period of several hours.

CHARACTERIZATION

Direct evidence that exfoliation has occurred to produce single layers and the distribution of particle thicknesses is difficult to obtain. Electron microscopy has generally been used to study samples allowed to dry on microscope grids. Because of their high aspect ratio layers lie on transmission electron microscope grids almost exclusively in plane view with the sheet normals along the beam direction and edge on views being very rare. Intensities from conventional bright-field and dark field images are difficult to interpret qualitatively. For example, measurements of platelet thickness in H_xNbS_2 were made by measuring image contrast with the sample in and out of the electron beam. Intensities were calibrated by optical density measurements on cleaved crystals but the measurement uncertainty is large and only qualitative evidence for single layers was obtained (10).

An alternative approach originally developed for studies of clay layers uses a low angle shadowing technique (27). A sample of the dispersion is sprayed onto a flat surface and then shadowed by evaporating platinum onto the specimen at low angle. Measurement of the shadow length is then used to calculate the thickness of the particle. Studies of the clay mineral, allevardite, which has a double 2:1 layer structure and of MNb_2S_5 (M=Pb, Sm) have shown that resolutions of 3-4Å can be obtained at shadowing angles of 11-15°, sufficient to resolve single layers of these composite layered materials. An inherent problem with the shadowing technique is the requirement that the particles lie flat on the substrate and are not buckled or crumpled. It is also limited by delineation of the shadow and crystal edges.

An alternative technique is to measure the distribution of Rutherford scattered intensities into a high angle annular detector in a scanning transmission electron microscope (STEM). The intensity of the image formed in the high angle annular detector is approximately proportional to the atomic number and can be directly related to the specimen thickness. This imaging technique is relatively insensitive to sheet buckling, it measures the thickness at all points across a sheet rather than only at an edge and it obtains thickness distributions directly from the intensity histogram of a single image. This technique was applied to a study of the layered perovskites $H[Ca_2Na_{n-3}Nb_nO_{10}]$ with different layer thicknesses. The intensities are quantized in units of the layer thickness though the peaks broaden more rapidly than expected. The broadening is due to the distinction between, for example, two layers formed by two single layers that overlap on the grid and two layers that were not exfoliated in the original preparation. Approximately 80% of the n=3 compound was observed in the form of single layers whereas the corresponding values for the n=4 and 5 compounds were 50% and 10%. The reason for the difference is not understood.

Most recently, exfoliated $HCa_2Nb_3O_{10}$ was examined in a Hitachi H9000NAR microscope equipped with a high angle electronic hollow cone unit (28). In this mode, the platelet contrast is significantly enhanced as can be seen in the Figure 1. The data also reveals some new features in the form of "blisters" between adjacent overlapping sheets. The blisters typically have dimensions of ~100Å, are randomly distributed and are believed to have been caused by trapped molecules when the sheets deposited. The specimens were baked prior to imaging and it is likely that no particle remains, but the contrast arises from the layer distortion which has not annealed out.

 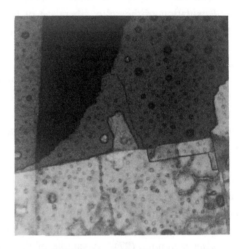

Figure 1 Electron micrographs of exfoliated layers of $HCa_2Nb_3O_{10}$

STEM or hollow cone microscopy data clearly can provide very detailed information on dispersions and how they aggregate under the conditions used to prepare specimens but for characterization in solution other techniques are required. One approach is to use dye adsorption if the optical absorption spectrum of a dye molecule is perturbed on interaction with the exfoliated layer surfaces. By measuring the concentration dependence of the spectrum and assuming an effective area for the molecule adsorbed, the layer surface area and hence the degree of dispersion can be estimated. The technique has been widely applied to clays (29) and was used to confirm single layer dispersion in H_xTaS_2 (8).

X-ray diffraction measurements of dispersions directly in solution also provides information. For example, in the case of MoS_2 exfoliated by the "explosion" method, diffraction data were compared with theoretical calculations of diffraction patterns at different layer thicknesses (30). As the number of layers in a stack decreases, the Bragg reflections broaden and for a single layer the 002 reflection disappears. Very good agreement was obtained between the observed data and the calculated single layer pattern (5, 31).

AGGREGATION/FLOCCULATION

Aggregation or flocculation of colloidal layered compound dispersions is induced by increasing the ionic strength of the solution or by decreasing the dielectric constant. In the case of added electrolyte different outcomes are possible depending on the conditions and can be summarized for the case of TaS_2 by the reaction: $A_x^+TaS_2^{x-}$ (dispersion) $\rightarrow A_{x-y}B_yTaS_2$ (solid)

In one limit, $y=0$, and the initial material is recovered in restacked form. The added electrolyte induces flocculation by compression of the double layer around each TaS_2 sheet but does not itself exchange with A^+ cations in the double layer. In general, reassembly can occur in several different ways depending on how individual particles connect. Edge to edge flocculation results in the formation of very open so called "house of cards" structures and occurs in clays at low electrolyte concentration. Edge defects with a different charge than the layers are thought to be the origin of the effect. At higher electrolyte concentrations both edge to edge flocculation and face to face aggregation occurs. The balance between the two effects is subtle in clays which have a relatively high concentration of edge sites. In other systems, with much larger lateral platelet dimensions, face to face aggregation is more prevalent.

In the other limit, addition of the cation B^{n+} induces flocculation accompanied by ion exchange for the A^+ cations to give the intercalation compound B_yTaS_2. The balance between the two limits is determined by the dispersion concentration, the concentration of electrolyte containing the intercalating cation and the ratio of the two. In order to assess the relative importance of these variables in a model system, flocculation of Na_xTaS_2 dispersions in NMF/H_2O with cobaltocenium cations was studied with the results shown in Figure 2.

At very low concentrations of Cp_2Co^+ (region D), flocculation does not occur irrespective of the concentration of TaS_2^{x-} colloid. Conversely, at relatively high concentrations of TaS_2^{x-} and Cp_2Co^+ (region I), flocculation and ion-exchange occur to give either a very well-ordered, completely ion-exchanged product (at higher $[TaS_2^{x-}]/[Cp_2Co^+]$ ratios), or a poorly-ordered product (at lower $[TaS_2^{x-}]/[Cp_2Co^+]$ ratios). The X-ray diffraction of the well-ordered product gives an interlayer spacing identical to that reported for direct intercalation of Cp_2Co into TaS_2 or by ion-exchange with $Cp_2Co^+I^-$ (32) In region II, diffraction patterns indicated two phases with distinctly different interlayer spacings. One phase corresponds to $[Cp_2Co]_xTaS_2$, and the other to a NMF or H_2O-solvated Na^+ phase with an interlayer spacing of 9.6Å.

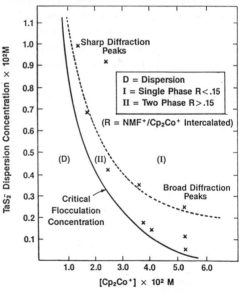

Figure 2 Flocculation of TaS_2 dispersions with cobaltocenium iodide (from reference 18)

The flocculation behavior established for cobaltocenium with TaS_2^{x-} dispersions is a good model for other large cations. The "flocculation diagram" is roughly applicable to a wide range of electrolytes, although the size and charge of the B cation can significantly shift the relative positions of the indicated regions. Intercalation of two other large cations, $Al_{13}O_4(OH)_{24}(H_2O)_{12}^{7+}$ (18) and $Fe_6(\mu_3\text{-}S)_8(PEt_3)_6^{2+}$ (11), have been achieved using Figure 2 as a guide.

The product of the reaction with solutions containing the Al_{13} cation prepared by hydrolysis of $AlCl_3$ solutions depends on the Al_{13}/TaS_2 ratio. For ratios between 0.04 and 0.06 a single phase product was obtained with a sharp diffraction pattern corresponding to an interlayer separation of 20.3Å ("20Å phase"). Prolonged aging of oriented films under ambient conditions showed a slower, but gradual decrease of the interlayer spacing. The peak widths initially broadened during this decrease, and then sharpened again to give a well ordered phase with a interlayer separation of 16.2Å. The change in layer spacing is essentially complete after 2h, and no further change occurs at ambient conditions after 2d. The 20Å phase and the 16Å phase could be interconverted by the addition of either NMF or H_2O and the changes are thought to be due to solvation of the interlayer Al_{13} cation.

The X-ray data for two samples of Al_{13} cation intercalated TaS_2, with interlayer separations corresponding to 16.1Å and 20.3Å, were analyzed to obtain the one dimensional projections of the electron density along the c axis by the method described above. The projection of the electron density for the 16.1Å phase is shown in Figure 3 together with a drawing of the structure of the $Al_{13}O_4(OH)_{24}(H_2O)_{12}^{7+}$ cation. Two unit cells are plotted. In addition to the scattering from the TaS_2 layers, four prominent peaks are apparent between the layers. Comparison with the structure of the Al_{13} cation suggests that these peaks (2.5Å apart) correspond to the four layers of oxygen atoms in the cation oriented as shown (or in the inverse configuration). A similar arrangement has been previously proposed to occur in Al_{13} pillared clays (33) but is different from that observed in MoO_3 where the C_2 axis of the cluster is perpendicular to the layers (6). A similar projection of the electron density was generated for the 20.3Å phase which contains both water and NMF molecules of solvation. The peaks in the electron density between the layers were less well defined, presumably reflecting contributions from the different possible orientations and locations of the NMF molecules.

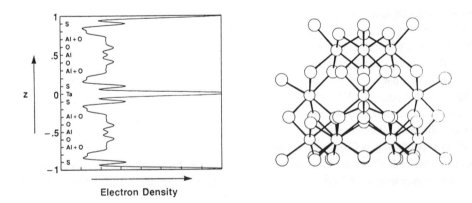

Figure 3. One dimensional electron density projection for Al_{13} cations intercalated in TaS_2 (left) and the cation structure (right) (adapted from reference 18)

Flocculation of the TaS_2^{x-} dispersion with a 50/1 molar excess of $Fe_6(\mu_3\text{-}S)_8(PEt_3)_6^{2+}$ (as the BF_4^- salt) resulted in incorporation of the cluster to give an interlayer spacing of d = 17.491Å. This corresponds to a lattice expansion of 11.5Å, consistent with the size of this cluster estimated by molecular modelling. The Fourier transform of the structure factors derived from the X-ray intensities gave the one-dimensional electron density map that was used to show that the cluster is oriented with the C_3 axis of the Fe_6 octahedron perpendicular to the layers (11).

Inclusion compounds of dispersions of MoS_2 have been prepared by two methods. One approach used pH to control the flocculation of aqueous metal nitrate dispersions of MoS_2 (34). Reactions were successful for Ni, Co, Cd, Zn, and Pb as determined by changes in the X-ray diffraction data for the restacked materials. Auger spectra indicated the presence of oxygen consistent with the presence of hydroxide layers between layers of MoS_2.

In the second approach, inclusion compounds of organic and organometallic cations were prepared by dipping a substrate into a two phase system of the aqueous MoS_2 dispersion and an imiscible organic liquid containing the guest species (*e.g.* ferrocene in CCl_4) (35). It was observed that the exfoliated layers segregated to the interface formed by the two liquids and could be transferred from the interface with inclusion of the guest species. For ferrocene, an X-ray diffraction pattern from a film prepared as described showed a high degree of orientation. The interlayer separation showed an expansion of 5.6Å consistent with similar values observed for metallocene cations in other dichalcogenides. The amount included, however, was significantly less (0.1 vs 0.25). Films containing other molecular species (dimethoxybenzene, t-butylbenzene, heptane, styrene) were also prepared. In general, the films have low thermal stability indicative of weak bonding. The nature of the interactions in these systems is not completely understood but it has been suggested that some aspects of the behavior can be explained by partial replacement of the sulfide ions at the edges of the MoS_2 layers by hydroxide ions.

COMPOSITE MATERIALS

Dispersions of layered compounds can be used in the synthesis of other types of materials. An interesting example is the work carried out at ICI on synthesis of inorganic films derived from clays (36). In one study, vermiculite was delaminated by exchange in sodium chloride and then n-butyl ammonium chloride. Stable dispersions were obtained by mechanical exfoliation in a high shear mixer. The dispersions were used to form films by evaporation. Surface tension forces on drying produce alignment of the particles and flat coherent films that could be stabilized in the presence of water by re-exchange with magnesium ions. Flexible films with high tensile strengths relative to organic films could be achieved. Suspensions were also used to form stable coatings on glass fibers (36).

Similar films have been prepared by Alberti and coworkers from dispersions of $Zn(HPO_4)_2 \cdot H_2O$ (13). Acid treatment to remove the amine, followed by filtration and drying, produces flat flexible films with some mechanical strength especially when large crystallites are used in the initial synthesis. High proton conductivity in samples prepared from these films has been reported (37). Films of $HCa_2Nb_3O_{10}$ have been prepared by the same technique. Other examples include deposition of MoS_2 onto high surface area alumina (38), the synthesis of high surface area $HCa_2Nb_3O_{10}$ by acid flocculation and the preparation of composites of metal hydroxides and MoS_2 discussed above.

Organic - inorganic composite materials can also be synthesized by flocculation of a dispersion in the presence of a monomer followed by polymerization. For example, thin films of MoS_2 incorporating styrene were prepared by the technique described above (39) and then polymerized by heating at 60°C. Before the thermal polymerization, the films reversibly incorporate water and the film resistivities are sensitive to the ambient humidity. The polymerized films show very high anisotropy (10^8) in the electrical conductivity as a result of the inclusion of an insulating polystyrene layer between layers MoS_2.

Dispersions of MoS_2 and WS_2 can both be prepared but flocculate under different pH conditions (MoS_2, pH=2; WS_2, pH = 0.8). When a mixture containing both types of layers is flocculated at a final pH of 1.3, a nearly single phase material is obtained with a *c* axis spacing of 6.84Å suggesting the formation of $MoWS_4$ with regularly alternating layers.(40).

CHAIN STRUCTURES

Dispersion reactions for only a small number of chain structures have been investigated. Two specific examples, $[Mo_3Se_3]$ and $[FeS_2]$, illustrate the general behavior which is similar to that observed in the layered systems.

(i) $[Mo_3X_3]$ chains:

The quasi one dimensional $M_2Mo_6X_6$ compounds (x = Se, Te) can be described as infinite linear chains of $[Mo_3X_3]$ units separated by the ternary element M, with large interchain separations. Direct reaction of the elements is used to form $In_2Mo_6Se_8$ or $In_2Mo_6Te_8$. The corresponding alkali metal compounds are obtained from the indium phases by ion exchange in molten halides. The resulting metastable alkali metal phases completely disperse in organic solvents, for example, $Li_2Mo_6Se_6$ swells in NMF, dimethylsulfoxide, and propylene carbonate, but not in tetrahydrofuran. In mixed phases, $Li_xIn_{2-x}Mo_6S_6$ the ease of exfoliation increases as the lithium content increases. Only DMSO and NMF disperse $Na_2Mo_6Se_6$; the heavier alkali metal compounds (K, Rb, Cs) do not disperse. Removal of the solvent results in the recovery of the starting materials. Similarly, addition of salts or acids also causes flocculation but the starting phase is recovered independent of the added alkali with no evidence of any ion exchange. Measurement of chain diameters by transmission electron microscopy showed smallest diameters of 20Å or about 3x the interchain distance in the reprecipitated solid. The fiber diameters in solution are smaller and were shown by EXAFS measurements to retain the solid structure (41). Films of chain like layers can be oriented by shear and showed anisotropy in their electrical conductivity.

(ii) $[FeS_2^-]$ chains:

The compound $KFeS_2$ has a structure that comprises infinite chains of edge-shared FeS_4^- tetrahedra (42). The K^+ ions occupy interchain sites and are eight coordinate in distorted square antiprismatic sites. The compound was prepared originally by reaction of iron metal, sulfur and potassium carbonate (43).

An interesting feature of the potassium iron sulfide structure is that the interchain potassium ions are readily exchangeable. Exchange with Cu^+ ions in aqueous solutions was first reported by O'Daniel (44) and subsequently, exchange reactions with copper and silver ions were studied in more detail. These reactions are not topochemical and exchange of potassium ions is accompanied by migration of Fe^{3+} ions from chain to interchain sites. The structure of the final products $MFeS_2$ (Cu, Ag) are chalcopyrite. In contrast to reactions with Ag^+ and Cu^+, other exchange reactions occur with preservation of the chain structure. Reactions with aqueous alkaline earth metal chlorides for example, leads to the formation of hydrated phases $M_{0.5}FeS_2 \cdot xH_2O$ (M=Ca, Sr) with variable water content and anhydrous α-$Ba_{0.5}FeS_2$ (45).

In the course of investigating electrochemical reactions of the alkali and alkaline earth metal thioferrates (46) we attempted to prepare $MFeS_2$ (M=Li, Na) by ion exchange of $KFeS_2$ in aqueous MCl solutions. Prolonged exchange in 1M aqueous NaCl leads to complete replacement of the potassium ions and the formation of a hydrated sodium phase with the composition $NaFeS_2 \cdot 2H_2O$. The sodium thioferrate is identical to the mineral erdite found at Coyote Park, Humboldt County, California (47), the structure of which has been determined (48).

Ion exchange of $KFeS_2$ with aqueous lithium ion solutions produces a dramatically different result. An intense green coloration is observed and the solid phase gradually disperses. The visible spectrum of the solution shows three characteristic peaks at 640, 530 and 435 mm. Similar intensely colored solutions are formed by the addition of excess NaSH or KSH to ferric nitrate solutions in the pH range 11 to 12 (16). Addition of acetone or 1M solutions of MOH or MCl causes irreversible precipitation, and the formation of $KFeS_2$ or $NaFeS_2 \cdot 2H_2O$ with broad diffraction patterns, but which clearly indicate that the structures are similar to the phases prepared by high temperature and ion exchange reactions. The direct precipitation reaction indicates that the intense green solutions are colloidal dispersions of FeS_2^- chains.

Lithium and tetramethyl ammonium ion containing dispersions can be prepared similarly, but are more stable with respect to flocculation (22). Colloidal $[FeS_2^-]$ is an interesting example where similar dispersions can be prepared both by direct synthesis and from a preformed solid.

SUMMARY

Colloidal dispersions can be prepared under appropriate conditions for a range of examples of layer and chain structures and used to synthesize intercalation compounds and other types of materials. At present, the general features of their properties and potential utility are appreciated but further detailed studies are needed for better control of the chemistry and to develop detailed characterization methods.

ACKNOWLEDGEMENTS
Partial support of this work was provided by the Robert A Welch Foundation.

REFERENCES
1. Legendre, J. J. and Livage, J.: J. Coll. Int. Sci., 1983, 94 (1), 75; Livage, J. and Lemerle, J.: Ann. Rev. Mat. Sci., 1982, 12, 103.
2. R'kha, C.; Vandenborre, M. T.; J. Livage; J.; Prost, R.; and Huard, E.: J. Solid State Chem., 1986, 63, 202.
3. Chianelli, R. R.and Dines, M. B.: Inorg. Chem., 1978, 17, 2758.
4. For example, Grim, R. E.: Clay Mineralogy, 2nd. ed. McGraw-Hill: New York, 1968; Nadeau, P. H.: Applied Clay Science, 1987, 2, 83.
5. Joensen, P.; Frindt, R. F.; and Morrison, S. R.: Mater. Res. Bull., 1986, 21, 457.
6. Nazar, L. F.; Liblong, S. W.; and Yin, X. T.: J. Amer. Chem. Soc., 1991, 113, 5889.

7. Weiss, A. and Sick, E.: Z. Naturforsch., 1978, 33b, 1087.
8. Murphy, D. W. and Hull, Jr., G. W.: J. Chem. Phys., 1975, 62(3), 967.
9. Lerf, A. and Schöllhorn, R.: Inorg. Chem., 1977, 16, 2951.
10. Liu, C.; Singh, O.; Joensen, P.; Curzon, A. E.; and Frindt, R. F.: Thin Solid Films, 1984, 113, 165.
11. Nazar, L. F. and Jacobson, A. J.: J. Chem. Soc. Chem. Commun., 1986, 570.
12. Clement, R.; Garnier, O.; and Jegoudez, J.: Inorg. Chem., 1986, 25, 1404.
13. Alberti, G.; Casciola, M.; and Costantino, U.: J. Colloid Interface Sci., 1985, 107, 256.
14. Treacy, M. M. J., Rice, S. B., Jacobson, A. J. and Lewandowski, J. T.: Chem. Mater. 1990, 2, 279.
15. Rebbah, H.; Borel, M. M.; and Raveau, B.: Mater. Res. Bull., 1980, 15, 317.
16. Taylor, P. and Shoesmith, D. W.: Can. J. Chem., 1978, 56, 2797.
17. Tarascon, J. M.; DiSalvo, F. J.; Chen, C. H.; Carroll, P. J.; Walsh, M.; and Rupp, L.: J. Solid State Chem., 1985, 58, 290.
18. Nazar, L. F. and Jacobson, A. J. submitted for publication 1993.
19. Clearfield, A.: *Inorganic Ion Exchange Materials*, CRC Press, Florida, 1982.
20. Raveau, B.: Rev. Chim. Miner., 1984, 21, 391.
21. Johnson, J. W.; Johnston, D. C.; Jacobson, A. J.; Brody, J. F.: J. Am. Chem. Soc., 1984, 106, 8123.
22. Jacobson, A. J., Lim, S. C. and Vaughey, J. T. unpublished results.
23. Ortiz-Avila, C. Y. and Clearfield, A.: Inorg. Chem. 1985, 24, 1733.
24. Jeffamine surfactants supplied by Texaco.
25. Chatakhondu, K.; Green, M.L.H.; Mingos, D. M. P.; and Reynolds, J.: Chem. Soc. Chem. Commun., 1987, 900.
26. Bonneau, P.; Mansot, J. L.; and Rouxel, J.: Mater. Res. Bull., 1993, 28, 757.
27. Weir, A. H.; Nixon, H. L.; and Woods, R. D.: Clays and Clay Minerals, 1962, 9, 419.
28. Treacy, M. M. J.; Bisher, M. E.; and Jacobson, A. J.: Proc 51st Annual Mtg.Microscopy Society of America, 1993.
29. Hang, P. T. and Brindley, G. W.: Clays and Clay Minerals, 1970, 18, 203.
30. Liang, K.S.; Chianelli, R.R.; Chien, F.Z.; and Moss, S.C.: J. Non Cryst. Solids 1986, 79, 251.
31. Joensen, P.; Crozier, E. D.; Albending, N.; and Frindt, R. F.: J. Phys. C.: Solid State Phys., 1987, 20, 4043.
32. Dines, M. B.: Science, 1975, 188, 1210.
33. Plee, D.; Borg, F.; Gatineau, L.; and Fripiat, J. J.: J.. Amer. Chem Soc., 1985, 107, 2362.
34. Gee, M. A.; Frindt, R. F.; Joensen, P.; and Morrison, S. R.: Mater. Res. Bull., 1986, 21, 543.
35. Divigalpitiya, W. M. R.; Frindt, R. F.; and Morrison, S. R.: Science, 1989, 246, 369.
36. Ballard, D. G. H. and Rideal, G. R.: J. Materials Science, 1983, 18, 545.
37. Casciola, M. and Costantino, U.: Solid State Ionics, 1986, 20, 69.
38. Miremadi, B. K. and Morrison, S. R. : J. Catal., 1987, 103, 334.
39. Divigalpitiya, W. M. R.; Frindt, R. F. and Morrison, S. R.: J. Mater. Res., 1991, 6(5), 1103.
40. Miremadi, B. K. and Morrison, S. R.: J. Appl. Phys., 1990, 67(3), 1515.
41. Holtman, D.A.; Teo, B. K.; Tarascon, J. M.; and Averill, B. A.: Inorg, Chem., 1987, 26, 1669.
42. Boon, J. W. and MacGillavry, C. H.: Rec. Trav. Chim. Pays-Bas., 1942, 61, 910.
43. Preis, K.: J. prakt. Chem., 1869, 107, 12.; Schneider, R.: J. prakt. Chem., 1869, 108, 16.
44. O'Daniel, H.: Z. Krist, 1933, 86, 192.; Boon, J. W.: Rec. Trav. Chim. Pays-Bas., 1943, 69,.
45. Boller, H.: "Solid Compounds of the Transition Elements", Uppsala Sweden, 1976, p. 93.
46. Jacobson, A. J.; Whittingham, S. J.; and Rich, S. M.: J. Electrochem. Soc., 1979, 126, 887.
47. Czamanske, G. K.; Leonard, B. F.; and Clark, J. R.: Amer. Mineral, 1980, 65, 509.
48. Konnert, J. A. and Evans, Jr., H. T.: Amer. Mineral, 1980, 65, 516.

Materials Science Forum Vols. 152 - 153 (1994) pp. 13-34

AMBIENT TEMPERATURE SOLID STATE REACTIONS IN BATTERY ELECTRODES

J.O. Besenhard

Institute of Chemical Technology of Inorganic Materials, Technical University of Graz, Stremayrgasse 16, A-8010 Graz, Austria

Keywords: Electrodes, Batteries, Electrochemical Insertion Reactions, Mixed Conductor Compounds

ABSTRACT

Electrochemical insertion of ions into mixed conductor electrode materials plays a key-role in almost any primary or secondary ambient temperature battery system. Electron transfer is usually coupled with insertion/removal of protons or Li^+-ions into/from host matrices such as oxides, chalcogenides or metals. This is illustrated by examples. Some recent developments such as lithium-ion batteries and rechargeable Zn/MnO_2 batteries are discussed in more detail.

INTRODUCTION

More than 10^{10} batteries, worth ca. 3 x 10^{10} US-$, are fabricated in the world every year and - apart from very few exceptions - each of them is an ambient temperature solid state electrochemical reactor. This is probably the largest scale technical application of soft solid state chemistry. Battery making has a long tradition and - at least in the case of traditional aqueous electrolyte batteries - theoretical understanding of reaction mechanisms was quite poor for a long time. This is also related with the fact that active materials in technical battery electrodes are usually poorly crystalline and also inhomogeneous and sometimes fairly difficult to study, even with sophisticated modern methods.

Solid active materials of battery electrodes can be discharged or charged by two principally different pathways: i) via soluble species or ii) by solid state redox reaction coupled with solid state transport of ions. This may be illustrated by the two reduction reactions:

$$Pb^{2+} SO_4^{2-} + 2 H^+ + 2 e^- \rightarrow \quad Pb + H_2SO_4 \tag{1}$$

$$TiS_2 \quad + x Li^+ + x e^- \rightarrow \quad Li_xTiS_2 \tag{2}$$

The first reaction can only proceed via soluble Pb^{2+}-cations, i.e., after dissolution of $PbSO_4$, as $PbSO_4$ is an insulator. The second reaction is a solid state reaction, insertion of Li^+-cations into the mixed conductor Li_xTiS_2. The different kinds of reduction mechanisms affect also the shape of the reaction zone and the completeness of material utilization (see figure 1).

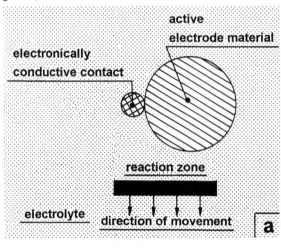

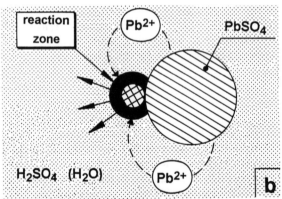

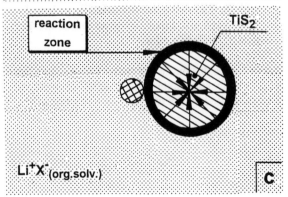

Fig. 1 a - c)
The active electrode material is in contact with an electronically conductive material (current collector). The reaction zone will develop depending on the properties of the active material (a). If the active material is a nonconductor (e.g. $PbSO_4$) its reduction is only possible via a solution mechanism. Deposition of metallic lead will be on the conductive material, i.e., the micromorphology of the electrode will change considerably (b). If the active material is a conductive host structure (e.g. TiS_2) the inter-calation reaction zone will proceed into the active material. Utilization of the active material can be almost 100 % as the reaction product is a mixed conductor. So long the volume changes and irreversible structural changes related with ion-insertion/removal are small the morphology of the electrode will remain unchanged for a large number of insertion/removal cycles (redrawn from ref. [1]).

Most of the successful primary and secondary batteries make use of ion insertion/removal into/from mixed conductors. In the case of aqueous electrolyte batteries electron transfer is usually coupled with proton insertion/removal whereas in nonaqueous electrolyte batteries the small Li^+-ions are the natural choice for the mobile cations in mixed conductor electrodes. Attempts to use Na^+- [2], K^+- [3] or Mg^{2+}-ion [4] insertion reactions in batteries have not been extremely promising so far. The development of lithium batteries and in particular, the rechargeable ones, was based from the very beginning on the concept of Li^+-insertion into mixed conductors. On the other hand, it took almost a century until it was understood and accepted that reduction of MnO_2 in aqueous electrolytes can proceed via an "electron-proton" mechanism [5,6].

PROTON INSERTION INTO OXIDE ELECTRODES

Manganese dioxide:
Manganese dioxide preparations are certainly the technically most important battery cathode materials. In combination with aqueous or nonaqueous electrolytes they are used in primary and more recently also in secondary batteries.

In the early stages of discharge of MnO_2 in aqueous electrolytes electrons and protons (originating from water molecules) are inserted into the crystal structure of MnO_2 in a homogeneous reaction, i.e., the discharge product is not formed as a new separate phase [6,7].

$$MnO_2 + x\,H^+ + x\,e^- \rightarrow MnO_{2-x}(OH)_x \tag{3}$$

$$MnO_2 + x\,H_2O + x\,e^- \rightarrow MnO_{2-x}(OH)_x + x\,OH^- \tag{4}$$

Equations 3 and 4 are valid in acidic as well as in alkaline electrolytes.

The electron-proton mechanism is schematically represented in figure 2. The potential of the partly discharged MnO_2 electrodes agrees usually quite well with that calculated from the Nernst equation, if $(OH)^-$ and (H_2O) are assumed constant.

$$E = E^0 - \frac{RT}{F}\ln\frac{\left[Mn^{3+}\right]_{(solid)}}{\left[Mn^{4+}\right]_{(solid)}} \tag{5}$$

It should be pointed out, however, that the Nernst equation is not a sufficient general approach to calculate electrochemical potentials of insertion compound electrodes, because it does not take into account the (limited) number and (different) quality of the ionic and electronic sites available in solids.

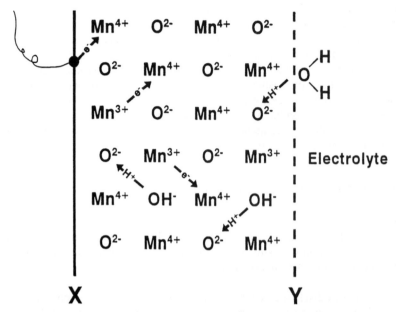

Fig. 2) Schematic representation of proton and electron movements in "MnO_2",
         X: Interface to electronic conductor, Y: interface to solution (redrawn from ref. [6]).

The discharge diagram of MnO_2 shows that the homogeneous part of the discharge is
followed by a heterogeneous one (figure 3). In terms of "x" in MnO_x the onset of the
heterogeneous discharge was observed at x < 1.875 in alpha-MnO_2 and at x < 1.6 in
gamma-MnO_2 [8].

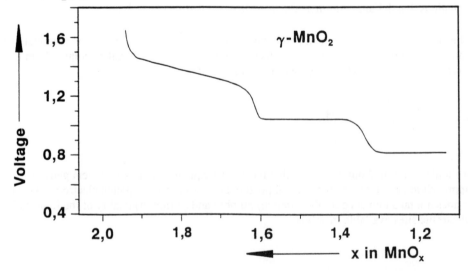

Fig. 3) Discharge diagram (open cell conditions) of gamma-MnO_2 in a Leclanché battery
         (redrawn from ref. [9]) .

The discharge of MnO_2 in aqueous electrolytes is a fairly reversible reaction and rechargeable alkaline Zn/MnO_2 batteries are now available commercially [10-12]. Reversibility of MnO_2-discharge is not strictly limited to the homogeneous part of the reaction, however, the "cycleability" of MnO_2 (i.e., the number of useful discharge-/charge-cycles and also their capacity) suffers if the discharge was extended too far [13]. Heterogeneous reoxidation of discharged MnO_2 is possible by oxygen molecules produced during charging [14,15]. Deep discharge of MnO_2 may cause failure of the cell due to the formation of Haeterolite $ZnO_xMn_2O_3$ [16].

The design of rechargeable alkaline manganese (RAM) batteries is only slightly different from the design of primary ones which should not be recharged mainly for reasons of safety, although they could be recharged to some extent.

The most important change in cell design is the limitation of the zinc anode (gelled anode made from zinc powder in immobilized KOH electrolyte) in order to prevent overdischarge of the MnO_2 cathode made from electrolytic manganese dioxide (EMD), i.e., gamma-MnO_2 and graphite powder. The zinc anode is mercury-free, corrosion and gassing is prevented by a combination of organic and inorganic inhibitors. The cathode formulations contain also some inorganic catalysts such as Ag_2O to promote the recombination of hydrogen gas formed at the zinc anode.

To avoid overcharging, special chargers are recommended, opening overflow circuits if a threshold voltage of ca. 1.65 V per cell is approached. This can easily be affected by Zener diodes. As zinc tends to deposit dendritically during charging, a two-component separator composed of i) a microporous cellulose barrier and ii) a nonwoven spacer and electrolyte reservoir made from rayon fibres is used [17].

Unlike Ni/Cd cells, RAM cells are sold in the charged state. Their initial capacity is only slightly (less than 20%) below that of a conventional single use alkaline cell. In the first few cycles, however, the capacity drops considerably until a plateau of ca 1/3 of the original capacity is reached. At this level, which is still about the capacity of a Ni/Cd cell, RAM cells may be used for several hundreds of cycles.

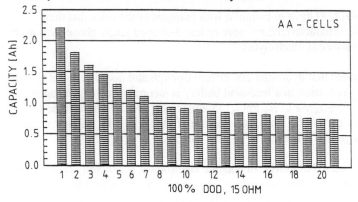

The typical capacity fading of a RAM cell during the first few cycles is shown in figure 4.

Fig. 4) Cycling curve of a AA-size RAM cell on continuous 15 Ohm load, cut-off voltage 0.9 V (100% DOD), ref. [16].

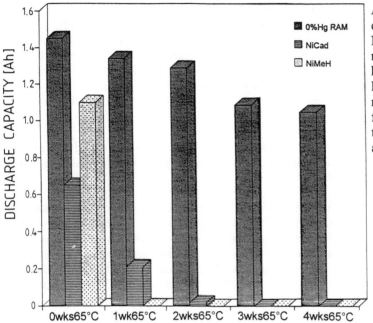

A dramatic advantage of RAM cells over Ni/Cd-cells or, even more, over Ni/metal hydride-cells is their low self discharge rate. This is shown in figure 5, comparing the charge retention at 65 °C.

Fig. 5) Charge Retention of AA-size RAM cells at 65 °C compared with the losses of Ni/Cd- and Ni/MeH-cells, discharge on 4 Ohm to 0.8 V cut-off. Note: 4 wk. at 65°C approximately corresponds to 2 yrs. at 20°C, ref [16].

Due to the low self-discharge at elevated temperatures, RAM batteries are expected to create significant new opportunities in a variety of applications using solar chargers.

Other Oxides:
V_2O_5, RuO_2 [7] and MoO_3 [18] have in common with manganese dioxides that their discharge diagrams are also characterized by more or less S-shaped single phase parts at the beginning of discharge in aqueous electrolytes.

It has also been demonstrated that hydrogen can diffuse through and dissolve in PbO_2 and that the first step of PbO_2-reduction in a lead-acid battery is incorporation of hydrogen, followed by formation of $PbSO_4$ via a "$H_x PbO_4$" intermediate [19-21]. There is, however, no significant single phase part in the discharge diagram of PbO_2, i.e., the stability range of the homogeneous discharge product seems to be very narrow.

The discharge of nickel oxides "NiO(OH)" in alkaline electrolytes is generally represented by simplified equations corresponding to equation (4), which was given for the reduction of MnO_2. Nevertheless, proton insertion into "NiO(OH)" is not generally considered to be the dominant pathway for the reduction reaction, despite strong evidence in favour of this mechanism [22-24]. Unfortunately, because of the mostly amorphous character of the reactants and products there is still some confusion about the processes at nickel oxide electrodes (cf. e.g. ref. [25]). Transfer of OH^- anions is commonly assumed to be coupled with the electron transfer [26]. Moreover, reduction of nickel oxides is also related with some insertion of alkali cations [24, 27].

PROTON INSERTION INTO METALLIC ELECTRODES

Most of the metallic elements will - at least to some extent - react directly and reversibly with hydrogen. In particular, hydrogen sorption by platinum group metal electrodes has been familiar to electrochemists since a very long time. Binary metal hydrides are usually too stable and/or too expensive to be considered for reversible hydrogen storage systems. However, shortly after intermetallic compounds such as $LaNi_5$ with rapid hydrogen sorption properties near room temperature were discovered in the late 1960s, several larger scale applications of metal hydrides have developped [28, 29].

Metal hydrides are now widely used as reversible hydrogen electrodes in rechargeable alkaline cells with metal oxide (mostly nickel oxide) cathodes. These cells operate according to the general equation (6)

$$MH + NiO(OH) \underset{charge}{\overset{discharge}{\rightleftharpoons}} M + Ni(OH)_2 \qquad (6)$$

"M" being the hydrogen storage material. Common electrolytes are similar to those used in nickel-cadmium cells, i.e., quite concentrated solutions of KOH containing some LiOH to improve the properties of the NiO(OH) electrode. The open cell voltage of fully charged cells is slightly above 1.3 V.

Pure $LaNi_5$ and also the widely studied TiNi-based alloys have too short cycle life times to be used in practical batteries. Formation of passive oxide layers, e.g. $La(OH)_3$, was found to be the main reason for capacity decay and failure [30].

A dramatic improvement of the cycle life of metal hydride electrodes was achieved when in alloys of the $LaNi_5$-type nickel was partly substituted by other elements [31, 32]. Cycle lifetimes of alloys $LaNi_{5-x}M_x$ (M = Mn, Cu, Cr, Al, Co) are obviously related with their mechanical properties such as the pulverizing character [33]. There are also promising results with MH-anodes based on AB_2-type "Laves alloys" such as $ZrMn_{0.6}Cr_{0.2}Ni_{1.2}$ [34] and $V_{22}Ti_{16}Zr_{16}Ni_{42}Cr_7$ [35].

As pure Lanthanum is quite expensive, so-called "Mischmetall" (Mm), i.e., unrefined rare earth metal mixtures containing larger fractions of La, Ce, Pr and Nd but also some Sm, Y and other elements, are commonly used in La-based hydrogen storage alloys (see e.g. refs. [36-38]).

To make up electrodes, powdered hydrogen storage alloys are usually mixed with PTFE binders. Microencapsulation of the alloy particles by "electroless" deposition of metals (mostly Ni or Cu) obviously improve their cycle lives [39,40].

The construction of cylindrical and prismatic nickel oxide metal hydride cells is shown in Fig. 6.

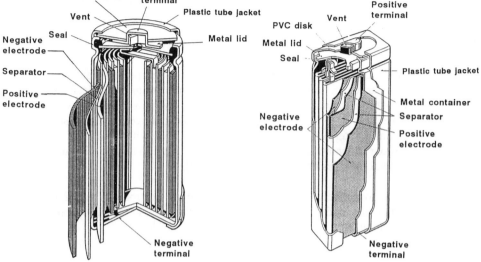

Fig. 6)   Construction of cylindrical and prismatic nickel oxide / metal hydride cells
            (by courtesy of VARTA)

The energy density of nickel oxide / metal hydride cells is in the order of 75 % higher than that of nickel / cadmium cells (see figure 7). For very low discharge rates the advantage of

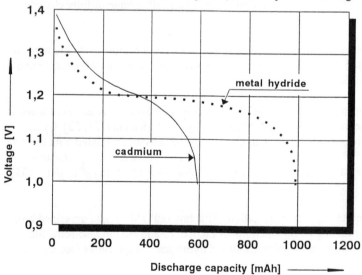

metal hydride batteries is even higher. On the other hand, at very high rates of charge and discharge metal hydride electrodes cannot compete with cadmium electrodes. A further drawback of metal hydride batteries is their very poor charge retention at elevated temperatures (see figure 5).

Fig. 7)   Typical discharge curves of AA-size nickel oxide /
            cadmium AA-size nickel oxide / metal hydride
            cells at 1 A   (by courtesy of VARTA)

Like conventional sealed nickel cadmium batteries, metal hydride batteries are also protected against overcharge and overdischarge. The capacity of metal hydride batteries is limited by the positive electrode. An "oxygen cycle" provides that oxygen produced during overcharging is consumed at the negative electrode.

As the performance of the nickel oxide electrode is improved by addition of some cadmium oxide, high duty metal hydride batteries are not always quite cadmium-free [41]. There are, however, also cadmium-free versions available.

Sanyo's Cylindrical Ni/MeH Rechargeable Cells		
MODEL	HR-4/3A	HR-AA
CAPACITY	2300 mAh (0.2C)	1100 mAh (0.2C)
VOLTAGE	1.2 V	1.2 V
CHARGE CURRENT	2300 mA	1100 mA
TIME	1.5 Hrs.	1.5 Hrs.
WEIGHT	51g	26g
EXTERNAL DIMENSIONS (D)x(H) (INCLUDING TUBE)	17 x 67 mm	14.2 x 50 mm
ENERGY VOLUME	196 Wh/l	178 Wh/l
WEIGHT	53 Wh/kg	51 Wh/kg

Tab. 1)   Specifications of Sanyo's cylindrical nickel oxide / metal hydride storage batteries [42]

INSERTION CATHODES IN LITHIUM BATTERIES

It was already observed in the early days of lithium battery research that many electrode materials which were completely insoluble in the electrolyte solution (e.g. MoO_3, TiS_2) could be reduced at high rates in Li^+-electrolytes, and much worse e.g. in K^+-electrolytes. When it was understood that this behaviour was related with the insertion of Li^+-ions (in some cases in a topotactic and fairly reversible process, sometimes immediately followed by considerable and more or less irreversible structural changes), an avalanche of scientific papers and patents on insertion cathodes went down.

Various types of host structures were studied, mostly transition metal oxides and chalcogenides but also conductive polymers and graphite compounds. As this field of research has already been covered by many reviews (see e.g. refs. [1, 43 - 50]) it will not be discussed in detail in this paper.

There is no doubt that manganese dioxide (gamma/beta-MnO_2 prepared by heat treatment of electrochemically fabricated gamma-MnO_2 [51]) has "made the race" for primary battery applications. The discharge of gamma/beta-MnO_2 is practically irreversible and yields Li_xMnO_2 ($0 < x < 1$) [52].

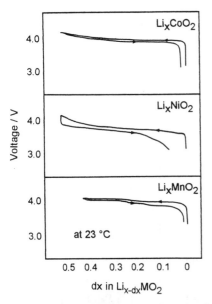

Fig. 8) Charge/discharge curves of cells Li/$Li_{x-dx}MO_2$ M = Co, Ni, Mn, in the discharged state x = 1.0 in Li_xCoO_2 and Li_xNiO_2 but 0.5 in Li_xMnO_2, $i_{charge} = i_{discharge}$ = 0.5 mA/cm^2, cut-off 4.2 V / 3.3 V, [59].

Many oxide host structures which are highly attractive in view of their energy densities calculated on the basis of a full capacity discharge are not useful for rechargeable batteries, because they can in fact only be cycled within very narrow ranges of stoichiometry. Cycling beyond these limits causes irreversible damage to the host lattice. On the other hand, most of the transition metal chalcogenides do resist long-term full capacity cycling. Unfortunately, their voltages vs. lithium and also their energy densities are relatively poor.

For the sake of reliabilty and safety of rechargeable liquid electrolyte lithium batteries metallic lithium anodes are more and more going to be replaced by lithium insertion anodes, at the expense of cell voltage and energy density. For compensation, the present trend is towards high voltage "4-volt" oxide cathodes such as Li_xCoO_2, Li_xNiO_2 and spinel-type $Li_xMn_2O_4$ [53-56]. These high voltage oxides are prepared in the lithiated (discharged) state and lithium is removed during charging [57, 58]. About 0.5 Li per transition metal M can be cycled (see figure 8).

INSERTION ANODES IN LITHIUM BATTERIES

Lithium alloy anodes:
As the use of metallic lithium anodes in organic solvent based rechargeable batteries is associated with severe problems, above all dendrite growth and corrosion, lithium alloy negatives Li_xM (e.g. Li_1Al) are attractive alternative anode materials [1, 60, 61].

$$Li_xM \underset{charge}{\overset{discharge}{\rightleftharpoons}} x\,Li^+ + x\,e^- + M \qquad (7)$$

For many applications the obvious disadvantages of Li-alloy anodes, i.e., lower rate capabilities and lower energy densities, are balanced by their improved cycling behaviour.

Like metallic lithium or lithium graphite intercalation compounds, Li-rich Li-alloys are thermodynamically instable in common battery electrolytes. Fortunately, the surfaces of all these highly reactive materials are protected by films of electrolyte decomposition products. There is general agreement that films formed on anodes in contact with organic electrolyte solutions comprise i) organic polymer and ii) inorganic decomposition products [62-64]. These films are still permeable to Li^+-ions but impermeable to solvent molecules.

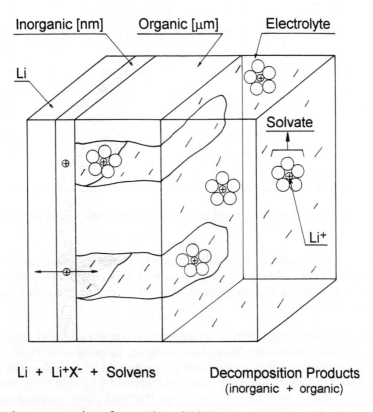

Fig. 9) Schematic representation of protection of lithium in organic electrolytes, inner layer: "solid electrolyte" layer permeable to unsolvated Li^+-cations outer layer: "polymer electrolyte" layer, permeable to solvated Li^+-cations

The main problems with Li-alloy electrodes are connected with the significant differences in volume between the Li-alloys Li_xM and the pure matrix metals M (see Tab. 2).

M	Li_xM	Δ volume	molar volume of Li in Li_xM
Al	LiAl	96.78	19.67
As	LiAs	91.63	24.88
	Li_3As	95.59	8.47
Bi	LiBi	75.88	29.06
	Li_3Bi	176.51	45.22
C	LiC_6	9.35	35.50
Cd	Li_3Cd	267.71	15.29
In	LiIn	52.29	23.53
Pb	LiPb	44.70	26.43
	$Li_{22}Pb_5$	233.66	13.85
Sb	Li_3Sb	147.14	14.99
Si	Li_2Si	175.12	15.15
	Li_4Si	322.57	11.64
Sn	$Li_{22}Sn_5$	76.31	28.75
Zn	LiZn	70.64	17.30
-	Li	-	13.00

Tab. 2) Volume changes related with insertion of lithium into matrix metals M [65-67].

Therefore, charge/discharge cycles induce mechanical stresses, cause cracks and finally the materials get crumbly. This is, of course worst with thick electrodes whereas thin layers of Li-alloys, corresponding to a charge of ca. 50 - 100 C / cm^2 can be cycled with high efficiencies [68].

There have been various attempts to reduce deterioration of Li-alloy electrodes during cycling in organic electrolytes. Most of this work was focussed on Al / beta-LiAl electrodes. Thin film aluminum-based electrodes have been suggested by several groups in order to minimize crack-formation due to insertion / extraction of lithium [69, 70]. Silicon-rich aluminium electrodes were shown to last longer than high purity aluminium electrodes in cycling experiments. This was attributed to a reinforcement of the active material by (much less reactive) silicon crystallites [71, 72] as silicon is not soluble in aluminium at room temperature. A similar concept (addition of a few % of Mn which is also insoluble in aluminium at room temperature) has been successfully used more recently [73, 74]. A patent was granted to aluminium-copper composite electrodes prepared by vapour phase deposition of thin layers of both metals [75] and "dimensionally stable" LiAl-copper electrodes prepared by compression of small particle size LiAl and Cu were also suggested [76].

The use of soft matrix metals or alloys of matrix metals (e.g. Wood's metal [77] or indium [78]) also reduces the problems related with crumbling. A very recent approach to stabilize Li-alloy anodes is incorporation of the matrix metals into the flexible microporous structure of polypropylene separator materials [79].

All-solid lithium electrodes with mixed conductor matrix, i.e., compact composite electrodes in which a reactant (e.g. Li_xSi) is finely dispersed in a solid mixed conducting matrix (e.g. Li_xSn) have been proposed by Huggins [80, 81]. These electrodes show excellent cycling behaviour, however, deep discharge will attack, i.e., also discharge, the mixed conductor matrix. Composite electrodes made from an intimate mixture of lithium and the solid Li^+-conductor Li_3N ("Linode") are also characterized by good cycling behaviour [82, 83].

Lithium-carbon anodes:
The electrochemical intercalation of alkali metal cations A^+ from organic solvent electrolytes into fairly crystalline graphitic materials usually yields solvated graphite intercalation compounds (GICs) $A^+(solv)_yC_n^-$ [84-87]. By contrast, alkali metal transfer in hydrocarbon solutions yields binary alkali metal GICs [88, 89]. The electrochemical potentials of these solvated compounds are more positive vs. the free alkali metals than those of the correponding binary GICs. Nevertheless solvated GICs are usually still thermodynamically instable vs. reduction of the intercalated polar solvents. The rate of the solvent reduction is mostly kinetically controlled. It is slow e.g for dimethyl sulfoxide (DMSO) [87] but very much faster e.g. for propylene carbonate (PC) [85, 90, 91].

The tendency to form solvated GICs increases with stage number, i.e. with the number of empty galleries in a GIC. It was shown that 1st stage binary alkali metal GICs do not tend to solvate in electron donor solvents such as benzene, whereas higher stage compounds much more readily co-intercalate solvent molecules [92, 93].

On the other hand, further electrochemical intercalation of alkali cations into 1st stage solvated GICs $A^+(solv)_yC_n^-$ may finally lead to alkali-rich binary GICs, i.e., the solvent is "squeezed" out and the vacant sites can be used to accomodate more alkali cations [94-96].

The degree of solvation in compounds $A^+(solv)_yC_n^-$ is not only controlled by the stoichiometry factor "n", the kind of solvent "solv" and the cation "A^+", but also very much by the crystallinity and morphology of the parent graphite material [97].

The formation of - at least essentially - binary Li-GICs during electrointercalation into p o o r l y crystalline "turbostratic" (= random stacking) graphitic materials from organic solutions of Li+-salts has been reported by many groups (see e.g. refs. [98-108]). The minimum stoichiometry factor "n" in these compounds is typically 8 - 10, whereas it is 6 in graphite-based 1st stage LiC_n [66]. Most of the parent graphites used in these studies were prepared by pyrolysis of organic polymer precursors at fairly low temperatures and it can be concluded that in these materials there will be not only random stacking but also considerable "crosslinking" of graphitic parts of the structure. Figure 10 shows schematically a "linear graphite hybrid", a material with ribbon-type and layer-type areas which was prepared by pyrolysis of crosslinked aromatic polymers.

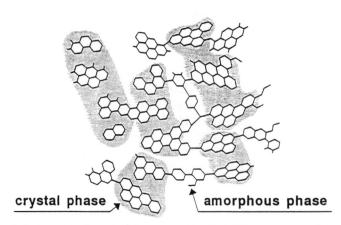

crystal phase   amorphous phase

Fig. 10) Schematic drawing of a "linear graphite hybrid" anode material
          (redrawn from ref. [109]).

Turbostratic carbon based binary GICs have found considerable interest as a substitute for
metallic lithium or lithium alloys as the anode material in rechargeable organic electrolyte
lithium batteries. Several companies have already commercialized batteries with LiC_n
anodes (see Tab. 3).

Sony's Lithium Ion Batteries			
Type of Cell	US-61 14500 (AA-size)	US-61 16530 (O 16 x 53 mm)	US-61 20500 (C-size) (O 20 x 50 mm)
Volume	7.5 cm^3	11.3 cm^3	16.5 cm^3
Weight	18 g	28 g	39 g
Operating Voltage	3.6 V	3.6 V	3.6 V
Capacity at 0.2C	400 mAh	640 mAh	1080 mAh
Energy Density	192 wh/l 78 wh/kg	204 wh/l 83 wh/kg	236 wh/l 99 wh/kg
Cycle Life in 100% DOD	1200 cycles	1200 cycles	1200 cycles
Drain Capability Capacity at 1C 2C	370 mAh 330 mAh	600 mAh 520 mAh	1000 mAh 880 mAh
Low Temp. 0,5C Dis. Capacity at 0°C -20°C	335 mAh 250 mAh	550 mAh 430 mAh	920 mAh 720 mAh

Tab. 3) Specifications of Sony's rechargeable organic electrolyte LiC_n / Li_xCoO_2 battery
        (lithium ion battery") [110]

Unlike metallic lithium or Li-alloys, LiC_n anodes can stand more than 1000 d e e p charge/discharge cycles and this advantage compensates for their relatively poor volumetric and gravimetric energy density. On the other hand, Li-alloy anodes would easily outperform LiC_n anodes, if there was a solution to the "pulverization" problem (cf. Tab. 2).

The phantastic cycling behaviour of LiC_n anodes is, of course, associated with their dimensional stability during insertion/removal of lithium. The volume increase caused by intercalation of lithium into highly oriented graphite (graphite $\rightarrow LiC_6$) is only 9.35 %. As the gravimetric densities of turbostratic carbon materials are significantly below that of crystalline graphite (because of imperfect structure and also some larger internal voids), the actual volume changes will be even lower.

More recently it was found by several groups that in ethylene carbonate (EC) based electrolytes b i n a r y GICs LiC_n are formed even if c r y s t a l l i n e graphite is used as a host material (see e.g. refs. [111, 112] and figure 11). In view of the fact that EC is an extremely polar solvent (dielectric constant ca. 95 !) this might be interpreted in terms of poor solvation of graphene layers by EC. On the other hand, it was shown that binary Li-GICs can also be obtained in mixed solvent electrolytes made up from 1,2-dimethoxyethane (DME) plus EC (50 : 50) [112, 113]. As DME (dielectric constant 5.5) is well known to be a typical solvent that is co-intercalated with alkali cations into graphite [87] the special effect of EC is very likely to be related with EC-based surface films on graphite which are only permeable to unsolvated Li^+-cations.

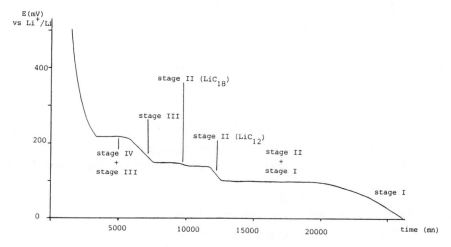

Fig. 11) Low-rate (1.02 μA mg^{-1}) discharge of highly oriented pyrolytic graphite in $LiClO_4$/EC [111].

In fact the (kinetical) stability of LiC_n-anodes in organic electrolytes is controlled by the more or less Li^+-selective permeability of protective films (consisting of electrolyte decomposition products) formed on their surface (see also figure 9). Many research activities are focussed on the improvement of these films by proper choice of electrolyte

[114, 115]. Some progress has been made with inorganic additives to promote the formation of thin inorganic films, such as CO_2, N_2O and S_x^{2-} [105, 107, 116-118]. Filming of carbon surfaces even renders possible reversible insertion/removal of Li^+-ions in $SOCl_2$ - electrolytes [119].

A serious drawback of LiC_n-anodes in organic electrolytes is their considerable self-discharge rate (in commercial batteries ca. 10 % per month). Unfortunately, in organic electrolyte batteries self-discharge is mostly associated with irreversible decomposition of the electrolyte, i.e., with a loss of material. By contrast, self-discharge in sealed aequous electrolyte batteries (e.g. nickel oxide / cadmium cells) means only a loss of charge, because recombination of decomposed water is possible.

Positive electrodes for batteries with LiC_n-anodes are usually chosen from the group of high voltage cathode materials, such as Li_xCoO_2 and related compounds. These cathode materials are fabricated in the fully lithiated, i.e., in the discharged state. Therefore cells of the type LiC_n/Li_xMO_2 can be mounted conveniently as only carbon (and not LiC_n) has to be handled to make up the negative electrodes. As overcharge and also overdischarge of LiC_n/Li_xMO_2 batteries may be harmful, commercial systems are provided with intelligent charging devices and additional protective electronics.

Aspects of carbon materials as anodes for "ion batteries" have also been covered by several recent comprehensive and review papers [112-122].

REFERENCES

1)   J.O. Besenhard in: W. Müller-Warmuth, R. Schöllhorn (eds.),
     Progress in Intercalation Research, Kluwer Academic Publishers, Dordrecht, in press.

2)   K.M. Abraham, Solid State Ionics, 7 (1982) 199.

3)   L.P. Kleman, G.H. Newman, J. Electrochem. Soc., 129 (1982) 230.

4)   P. Novak, J. Desilvestro, J. Electrochem. Soc., 140 (1993) 140.

5)   J.J. Coleman, Trans. Electrochem. Soc., 90 (1946) 545.

6)   A. Kozawa, R.A. Powers, J. Electrochem. Soc., 113 (1966) 870.

7)   K.V. Kordesch in: J.O'M. Bockris, B.E. Conway, E. Yeager, R.E. White (eds.),
     Comprehensive Treatise of Electrochemistry, Vol. 3, Electrochemical Energy Conversion and
     Storage, Plenum Press, New York, 1981, p. 219.

8)   K. Wiesener, J. Garche, W. Schneider,
     Elektrochemische Stromquellen, Akademie-Verlag, Berlin, 1981.

9)   R. Huber, Trockenbatterien, Verlag VDI, Düsseldorf, 1968.

10)   Batteries International, October 1993, p. 12.

11)   Advanced Battery Technology, 29(8) (1993).

12)   RENEWAL Reusable Alkaline, Product Information, (1993),
     Rayovac Corporation, 601 Rayovac Drive, Madison, WI 53711, USA.

13)   K. Kordesch, J. Gsellmann, M. Peri, K. Tomantschger, R. Chemelli,
     Electrochim. Acta, 26 (1981) 1495.

14) S. Llompart, L.T. Yu, C. J. Mas, A. Mendiboure, R. Vignaud,
J. Electrochem. Soc., 137 (1990) 371.

15) S. Llompart, H. Ouboumour, L.T. Yu, J. Electrochem. Soc., 138 (1991) 665.

16) K. Kordesch, Ch. Faistauer,
8th International Battery Materials Symposium, Brussels, May 1993.

17) W. Taucher, K. Kordesch in: C.A.C. Sequeira (ed.),
Environmental Oriented Electrochemistry, Elsevier Science Publishers, in press.

18) R. Schöllhorn, R. Kuhlmann, J.O. Besenhard, Mat. Res. Bull., 11 (1976) 83.

19) J.P. Pohl, H. Rickert, in: A. Kozawa, K, Kordesch (eds.),
Progress in Batteries and Solar Cells, Vol. 2, JEC Press, Cleveland, 1979, p. 179.

20) J.P. Pohl, H. Rickert, Z. Phys. Chem. N.F., 112 (1978) 117.

21) J.P. Pohl, H. Rickert, in: S. Trasatti (ed.),
Electrodes of Conductive Oxides, Elsevier, Amsterdam, 1980, p. 183.

22) J. Douglade, A. Metrot, P. Willmann, Mat. Sci. Forum, 91-93 (1992) 653.

23) P. Willmann, J. Douglade, V. Mancier, A. Metrot,
European Space Power Conference, Graz, 1993, Proceedings p. 581.

24) C. Delmas, International Symposium on Soft Chemistry Routes to New Materials,
Nantes, September 6 - 10, 1993.

25) K.I. Pandya, R.W. Hoffman, J. McBreen, W.E. O'Gady,
J. Electrochem. Soc., 137 (1990) 383.

26) R.D. Armstrong, E.A. Charles, J. Power Sources, 27 (1989) 15.

27) D.A. Corrigan, S.L. Knight, J. Electrochem. Soc., 136 (1989) 613.

28) F.E. Lynch, J. Less-Common Met., 172-174 (1991) 943.

29) H.F. Bittner, C.C. Badcock, J. Electrochem. Soc., 130 (1983) 193C.

30) T. Sakai, T. Hazama, H. Miyamura, N. Kuriyama, A. Kato, H. Ishikawa,
J. Less-Common Met., 172-174 (1991) 1175.

31) J.J.G. Willems, Philips J. Res., 39 (1984) 1.

32) J.J.G. Willems, K.H. Bushow, J. Less-Common Met., 129 (1987) 13.

33) T. Sakai, K. Oguro, H. Miyamura, N. Kuriyama, A. Kato, H. Ishikawa, C. Iwakura,
J. Less-Common Met., 161 (1990) 193.

34) Y. Moriwaki, T. Gamo, A. Shintani, T. Iwaki, Denki Kagaku, 57 (1989) 488.

35) S. Venkatasean, B. Reichman, M.A. Fetcenko, U.S. Pat. 4 728 586.

36) Y. Osumi, H. Suzuki, A. Kato, N. Nakane, J. Less-Common Met., 84 (1972) 99.

37) H. Ogawa, M. Ikoma, H. Kawano, I. Matsumoto, J. Power Sources, 12 (1988) 393.

38) Yong-Quan Lei, Zhou-Peng Li, Chang-Pin Chen, Jing Wu, Qi-Dong Wang,
J. Less-Common Met., 172-174 (1991) 1265.

39) T. Sakai, A. Yuasa, H. Ishikawa, H. Miyamura, N. Kuriyama,
J. Less-Common Met., 172-174 (1991) 1194.

40) T. Sakai, A. Takagi, K. Kinoshita, N. Kuriyama, H. Miyamura, H. Ishikawa, J. Less-Common Met., 172-174 (1991) 1185.

41) VARTA Batterie AG, Technical Information.

42) JEC Battery Newsletter, JEC Press, Brunswick, Ohio, (1991) No. 1, p. 17.

43) K.M. Abraham, J. Power Sources, 7 (1981) 1.

44) J.-P. Gabano, Lithium Batteries, Academic Press, London, 1983.

45) M. Hughes, N.A. Hampton, S.A.G.R. Karunathilaka, J. Power Sources, 12 (1984) 83.

46) R.V. Moshtev, J. Power Sources, 11 (1984) 93.

47) J. Desilvestro, O. Haas, J. Electrochem. Soc., 137 (1990) 5C.

48) R. Brec, E. Prouzet, G. Ouvrard, J. Power Sources, 43-44, (1993) 277.

49) J.B. Goodenough, A. Manthiram, B. Wnetrzewski, J. Power Sources, 43-44 (1993) 269.

50) D.W. Murphy, Adv. Synth. React. Solids, 1 (1991) 237.

51) M.M. Thackeray, A. de Kock, L.A. de Picciotto, J. Power Sources, 16 (1989) 355.

52) H. Ikeda, S. Narukawa, J. Power Sources, 9 (1983) 329.

53) K. Mizushima, P.C. Jones, P.J. Wiseman, J.B. Goodenough, Mat. Res. Bull., 15 (1980) 783.

54) L.A. de Picciotto, M.M. Thackeray, Mat. Res. Bull., 18 (1983) 1497.

55) J.M. Tarascon, D. Guyomard, J. Electrochem. Soc., 138 (1991) 2864.

56) M.M. Thackeray, M.H. Rossouw, A. de Kock, A.P. de la Harpe, R.J. Gummow, K. Pearce, D.C. Liles, J. Power Sources, 43-44 (1993) 289.

57) A. Mendioure, C. Delmas, P. Hagenmuller, Mat. Res. Bull., 19 (1984) 1383.

58) T. Ohzuku, M. Kitagawa, T. Hirai, J. Electrochem. Soc., 137 (1990) 769.

59) K. Sekai, H. Azuma, A. Omaru, S. Fujita, H. Imoto, T. Endo, K.Yamaura, Y. Nishi, J. Power Sources, 43-44 (1993) 241.

60) B.M.L. Rao, R.W. Francis, H.W. Christopher, J. Electrochem. Soc., 124 (1977) 1490.

61) J. Wang, I.D. Raistrick, R.A. Huggins, J. Electrochem. Soc., 133 (1986) 185.

62) M. Garreau, J. Thevenin, B. Milandou, in: A.N. Dey (ed.), Lithium Batteries, The Electrochemical Society, Pennington, N.J., 1984, p. 28.

63) D. Aurbach, M.L. Daroux, P.W. Faguy, E. Yeager, J. Electrochem. Soc., 134 (1987) 1611.

64) J. Thevenin, R.H. Muller, J. Electrochem. Soc., 134 (1987) 273, ibid, 134 (1987) 2650.

65) T. Landolt-Börnstein, Vol. 6, Structure Data of Elements and Intermetallic Phases, Springer, Berlin 1971.

66) D. Billaud, E. McRae, A. Herold, Mat. Res. Bull., 14 (1979) 857.

67) E. Zintl, G. Bauer, Z. Phys. Chem., B20 (1933) 245.

68) M. Garreau, J. Thevenin, M. Fekir, J. Power Sources, 9 (1983) 235.

69) J.R. Owen, W.C. Maskell, B.H.C. Steele, T. Sten Nielsen, O. Toft Soerensen,
Solid State Ionics, 13 (1984) 329.

70) P. Zlatilova, I. Balkanov, Y. Geronov, J. Power Sources, 24 (1988) 71.

71) J.O. Besenhard, H.P. Fritz, E. Wudy, K. Dietz, H. Meyer,
J. Power Sources, 14 (1985) 193.

72) J.O. Besenhard, P. Komenda, A. Paxinos, E. Wudy, M. Josowicz,
Solid State Ionics, 18 & 19 (1986) 823.

73) Sanyo Electric, U.S. Pat. 4 820 599.

74) JEC Battery Newsletter, JEC Press, Brunswick, Ohio, (1989) No.3, p. 17.

75) K. Fusaji, Y. Kazumi, K. Kozo,
Jap. Kokai, 63 132 64 A2, Chem. Abstr., 108 (1988) 170814j.

76) J.O. Besenhard, M. Hess, P. Komenda, Solid State Ionics, 40 & 41 (1990) 525.

77) Y. Matsuda, Proc. Int. Power. Sources Symp., 32nd (1988) 124.

78) Y. Nakane, K. Terashi, S. Furukawa,
Jap. Kokai, 61 193 360, Chem. Abstr., 106 (1987) 70349b.

79) J.O. Besenhard, M. Heß, J. Huslage, U. Krebber, K. Jurewicz,
J. Power Sources, 43-44 (1993) 493.

80) B.A. Boukamp, G.C. Lesh, R.A. Huggins, J. Electrochem. Soc., 128 (1981) 725.

81) R.A. Huggins, J. Power Sources, 26 (1989) 109.

82) C.D. Desjardins, G.K. MacLean,
in: S. Subbarao, V.R. Koch, B.B. Owens, W.H. Smyrl (eds.), Rechargeable Lithium Batteries,
The Electrochemical Society, Pennington, N.J., 1990, p. 108.

83) M. Maxfield, T.R. Jaw, S. Gould, M.G. Sewchok, L.W. Shacklette,
J. Electrochem. Soc., 135 (1988) 299.

84) J.O. Besenhard, H. P. Fritz, Angew. Chem. Int. Ed. Engl., 22 (1983) 950.

85) J.O. Besenhard, H.P. Fritz, J. Electroanal. Chem., 53 (1974) 329.

86) J.O. Besenhard, Carbon, 14 (1976) 111.

87) J.O. Besenhard, H. Möhwald, J.J. Nickl, Carbon 18 (1980) 399.

88) J.O. Besenhard, I. Kain, H.-F. Klein, H. Möhwald, H. Witty,
in: M.S. Dresselhaus G. Dresselhaus, J.E. Fischer, M.J. Moran (eds.),
Intercalated Graphite, North-Holland, New York, 1983, p. 221.

89) J.O. Besenhard, H. Witty, H.-F. Klein, Carbon, 22 (1984) 97.

90) A.N. Dey, J. Electrochem. Soc., 117 (1970) 222.

91) G. Eichinger, J. Electroanal. Chem., 74 (1976) 183.

92) L. Bonnetain, Ph. Touzain, A. Hamwi, Mater. Sci. Eng., 31 (1977) 45.

93) F. Beguin, R. Setton, L. Facchini, A.P. Legrand, G. Merle, C. Mai,
Synth. Metals, 2 (1980) 161.

94) Ph. Touzain, B. Marcus, Y. Maeda, L. Bonnetain, International Colloquium on Layered
Compounds, Pont-a-Mousson, 1988, Proceedings, p. 131.

95) B. Marcus, Ph. Touzain, J. Solid State Chem., 77 (1988) 223.

96) C. Fretigny, D. Marchand, M. Lagues,
Proc. 4th Int. Conf. Carbon, Baden-Baden, 1986, p. 204.

97) J.R. Dahn, A.K. Sleigh, Hang Shi, J.N. Reimers, Q. Zhong, B.M. Way,
Electrochim. Acta, 38 (1993) 1179.

98) R. Kanno, Y. Takeda, Y. Ichikawa, T. Nakanishi, O. Yamamoto,
J. Power Sources, 26 (1989) 535.

99) M. Mohri, N. Yanagishaw, Y. Tajima, H. Tanaka, T. Mitate, S. Nakajima,
M. Yoshida, Y. Yoshimoto, T. Suzuki, H. Wada,
J. Power Sources, 26 (1989) 545.

100) F. Fong, U. von Sacken, J.R. Dahn, J. Electrochem. Soc., 137 (1990) 2009.

101) T. Nagaura, K. Tozawa, Progress in Batteries and Solar Cells, 9 (1990) 209.

102) J.R. Dahn, U. von Sacken, M.W. Juzkow, H. Al-Janaby,
J. Electrochem. Soc., 128 (1991) 2207.

103) M. Sato, T. Iijima, K. Suzuki, K.I. Fujimote,
in: K.M. Abraham, M. Salomon (eds.), Primary and Secondary Lithium Batteries,
The Electrochemical Society, Pennington, NJ, 1991, p. 407.

104) Y. Yoshimoto, H. Wada, H. Nakaya, M. Yoshida, S. Nakajima,
Synth. Metals, 41-43 (1991) 2707.

105) B. Simon, J.P. Boeuve, M. Broussely, J. Power Sources, 43-44 (1993) 65.

106) P. Schoderböck, H.P. Boehm, Mater. Sci. Forum Vols., 91-93 (1992) 683.

107) O. (Youngman) Chusid, Y. Ein Ely, D. Aurbach, M. Babai, Y. Carmeli,
J. Power Sources, 43-44 (1993) 47.

108) R. Yazami, D. Guérard, J. Power Sources, 43-44 (1993) 39.

109) K. Inada, K. Ikeda, S. Inomata, T. Nishii, M. Miyabayashi, H. Yui,
in: Practical Lithium Batteries, JEC Press, Brunswick, Ohio, p. 96.

110) JEC Battery newsletter, JEC Press, Brunswick, Ohio, (1992) No. 3, p. 23.

111) P. Willmann, D. Billaud, F.X. Henry,
European Space Power Conference, Graz 1993, Proceedings, p. 789.

112) T. Ohzuku, T. Hirai, 44th Meeting of the International Society of Electrochemistry, Berlin,
1993, Abstracts, p.409.

113) T. Ohzuku, Y. Iwaskoshi, K. Sawari, J. Electrochem. Soc., 140 (1993) 2490.

114) C.-K. Huang, S. Surampudi, D.H. Shen, G. Halpert, The Electrochemical Society,
Fall Meeting, New Orleans, October 1993, Extended Abstracts p. 26.

115) S. Passerini, F. Croce, B. Scrosati, The Electrochemical Society,
Fall Meeting, New Orleans, October 1993. Extended Abstracts p. 39.

116) J.O. Besenhard, P. Castella, M.W. Wagner, 6th International Symposium on Intercalation
Compounds, ISIC-6, Orleans, 1991, Poster P-1.

117) J.O. Besenhard, P. Castella, M.W. Wagner,
3Mater. Sci. Forum Vols., 91-93 (1992) 647.

118) J.O. Besenhard, M.W. Wagner, M. Winter, A.D. Jannakoudakis, P.D. Jannakoudakis, E. Theodoridou, J. Power Sources, 43-44 (1993) 413.

119) J.O. Besenhard, M. Winter, in preparation.

120) K. Tatsumi, A. Mabuchi, T. Maeda, S. Higuchi, Osaka Kogyo Gijutsu Shikesho Kiho, 42 (1991) 150, Chem. Abstr., 116 (1992) 217957r.

121) T. Nagaura, Progress in Batteries and Solar Cells, 10 (1991) 218.

122) B. Scrosati, J. Electrochem. Soc., 139 (1992) 2776.

Materials Science Forum Vols. 152 - 153 (1994) pp. 35-42
© *1994 Trans Tech Publications, Switzerland*

REACTIVE FLUX SYNTHESES AT LOW AND HIGH TEMPERATURES

J.A. Cody, M.F. Mansuetto, S. Chien and J.A. Ibers

Department of Chemistry, Northwestern University, Evanston, IL 60208-3113, USA

Keywords: Reactive Flux Method, Chalcogenide, Polychalcogenide, Ternary, Quaternary, Alkali Polychalcogenide, $K_4Ti_3S_{14}$, $Na_2Ti_2Se_8$, $K_4Hf_3Te_{17}$, $Cs_4Zr_3Te_{16}$, Cu_3NbSe_4, KCu_2NbSe_4, $K_2CuNbSe_4$, K_3NbSe_4, $KCuZrQ_3$, $NaCuZrQ_3$, $Cs_{0.68}CuTiTe_4$, $CsTiUTe_5$

ABSTRACT

This paper reviews the reactive flux method for the synthesis of ternary and quaternary metal polychalcogenides, starting with the early synthesis of $K_4Ti_3S_{14}$ and $Na_2Ti_2Se_8$. It is shown that a variety of interesting solid-state metal chalcogenides, many of which show low dimensionality, may be synthesized over a range of temperatures. Compounds discussed include $K_4Hf_3Te_{17}$ and $Cs_4Zr_3Te_{16}$, the series Cu_3NbSe_4, KCu_2NbSe_4, $K_2CuNbSe_4$, and K_3NbSe_4, $KCuMQ_3$ and $NaCuMQ_3$ (M = Ti, Zr, Hf; Q = S, Se, Te), $Cs_{0.68}CuTiTe_4$, and $CsTiUTe_5$. The unusual structural features of some of these newly synthesized compounds are discussed.

INTRODUCTION

The reactive flux method [1] has proven to be widely applicable to the preparation of ternary and quaternary metal polychalcogenides containing an alkali metal or copper or both. It has produced an amazing variety of low-dimensional materials with novel structures.

In the reactive flux method one or more elemental powders and the components of the flux are heated together. Upon slow cooling of the resultant solution to room temperature a material, often crystalline, is formed that incorporates the elements added as well as the elements of the flux. Usually, the excess flux is washed away with water. The flux itself consists either of an A_2Q / Q mixture, where A = alkali metal and Q = S, Se, or Te (chalcogen), or of Cu / Te.

Following its initial demonstration, the method has been applied by several groups, most notably by Kanatzidis and coworkers [2], but this paper will be limited to a discussion of syntheses from our own laboratory.

EARLY SYNTHESES

While molten alkali-metal polysulfides had previously been used as fluxes in the growth of solid-state sulfides [3-5] none of the compounds prepared contained polysulfide ions. In the initial demonstration of the reactive flux method [1] Ti, K_2S, and S in the molar ratio 1 : 2 : 6 were kept at 375 °C for 50 hr and then allowed to cool to room temperature. The excess melt was dissolved in distilled water and crystals of the new one-dimensional material $K_4Ti_3S_{14}$ were obtained. The crystal structure of $K_4Ti_3S_{14}$ comprises well separated K^+ cations and one-dimensional anionic chains. The chain (figure 1) contains two crystallographically unique Ti atoms, one of which is seven coordinate, the other eight coordinate. The chain contains both isolated S^{2-} and S_2^{2-} ions, the latter being readily identified by S–S distances of 2.065(2) to 2.071(2) Å.

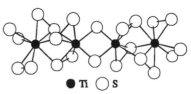

● Ti ○ S

Figure 1. View of the $\frac{1}{\infty}[Ti_3(S_2)_6(S)_2^{4-}]$ chain in $K_4Ti_3S_{14}$.

The anionic chain may thus be formulated as $\frac{1}{\infty}[Ti_3(S_2)_6(S)_2^{4-}]$. With the usual convention of describing a polychalcogenide species as possessing a II– charge, Q_n^{2-}, Q = S, Se, Te, this anionic chain contains Ti^{IV} centers. In a similar synthetic procedure the compound $Na_2Ti_2Se_8$ was synthesized through the reaction of Ti metal with a Na_2Se/Se flux at 345 °C [6]. This compound contains the anionic chain $\frac{1}{\infty}[Ti_2(Se_2)_3(Se)_2^{2-}]$ (figure 2), as deduced from the Se–Se distances of 2.347(2) to 2.356(2) Å characteristic of the Se_2^{2-} species. Once again, this chain contains Ti^{IV} centers.

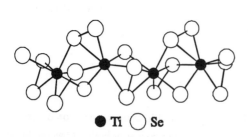

● Ti ○ Se

Figure 2 View of the $\frac{1}{\infty}[Ti_2(Se_2)_3(Se)_2^{2-}]$ chain in $Na_2Ti_2Se_8$.

Two observations were made about these

syntheses. The first was that better crystals of $K_4Ti_3S_{14}$ resulted if the solution were first heated to 890 °C, before being cooled to 375 °C for extended heating. This suggests that $K_4Ti_3S_{14}$ is stable at higher temperatures and that crystals grew when the reaction solution was cooled from 890 °C to 375 °C. That a polychalcogenide should be stable at high temperatures is not surprising. For example, Nb_2Se_9 on the basis of its structure [7, 8] may be formulated as $Nb_2(Se_5)(Se_2)_2$; it is synthesized around 700 °C. The second early observation was that when mixtures of $Na_2Se + \frac{1}{2}$ Ti + n Se (n = 3–7) were heated at 375 °C for 100 hr, reactions where n = 3 or 4 afforded no ternary product; reactions where n ≥ 5 gave crystals of $Na_2Ti_2Se_8$ in very low yield. A reaction at 345 °C where n = 5 afforded the same product in about 18% yield. Thus, the product obtained in this instance (and in general) is strongly dependent on the composition of the flux.

The synthesis of $K_4M_3Te_{17}$ (M = Zr, Hf) [9] was important in that it established that the reactive flux method could be extended to tellurides and that polytellurides could be made at high temperatures, in this instance 900 °C. $K_4M_3Te_{17}$ is once again a one dimensional material. The anionic chain (figure 3) is more difficult to describe because, as opposed to S–S and Se–Se distances which tend to be characteristic of Q^{2-} or Q_n^{2-}, Q = S, Se, Te–Te distances can vary continuously from about 2.7 Å to 4.0 Å [10-12]. In the $K_4Hf_3Te_{17}$ anionic chain the "short" Te–Te distances range from 2.756(5) to 3.260(4) Å. If we take as bonding Te–Te distances less

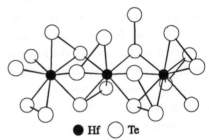

● Hf   ○ Te

Figure 3. View of the anionic chain in $K_4Hf_3Te_{17}$.

than 2.95 Å, then the anion may be formulated as $\frac{1}{\infty}[Hf_3(Te_3)(Te_2)_7{}^{4-}]$ and it contains Hf^{IV} centers. Alternatively, if we selected all Te–Te distances less than 3.26 Å as bonding, then the anion becomes $\frac{1}{\infty}[Hf_3(Te_{14})(Te_2)(Te)^{4-}]$ and it contains $Hf^{0.67}$ centers! As we shall see, no obvious formulation of a closely related Cs compound is possible even with this rather arbitrary method of counting.

RECENT SYNTHESES

From a loading of Cu : Zr : Cs_2Te_3 : Te of 2 : 1 : 2 : 4 at 900 °C the compound $Cs_4Zr_3Te_{16}$ was synthesized. (The target compound was actually $CsCuZrTe_3$!) $Cs_4Zr_3Te_{16}$ contains the anionic infinite chain shown in figure 4. Here the Te–Te distances and their numbers are: 2.76 Å (4); 2.93 – 2.97 Å (6); 3.05 – 3.07 Å (2). Thus the only formulation of this anionic chain we offer is $\frac{1}{\infty}[Zr_3(Te_2)_4??^{4-}]$; all other possibilities are rather arbitrary in view of the distribution of Te–Te distances. The trigonal prismatic frameworks about the Hf or Zr centers in $K_4Hf_3Te_{17}$ and $Cs_4Zr_3Te_{16}$ are contrasted in figure 5. A difference in cation radius of 0.33 Å [13] moving from K^+ to Cs^+ induces many subtle changes in the structure of the one-dimensional M/Te chains. The coherency

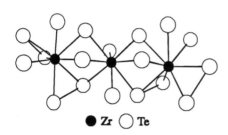

● Zr   ○ Te

Figure 4. View of the anionic chain in $Cs_4Zr_3Te_{16}$.

of the trigonal prismatic framework is broken (gray boxes represent missing vertices in figure 5) and the $\mu_2-\eta^1-Te_2^{2-}$ ligand is not present.

In solid-state chalcogenides, Cu is invariably found in the I+ oxidation state, as, of course, are the alkali metals. The compounds Cu_3NbSe_4 [14] and K_3NbSe_4 [15] are readily made. Cu_3NbSe_4 has a three-dimensional structure comprising edge- and corner-sharing $CuSe_4$ and $NbSe_4$ units to form a structure in which there are Cu–Se and Nb–Se bonding distances. On the other hand, the structure of K_3NbSe_4 is composed of isolated K^+ and $NbSe_4^{3-}$ ions. Can we make the intermediates, in which we progressively substitute K for Cu? Indeed we can through the use of the reactive flux method. The structure of KCu_2NbSe_4 [16] comprises edge-sharing and corner-sharing tetrahedra that form layers

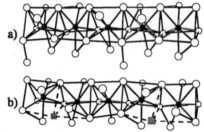

Figure 5. The trigonal prismatic framework in a) $K_4Hf_3Te_{17}$ and b) $Cs_4Zr_3Te_{16}$

separated by K^+ cations. This is a two-dimensional structure. The structure of $K_2CuNbSe_4$ [17] consists of infinite chains of edge-sharing, alternating $CuSe_4$ and $NbSe_4$ tetrahedra separated from one another by K^+ cations. This is a one-dimensional structure. The decreasing dimensionality from three to two to one to isolated of the non-alkali metal portion of these structures as the alkali metal is added is a general trend [18].

The sensitivity of resultant compound to flux composition is well illustrated by the synthesis of $K_3Cu_3Nb_2Se_8$ [19] and $K_3CuNb_2Se_{12}$ [17], which were produced by small variations in conditions used to obtain KCu_2NbSe_4 and $K_2CuNbSe_4$. Both of these compounds display one-dimensional structures.

Above 50 mole% Te the Cu/Te phase diagram shows a low-melting region with a wide liquid range [20]. It therefore seemed feasible to apply the reactive flux method to the synthesis of Cu/M/Te systems. We have synthesized the series of compounds Cu_2MTe_3, M = Ti, Zr, Hf [21], as well as $Cu_{1.85}Zr_2Te_6$ [22]. There is no Te–Te bonding in Cu_2MTe_3, and so these may be formulated as containing Cu^I, M^{IV}, and Te^{II-}. As such, they should be insulators or semiconductors, but instead they are metallic conductors. It has been suggested from band-structure calculations [23] that the conductivity arises from non-stoichiometry, but the carrier concentration derived from the Hall effect measurements on Cu_2ZrTe_3 is such that any nonstoichiometry is below the limit detectable by X-ray diffraction techniques [23].

In view of the existence of the Cu_2MTe_3 (M = Ti, Zr, Hf) series and our interest in substituting alkali metal for copper, the obvious study of potential quaternary compounds of the type A/Cu/M/Q (A = alkali metal; M = Ti, Zr, Hf; Q = S, Se, Te) was made. Thus far, a number of compounds in this series have been prepared and characterized [24, 25]. The compounds $KCuZrQ_3$ (Q = S, Se, Te) are isostructural and are composed of $^2_\alpha[CuZrQ_3^-]$ layers (figure 6) separated by K^+ cations. In the layers the Cu atoms are tetrahedrally coordinated and the Zr atoms are octahedrally coordinated by Q atoms. Within the $^2_\alpha[CuZrQ_3^-]$ layer, the tetrahedra share edges with four adjacent octahedra. In this structure there is a zigzag chain of alternating ZrQ_6 octahedra and CuQ_4 tetrahedra. This series of compounds does not exhibit Q–Q interactions and hence is described formally as

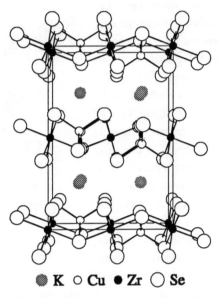

● K ○ Cu ● Zr ○ Se

Figure 6. View of KCuZrSe₃ down [100].

containing Cu^I, Zr^{IV}, and Q^{II-}. As such we would expect these compounds to be insulators or semiconductors. Conductivity measurements [24] indicate that $KCuZrTe_3$ is a metal from 4 to 300 K, $KCuZrSe_3$ displays metallic behavior down to 50 K, where there is a transition to semiconducting behavior, while $KCuZrS_3$ is an insulator. These varied conductivity properties need further investigation.

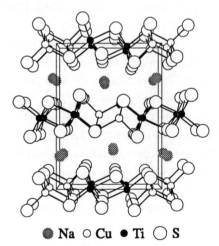

● Na ○ Cu ● Ti ○ S

Figure 7. View down [010] of the structure of NaCuTiS₃.

Closely related to the $KCuZrQ_3$ series are the compounds $NaCuZrQ_3$ and $NaCuTiS_3$ [25]. While $NaCuZrS_3$ has the $KCuZrQ_3$ structure, the other Na-containing compounds, which are isostructural, have a different structure (figure 7). The basic difference is the existence of alternating *pairs* of CuQ_4 and ZrQ_6 octahedra.

Since the structure of $ACuMQ_3$ changes on going from A = K to A = Na, a reasonable extension is to see what the structure of A = Cs might be. As is so often the case in synthetic solid-state chemistry, the expected compound was not obtained. Rather $Cs_{0.68}CuTiTe_4$ resulted (figure 8). $Cs_{0.68}CuTiTe_4$ is a three-dimensional channel structure composed of $^2_\alpha[Cu_2Ti_2Te_8^{1.36-}]$ units (figure 9) with pairs of Cs^+

cations occupying the channels. Within the unit there are CuTe$_4$ tetrahedra and TiTe$_6$ octahedra. The units contain *pairs* of Cu tetrahedra bordered by Ti octahedra (figure 9) that are connected through Te–Te bonds to complete the channel structure. The occurrence of short Te–Te distances of 2.923(1) and 3.084(2) Å and the nonstoichiometry on the Cs position make assignment of formal

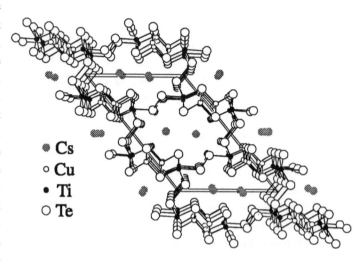

* Cs
○ Cu
• Ti
○ Te

Figure 8. View of Cs$_{0.68}$CuTiTe$_4$ down [010].

oxidation states difficult. Whether the unexpected product Cs$_{0.68}$CuTiTe$_4$, rather than CsCuTiTe$_3$, resulted from the use of Cs in place of K or Na or the use of a lower temperature (350 °C) rather than 700 °C or above, is a subject that needs investigation.

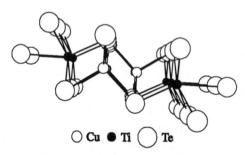

○ Cu • Ti ○ Te

Figure 9. The $_\infty^2[Cu_2Ti_2Te_8^{1.36-}]$ unit in Cs$_{0.68}$CuTiTe$_4$.

The new layered compound CsTiUTe$_5$ was synthesized at 900 °C from the reactive flux Cs$_2$Te$_3$/Te. The unit cell of CsTiUTe$_5$ viewed down [100] shows the layers of $_\infty^2[UTiTe_5^-]$ separated by Cs$^+$ cations (figure 10). The U atom is in a bicapped trigonal prism and the Ti atom is in the center of an octahedron of Te atoms (figure 11). Each TiTe$_6$ octahedron shares a *face* with two adjacent octahedra to form a $_\infty^1[TiTe_3^{2-}]$ chain. Interestingly, the Cs atom is in a pentagonal prism

of Te atoms. The chain of face-sharing TiTe$_6$ octahedra is unusual (figure 12). TiTe$_2$ has the CdI$_2$ structure and is composed of edge-sharing octahedra. Other layered compounds have TiQ$_6$ octahedra, but other than in the present compound apparently only in BaTiS$_3$ [26] do the octahedra face-share. From an empirical study of the relationship between the oxidation state of U and the ionic radius of the species [27] we conclude that eight-coordinate U^{IV} should exhibit U–Te distances greater than 3.1 Å. The U^{III}–Te distances should be even longer. U–Te distances of 3.058(1), 3.112(5), and 3.262(1) Å are found in CsTiUTe$_5$. The bicapped trigonal prisms of U atoms share an edge and have a short Te(2)–Te(2) distance of 3.065(1) Å.

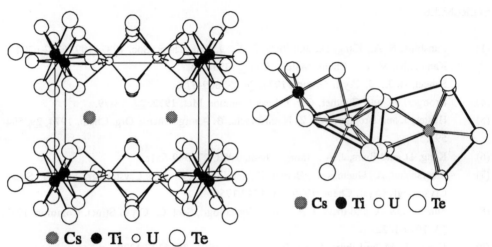

● Cs ● Ti ○ U ○ Te

Figure 10. The unit cell of CsTiUTe₅ viewed down [100].

● Cs ● Ti ○ U ○ Te

Figure 11.   The unique metal atom environments in CsTiUTe₅.

SUMMARY AND THE FUTURE

The reactive flux method has proved very useful in the synthesis of new ternary and quaternary metal chalcogenides. A wide variety of new compounds with unexpected compositions and structures and interesting physical properties, especially many one-dimensional materials, have been synthesized.

To make reactive flux syntheses more predictable and even more useful the effects of temperature and composition of the flux need to be investigated systematically. Can the method be extended to other systems? While the limit for quartz is around 1200 °C, if suitable containers can be found it would certainly be feasible to use fluxes whose liquid range extends beyond that temperature. Undoubtedly, entirely new materials could be synthesized in that way.

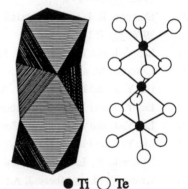

● Ti ○ Te

Figure 12.   Chain of TiTe₆ face-sharing octahedra in CsTiUTe₅.

ACKNOWLEDGMENT

This research was supported by the U. S. National Science Foundation through Grant DMR91-14934.

REFERENCES

(1)   Sunshine, S. A., Kang, D., and Ibers, J. A.: J. Am. Chem. Soc. 1987, 109, 6202-6204.

(2)   Kanatzidis, M. G.: Chem. Mater. 1990, 2, 353-363.

(3)   Scheel, H. J.: J. Cryst. Growth 1974, 24/25, 669-673.

(4)   Bronger, W. and Günther, O.: J. Less-Common Met. 1972, 27, 73-79.

(5)   Huster, J. and Bronger, W.: Z. Naturforsch., B: Anorg. Chem., Org. Chem. 1974, 29, 594-595.

(6)   Kang, D. and Ibers, J. A.: Inorg. Chem. 1988, 27, 549-551.

(7)   Meerschaut, A., Guémas, L., Berger, R., and Rouxel, J.: Acta Crystallogr., Sect. B: Struct. Crystallogr. Cryst. Chem. 1979, 35, 1747-1750.

(8)   Sunshine, S. A. and Ibers, J. A.: Acta Crystallogr., Sect. C: Cryst. Struct. Commun. 1987, 43, 1019-1022.

(9)   Keane, P. M. and Ibers, J. A.: Inorg. Chem. 1991, 30, 1327-1329.

(10)   Mar, A., Jobic, S., and Ibers, J. A.: J. Am. Chem. Soc. 1992, 114, 8963-8971.

(11)   Canadell, E., Jobic, S., Brec, R., and Rouxel, J.: J. Solid State Chem. 1992, 98, 59-70.

(12)   Canadell, E., Jobic, S., Brec, R., Rouxel, J., and Whangbo, M.-H.: J. Solid State Chem. 1992, 99, 189-199.

(13)   Shannon, R. D.: Acta Crystallogr., Sect. A: Cryst. Phys., Diffr., Theor. Gen. Crystallogr. 1976, 32, 751-767.

(14)   Lu, Y.-J. and Ibers, J. A.: J. Solid State Chem. 1993, in press.

(15)   Latroche, M. and Ibers, J. A.: Inorg. Chem. 1990, 29, 1503-1505.

(16)   Lu, Y.-J. and Ibers, J. A.: J. Solid State Chem. 1991, 94, 381-385.

(17)   Lu, Y.-J. and Ibers, J. A.: Inorg. Chem. 1991, 30, 3317-3320.

(18)   Lu, Y.-J. and Ibers, J. A.: Comments Inorg. Chem. 1993, 14, 229-243.

(19)   Lu, Y.-J. and Ibers, J. A.: J. Solid State Chem. 1992, 98, 312-317.

(20)   Blachnik, R., Lasocka, M., and Walbrecht, U.: J. Solid State Chem. 1983, 48, 431-438.

(21)   Keane, P. M. and Ibers, J. A.: J. Solid State Chem. 1991, 93, 291-297.

(22)   Keane, P. M. and Ibers, J. A.: Inorg. Chem. 1991, 30, 3096-3098.

(23)   Mitchell, J. F., Burdett, J. K., Keane, P. M., Ibers, J. A., DeGroot, D. C., Hogan, T. P., Schindler, J. L., and Kannewurf, C. R.: J. Solid State Chem. 1992, 99, 103-109.

(24)   Mansuetto, M. F., Keane, P. M., and Ibers, J. A.: J. Solid State Chem. 1992, 101, 257-264.

(25)   Mansuetto, M. F., Keane, P. M., and Ibers, J. A.: J. Solid State Chem. 1993, 105, 580-587.

(26)   Clearfield, A.: Acta Crystallogr. 1963, 16, 135-142.

(27)   Noel, H.: J. Solid State Chem. 1984, 52, 203-210.2

Materials Science Forum Vols. 152 - 153 (1994) pp. 43-54

THE SOL-GEL ROUTE TO ADVANCED MATERIALS

J. Livage

Chimie de la Matière Condensée, Université Pierre et Marie Curie,
4, place Jussieu, F-75252 Paris, France

Keywords: Molecular Precursors, Sol-Gel, Akoxides

ABSTRACT

The sol-gel route to oxide materials is based on inorganic polymerization reactions. A solution of molecular precursors is converted by a chemical reaction into a sol or a gel which on drying and densification give a solid material. This allows the production of single or multicomponent materials with high purity, novel compositions, tailored microstructures and potentially greater chemical homogeneity at lower temperature. Moreover films or fibres can be obtained directly from sols or gels by such techniques as dip-coating, spin-coating, spray or drawing.

Condensation reactions can be chemically controlled via the molecular design of alkoxide precursors with nucleophilic species such as carboxylates or β-diketones. Non hydrolyzable organic ligands lead to the formation of hybrid materials in which organic and inorganic moities are chemically bonded.

The sol-gel process is highly amenable to incorporating organic or biological species into oxide matrices. These hybrid nanocomposites open new possibilities for producing advanced materials for optical devices or chemical sensors.

INTRODUCTION

The first sol-gel synthesis of an inorganic compound was published almost 150 years ago when a french chemist, M. Ebelmen, observed that tetraethylorthosilicate (TEOS) slowly transformed into glassy silica in the presence of moisture [1]. Hundreds of scientific papers are now published yearly throughout the world and the first sol-gel products appeared on the market in the fifties. Large scale production started with automotive rear-view mirrors and anti-reflecting coatings. A wide range of products such as fibres, fine powders, porous solids and coatings, are now produced via the so-called "sol-gel" process [2].

The sol-gel synthesis of oxide materials provides a new approach to the preparation of glasses and ceramics [3]. A solution of molecular precursors is progressively transformed into an oxide network via inorganic polymerization reactions. Most versatile precursors are metal alkoxides $M(OR)_z$ where R is an alkyl radical (R=CH_3, C_2H_5,...) [4]. Hydrolysis and condensation give oligomers, oxopolymers and sols which can be shaped, gelled, dried and densified in order to get powders, films, fibres or monolithic glasses [2].

CHEMICALLY CONTROLLED CONDENSATION

The hydrolysis and condensation of metal alkoxides leads to the formation of an oxide network directly from a solution at room temperature. It can therefore be described as a soft chemistry route to glasses and ceramics.

The hydrolysis and condensation of coordinatively saturated metal alkoxides correspond to the nucleophilic substitution of alkoxy ligands by hydroxylated species XOH [5]:

$$M(OR)_z + x\ XOH \Rightarrow [M(OR)_{z-x}(OX)_x] + x\ ROH \qquad (1)$$

where X stands for Hydrogen (hydrolysis), a Metal atom (condensation) or even an organic or inorganic Ligand (complexation) .

These reactions can be described by a S_N2 mechanism as follows :

$$
\begin{array}{c}
H \\
\overset{|}{\underset{X}{O}}{}^{\delta-} + M^{\delta+}\text{-}O^{\delta}\text{-}R \Rightarrow
\end{array}
\begin{array}{c}
H^{\delta+} \\
\overset{\diagdown}{\underset{X}{O}}\text{-}M\text{-}O^{\delta-}\text{-}R \Rightarrow
\end{array}
\begin{array}{c}
H^{\delta+} \\
XO\text{-}M\text{-}\overset{\diagup}{\underset{R}{O}} \Rightarrow XO\text{-}M + ROH \qquad (2)
\end{array}
$$

The nucleophilic addition of $HO^{\delta-}$ groups onto $M^{\delta+}$, increases the coordination number of the metal atom in the transition state. The positively charged proton is then transfered toward an alkoxy group and the protonated ROH ligand is finally removed.

According to this mechanism the chemical reactivity of metal alkoxides toward hydrolysis and condensation mainly depends on the positive charge of the metal atom δ_M and its ability to increase its coordination number "N". As a general rule, the electronegativity of metal atoms decreases and their size increases when going down the periodic table (Table 1). The corresponding alkoxides become progressively more reactive toward hydrolysis and condensation. Silicon alkoxides are rather stable while alkoxides of electropositive metals must be handled with care under a dry atmosphere in order to avoid precipitation. Alkoxides of highly electronegative elements such as $PO(OEt)_3$ cannot be hydrolyzed under ambient conditions, whereas the corresponding vanadium derivatives $VO(OEt)_3$ are readily hydrolyzed into vanadium pentoxide gels [6].

TABLE 1 : Electronegativity "χ", partial charge "δ", ionic radius "r" and maximum coordination number "N" of some metal alkoxides

alkoxide	χ	δ	r(Å)	N
$Si(OPr^i)_4$	1.74	+0.32	0.40	4
$Ti(OPr^i)_4$	1.32	+0.60	0.64	6
$Zr(OPr^i)_4$	1.29	+0.64	0.87	7
$Ce(OPr^i)_4$	1.17	+0.75	1.02	8
$PO(OEt)_3$	2.11	+0.13	0.34	4
$VO(OEt)_3$	1.56	+0.46	0.59	6

The sol-gel process does not only lead to the formation of metal-oxygen bonds in the solution. It also makes it possible to control the formation of the oxide network. The strong electronegativity of oxygen (χ_O=3.5) makes M-OR bonds strongly polar and the metal atom highly prone to nucleophilic reagents. Metal alkoxides react with hydroxylated ligands XOH giving $[M(OR)_{z-n}(OX)_n]$ species. Chemical additives such as carboxylic acids or β-diketones are therefore currently employed in order to stabilize highly reactive metal alkoxides and control condensation reactions. They behave as bidentate, bridging or chelating, ligands. Such reactions lead to important modifications of the molecular structure of the alkoxide (Fig.1). A new precursor is formed which exhibits different molecular structure, chemical reactivity and functionality [7].

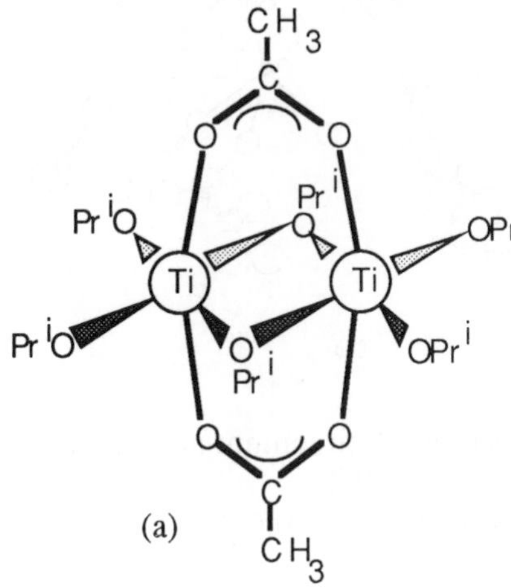

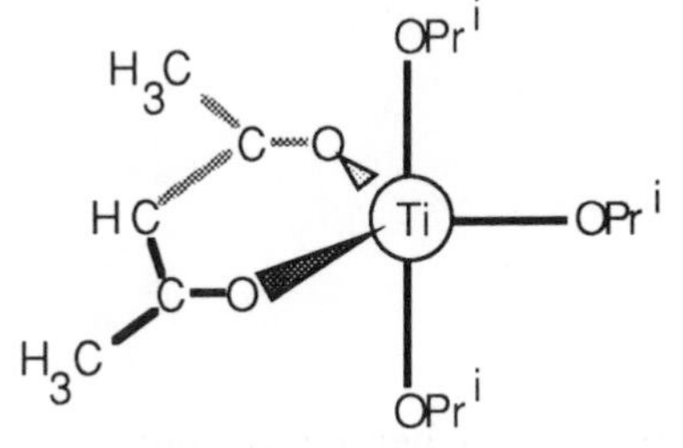

Figure 1. Chemical modification of $Ti(OPr^i)_4$ with :
a : acetic acid $[Ti(OPr^i)_3(OAc)]_2$, b : acetylacetone $[Ti(OPr^i)_3(acac)]$

The hydrolysis of metal alkoxides gives reactive M-OH bonds which lead to condensation and favor the formation of larger species. Complexation leads to non reactive M-OX bonds which act as polymerization lockers and favor the formation of smaller species. A large variety of oligomeric species can then be obtained upon hydrolysis and condensation. Molecular clusters or colloidal particles can be synthesized depending on the relative amount of hydrolysis ($h=H_2O/M$) and complexation ($x=X/M$).

In the presence of a small amount of water ($h<1$), only few alkoxy groups are removed leading to molecular clusters which sometimes give rise to single crystals when the mother solution is left for aging in a closed vessel (Fig.2). Several alkoxy groups are hydrolyzed giving rise to μ-oxo or μ-OH bridges, while complexing ligands are still bonded to the metal atom. They are located outside the oxo rich core of the molecular species (i.e. at the surface) and act as terminal ligands preventing further condensation. More condensed species are obtained as x decreases and h increases [8].

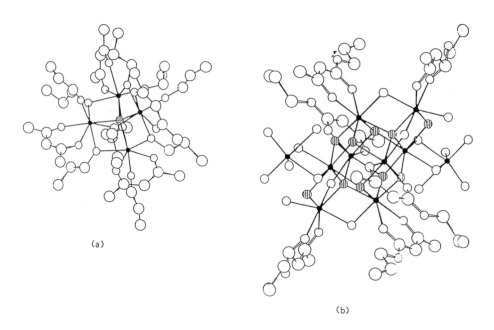

(a)

(b)

Figure 2. Chemically controlled condensation of zirconium alkoxides
a. $Zr_4O(OPr)_4(acac)_4$ ($x=1$, $h=0.2$)
b. $Zr_{10}O_6(OH)_4(OPr)_{18}$(allylacetoacetate)$_6$ ($x=0.6$, $h=1$)
● = M, ◉ = μ-oxo, O = other atoms

Polymeric sols are obtained when more water is added to zirconium alkoxide solutions. They are made of amorphous zirconium oxo-species in which most acac ligands remain bonded to zirconium preventing condensation and the formation of an oxide network. Crystalline zirconia particle are never obtained when hydrolysis is performed at room temperature. For crystallization to occur, zirconium bonded complexing ligands have to be removed so that Zr-O-Zr bonds can be formed. A crystalline oxide network can then be formed upon heating as for the thermo-hydrolysis of aqueous solutions of zirconium oxychloride. This has been obtained when reflux is performed in the presence of an organic acid such as PTSA (Para Toluene Sulfonic Acid). Reversible dissociation of Zr-acac bonds then occurs followed by preferential recoordination at the surface of the growing cluster. These surface acetylacetonato ligands, in equilibrium with free acac in the solution, slow down the growth process of the colloidal particles and prevent their aggregation via steric hindrance effects. Kinetically stable non aggregated crystalline zirconia nano-particles are then obtained [9].

MOLECULAR BUILDING BLOCKS

Most advanced ceramics are multicomponent materials having two or more types of cations in the lattice. Since alkoxide precursors are mixed at the molecular level in the solution, a high degree of homogeneity can be expected. However a major problem in forming homogeneous multicomponent gels is the unequal hydrolysis and condensation rates of the metal alkoxides. This may result in phase separation, during hydrolysis or thermal treatment, leading to higher crystallization temperatures or even undesired crystalline phases. The sol-gel synthesis of high-temperature superconductors such as yttrium barium cuprates for instance leads to the formation of $BaCO_3$, CuO and Y_2O_3 which gives few advantages over conventional solid-state syntheses [10]. It is therefore necessary to prepare gels of high homogeneity in which cations of various kinds are uniformly distributed at an atomic scale through M-O-M' bridges. Several approaches have been attempted to overcome this problem, including partial prehydrolysis or matching of hydrolysis rates by chemical modification with chelating ligands. In order to prepare crystalline materials at low temperatures it is necessary to design metal-organic precursors such that the metal ions are dispersed at the molecular level and the ligands undergo facile elimination during the transformation from molecular to bulk material. The development of precursor solutions which consist of polynuclear complexes with the metals in the stoichiometry of the desired oxide product will have a beneficial effect by lowering processing temperatures and times. Heterometallic alkoxides, which contain two or more different metal atoms linked by μ-OR bridges are claimed to provide best precursors for the sol-gel synthesis of multicomponent materials [11].

Most metal alkoxides actually exhibit oligomeric molecular structures. This is due to the tendency of the metal atom to expand its coordination number via the nucleophilic addition of other alkoxide molecules and the formation of alkoxy bridges. The molecular complexity of metal alkoxides increases when the size of the metal atom increases and the steric hindrance of alkoxy ligands decreases [4][12]. Oligomeric species $[Ti(OEt)_4]_n$ have been evidenced for titanium ethoxide (Fig.3b), both in the solid state (n=4) and in the liquid state (n=3) whereas titanium iso-propoxide $Ti(OPr^i)_4$ remains monomeric (Fig.3a). Solvent molecules can also be used for coordination expansion leading to the formation of solvates. Zirconium iso-propoxide, for instance, gives solvated dimers $[Zr(OPr^i)_4(Pr^iOH)]_2$ when dissolved in its parent alcohol (Fig.3c) [7].

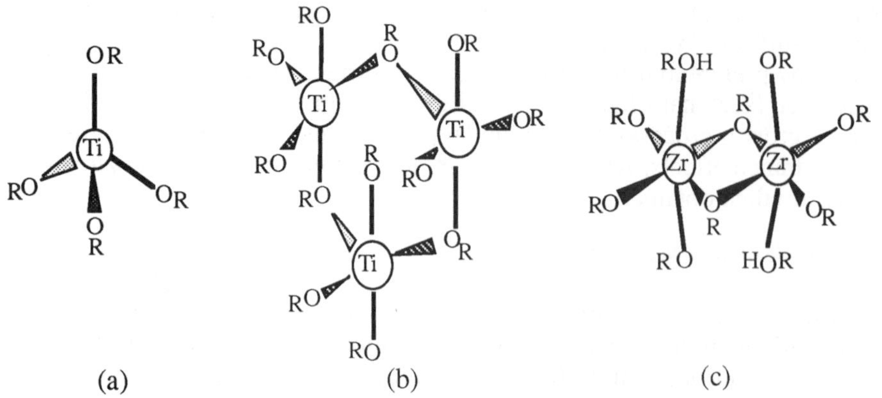

Figure 3. Molecular structure of metal alkoxides
$a = Ti(OPr^i)_4$, $b = [Ti(OEt)_4]_3$, $c = [Zr(OPr^i)_4(Pr^iOH)]_2$

The formation of heteroalkoxides is governed by coordination expansion and acid-base properties [11]. The nucleophilic addition of alkoxide groups between two different alkoxides can be described as a Lewis acid-base reaction. It is favored by a large difference in the electronegativity of metal atoms. Heteroalkoxides are therefore easily formed by simple mixing of two alkoxides of low and high electronegativity. Heteroalkoxides incorporating an alkali metal such as $LiNb(OEt)_6$ represent so far the most important class of heteroalkoxides. They are often used as "building blocks" for the sol-gel synthesis of multicomponent ceramics. The $LiNbO_3$ perovskite or the spinel $MgAl_2O_4$ for instance have been obtained via the hydrolysis of the bimetallic alkoxides $LiNb(OEt)_6$ and $Mg[Al(OR)_4]_2$ respectively [13][14]. Heterometallic alkoxides are not limited to two metals. A wide range of polymetallic alkoxides containing up to five different metals in the same molecular species have been reported [15]. This explain the emphasis on heterometallic alkoxides. However polar M-OR-M' bonds can be broken upon hydrolysis and the question arises as to how the sol-gel synthesis really leads to the formation of an homogeneous oxopolymer or oxide network rather than in nanophase materials.

Oxo-alkoxides would therefore be better precursors than heterometallic alkoxides. Condensation, via ether elimination, leads to the formation of oxo-bridges. This normally occurs upon heating and μ-oxo complexes are often formed during the purification of metal alkoxides via distillation [4]. Condensation via ether elimination is favored by the smaller size of μ-oxo ligands and their ability to exhibit higher coordination numbers, up to six, favoring coordination expansion of the metal atom. Oxo-alkoxides are more stable than the corresponding alkoxides and less reactive toward hydrolysis and condensation. The tendency to form oxo bridges increases with the size and charge of metal ions. Large electropositive metals are known to give oxoalkoxides such as $Pb_4O(OEt)_6$, $Pb_6O_4(OEt)_4$, $Bi_4O_2(OEt)_8$, $Y_5O(OPr^i)_{13}$ or $Nb_8O_{10}(OEt)_{20}$. They are usually built of edge sharing $[MO_6]$ octahedra and their molecular structure is close to that of the corresponding polyoxoanions formed in aqueous solutions [12].

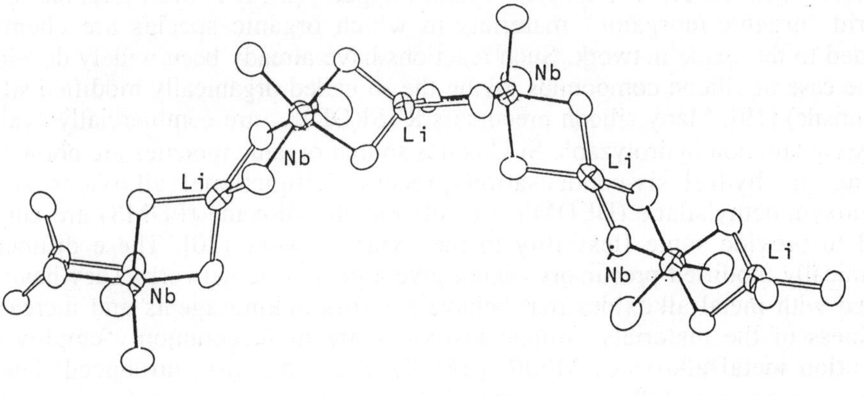

Figure 4. Solid state structure of the bimetallic alkoxide $LiNb(OEt)_6$
(according to ref.13)

As for homometallic alkoxides, oxo-bridges can also be formed when two different metal alkoxide solutions are refluxed leading to heterometallic oxo-alkoxides. $Pb_4O(OEt)_6$ for instance undergoes complete dissolution in ethanol when $Nb(OEt)_5$ is added giving rise to $[Pb_6O_4(OEt)_4][Nb(OEt)_5]_4$. Hetero-alkoxides can also provide molecular precursors with the correct M'/M stoichiometry in which some M-O-M' bonds are already formed. Whereas alkoxy bridges are usually hydrolyzed during the sol-gel synthesis, oxo bridges are strong enough not to be broken [11].

The crystallization of the perovskite $BaTiO_3$ phase synthesized via the sol-gel route occurs around 600°C when $Ti(OPr^i)_4$ and $Ba(OPr^i)_2$ are used as precursors. This crystallization temperature decreases to 60°C when the mixture of alkoxides is refluxed prior to hydrolysis. This should be due to the *in-situ* formation of the bimetallic oxo-alkoxide $[BaTiO(OPr^i)_4(Pr^iOH)]_4$. Such a compound has been isolated as single crystals and characterized by X-ray diffraction. This bimetallic precursor exhibits the correct Ba/Ti stoichiometry and Ba-O-Ti bonds are already formed in the solution [16].

Interesting attempts to make M-O-M' bonds via the non hydrolytic condensation between metal halides and metal alkoxides have been published recently by Corriu et al. [17]. They could offer a new soft chemistry route to multicomponent materials.

HYBRID ORGANIC-INORGANIC GELS

The sol-gel synthesis of metal oxides is based on the hydrolysis and condensation of metal-organic precursors in organic solvents. Little or no heating is required and oxide gels may be doped with organic molecules. The basic idea is very simple, organic species are dissolved in a common solvent with the alkoxide precursor. Hydrolysis and condensation then lead to the formation of an oxide network in which organic species remain trapped [18]. It is even possible to build hybrid "organic-inorganic" materials in which organic species are chemically bonded to the oxide network. Such reactions have already been widely developped in the case of silicon compounds giving the so-called organically modified silicates (Ormosils) [19]. Many silicon precursors $R'_xSi(OR)_{4-x}$ are commercially available. They contain non hydrolyzable Si-C bonds so that organic moieties are not removed during the hydrolysis-condensation process. Difunctional alkoxides such as diethoxydimethylsilane (DEDMS) or polydimethylsiloxane (PDMS) are currently used to provide some flexibility to the oxide network [20]. These difunctional organically modified precursors cannot give a polymeric network. They have to be mixed with metal alkoxides that behave as cross-linking agents and increase the hardness of the materials. Silicon alkoxides are more commonly employed but transition metal alkoxides $M(OR)_n$ (M=Ti, Zr,...) are also introduced. They not only serve as cross-linking reagents but can also increase the refractive index or catalyze the condensation reactions of siloxanes [21].

Polymerizable organic monomers can also be directly bonded to the silica framework via trifunctional silicon alkoxides $R'Si(OR)_3$ in which R' is a organic group that could undergo further polymerization. The most common polymerizable groups introduced thus far are vinyl, epoxy or methacrylate groups. A double polymerization process is then performed via hydrolysis for the inorganic part and photochemical or thermal curing for the organic polymer. Depending on the nature and amount of organic and inorganic components a whole family of new materials ranging from rubbery ormosils to brittle glasses can be synthesized. They open new possibilities in the field of materials science, mainly as transparent matrices for optical applications.

Ormosils lead to the formation of hybrid materials which are specially suitable for the sol-gel synthesis of non-porous thick films. Hard coatings and matrices for patterning have been developed for optical applications [22]. Transparent coatings, several tens of microns in thickness, have been deposited from DEDMS precursors crosslinked by metal alkoxides. DEDMS $[Si(CH_3)_2(OEt)_2]$ contains hydrophobic methyl groups so that only very small amounts of solvent molecules remain in the gel when dried at room temperature. Cracks are therefore not formed upon drying. These films exhibit good transparency and mechanical properties. They can be polished for making optical devices. Their refraction index can be tailored by adding transition metal alkoxides such as $Ti(OR)_4$ or $Zr(OR)_4$ as crosslinking agents [21].

A wide range of organic dyes have been incorporated into these gels for the fabrication of luminescent materials and the number of papers describing the optical properties of such coatings is constantly expanding. Significant results have been achieved in the areas of solid state lasers, nonlinear optical materials and photochromics [23]. A variety of laser dyes (rhodamine, coumarin) doped in sol-gel matrices have demonstrated laser action. Their properties are comparable to those of the dye in solution and their stability appears to be significantly greater than in organic polymeric matrices. Solid-state tunable lasers can be fabricated throughout the visible spectrum and into the near infrared. The photostability of dye-doped Ormosils have greatly surpassed the laser lifetimes obtained with polymer hosts [23].

Since the non-linear optical properties (NLO) of organics are frequently the highest known, the sol-gel synthesis of NLO materials is quite promising. One way is to introduce conjugated polymers or organic molecules in the inorganic gel [24]. However most studies deal with third order rather than second order effects which require specific orientation of organic species. Molecular orientation is currently performed via the poling process by applying an external electrical field. However the fast relaxation of NLO molecules back to random orientation is rather fast and $\chi^{(2)}$ properties vanish quite rapidly. To prevent the randomization of the poled molecules, the NLO chromophores are usually incorporated in a polymer which has a high glass transition temperature. Second harmonic generation was first demonstrated in doped sol-gel films by C. Sanchez et al. by using a chromophore group bonded to a functionalized silicon alkoxide such as N-(3-triethoxysilyl-propyl)-2,4 dinitrophenyl amine, $SiN\Phi(OEt)_3$ [25]. The NLO molecule is then grafted onto the oxide network rather than simply embedded within the sol-gel matrice. Such chemically bonded chromophores can be introduced in higher concentration and offer better orientational stability [26].

Porous hydrophilic matrices are required for the sol-gel synthesis of chemical sensors. Such materials can be easily made from silicon alkoxide precursors and coated onto glass plates or optical fibers. They are almost ideal matrices for optical chemical sensors. The indicator molecules can be encapsulated within the pores of the glass. They cannot be leached out, while smaller species can diffuse in and out of the porous sol-gel matrice [27]. The simplest chemical sensors which have been demonstrated to date are pH sensors [28], but sensors which respond to metal ions have also been reported [29].

Even biological species such as enzymes or proteins can be encapsulated into porous sol-gel glasses [30][31]. However the entrapment of biological species is more difficult as these species can be very sensitive to ethanol or pH. The usual sol-gel procedure has then to be modified in order to meet these requirements [31]. Pure TMOS, which is not soluble in water, is used as a precursor. The diphasic water-TMOS mixture is then sonicated in the presence of HCl (acid catalysis) for hydrolysis to occur. Small quantities of methanol are released during this hydrolysis reaction. They behave as a co-solvent so that a transparent solution is finally obtained. A buffer is then added in order to increase the pH and biological species are introduced into the solution before gelation occurs. They remain trapped into the growing oxide silica network. Many cell-free proteins or enzymes such as glucose oxidase [32-34], trypsin [35] or bacteriorhodopsin [36] have been embedded within sol-gel matrices where they have been shown to retain their characteristic reactivities.

More recently sol-gel synthetic techniques were used for the entrapment of parasitic protozoa (*Leishmania*) into porous silica glasses [37]. One remarkable point is that protozoa keep their integrity into the dried gel. Transmission electron microscopy shows that the cellular organization of the parasite remains clearly recognizable (Fig.5). The nucleus and numerous ribosomes can still be seen in the cytoplasm and the regular pattern of the granules underlying the plasma membrane suggests that the cytoskeletal structures are not destroyed.

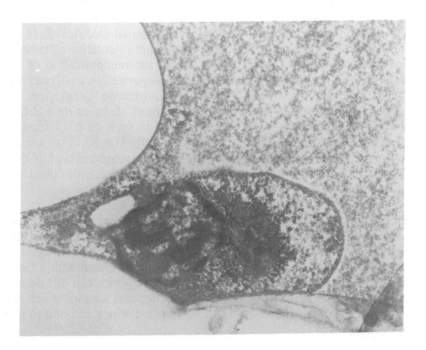

Figure 5. Transmission electron microscopy of *Leishmania* entrapped
within a sol-gel silica matrix showing that the cellular organization is preserved.

Moreover, *Leishmania*s retain their antigenic properties in the macropores of the gel. They react with antibodies in infected sera. This antigen-antibody reaction was detected by the so-called ELISA test which is now widely used in studying serological responses to parasitic infections [38]. A clear-cut difference in optical density has been measured between positive and negative sera allowing the easy detection of *Leismania* antibodies. The entrapment of antigenic species into porous sol-gel matrices avoids most problems due to non specific binding and could be advantageously used in diagnostic kits.

CONCLUSION

Sol-gel syntheses offer a very versatile soft chemistry route to advanced materials. They lead to the formation of an oxide network directly from the solution at room temperature. The powderless processing of ceramics becomes possible, allowing the formation of fibers or coatings directly from sols or gels. The molecular engineering of metal alkoxide precursors provide a chemical control of the nanostructure of sol-gel materials. Moreover, hybrid organic-inorganic materials can be easily made which open new possibilities in the field of materials science. Even biological species can be encapsulated in inorganic matrices leading to the formation of "biogels".

REFERENCES

1. Ebelmen, Comptes-Rendus Acad. Sci. Fr., 1845, 21, 502
2. Klein L.C., "Sol-Gel Technology", Noyes, Park Ridge, 1988
3. Brinker C.J., Scherer G.W., "Sol-Gel Science", Academic Press, San Diego, 1990
4. Bradley D.C., Mehrotra R.C., Gaur D.P., "Metal Alkoxides", Academic Press, London, 1978
5. Livage J., Henry M., Sanchez C., Prog. Solid State Chem., 1988, 18, 2596. 6. Livage J., Chem. Mater., 1991, 3, 578
7. Livage J., Sanchez C., J. Non-Cryst. Solids, 1992, 145, 11
8. Livage J., Sanchez C., Toledano P., Mat. Res. Soc. Symp. Proc. "Chemical Processes in Inorganic Materials, 1992, 272, 3
9. Chatry M., Henry M., Sanchez C., Livage J., Mater. Res. Bull. (in press)
10. Lee G.R., Crayston J.A., Adv. Mater., 1993, 5, 434
11. Caulton K.G., Hubert-Pfalzgraf L.G., Chem. Rev., 1990, 90, 969
12. Hubert-Pfalzgraf L.G., New J. Chem., 1987, 11, 663
13. Eichorst D.J., Payne D.A., Wilson S.R., Howard K.E., Inorg. Chem., 1990, 29, 1458
14. Jones K., Davies T.J., Emblem H.G., Parkes P., Mater. Res. Soc. Symp. Proc., 1986, 73, 111
15. Chandler C.D., Roger C., Hampden-Smith M.J., Chem. Rev., 1993, 93, 1205
16. Yanovsky A.I., Yanovskaya M.I., Limar V.K., Kessler V.G., Turova N.Y., Struchkov Y.T., J. Chem. Soc., Chem. Commun., 1991, 1605
17. Corriu R., Leclercq D., Lefèvre P., Mutin P.H., Vioux A., Chem. Mater., 1992, 4, 961
18. Avnir D., Levy D., Reisfeld R., J. Phys. Chem., 1984, 88, 5956
19. Schmidt H., J. Non-Cryst. Solids, 1985, 73, 681
20. Wilkes G.L., Orter B., Huang H., Polymer Preprints, 1985, 26, 300
21. Diré S., Babonneau F., Sanchez C., Livage J., J. Mater. Chem., 1992, 2, 239
22. Schmidt H., Seiferling B., Mat. Res. Soc. Symp. Proc. "Better Ceramics through Chemistry", 1986, 73, 739
23. Dunn B., Zink J.I., J. Mater. Chem., 1991, 1, 903
24. Prasad P.N., Mat. Res. Soc. Symp. Proc., 1990, 180, 741
25. Toussaere E., Zyss J., Griesmar P., Sanchez C., Nonlinear Optics, 1991, 1, 349

26. Jeng R.J., Chen Y.M., Jain A.K., Kumar J., Tripathy S.K., Chem. Mater., 1992, 4, 972

27. Lev O., Analusis, 1992, 20, 543

28. Rottman C., Ottolenghi M., Zusman R., Lev O., Smith M., Gong G., Kagan M.L., Avnir D., Mater. Lett., 1992, 13, 293

29. Zusman R., Rottman C., Ottolenghi M., Avnir D., J. Non-Cryst. Solids, 1990, 122, 107

30. Braun S., Rappoport S., Zusman R., Avnir D., Ottolenghi M., Mater. Lett., 1990, 10, 1

31. Ellerby L.M., Nishida C.R., Nishida F., Yamanaka S.A., Dunn B., Valentine J.S., Zink J.I., Science, 1992, 255, 1113

32. Braun S., Shtelzer S., Rappoport S., Avnir D., Ottoenghi M., J. Non-Cryst. Solids, 1992, 147-148, 739

33. Yamanaka S.A., Nishida F., Ellerby L.M., Nishida C.R., Dunn B., Valentine J.S., Zink J.I., Chem. Mater., 1992, 4, 495

34. Audebert P., Demaille C., Sanchez C., Chem. Mater., 1993, 5, 911

35. Shtelzer S., Rappoport S., Avnir D., Ottolenghi M., Braun S., Biotech. Applied Biochem., 1992, 15, 227

36. Wu S., Ellerby L.M., Cohan J.S., Dunn B., El-Sayed M.A., Valentine J.S., Zink J.I., Chem. Mater., 1993, 5, 115

37. Barreau J.Y., Da Silva J.M., Desportes I., Monjour L., Gentillini M., Livage J., Science (in press)

38. Venkatesan P., Wakelin D., Parasitology Today, 1993, 9, 228

Materials Science Forum Vols. 152 - 153 (1994) pp. 55-68
© 1994 Trans Tech Publications, Switzerland

SOFT CHEMISTRY: THERMODYNAMIC AND STRUCTURAL ASPECTS

M. Figlarz

Laboratoire de Réactivité et de Chimie des Solides - URA CNRS 1211
Université de Picardie Jules Verne, 33, rue Saint-Leu, F-80039 Amiens Cédex, France

Keywords: Soft Chemistry, Thermodynamic Aspects of Soft Chemistry, Structural Aspects of Soft Chemistry, WO_3 Polymorphs

ABSTRACT

Soft chemistry is defined as the set of mild chemical operations, which allow to generate new metastable phases, which cannot be obtained from the thermodynamic stable polymorphs, by some structural filiation between mother-daughter phases. The thermodynamic aspects are discussed in relation to the metastability. It is shown that the phases prepared by soft chemistry routes do not possess the lowest free energy, which characterizes thermodynamic stable phases. Different degrees of metastability can occur: metastable, supermetastable, hypermetastable... polymorphic forms with increasing free energy can be obtained by different chimie douce routes. The structural aspects are discussed in relation to the structural relationship between mother and daughter phases observed in soft chemistry transformations. The non-reconstructive and reconstructive transformation situations are discussed. In non-reconstructive transformations the mother structural skeleton is maintained during the course of the transformation. In reconstructive transformations the mother phase framework has to be broken; the chemical composition changes, and new bonds are formed leading to a new structure. In these reconstructive transformations oriented nucleation and growth are responsible for the occurrence of mother-daughter structural relationship. Therefore, in both situations, the soft chemistry phases possess higher free energy, in comparison with thermodynamic stable phases, induced by the mother phases free energy in relation to mother-daughter structural relationship. As an illustration of these general principles the WO_3 case is presented and discussed with four new phases: metastable cubic WO_3 with ReO_3 structure, metastable cubic pyrochlore-type WO_3, supermetastable orthorhombic $WO_3.1/3H_2O$-type WO_3 and metastable hexagonal HTB-type WO_3.

I. INTRODUCTION

In an address given at the Colloquium on Preparative Solid State Chemistry in 1969, H. Schäfer distinguished two types of preparative reactions: those in which high mobility of the reactants is aimed and mainly achieved by the use of high temperatures and those in which the course of the reaction and the nature of the products is influenced by the structure of the reactants (1). He emphasized that using the first methods the question of the reaction mechanism is left in the background and many interesting synthetic pathways are thus left unexplored and unused. Indeed in classical preparative chemistry the use of high temperatures leads to the phases expected from thermodynamic considerations: in these hard conditons all the structures out of thermodynamic equilibrium cannot be achieved. On the contrary the use of mild reactive conditions, i.e. soft chemistry, allows the preparation of metastable compounds. It is mainly during the last two decades that the "chimie douce" or "soft chemistry" methods have been increasingly exploited for the preparation of numerous solid state compounds. The chimie douce processes have now assumed special significance exemplified by this first international conference devoted to the subject.

Soft chemistry appears to provide new routes for the preparation of new materials, which often cannot be obtained using classical synthesis methods. A wide range of chemical reactions is involved in chimie douce processes as intercalation, deintercalation, cationic exchange, dehydration, dehydroxylation, hydrolysis, redox, etc... The potential range of application of the chimie douce approach for making new compounds will be first exemplified to introduce the subject.

A first typical example is the preparation of hexagonal VS_2. The precursor $LiVS_2$ is obtained by the classical ceramic method and by deintercalation of Li the V-S matrix of the precursor is preserved and VS_2 is obtained (2). Li deintercalation is chemically produced by I_2 in acetonitrile. This kind of reaction has led to the concept of chemical reactions which mimic electrode reactions (3, 4) and has been very useful in soft chemistry preparation.

The preparation of a new form of titanium dioxide, $TiO_2(B)$ by M. Tournoux et al. is another interesting example (5). The starting material is the tetratitanate $K_2Ti_4O_9$ which is obtained by direct synthesis. In a first step K^+/H^+ cationic exchange is obtained: in this cationic exchange reaction the structure of the precursor is maintained. The exchange product undergoes progressive dehydration through a series of hydrated intermediate forming $TiO_2(B)$ as the last step by thermolysis at about 500°C. The product of the reaction $TiO_2(B)$, presents some structural similarity with the precursor.

McCarron prepared a novel metastable monoclinic βMoO_3 from molybdic acid solution (6). By spray-drying the molybdic acid solution he obtained an amorphous powder, which by thermal treatment at 300°C leads to this β-MoO_3 with a ReO_3-type structure whereas the thermodynamic stable MoO_3 form adopts a double-layered structure with an orthorhombic cell. The formation of β-MoO_3 is a phase transition amorphous $MoO_3 \rightarrow$ new crystallized MoO_3 form in which the short range order in the amorphous phase induces the nucleation of the new ReO_3-type MoO_3.

It would be possible to give numerous examples of such *chimie douce* reactions leading to the obtention of new compounds which are very often metastable phases. In all these examples it is clear that there are some structural relationships between the starting material and the product of the reaction(s). In this way it is possible to define soft chemistry as the set of mild chemical operations which allow to generate new metastable phases by some structural filiation between mother-daughter phases (7, 8).

Therefore we will discuss two main aspects of soft chemistry: the thermodynamic aspects related to the metastable phases and the structural aspects related to the structural filiation .

In a last part we will exemplify these thermodynamic and structural aspects with the different metastable WO_3 phases we have prepared in our laboratory using the soft chemistry approach.

II.   THERMODYNAMIC ASPECTS OF SOFT CHEMISTRY

To discuss the thermodynamic aspects of *chimie douce* we will consider the phase transitions - i.e. without any changes in composition, restricted to changes in structure only - in the case of two polymorphic phases under equilibrium conditions (classical preparative or hard chemistry) and in the case of metastable polymorphic forms obtained by soft chemistry.

II.1   Phase transitions and metastability for phases at thermodynamic equilibrium

Let us consider a first-order phase transition in the Ehrenfest thermodynamic classification (9). The temperature dependences of the free energy and enthalpy are shown in figure 1a. Each polymorph has its own G-T and H-T curve. By definition the polymorph that is stable under a particular set of conditions is the one with lowest free energy. The two free energy curves cross over at the equilibrium transition temperature T_t at which $G_1=G_2$. The phase transition $1 \rightleftarrows 2$ is reversible. Discontinuity in H occurs at T_t, ΔH is positive on heating and the transition from one form to another is endothermic on heating ($1 \rightarrow 2$) and exothermic on cooling ($2 \rightarrow 1$) as schematized in Fig.1b. The dashed lines represent extensions of stable states: superheated and supercooled metastable phases obtained from the stable phases.

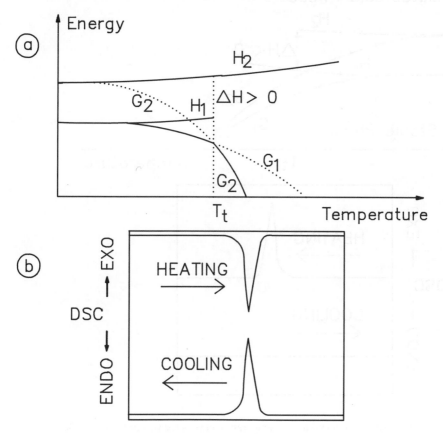

Figure 1:   a) Free energy and enthalpy - temperature diagram for a first order polymorphic phase transition; b) endothermic and exothermic peaks associated with the phase transitions on heating and on cooling respectively.

<u>II.2   Phase transitions and metastability for phases obtained by soft chemistry</u>

The situation is quite different for phases obtained by soft chemistry. The free energy and enthalpy temperature curves in the case of a metastable phase obtained by soft chemistry and in the case of the thermodynamic stable phase as a reference are given in figure 2a. The metastable phases obtained by soft chemistry are still those with higher free energy but cannot be obtained from the thermodynamic stable phase. The two free energy curves do not cross over. The soft chemistry system is not at thermodynamic equilibrium. The phase transition is only possible in one way from metastable to thermodynamic stable phases: it is a monotropic non-reversible phase transition. ΔH is negative on heating and the corresponding phase transition is exothermic (Fig. 2b). It is important to recognize that this monotropic phase transition is distinctly different from equilibrium phase transition since the transition involving metastable phase are solely the result of reaction kinetics: on Fig. 2a the two free energy curves do not cross over and T_t depends on kinetic factors.

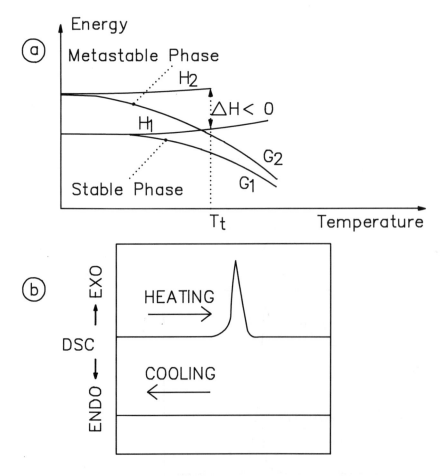

Figure 2:   a) Free energy and enthalpy - temperature curves for two polymorphic forms of a metastable phase obtained by soft chemistry and the thermodynamic stable phase; b) exothermic peak associated with the monotropic transition metastable $\rightarrow$ stable (the reverse transition does not occur).

Finally it is worth noticing, as the phases prepared by soft chemistry routes do not possess the lowest free energy which characterizes thermodynamic stable phases, that different degrees of metastability can occur and metastable, super-metastable, hyper-metastable polymorphic forms with increasing free energy (Fig.3) can be obtained by different soft chemistry routes. This will be exemplified in the last part of this paper in the case of WO_3 polymorphs.

Two remarks will conclude this thermodynamic discussion. It is unfortunate that the same denomination namely "metastable" is used for systems at thermodynamic equilibrium and soft chemistry systems not at thermodynamic equilibrium, with different meanings leading to a great deal of confusion. The second remark is more fundamental. The initial formation of metastable phases in soft chemistry conditions is not predicted from thermodynamics. Indeed the above thermodynamic discussion is simply a qualitative and comparative treatment of soft chemistry metastable phases compared with thermodynamic stable phases. It does not provide an understanding of the occurence of such metastable phases in a true thermodynamic sense. The correct interpretation must be searched for in the structural aspects of chimie douce that we will discuss now.

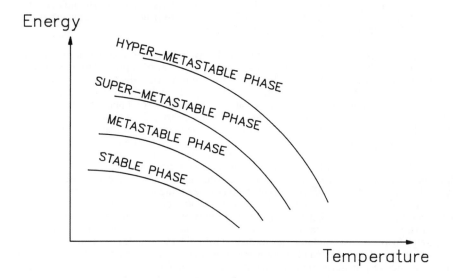

Figure 3:   Free energy - temperature curves for different metastable, supermetastable, hypermetastable polymorphic forms with increasing free energy obtained by different soft chemistry routes.

III.   STRUCTURAL ASPECTS OF SOFT CHEMISTRY

In this part we will deal with the chemical transformation, i.e. with changes in composition, from the precursor (mother phase) to the product of the transformation (daughter phase).
As we have emphasized in the introduction, in all the chimie douce reactions there is a structural relationship between mother and daughter phases. Two situations can be considered for the structural relationship between mother and daughter phases: non-reconstructive and reconstructive transformations. We use here the same denomination as that introduced by Buerger in his structrual classification of phase transitions (10). The main difference is that Buerger considered phase transitions restricted to changes in structure only without any changes in composition, whereas we consider transformation with changes in composition.

III.1 Non-reconstructive transformations

On one hand there are reactions in which most of the transformation product atoms move only slightly, if at all, in comparison with the mother phase: these are displacive or non-reconstructive transformations. This kind of transformation is well exemplified with cationic exchange reactions, intercalation-deintercalation reactions, some dehydration reactions when the water molecules are very weakly bonded as for zeolithic-type water, etc... This is the simplest situation in which a three or bi-dimensional structural skeleton is really maintained during the course of the reaction. The effect of the transformation is to produce distortions which lead to a variation of the lattice parameters and possibly a modification of the cell symmetry.

III.2 Reconstructive transformations

On another hand there are reactions which involve the disruption of the mother phase structure to form a new and different structure for the daughter phase: these are *reconstructive transformations*. Dehydration and condensation reactions are typical examples of such reconstructive transformation. In this case the entire framework has to be broken, chemical composition changes and new bonds formed leading to a new structure. In general even if there is a complete change in crystal structure some parts of the mother and daughter phases are "similar in structure" but that does not mean those structural elements are really maintained sensu stricto during the course of the reaction.

There is an apparent contradiction between the two ideas of reconstructive phase transformation with breaking of many strong bonds and that of mother/daughter phases structural relationships. This contradiction can be removed by discussing the way the daughter phase is formed from the mother one through a nucleation-growth mechanism. Indeed, providing that the reaction conditions are soft enough, in solid phase transformations the nucleus of the product of reacion (daughter phase) appearing in the lattice of the mother phase could hardly escape the influence of the mother lattice in its formation (11). This is due to the substantial contribution of surface free energy to the total nucleation free energy. Under such conditions the oriented daughter/mother nuclei are the thermodynamically more stable nuclei. Consequently their formation and subsequent oriented growth, with displacement of the daughter/mother interphase boundary, favour the formation of the metastable daughter phase. This is the explanation to the occurence of structural relationships in reconstructive transformations. These aspects are correlated with the topotaxy phenomenon and have been discussed in details elsewhere (11). It must be emphasized that in order to obtain by soft chemistry such metastable materials, the reaction conditions must be mild enough to avoid any structural transformation from the metastable state to the stable one which would cancel the oriented nucleation and growth effect or the structural preservation in non-reconstructive transformations i.e. the mother/daughter structural relationships. This implies the use of relatively low reaction temperatures and/or sufficiently soft reaction conditions. It is therefore often necessary to use finely divided compounds in order to increase the reactivity of the mother phase and therefore to obtain sufficiently soft reaction conditions.

To conclude this structural discussion it is interesting to join the thermodynamic and structural aspects of *chimie douce*. Indeed from the above structural aspects it becomes clear that the phases prepared by *chimie douce* routes do not possess the lowest free energy values which characterize thermodynamic stable phases. They possess higher free energy values induced by the mother phase free energy in relation to the mother/daughter structural relationships. This is quite obvious for non-reconstructive transformations in which the structural framework is the same for mother and daughter phases. For reconstructive transformations we have shown the predominant role of oriented nucleation of the daughter on the mother phase and the subsequent oriented growth which induce the free energy dependence between mother and daughter phases. From that point of view this structural discussion brings an a posteriori justification of the thermodynamic results.

IV. THE WO₃ CASE AS AN EXAMPLE

As an illustration of these general principles some results concerning the stable and metastable WO_3 oxides are presented and discussed. Before presenting the preparations and characteristics of four metastable WO_3 oxides we will recall the structure of WO_3 thermodynamic stable forms.

The thermodynamic stable forms of WO_3 have a ReO_3-type structure and can be described by a three dimensional array of (WO_6) octahedra sharing all their corners (Fig.4). The octahedra are distorted and the type and magnitude of the distortions are dependent on the temperature, which explains the existence of several distorted ReO_3-type polymorphic forms which are stable within well-defined temperature ranges and transform reversibly into each other (12).

We will show that completely new WO_3 modifications, which cannot be obtained from any of the thermodynamic stable forms, can be prepared by *chimie douce* routes (13-15).

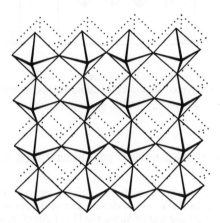

Figure 4:   The ReO_3-type structure of the different stable thermodynamic forms of WO_3.

IV.1 Metastable cubic WO₃ with ReO₃ structure (prepared via a reconstructive transformation)

Metastable cubic WO_3 is prepared by dehydration in mild conditions of the hydrate $WO_3.1H_2O$ (15). Indeed the WO_3 oxide which appears at 200°C has a cubic ReO_3 structure without any distortion of the (WO_6) octahedra (Fig. 5). The refinement of the lattice parameter leads to a=3.71Å. This metastable cubic WO_3 remains stable up to 300°C. At this temperature it changes into the stable thermodynamic monoclinic WO_3 with an exotherm on DSC corresponding to a small change in enthalpy $\Delta H=0.6$ kcal.mole^{-1}, associated with the displacive non-reconstructive monotropic phase transition metastable cubic $\rightarrow$ stable monoclinic WO_3.

The formation of cubic WO_3 is a reconstructive transformation from $WO_3.1H_2O$ schematized in Fig. 5. The hydrate has a layered structure (16) and in the (001) plane the (WO_6) octahedra arrangement is of the ReO_3 type but in the dehydration transformation W-OH₂ bond are disrupted and new W-O strong bonds formed. It was generally accepted that the dehydration of $WO_3.1H_2O$ hydrate leads to the formation of WO_3 with a distorted ReO_3-type structure but the intemediary metastable WO_3 formation was not detected. L. Marty reported for the first time the occurence of the cubic WO_3 form in the dehydration of $WO_3.1H_2O$ (17). More recently Yamoguchi et al. prepared cubic WO_3 by hydrolysis of tungsten ethoxide with $WO_3.1H_2O$ as an intermediary product (18). A cubic WO_3 form was also reported by Palatnick et al. in thin condensed films with a thickness lower than 100Å (19).

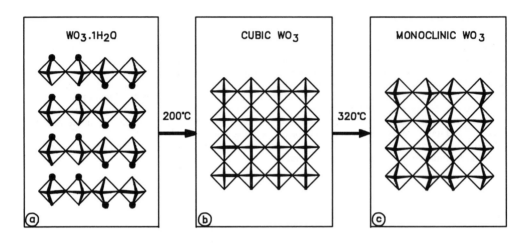

Figure 5:   Metastable cubic WO_3 with ReO_3 structure:
             a) the $WO_3.1H_2O$ layered structure; b) cubic WO_3 obtained by dehydration; c) the
             monotropic cubic → monoclinic WO_3 phase transition.

IV.2 Metastable cubic WO_3 with pyrochlore-type structure (prepared via non-reconstructive transformations)

The structure of pyrochlore-type WO_3 can be described in terms of distorted hexagonal WO_3 layers . In the (111) plane the (WO_6) octahedra are arranged as in the hexagonal WO_3 (001) plane (Fig. 7a). These layers are linked along the [111] direction by intermediary (WO_6) octahedra (Fig. 6); three dimensional interconnected tunnels are thus formed.

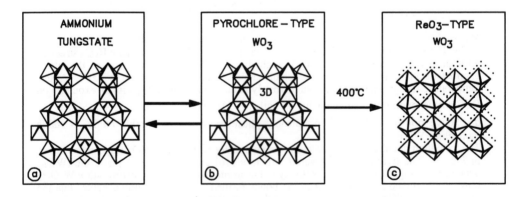

Figure 6:   Metastable cubic WO_3 with a pyrochlore-type structure:
             a) pyrochlore-type ammonium tungstate precursor;   b) pyrochlore-type WO_3
             obtained after NH_4^+/H^+ exchange and dehydration of the hydronium compounds at
             200°C;   c) the metastable pyrochlore → stable monoclinic WO_3 monotropic phase
             transition.

This metastable cubic WO_3 with a pyrochlore-type structure (p.WO_3) was prepared from a pyrochlore-type ammonium tungstate after NH_4^+/H^+ exchange and dehydration of the hydronium compound (13, 20). These two-step transformations are non-reconstructive reactions with conservation of the three-dimensional W-O network from ammonium tungstate to oxide. All the structural features of the mother phase are to be found again in the daughter phase. This pyrochlore WO_3 presents a lacunar structure with W and O vacancies and protons to maintain global electroneutrality with the formula $H_{0.65}W_{1.635}O_{5.23}$ (l.p.WO_3) (21). H and O contents can vary under thermal treatment in the temperature range 100-350°C due to the reaction of protons with oxygen of the framework. Unfortunately stoichiometric p.WO_3 can never be obtained due to the transformation l.p.$WO_3 \rightarrow$ m.WO_3 which occurs at 400°C. As expected this monotropic transition metastable l.p.$WO_3 \rightarrow$ stable m.WO_3 is exothermic with $\Delta H = 2.7$ kcal.mole^{-1} and is reconstructive (22). Fig. 6 schematizes these different transformations.

IV.3 Supermetastable orthorhombic WO_3 with a $WO_3.1/3H_2O$-type structure (prepared via a non-reconstructive transformation)

Orthorhombic WO_3 is prepared by dehydration in mild conditions of the hydrate $WO_3.1/3H_2O$ (14, 15). The hydrate is prepared under hydrothermal treatment at 120°C of a tungstic acid gel (23). The hydrate structure can be described as follows (23). The basic structural element is an infinite plane of (WO_6) octahedra sharing their corners and forming six-membered rings in the (001) plane (Fig. 7a). The complete structure arises from a stacking of such layers along the [001] axis, every other layer being shifted by a/2 (Fig. 7b). All the tungsten ions are bonded to six oxygens in a slightly distorted octahedral coordination. As schematized in Fig. 7b W(2) is bonded to four O(2) atoms and to O(4) and O(5) atoms which are coordinated only to W(2). O(4) is the oxygen of the water molecule with a rather long distance W(2)-O(4) = 2.1Å and conversely W(2)-O(5) = 1.8Å is rather short and indicates a double-bond character. In mild conditions dehydration of $WO_3.1/3H_2O$ takes place around 200°C, leading to a new anhydrous WO_3 with the same structure as the mother hydrate, but in the oxide W(2) is then five-coordinate. Neutron diffraction study of dehydration using the H diffusion of water molecules, shows that the water has completely disappeared even though the hydrate structure is maintained (24). IR, TG and DSC studies

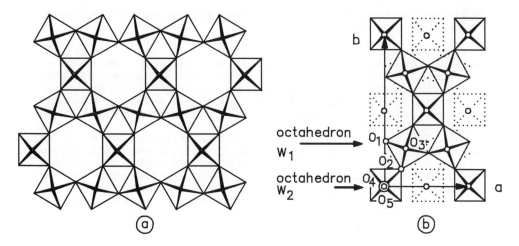

octahedron $W_1 \longrightarrow$

octahedron $W_2 \longrightarrow$

(a) (b)

Figure 7: Structure of $WO_3.1/3H_2O$ hydrate: a) the six-membered rings formed by the (WO_6) octahedra in the (001) plane; b) projection of the structure along the [001] axis.

(14, 15) confirm the existence of this WO_3 phase. As the dehydration reaction takes place without any structural transformation, the water molecule must be very weakly bonded; this is confirmed by the dehydration enthalpy measured from DSC, $\Delta H=7$ kcal.mole^{-1}. Due to the super-zeolitic behaviour of this structural water the dehydration is reversible.

Moreover according to our thermodynamic classification this orthorhombic WO_3 oxide is a supermetastable phase. Indeed it transforms irreversibly to a metastable hexagonal WO_3 at 300°C and this monotropic transition is exothermic with $\Delta H=0.5$ kcal mole^{-1}. These transformations are schematized in Fig. 8. The supermetastable orthorhombic → metastable hexagonal WO_3 transition will be now discussed as it is the preparation method of hexagonal WO_3.

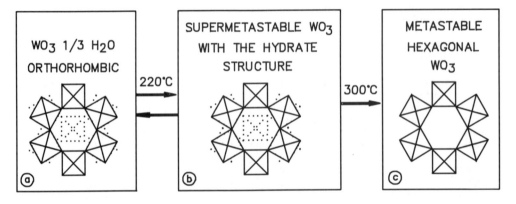

Figure 8:   Supermetastable orthorhombic WO_3 with $WO_3.1/3H_2O$-type structure:
a) $WO_3.1/3H_2O$ hydrate precursor; b) supermetastable WO_3 obtained by dehydration; c) the monotropic supermetastable orthorhombic → metastable hexagonal WO_3 phase transition.

<u>IV.4 Metastable hexagonal WO_3 with a hexagonal tungsten bronze-type structure (prepared by a reconstructive transformation)</u>

In the previous publications on the preparation of hexagonal WO_3 the existence of the intermediary supermetastable orthorhombic WO_3 with the hydrate-type structure was not pointed out (25, 13). Indeed, X-ray diffraction study of the dehydration does not show any evidence of its occurrence because its structure is similar to the $WO_3.1/3H_2O$ mother phase. So we concluded that dehydration of $WO_3.1/3H_2O$ leads directly to hexagonal WO_3, but all the previous conclusions remain correct as the structure of the supermetastable oxide and that of the hydrate are the same.

Hexagonal WO_3 has the empty tunnel structure of the hexagonal tungsten bronzes (26). It is built-up of slighly distorted (WO_6) octahedra arranged in six-membered rings in layers perpendicular to the hexagonal axis; the stacking of such layers leads to the formation of hexagonal tunnels running along the c axis (Fig. 9). Hexagonal oxide is prepared from super metastable orthorhombic WO_3 by a reconstructive transition occurring at 300°C. In this transition the W-O bonds of the mother phase are disrupted and a new hexagonal W-O network is formed. Nevertheless there is a strong strutural filiation between mother and daughter phases. We have shown that this structural filiation is related to the oriented nucleation/growth process of hexagonal on orthorhombic WO_3 (7, 11 ,13). Indeed h.WO_3 appears by oriented nucleation on the (001) orthorhombic oxide plane: in this plane the W-O arrangement is of hexagonal-type. This oriented nucleation is thermodynamically favoured because it diminishes the contribution of the nucleation surface free energy term leading to the lowest nucleation free energy. The mother/daughter structural relationship is maintained during the course of the dehydration by oriented growth with displacement of the reaction

interphase boundary. Despite the fact that, from a structural viewpoint, the (001) planes are
the same in the orthorhombic and hexagonal oxide phases, these structural elements are not
maintained sensu stricto during the course of the dehydration reaction.

In comparison with the supermetastable orthorhombic WO_3 hexagonal WO_3 is a metastable
phase which transforms irreversibly at 400°C by a reconstructive transition to the
thermodynamic stable monoclinic WO_3. This transition is exothermic with

$\Delta H = 3 Kcal.mole^{-1}$. These transformations are schematized in Fig. 9.

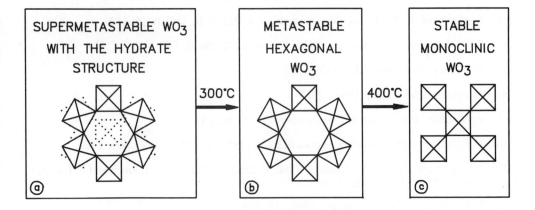

Figure 9:   Metastable hexagonal WO_3:
               a) the supermetastable orthorhombic WO_3 precursor; b) metastable hexagonal WO_3
               obtained by monotropic phase transition;   c) the hexagonal $WO_3 \rightarrow$ stable
               monoclinic WO_3 monotropic phase transition.

V.   CONCLUSION

During the last two decades the solid state chemists have answered H. Schäfer's hope expressed
about 25 years ago of paying more attention to reaction mechanism in preparative solid state
chemistry (1). Soft chemistry strategies bring up to date the main role of chemistry in the synthesis
of materials. Soft chemistry has opened new fields in preparative solid state chemistry with the
possibility of designing tailor-made new metastable materials which cannot be prepared by the hard
chemistry strategy, illustrating Machiavelli's concept "there is more power in softness than in
violence or barbary". Nevertheless the opposition between hard and soft chemistry should not be
understood as antinomics terms but as complementary tools: both still have a promising future.
In this soft chemistry new space, in full growth, we have developed these last years a conceptual
approach in order to account for the peculiar nature of soft chemistry mainly the metastability of the
phases so prepared (7, 8, 11, 13, 27). The thermodynamic and structural aspects which have been
dealt with here are an illustration of this approach of the soft chemistry exemplified with the new
phases obtained in the WO_3 system.

ACKNOWLEDGEMENTS

I wish to acknowledge the referenced works of my co-workers who contributed so greatly to the
development of the suject. In particular I thank Drs A. Coucou, A. Driouiche, B. Gérand,
F. Harb, G. Nowogrocki, J. Pannetier and Mr L. Seguin.

REFERENCES

(1)   H. Schäfer, Angew. Chem. internat. ed., (1971), 10, 43 .

(2)   D.W. Murphy, C. Cros, F.J. Di Salvo and J.V. Waszczak, Inorg. Chem., (1977), 16, 3027.

(3)   D.W. Murphy in Proceedings of Workshop on Lithium Non-aqueous Battery Electrochemistry, (1980), vol. 7 p. 197 Electrochemical Society.

(4)   D.W. Murphy and P.A. Christian, Science, (1979), 205, 651.

(5)   R. Marchand, L. Brohan and M. Tournoux, Mat. Res. Bull., (1980), 15, 1129.

(6)   E.M. Mc Carron, J; Chem. Soc. Chem. Commun., (1986), 4, 336.

(7)   M. Figlarz, Rev. Chim. Min., (1985), 22, 177.

(8)   M. Figlarz, Chem. Scr., (1988), 28, 3.

(9)   P. Ehrenfest, Proc. Acad. Sci. Amsterdam, (1933), 36, 153.

(10) M.J. Buerger in Phase transformations in solids (1951), p. 183, J. Wiley New York.

(11) M. Figlarz, B. Gérand, A. Delahaye-Vidal, B. Dumont, F. Harb, A. Coucou and F. Fievet, Solid State Ionics, 1990, 43, 143.

(12) S. Tanisaki, J. Phys. Soc. Japan, (1960), 15, 573.

(13) M. Figlarz, Prog. Solid State Chem., (1989), 19, 1.

(14) L. Seguin, J. Pannetier and M. Figlarz, Solid State Ionics (in the press).

(15) L. Seguin, B. Gérand, F. Portemer and M. Figlarz (this symposium).

(16) J.T. Szymanski, Canadian Mineralogist, (1984), 22, 681.

(17) L. Marty, thèse de 3ème cycle Université de Paris 6, Oct. 1972.

(18) O. Yamaguchi, D. Tomihisa, H. Kawabata and K. Shimizu, J. Am. Ceram. Soc., (1987), 70, C94 .

(19) L.S. Palatnik, O.A. Obolyaninova, M.N. Naboka and N.T. Gladkikh, Izv. Akad. Nauk SSSR, Neorg. Mater., (1973), 9, 801.

(20) A. Coucou and M. Figlarz, Solid State Ionics, (1988), 28-30, 1762.

(21) A. Coucou, A. Driouiche, M. Figlarz, M. Touboul and G. Chevrier, J. Solid State Chem., (1992), 99, 283.

(22) M. Figlarz, B. Gérand, B. Dumont, A. Delahaye-Vidal and F. Portemer, Phase Transitions, (1991), 31, 167.

(23) B. Gérand, G. Nowogrocki and M. Figlarz, J. Solid State Chem., (1981), 38, 312.

(24) J. Pannetier, Chem. Scr., (1986), 26, 131.

(25) B. Gérand, G. Nowogrocki, J. Guenot and M. Figlarz, J. Solid State Chem., (1979), 29, 429 .

(26)` A. Magneli, Acta. Chem. Scand., (1953), 7, 315.

(27) M. Figlarz, plenary lecture given at the 12th International Symposium on the Reactivity of Solids. Madrid September 1992.

Materials Science Forum Vols. 152 - 153 (1994) pp. 69-80
© 1994 Trans Tech Publications, Switzerland

NEW FUNCTIONAL SOLIDS DERIVED FROM LAYER STRUCTURED CRYSTALS BY PILLARING AND GRAFTING

S. Yamanaka

Department of Applied Chemistry, Faculty of Engineering, Hiroshima University,
Higashi-Hiroshima 724, Japan

Keywords: Layer Structure, Pillaring, Grafting, Clay, Zirconium Phosphate, Botallackite

ABSTRACT

Soft chemical processes developed mainly for the structural modification of organic polymers, cross-linking and grafting, were applied to two-dimensional inorganic layered crystals. The silicate layers of clay were cross-linked, or pillared with oxide sol particles, which were packed between the silicate layers, and micropores were formed in the interstices between the packed sol particles and the silicate layers. Supercritical drying of the pillared clay resulted in the preservation of a card-house type unique porous structure. The interlayer surface of zirconium phosphate was modified by grafting organic functional groups by ester bonds. Similar organic derivatives of zirconium phosphate layers were obtained by a direct ion exchange of the interlayer phosphate groups with organic phosphoric ester groups. Basic copper salts with the botallackite type layered structure are new anion-exchangeable layered crystals which can form organic derivatives.

INTRODUCTION

Two-dimensional layer structured crystals which can experience intercalation are often regarded as ideal nanocomposite matrices [1], because the structure consists of thin crystalline layers with a thickness of molecular scale, and a variety of foreign chemical species are inserted as guests into the interlayer space, forming a uniform nanomixture. However, if we look over the crystalline layer more carefully, we find that the layer is very different from a fragment made by simply dividing or grinding three dimensional crystals; the chemical bonds of each crystalline layer are completed within the layer, and cleaving the crystal along the plane parallel to the layers does not create dangling bonds. In this context of understanding, a layer structured crystal could be better-characterized as a stack of two-dimensional inorganic polymer molecules. It will be very interesting to apply soft chemical processes of cross-linking and grafting to inorganic polymer layers to derive new functional solids, though these processes have been developed in organic polymer chemistry. The term pillaring is used as cross-linking between layers with supposition of making free space between the layers. The term grafting will be used only when organic functional groups are bonded onto the interlayer surface by well-defined covalent bonds.

PILLARING

Two-dimensional crystalline layers are used as building units to construct porous pillared structures. The requirements for such building units are high swellability, low ion exchange capacity, proper stiffness of the structure, and easy availability. Several types of layered host structures have been extensively studied, such as clays [2-8], zirconium phosphate, layered double hydroxides [9,10], and graphite [11]. Among them, smectite clays are the most versatile host layers. Figure 1 shows the structure of montmorillonite, a typical smectite clay mineral. The 2:1 silicate layers (two tetrahedral and one octahedral sheets) are negatively charged by the substitution of the Al^{3+} with lower valent cations such as Mg^{2+} in the octahedral sheets; the resulting negative charge is balanced by the exchangeable interlayer cations.

General procedure for the preparation of pillared clays are schematically illustrated in Fig. 2 [1]. The first and most important reaction for the introduction of pillars is ion-exchange; the hydrated interlayer cations of montmorillonite are exchanged with precursory polynuclear metal hydroxy cations such as $[Al_{13}O_4(OH)_{24}]^{7+}$ and $[Zr_4(OH)_{14}]^{2+}$. After the ion-exchange, the montmorillonite is separated by centrifugation and washed with water to remove excess metal hydroxy ions. The interlayered metal hydroxy cations are then converted into the respective oxide pillared by calcination. Various kinds of metal oxides such as Al_2O_3, ZrO_2, Fe_2O_3, Cr_2O_3, and TiO_2 have been introduced by this procedure. The pillar heights obtained so far are in the range of 0.7 to 1.0 nm. Most of the pillared structured are thermally stable up to 500°C, and keep the specific surface area as large as 300-500 m^2/g.

Sol pillared clays

Metal oxide sol particles can also be used as pillar precursors, as long as those are positively charged. The silica sol particles obtained by the hydrolysis of silicon tetra-ethoxide are negatively charged, which can be converted into positively charged sols by the modification with small amount of titanium ions [12]. The modified sol particles can be ion exchanged with the interlayer cations of montmorillonite. The spacing increased rapidly soon after the clay was mixed with the sol

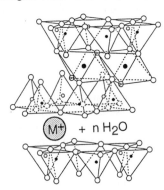

Fig. 1. Structure of montmorillonite.

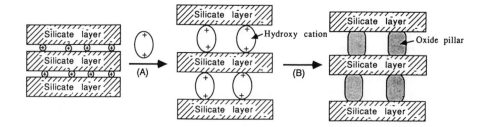

Fig. 2. Schematic illustration of the pillaring process in clay: (A) ion exchange with precursory cations and (B) conversion to oxide by calcination.

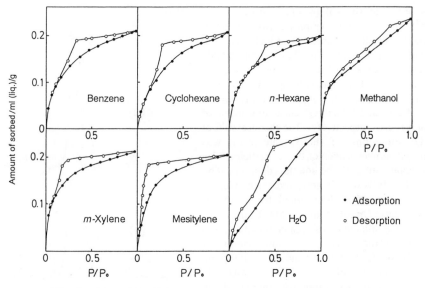

Fig. 3. Adsorption-desorption isotherms of SiO_2-TiO_2 sol pillared clay (30/3/1) for different solvent vapors.

solutions and then attained constant values at ~4 nm after about 6 h. The mixed sol composition will be designated like 30/3/1 for example, where the first and the second numbers refer to the molar ratios of silica and titania to the cation exchange capacity (CEC) equivalent (the third number) of the clay. The nitrogen adsorption isotherm of the sol pillared clay is fitted on the BET plot for the limited number of adsorption layers, suggesting that the pore sizes are much smaller than the pillar heights. The large basal spacing and a high surface area of about 400 m^2/g were maintained at least up to 500°C. Adsorption-desorption isotherms for various kinds of solvent vapors are shown in Fig. 3. The adsorption isotherms for large

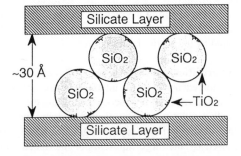

Fig. 4. Structural model of the arrangement of the SiO_2-TiO_2 sol particles in the interlayer space of montmorillonite.

molecules such as xylenes and mesitylene fitted on the Langmuir linear plot, whereas those for smaller molecules such as water and methanol fitted rather on the BET plot. This finding also supports that the pore sizes are of the order of the molecular dimensions examined. It is interesting to note that the hystereses observed in the adsorption of large molecules such as *m*-xylene and mesitylene persist to an unusually low relative pressure. A structural model for the SiO_2-TiO_2 sol pillared clays is proposed in Fig. 4, where the average size of the SiO_2-TiO_2 sol particles must be much smaller than the interlayer spaces and such small sol particles are packed uniformly so as to form small pores in between the sols and silicate layers. The SiO_2 sols are modified by the TiO_2 on the surface and positively charged.

NMR study of the sol pillared clays

^{29}Si magic angle spinning (MAS) NMR spectra of the sol pillared clay (30/1.5/1) were measured, and shown in Fig. 5 [13]. The sample dried at room temperature in air had three resonance peaks at -93.5, -101.5 and -110 ppm, which could be assigned to Q^3 silica component of the silicate layers of clay, and Q^3 and Q^4 silica components of the sol particles, respectively. The chemical shift of the Q^3 silica unit of the silicate layers was unchanged on intercalation of the sol particles. The silica molar ratios of the sol/clay determined by chemical analysis were in good agreement with the corresponding ratios determined by the NMR measurements. On calcination at 500°C, the NMR spectra of the pillared clays became broad and was decomposed into two silica components by deconvolution; Q^3 at -97 ppm and Q^4 at -109 ppm. The chemical shift of the Q^4 was essentially unchanged, but its intensity increased on calcination. The comparison of the intensities of the NMR signals suggested that a part of the Q^3 silica units of sol particles are converted into the Q^4 by condensation, and the rest of the Q^3 remain as Q^3 with showing a slight shift. Evidence for the formation of strong interactions between the silicate layers of clay and the sol particles could not be found by the NMR spectroscopy.

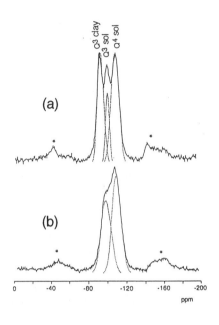

Fig. 5. ^{29}Si MAS NMR spectra of SiO_2-TiO_2 sol pillared clay (30/1.5/1); (a) dried in air and (b) calcined at 500°C.

Supercritical drying of the pillared clays

Ion exchange of the interlayer cations of montmorillonite with precursory cations for pillars is usually carried out in aqueous media, and the samples separated after the ion exchange are extremely swelled with water. The swelled sample experiences remarkable shrinkage during drying, as in the case of the drying of aqueous gels formed by solution-sol-gel route. The shrinkage of the gel is caused by the large liquid-vapor interfacial (capillary) forces, which act to disrupt the delicate microporous gel structure during an evaporative procedure. In order to avoid the formation of liquid-vapor interfaces, a supercritical drying technique was applied to the drying of SiO_2-TiO_2 sol pillared clays, and the resulting pore structure was investigated in comparison with that of the sample prepared by conventional air-drying procedures [14].

The SiO_2-TiO_2 pillared clay (4/0.4/1) was rinsed with ethanol in order to replace the water with ethanol, which was then extracted with supercritical fluid of CO_2 under a pressure of 120 atm at 40°C. For comparison, a separate sample of SiO_2-TiO_2 pillared clay was similarly prepared and air-dried at 60°C. The samples thus prepared were calcined at 500°C for 3 h in air.

The total pore volumes measured by the nitrogen adsorption study and the mercury porosimetry are compared for the two dried samples in Table 1. It is clearly shown that the air-dried sample has few macropores, while the supercritically dried (SCD) sample has an extremely large pore volume of 9.82 ml/g, the most part of which comes from the pores ranging from 5-0.5 μm in diameter. It is interesting to note that the total mercury intrusion volume for the air-dried sample is only 1% of that for the SCD sample. The micro and total pore volumes measured by nitrogen adsorption did not differ so much between the two samples. The observation by a scanning electron microscopy

Table 1. BET surface areas and the pore volumes measured by nitrogen adsorption and mercury porosimetry.

Drying procedure	BET surface area, $m^2 g^{-1}$	Pore volumes	
		N2 adsorption[1] $ml g^{-1}$	Hg porosimetry $ml g^{-1}$
Supercritical drying	459	1.13 (0.09)	9.82
Air drying	325	0.25 (0.19)	0.098

1) Micropore volumes are shown in parentheses.

indicated that the SCD sample had a well developed card-house structure, and macropores are constructed by thin silicate sheet walls. In contrast to the SCD sample, the silicate layers of the air-dried sample are stacked in a face-to-face fashion, and only a small number of large pores are observed. The macropores which were measured by mercury intrusion had been filled with water before drying. The pore-volume data given in Table 1 show that the supercritical drying can preserve the swelled pillared structure. During a conventional air drying procedure, the macropores filled with water shrink even to 1% of the original volume. A schematic structural model showing the dual-pore structure of the SCD sample is shown in Fig. 6. The micro- and mesopores are located in between the packed sol particles in the interlayer space of the pillared layers. The macropores are included in the highly porous card-house structure developed by the connection of the pillared layers.

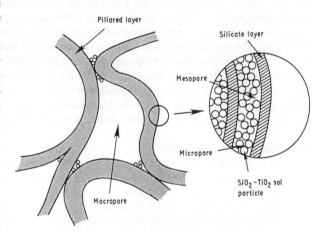

Fig. 6. Schematic structural model of the pillared clay supercritically dried.

GRAFTING

The interlayer surface of most of layered crystals consists of closely packed atom layers. Generally, it is very difficult to graft orgnic functional groups onto such surface. Among a number of layered crystals, zirconium phosphate is very special from a structural view point that the interlayer surface has reactive functional groups directing outside the layers; the tips of phosphate groups are directing toward the adjacent layers, which can be modified by organic functional groups through ester bonds.

Interlayer surface of zirconium phosphates

There are two types of layer structured zirconium phosphates, $Zr(HPO_4)_2 \cdot nH_2O$: a monohydrtate, α and a dihydrate, γ [15,16]. The γ phase having a larger basal spacing is more reactive for intercalation reactions. In a previous study on the preparation of organic derivatives of zirconium phosphate, the γ phase was used, and the phosphate groups on the interlayer surface were subjected to reaction with ethylene oxide [17]. It was concluded that all of the monohydrogenp-hosphate groups in the structure were involved in the formation of P-O-C ester bonds, since about two moles of ethylene oxide reacted with one mole of zirconium phosphate:

$$Zr(HPO_4)_2 \ + \ 2\overline{C_2H_4O} \ \longrightarrow \ Zr(HOC_2H_4OPO_3)_2.$$

Recently, Clayden [18] investigated the ^{31}P MAS NMR spectra of α- and γ-zirconium phosphates. The former showed only one ^{31}P resonance, whereas the latter two resonances. After the detailed analysis of the spectra, he concluded that there were two chemically distinct types of phosphate groups in the γ phase, and that it should be formulated as $Zr(PO_4)(H_2PO_4) \cdot 2H_2O$ rather than $Zr(HPO_4)_2 \cdot 2H_2O$. The crystal structure of γ-titanium phosphate, which is isomorphous with γ-zirconium phosphate was analyzed on the basis of the X-ray powder diffraction data by Christensen et al. [19]. The structural model proposed is compared with that of α-zirconium phosphate in Fig. 7. In the α phase, phosphate groups are all monohydrogenphosphate ions; each phosphate ion is bonded to three different zirconium atoms from above and below the zirconium atom layer [20]. The rest of the tips of the tetrahedral phosphate ion bears a hydrogen ion, which is exchangeable with various cations. In the γ phase the phosphate is present as equal number of PO_4 and H_2PO_4 groups, only the latter phosphate groups being on the interlayer surface.

Grafting of ethylene oxide

γ-Zirconium phosphate was dispersed in ethylene oxide aqueous solutions, and samples of three different degrees of reactions with basal spacings, d = 1.63, 1.85 and 1.94 nm were separated. The molar ratios of ethylene oxide taken up by γ-zirconium phosphate (n = EO/Zr) were determined to be 0.70, 1.75 and 2.30, respectively.

^{31}P cross-polarization (CP) MAS-NMR spectra were measured on the samples of different degrees of reaction, and are shown in Fig. 8 [21]. Two sharp ^{31}P resonance peaks are clearly seen in the spectrum of the starting γ-zirconium phosphate at -8.90 and -26.83 ppm, which are

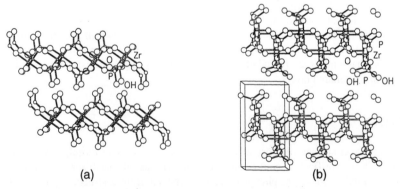

(a)   (b)

Fig. 7. Comparison of the two layer structures of zirconium phosphates; (a) α, and (b) γ. The interlayer water molecules are not shown.

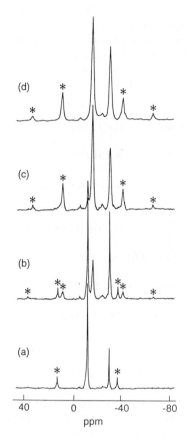

Fig. 8. ^{31}P CP MAS NMR spectra of (a) γ-zirconium phosphate and its organic derivatives with different degrees of reaction, (b) n = 0.70, (c) n = 1.75, and (d) n = 2.30.
* indicates spinning side bands.

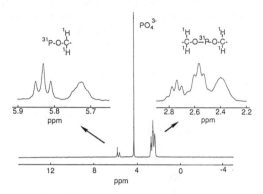

Fig. 9. ^{31}P NMR non-decoupled spectrum of the supernatant solution of the hydrolyzed product.

respectively assigned to H_2PO_4 and PO_4 by Clayden [18]. As the reaction with ethylene oxide proceeded, the intensity of the former peak due to H_2PO_4 was reduced and a new peak appeared at -13.24 ppm; the peak due to PO_4 remained unshifted at -26.9 ppm. This finding suggests that only half of the phosphate groups are used for the grafting reaction, though about two moles of ethylene oxide reacted with one mole of γ-zirconium phosphate. It is reasonable to assign the new peak to the ^{31}P of the phosphate groups with P-O-C ester bonds. The increased linewidth of the spectra of the organic derivatives are interpreted in terms of a reduction in the crystallinity of the samples.

The reaction product (n = 1.75, d = 1.85 nm) was hydrolyzed with a 1 M NaOH solution at 323 K for 2 h. Figure 9 shows ^{31}P non-decoupled NMR spectra of the supernatant solution of the hydrolyzed product. A sharp singlet peak at 4.31 ppm is assigned to orthophosphate ions. The presence of multipulet peaks can be interpreted in terms of the formation of several kinds of phosphate esters; more than two kinds of monoesters (triplets in the lower field), and more than three kinds of diesters (quintets in the higher field). The peak intensity of the diesters is much larger than that of the monoesters. If a small amount of NaOH (molar ratio of NaOH/Zr < 2) was used for the hydrolysis, the formations of monoesters and orthophosphate ions in the hydrolyzed products were much smaller.

^{13}C NMR spectra of the supernatant solution obtained by a similar hydrolysis had a large number of doublet peaks which could be assigned to ^{31}P-O-^{13}C and ^{31}P-O-C-^{13}C ester groups. The singlet ^{13}C resonance peaks due to ethylene and diethylene glycols were observed, the intensities of which greatly increased on further hydrolysis of the solution with NaOH at 373 K. Though we have more doublet peaks which cannot be assigned by standard samples, it is likely that those are doublet resonance peaks attributable to ^{31}P-O-^{13}C-C-O- and ^{31}P-O-C-^{13}C-O- groups of a mixture of mono- and di-esters of diethylene glycols and/or ethylene glycols.

From the NMR spectroscopic data, the following structural model is proposed for the reaction of ethylene oxide with the interlayer surface of γ-zirconium phosphate. γ-Zirconium phosphate has two types of phosphate groups; PO_4 and H_2PO_4, and only the interlayer H_2PO_4 groups react with ethylene oxide. The reason why one mole of γ-zirconium phosphate can react with two moles of ethylene oxide is found in the formation of diesters;

$$\begin{array}{c}
\text{Zr} \!\!-\!\! O \quad OH \\
\diagdown \! P \! \diagup \\
\text{Zr} \!\!-\!\! O \quad OH
\end{array}
\quad + \quad 2\ \overline{CH_2CH_2O}
\quad \longrightarrow \quad
\begin{array}{c}
\text{Zr} \!\!-\!\! O \quad O\text{-}CH_2CH_2OH \\
\diagdown \! P \! \diagup \\
\text{Zr} \!\!-\!\! O \quad O\text{-}CH_2CH_2OH
\end{array}$$

In parallel, diethylene glycol esters are formed, and the formation of more complicated esters will also be possible;

$$\begin{array}{c}
\text{Zr} \!\!-\!\! O \quad OH \\
\diagdown \! P \! \diagup \\
\text{Zr} \!\!-\!\! O \quad OH
\end{array}
\quad + \quad 2\ \overline{CH_2CH_2O}
\quad \longrightarrow \quad
\begin{array}{c}
\text{Zr} \!\!-\!\! O \quad O\text{-}CH_2CH_2OCH_2CH_2OH \\
\diagdown \! P \! \diagup \\
\text{Zr} \!\!-\!\! O \quad OH
\end{array}$$

$$\begin{array}{c}
\text{Zr} \!\!-\!\! O \quad OH \\
\diagdown \! P \! \diagup \\
\text{Zr} \!\!-\!\! O \quad OH
\end{array}
\quad + \quad (n+m)\ \overline{CH_2CH_2O}
\quad \longrightarrow \quad
\begin{array}{c}
\text{Zr} \!\!-\!\! O \quad O\text{-}(CH_2CH_2O)_nH \\
\diagdown \! P \! \diagup \\
\text{Zr} \!\!-\!\! O \quad O\text{-}(CH_2CH_2O)_mH
\end{array}$$

Anion exchange in zirconium phosphate

Ester derivatives of zirconium phosphate can also be prepared by an ion exchange of the interlayer phosphate groups with various phosphoric ester groups. Rahman and Barrett [22] first investigated the exchange of phosphate ions of α-zirconium phosphate with phosphate ions in solutions by using ^{32}P isotope. They found that the phosphate groups internal as well as outer surfaces of α-zirconium phosphate were exchanged with the labeled phosphate groups in the contacting solution:

$$Zr(HPO_4)_2 \ + \ H^{32}PO_4 \ \longrightarrow \ Zr(H^{32}PO_4)_2 \ + \ HPO_4$$

We have found that similar ion exchanges occur more easily in γ-zirconium phosphate; the H_2PO_4 groups are exchanged with various phosphate ester groups, where the phosphate groups are labeled with ester groups $(ROPO_3H)$ in stead of by ^{32}P isotope;

The reactions are carried out at 70°C in aqueous solutions or acetone + water mixed solutions containing phosphoric acid esters. The esters should be hydrogen form. If sodium salt forms are used, γ zirconium phosphate is changed into a stable Na ion-exchanged form, $Zr(PO_4)(NaHPO_4)$, which is innert against the exchange reaction [23]. The reactions are applicable a large number of phosphoric esters, as long as the esters are stable in aqueous solutions at about 70°C. Exchange with phosphonate ions are also possible. In our previous studies, we assumed that γ zirconium phosphate had a structure similar to that of the α phase and HPO_4 groups on the interlayer surface [24]. All the formulae reported in previous studies should be revised.

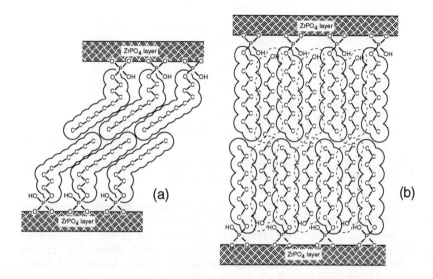

Fig. 10. Structural models of the arrangements of the alkyl chains bonded through ester bonds on to the interlayer surface of γ-zirconium phosphate; (a) before and (b) after swelling with alcohols.

A series of n-alkyl ester derivatives of γ-zirconium phosphate were prepared by the exchange method [25]. A systematic expansion in the basal spacing as a function of the number of carbon atoms in the alkyl chain suggested that the alkyl chains were arranged in a bimolecular layer with a bending in the middle of the chain as shown in Fig. 10a. The derivatives were swelled with alkanol with the same length of alkyl chains, the alkyl chains being straightened and oriented perpendicular to the phosphate layers (Fig. 10b).

The exchange rate of phenyl phosphate with the interlayer phosphate groups was measured [26]. The data shown in Fig. 11 can be explained by a diffusion model with a long cylinder where diffusion takes place only from the side edges. The diffusion coefficient was calculated to be $10^{-13} \sim 10^{-12}$ cm^2/s at 70°C. The exchange with phenyl phosphate was completed if the molar ratio of phenyl phosphate to the total amount of phosphate ions in the solution was larger than 0.8.

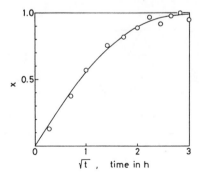

Fig. 11. The exchanged amount x in the product $Zr(PO_4)(H_2PO_4)_{1-x}(C_6H_5PO_4)_x$ as a function of the square root of the reaction interval (time in hours) at 70°C.

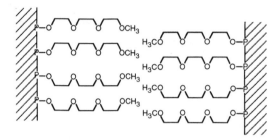

Fig. 12. A schematic illustration of the arrangement of oxyethylene chains grafted onto the interlayer surface of zirconium phosphate.

It is interesting to note that the resulting exchanged products are organic derivatives of inorganic layers. The organic functional groups are grafted onto the interlayer surface, and arranged in a regular manner. The derivatives obtained by the ion exchange with phosphoric esters having oxyethylene chains exhibit properties characteristic for crown ethers (Fig. 12); the interlayer oxyethylene chains can take up alkali metal salts such as $LiClO_4$, iodides, and thiocyanides [27]. The derivative intercalated with $LiClO_4$ is an interesting solid electrolyte [28].

New anion-exchangeable layered crystals
As described above, organic derivatives of inorganic layers can be prepared by grafting or exchanging framework anions with organic anions. However, anion-exchangeable layer structured crystals like zirconium phosphate are very rare. Recently, we have found that basic copper acetate of the botallackite type layer structure is a new anion-exchangeable crystals [29].
Basic copper acetate is prepared by titrating a copper acetate solution with a NaOH solution up to OH/Cu = 1. Green-colored platelet crystals with a composition of $Cu_2(OH)_3(OCOCH_3) \cdot H_2O$ were obtained. The structures of the botallackite and the hydrotalcite type layered double hydroxide (LDH) are schematically compared in Fig. 13. Both structures can be derived from the CdI_2 layer structure. In the LDH layer, the hydroxide layers are completed by hydroxy groups; the divalent metal ions are partially substituted with trivalent metal ions. The layers are positively charged. This excess charge is balanced by anions located between the layers. On the other hand, the botallackite has neutral hydroxy layers. A quarter of the hydroxy groups are substituted with acetate ions. Though the acetate framework ions are directly bound to Cu^{2+} ions, these are easily exchangeable with various anions.

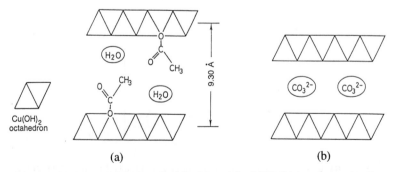

Fig. 13. Comparison of the botallackite (a) and the LDH (b) type layer structures.

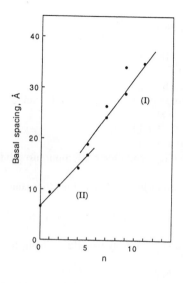

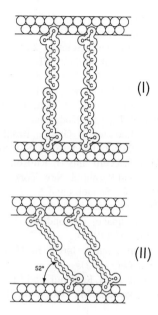

Fig. 14. Basal spacings of basic copper acetate after reaction with carboxylate ions with different number of carbon atoms (n) in the alkyl chains.

Fig. 15. Schematic structural models of the orientation of interlayer alkyl chains of the exchanged products I and II.

The acetate ions of basic copper acetate are exchanged with various anions merely by dispersing in aqueous solutions of NaX (X = Cl, Br, I, NO_3, ClO_4), and Na_2SO_4 at room temperature, the basal spacing being changed to those of the corresponding basic salts already reported;

$$Cu_2(OH)_3(CH_3COO) \ + \ X^- \longrightarrow Cu_2(OH)_3X \ + \ CH_33COO^-$$

The ions CH_3COO, NO_3, and ClO_4 are reversibly exchanged with each other, while the exchange with small size ions such as chloride and bromide ions are irreversible. The competitive ion exchange reactions studied on several pairs of anions showed that the selectivity of the anions by the basic copper layers were in the following order; $Cl^- > Br^- > NO_3^- > CH_3COO^-$, ClO_4^-.
The acetate ions are exchanged with various sodium salts (n-C_nH_{2n+1}COONa, n = 0-11) [30]. The basal spacings of the exchanged crystals are shown in Fig. 14 as a function of the number of carbon atoms (n) in the alkyl chains. Some products have more than one kinds of basal spacings depending on the preparation conditions. Two linear relationships are observed with slopes corresponding to 2.55 (I) and 2.0 (II) Å/carbon atom. These slopes suggest that the alkyl chains are oriented in bimolecular layers almost perpendicular and inclined at an angle of about 56° to the layers, respectively as shown in Fig. 15. Chemical and thermogravimetric analyses showed that the samples with the higher slope (I) have a composition of x = about 1, and the ones with the lower slope (II) have a composition of x = about 0.85 in $Cu_2(OH)_{4-x}(C_nH_{2n+1}COO)_x$.

Acknowledgments
This study was partly defrayed by Grant-in-Aid for Scientific Research from Ministry of Education, Science and Culture.

References
1. S. Yamanaka, Am. Ceramic Soc. Bull., **70**, 1056 (1991).
2. T. J. Pinnavaia, Science, **220**, 365 (1983).
3. R. Burch, ed., Pillared Clays, Catalysis Today, **2**, 1-185 (1988).
4. F. Figueras, Catal. Rev. Sci. Eng., **30**, 457 (1988).
5. E. M. Farfan-Torres and P. Grang, J. Chim. Phys., **87**, 1547 (1990).
6. S. Yamanaka and M. Hattori, in Chemistry of Microporous Crystals, (eds.) T. Inui, S. Namba and T. Tatsumi, Kodansha/Elsevier, Tokyo (1991) p.89.
7. B. Delmon and P. Grange, Erdol Erdgas Kohle, **107**, 376 (1991).
8. R. A. Schoonheydt, Stud. Surf.Sci. Catal., **58**, 201 (1991).
9. M. L. Occelli and H. Robson, (eds.) Expanded Clays and Other Microporous solids, Van Nostrand Reinhold, New York (1992).
10. C. A. C. Sequeira and M. J. Hudson, (eds.) Multifunctional Mesoporous Inorganic Solids, Kluwer Academic Publishers, Dordrecht (1993).
11. K. Watanabe, T. Kondow, M. Soma, T. Onishi, and K. Tamaru, Proc. Roy. Soc. London, **A333**, 51 (1973).
12. S. Yamanaka, Y. Inoue, M. Hattori, F. Okumura, and M. Yoshikawa, Bull. Chem. Soc. Jpn., **65**, 2494 (1992).
13. S. Yamanaka, K. Kunii, and M. Ohashi, to be published.
14. K. Takahama, M. Yokoyama, S. Hirao, S. Yamanaka, and M. Hattori, J. Mat. Sci., **27**, 1297 (1992).
15. A. Clearfield (ed.) Inorganic Ion Exchange Materials, CRS Press, Boca Raton, Fl., (1982).
16. S. Yamanaka and M. Hattori, in Inorganic Phosphate Materials, T. Kanazawa (ed.) Kodansha/Elsevier, Tokyo/Amsterdam, (1989) p.131.
17. S. Yamanaka, Inorg. Chem., **15**, 2811 (1976).
18. N. J. Clayden, J. Chem. Soc. Dalton Trans., 1877 (1987).
19. A. N. Christensen, E. K. Andersen, I. G. K. Andersen, G. Alberti, M. Nielsen, and M. S. Lehmann, Acta Chem. Scand., **44**, 865 (1990).
20. A. Clearfield, and G. D. Smith, Inorg. Chem., **8**, 431 (1969).
21. S. Yamanaka, T. Ohno, and H. Nakano, Chem. Lett., submitted.
22. M. K. Rahman and J. Barrett, J. Chromatogr., **69**, 261 (1972).
23. S. Yamanaka, K. Yamasaka, and M. Hattori, J. Inorg. Nucl. Chem., **43**, 1659 (1981).
24. S. Yamanaka and M. Tanaka, J. Inorg. Nucl. Chem., **41**, 45 (1979).
25. S. Yamanaka, M. Matsunaga, and M. Hattori, J. Inorg. Nucl. Chem., **43**, 1343 (1981).
26. S. Yamanaka, K. Sakamoto, and M. Hattori, J. Phys. Chem., **88**, 2067 (1984).
27. S. Yamanaka, K. Yamasaka, and M. Hattori, J. Inclusion Phenomena, **2**, 297 (1984).
28. S. Yamanaka, M. Sarubo, K. Tadanobu, and M. Hattori, Solid State Ionics, **57**, 271 (1992).
29. S. Yamanaka, T. Sako, K. Seki, and M. Hattori, Solid State Ionics, **53-56**, 527 (1992).
30. S. Yamanaka, T. Sako, and M.Hattori, Chem. Lett., 1869 (1989).

Materials Science Forum Vols. 152 - 153 (1994) pp. 81-86
© 1994 Trans Tech Publications, Switzerland

APPLICATION OF HOST-GUEST REACTIONS TO SYNTHESIS OF NEW MATERIALS: THE CASE OF S.T.A. PROJECT IN JAPAN

M. Watanabe

National Institute for Research in Inorganic Materials,
Namiki 1-1, Tsukuba, Ibaraki 305, Japan

Keywords: Shape-Control, Secondary Structure-Control, Li-Selectivity, Titanate-Fibers, Graphite-Thin Films, Mesoporous Materials

ABSTRACT

In the S.T.A. project materials with unique functions which can not be obtained or at least not easily through conventional methods are developed using host-guest reactions. On the basis of the characteristics that host compounds hardly suffer structural changes, the conversion of compounds is treated by dividing a whole conversion process into two components. The first is called a composition changing process in which the factors other than composition are basically kept unchanged, and the second a structure changing process where the factors other than structure remain approximately unchanged. Synthetical procedures developed in application of this policy are described in some concrete examples, e.g., $HSbO_3$ with high Li-selectivity, hexatitanate fibers, highly oriented graphite thin films, and so on. Finally the way and the definition of soft chemistry will be discussed from the viewpoint of materials synthesis.

INTRODUCTION

In general, inorganic compounds with ionic and/or covalent character are sensitive to thermodynamical stability and as a result tend to form according to phase rule. Therefore, one can often obtain desired compounds directly from starting materials by conventional methods; here the term " conventional method" is defined as the method which makes a whole reaction system reaching thermal equilibrium during process. Simultaneously it suggests that inorganic compounds would not easily suffer such local modifications as usually routine preparation of derivatives in organic field. The development of host-guest chemistry so far and that of soft chemistry in the last decade or more, however, has presented a clue to challenge this unadaptability of inorganic materials. In response we have been studying the application of host-guest chemistry or soft chemistry with the approach of material synthesis.

We are developing a research project entitled "Application of Host-Guest Reactions to Synthesize Novel Functional Materials", which has been promoted since 1990 with the support of Science Technology Agency Japan(hereafter, S.T.A.). In this project reaction-techniques in host-guest chemistry are regarded as one of the means which can realize a partial reaction, in other words prevent a whole body of the system from going to thermal equilibrium. We are focusing on the practical use of host-guest reactions for material synthesis rather on the investigation of the reactions themselves and expecting to develop some materials with unique functions that can not be obtained or at least not easily by conventioal methods. The outline of our project is introduced here as one of the materials for the meeting. We discuss the soft chemistry way referring to a series of advanced researches in the last decade and to some themes of our project.

S.T.A. PROJECT

The whole structure of the project is constructed with 11 themes, and 7 are related to material synthesis and to characterization; the titles of the themes, organizations, and their representative researchers are listed in Appendix. The research period is the first stage of 3 years(1990 to 1993) and the second stage of 2 years(1993 to 1995). In the first stage, which has finished in March 1993, emphasis was put on development of non-conventional reaction routes to individual objects and analysis of the processes.

AIMS
Basically we are interested in creating new compounds that can not be obtained conventionally or compounds that are known but can not be easily produced through conventional methods, but the general aim of the project is to develop unique materials. From this point of view, more concrete goals are described as follows;
(a) preparing compounds showing unique properties when particular framework ions or counter ions are replaced with negligible structural changes,
(b) obtaining materials in forms(e.g., fibers and thin films) which can not be elaborated in their usual preparation routes,
(c) achieving products where secondary structures (e.g., pores and texture) are controlled by host-guest reaction techniques.

POLICY
The conversion of compounds takes place in general along with compositional and/or structural changes. However, it can be assumed in principle that the host compound hardly suffers a structural change during a host-guest reaction. Accordingly we consider the conversion of compounds by dividing a whole process of conversion into two components. The first is what changes composition and basically holds other factors such as structure and crystal shape. The second is what converts the structure, other factors remaining approximately unchanged. The composition-change is performed effectively with host-guest reaction techniques like intercalation, deintercalation, and ion-exchange. The preparations of the (a) type are attained by this operation. In some cases unique compounds could only be obtained by the change of compositions[1,2,3]. The structure-change is usually performed by heat treatment at relevant temperatures. The materials of the (b) and (c) types could be obtained applying both of the composition- and structure-changing operations. However, there may be some cases that the operation combination results in unique compounds such as in the (a) type. Typical examples could be given by the references[4,5].

For the structure-changes the mild heat treatment has conveniently used, but it has often influenced the nature of compounds and leads to formation of thermal equilibrium phases. So we are planning to select other structure-changing ways, for example microwave-heating and high pressuring at relevant temperatures where ion diffusion is allowed locally, e.g., below a few hundred °C. The former may lower total energy supply and perform selective agitation of particular bonds. The latter may induce a bonding transition.

The functions expected for our products can be divided into three categories. The first is the functions remarkably enhanced by controlling the shape of products, which is related to mechanical or thermal properties such as heat insulating. The second is unique functions which are induced due to steric factors like intrinsic cavities in a crystal structure or micropores. Especially we expect the appearance of ion selectivity, molecular seiving effect and calatytic property. The third is the functions which are directly connected to the physical property of products, where dielectricity, photochromism, and so on are mainly examined.

SIGNIFICANT EXAMPLES
Some examples related to materials synthesis will be described:

1. Synthesis of HSbO$_3$ with high Li$^+$ ion-selectivity

The first example, corresponding to aim (a), is given by the elaboration of the monoclinic HSbO$_3$ which shows a very high ion-selectivity for Li$^+$ ion in comparison with the other alkaline metal ions. This material can be prepared by ion-exchanging the Li of orthorhombic LiSbO$_3$ with protons in aqueous solutions of nitric acid, that is the process is only performed by the composition-changing operation[6]. LiSbO$_3$ is obtained by dehydration of Li[Sb(OH)$_6$]. This conversion is characterized by a small structural distortion and hydration-free ion-exchange, which contributes to the high Li-selectivity as would the material have a memory effect for Li$^+$ ions. A similar modification was found for LiNbO$_3$ and LiTaO$_3$, although Li-selectivity was not mentioned[7], and their structural distortions can be regarded to be still displacive, but relatively more than that of LiSbO$_3$.

2. Synthesis of hexatitanate fibers from K$_2$Ti$_4$O$_9$

The second example, corresponding to aim (b), is given for the conversion from tetra- to hexatitanate fibers. K$_2$Ti$_6$O$_{13}$ fibers are excellent for heat insulation and resistance, and recently have been used for industrial products instead of asbestos. Its good fibers cannot be prepared easily by a conventional method, i.e. heating TiO$_2$ and K$_2$O mixtures. Fujiki et al.[8] showed that the fibers could be obtained by way of fibrous crystals of K$_2$Ti$_4$O$_9$ which were easily produced by flux growth method. K$^+$ ions in tetratitanate fibers are first replaced partially by proton or oxonium ions by ion-exchange, while the fibrous shape and the crystal structure are kept unchanged. Composition-changed fibers are then heated to convert the crystal structure of tetra- into hexatitanate, where the fibrous form and the chemical composition remain unchanged.

In relation to tetratitanate fibers, a new preparation process of fibrous K$_3$Ti$_8$O$_{17}$ crystals is described, which is attained by composition change only. This compound was of bronze-type and semiconducting, which was first prepared by the electrode reaction in a K$_2$O-TiO$_2$ melt with a small amount of Nb$_2$O$_5$ at about 1000°C[9]. In the new process, the fiber crystals can be derived by treating K$_2$Ti$_8$O$_{17}$ fibers in K vapor at about 260°C[10]. The starting K$_2$Ti$_8$O$_{17}$ absorbs one K atom per mole into the tunnel structure, but the crystal lattice is hardly distorted. This additional K atom reduces Ti^{4+} to Ti^{3+} in the framework, and as a result the fibers become semiconducting. The starting octatitanate fibers can be converted from tetratitanate fibers by using the soft chemical procedure described in reference[5].

3. Synthesis of highly oriented graphite thin films

A new preparation process of highly oriented graphite thin films is described below. The process can be regarded as the template method[11] using clay minerals for host materials[12], but here the host materials play the role of templates. Organic compounds such as furfuryl alcohol or acrylonitrile with polymeric property are intercalated into interlayer spaces of the host material, and then their polymerization is performed by heating or applying γ-rays. Those polymers are carbonized in the interlayer spaces at about 700°C. Precursor carbon is obtained by dissolving the host material in an acidic solution. The final product, i.e. highly oriented graphite thin film, can be obtained by graphitization of the precursor carbon at more than 2000 °C. Zeolite and sepiolite can be used as host materials. In those cases mesoporous carbon and fibrous carbon could be prepared.

4. Synthesis of mesoporous materials from hectorite

An example of (c) type, given by the preparation process of mesoporous materials from hectorites[13], is described. Starting materials are various precursor hectorites which have different layer charges and interlayer silicate contents. They are synthesized by a hydrothermal method. Precursor hectorites first suffer composition changes from interlayer cations like Na$^+$ to some alkyl ammonium ions at about 80 °C. In the following structure change, the organophilic hectorites are converted to mesoporous materials by heating between 300 and 900 °C. Mesoporous materials on clays have been studied in different ways, for example pillaring by heat treatment of polymeric hydroxyl cations or charged sols[14, 15]. It can be noted that this method gives mesoporous materials in which the average pore size is significantly larger than those of conventional pillared clays and the pore size distribution sharper.

DISCUSSION

In principle host-guest reactions do not bring about reconstruction of host structures although they actually result in displacive distortions. This feature is linked to the concept of "topotaxy" in structural chemistry. We use the word "soft chemistry" in close connection with host-guest chemistry. This word originally involved conversion of inorganic compounds by the topotactic modification of crystal structures performed by reactions such as ion-exchange or intercalation [16]. One may expect that soft chemistry with the time will be based on more comprehensive concepts. One possibility is to emphasize another aspect of host-guest reactions: only particular ions, usually guest species, are allowed to diffuse and hence a whole system of host and guest species is prevented from reaching thermal equilibrium. The system, either a compound or a complex of host and guest, remains metastable. One is able to create a metastable state with the help of a host structure, which is just the most important characteristic of host-guest reactions. Host-guest chemistry, however, does not require that the metastable state is a compound, i.e., a single phase. From this point it is expected that the basic object of soft chemistry may be put on the conversion of a system to a metastable phase.

There are various reaction techniques which do not bring about thermal equilibrium in a whole system. Host-guest reaction is an important technique among them. Some typical examples directly obtained by host-guest reactions are given below;
1) the conversion of $LiSbO_3$ to $HSbO_3$[6] and of $LiNbO_3$ to $HNbO_3$[7] by ion-exchange,
2) the conversion of $LiCoO_2$ to CoO_2[16], of $LiVS_2$ to VS_2[1], and of $CaSi_2$ to Si[2]
   by deintercalation,
3) the conversion of ReO_3 to $LiReO_3$[17] and of $K_2Ti_8O_{17}$ to $K_3Ti_8O_{17}$[10] by insertion.
These conversions are not accompanied by large structural changes and attained in one stage by a host-guest reaction. On the other hand, one can expect that host-guest reactions may be used at multi-steps. Such an example is given by the conversion of $NaNi_{1-y}Co_yO_2$ to $Ni_{1-y}Co_y(OH)_2X_{y/n}$ zH_2O, $X^{h-} = CO_3^{2-}$, etc. in two stages[3].

The most conventional reaction technique without thermal equilibrium in a whole system, other than host-guest reactions, would be a heat treatment under mild conditions. In the following, three typical processes are shown with simple flow charts; the conversion of $K_2Ti_4O_9$ to $K_2Ti_8O_{17}$[5] in 1) and of TiO_2 to $LiTi_2O_4$[4] in 2), and the template reaction of $AlPO_4$[18] in 3). They all suffer distinct changes of both composition and structure and lead to the products which can not be obtained conventionally or at least not easily.

1)

$$H_2Ti_4O_9 \ nH_2O \ \underset{\text{ion-exchange}}{\rightleftharpoons} \ KHTi_4O_9 \ n'H_2O \ \xrightarrow{500\ °C} \ K_2Ti_8O_{17}$$

(layer struct.)   (layer struct.)   (tunnel struct.)

2)

$$TiO_2 \ + \ x\ n\text{-BuLi} \ \underset{\text{insertion}}{\rightleftharpoons} \ Li_xTiO_2 \ + \ x/2\ C_8H_{18}$$

(anatase-type)   (anatase-type)

$$500°C \ \bigg| \ \text{irrevers.} \ \downarrow$$

$LiTi_2O_4$
(spinel-type)

3)

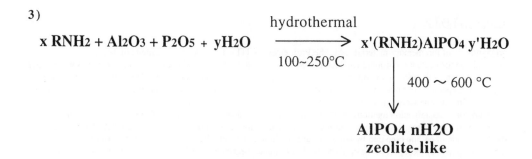

$$x \text{ RNH}_2 + \text{Al}_2\text{O}_3 + \text{P}_2\text{O}_5 + y\text{H}_2\text{O} \xrightarrow[100{\sim}250°\text{C}]{\text{hydrothermal}} x'(\text{RNH}_2)\text{AlPO}_4 \, y'\text{H}_2\text{O}$$

$$\downarrow 400 \sim 600 \,°\text{C}$$

AlPO4 nH2O
zeolite-like

In those examples mild heat treatment is used as a technique for realizing nonequilibrium reactions. The same happens also in our project. This treatment is very convenient, but simultaneously has the inconvenience of supplying easily an energy excess and of reaching thermal equilibrium. The use of microwave heating, high pressure at mild temperatures and so on would be interesting.

In conclusion it is expected that soft chemistry can be considered as dealing with the conversion of a system to a metastable phase along reaction routes realized by various ways, for example host-guest reactions, mild heating, etc., and their combinations. Finally we expect that soft chemical techniques are going to steadily develop as a chemical nonequilibrium preparation method as well as the physical nonequilibrium preparation ones like MBE supported by beam techniques.

REFERENCES

1)Murphy, D.W., Cros, C., Disalvo, F.J., and Waszczak, J.V.:*Inorg.Chem.*, 1977,**16**, 3027
2)Yamanaka, S., Suehiro, F., Sasaki, K., and Hattori, M.: *Physica,* 1981, **105B**, 230
3)Delmas, C. and Borthomieu, Y.,: *J.Solid State Chem.,* 1993,**104,** 345
4)Murphy, D.W., Greenblatt, M., Zahurak, S.M., Cava, R.J., Waszczak, J.V., Hull,Jr., G.W., and Hutton, R.:*Rev. de Chim.miner.*, 1982,**19**, 441
5)Marchand, R., Brohan, L.,and Tournoux, M.: *Mat.Res.Bull.*, 1980,**15**, 11296)Chitrakar, R. and Abe, M.: *Solvent Extr.Ion Exch.*,1989,**7**, 721
7)Rice, C.E. and Jackel, J.L.:*J. Solid State Chem.*, 1982,**41**, 308
8)Fujiki, Y.:*J.S.A.E.Rev.*, 1981,91
9)Watts, J.A.: *J.Solid State Chem.*, 1970,**1**, 319
10)Sasaki, T., Komatsu, Y., and Fujiki, Y.: *J.Solid State Chem.*,1989, **83**, 45
11)Kokotailo, G.T., Lawton, S.L., Olson, P.H., and Meier, W.M.: *Nature*, 1978, **272**, 437
12)Kyotani, T., Sonobe, N., and Tomita, A.: *Nature*, 1988, **331**, 331
13)Torii, K., Iwasaki, T., Onodera, Y., and Hatakeda, K.: *Proc.Chem.Microporous Crystals*, Ed. by T. Inui, et.al., 81(1990) Elsevier.
14)Brindley, G.W. and Yamanaka, S.: *Am.Min.*,1979, **64**, 830
15)Yamanaka, S., Suzuki, Y., Nishihara, T., and Hattori, M.: *Proc.Ceram.Soc.Jpn.*, 1986, Vol.**1**, 161
16)Delmas, C., Branconnier, J-J., Maazaz, A., and Hagenmuller, P.: *Rev. de Chim.miner.*, 1982, **19**, 343
17)Murphy, D.W., Greenblatt, M., Cava, R.J., and Zahurak, S.M.: *Solid State Ionics*,1981, **5**, 327

APPENDIX

(a1)Synthesis of ion memory materials., M.Abe(Prof. emer. of TIT), Tsuruoka Natl. College of Tech.
(a2)Study on the evaluation of ionic conductors., M.Amano, NRIM, STA./ (a3)Development of new inorganic water purification ion-exchangers with both cation and anion capacities. T.Suzuki, Dept. Appl.Chem. and Biotech., Facul. of Eng./ (a4)Preparation of novel layer compounds by the functionalization of their interlayer host spaces., T.Uematsu, Facul. of Eng., Chiba Univ./ (b1)Synthesis of fibrous materials with multi-functions., M.Watanabe, NIRIM,STA./ (b2)Study of characterization methods on electrical and optical properties., T.Ohachi, Dept.Elec.Eng., Doshisha Univ./ (b3)Establishment of template carbonization as a new method for material design., A.Tomita, ICRS, Tohoku Univ./ (b4)Evaluation of carbon prepared by template carbonization., Y.Chida, Res.Cent., Mitsubishi Kasei Co./ (b5)Development of mesoporous materials derived fom silicate-bearing layer silicates., Y.Onodera, Chem.Dept., Gover.Indust.Res Inst. Tohoku, MITI./ (b6)Department of Novel synthetic methods., K.Kuroda, Dept.Appl.Chem., School of Sci. & Eng., Waseda Univ./ (b7)Macroscopic and microscopic analysis of the crystal structure and properties., K.Fukuda, Dept.Inorg.Mater. & Dept.Phys.Chem., NIMCR, Agen. of Industr. Sci. & Tech.

Materials Science Forum Vols. 152 - 153 (1994) pp. 87-98
© *1994 Trans Tech Publications, Switzerland*

LAYERED AND PILLARED ZIRCONIUM PHOSPHATES WITH
α- AND γ-STRUCTURES

G. Alberti, S. Murcia Mascarós and R. Vivani

Department of Chemistry, University of Perugia, I-06100 Perugia, Italy

ABSTRACT
After the synthesis in 1978 of the first zirconium phosphonates and organo-phosphates with a layered structure closely related to that of α-zirconium phosphate, a large number of zirconium phosphonates have been prepared and investigated. Even covalently pillared M(IV) diphosphonates with a regular interlayer microporosity have recently been obtained. Furthermore, recent advances in the structure and chemistry of γ-zirconium phosphate dihydrogen phosphate have allowed the preparation of many organic derivatives with γ-layered structure.
An account of the chemistry of this wide class of compounds, largely based on the results obtained in our laboratory, is presented and a critical analysis of the present state of the art, with special regard to the aspects connected with soft chemistry, is attempted.

1. INTRODUCTION
Although amorphous M(IV) acid phosphates have been known for a long time and studied for their ion-exchange properties, it may be considered that the chemistry of layered phosphates began in 1964 when the first crystalline member of this class, α-Zr(HPO$_4$)$_2 \cdot$H$_2$O, was obtained by Clearfield and Stynes and its layered structure was clearly established [1].
The possibility of relating the properties of α-M(IV) acid phosphates to their layered structure and the many potential applications in the fields of ion-exchange, intercalation, catalysis and proton conductivity, stimulated a variety of researches on this class of compounds and a large number of papers were published in the following years. The interested reader is referred to some general reviews and references therein [2-5].
An important step in the development of this class of compounds was made in 1978 when the first M(IV)-phosphonates and organic phosphates with layered structures closely related to that of zirconium bis monohydrogen phosphate (α-ZrP) were reported by Alberti et al. [6]. Today we know that it is possible to obtain a large number of organic derivatives of α-ZrP. Even compounds containing -SO$_3$H groups (therefore exhibiting good protonic conductivity and catalytic activity) and

covalently pillared M(IV) diphosphonates with a regular interlayer microporosity have recently been obtained [7].
A further development on layered phosphates and phosphonates was made in the years 1987-1990 when it was realized that γ-ZrP, differently from zirconium bis monohydrogen phosphate, must be formulated as $ZrPO_4H_2PO_4 \cdot 2H_2O$ and that its interlayer $O_2P(OH)_2$ groups can be replaced by O_2PRR' groups by simple topotactic reactions [8-10]. Many organic derivatives have already been prepared and even pillared compounds with regular interlayer porosity have been obtained [11-15].
Layered and pillared M(IV) phosphonates can therefore be considered as a very large and versatile class of materials in which many different geometrical arrangements can be realized by the proper choice of the layered structure and/or the organic group.
An account of the chemistry of this wide class of compounds, largely based on the results obtained in our laboratory, is presented and a critical analysis of the present state of the art, with a special regard to the aspects connected with the soft chemistry, is attempted.

2. M(IV) PHOSPHONATES WITH α-LAYERED STRUCTURE

2.1. Preparation.

The preparation of the α-M(IV) phosphonates is closely related to the methods employed for the preparation of α-ZrP, i.e. the refluxing of amorphous or semicrystalline precipitates [1] and the direct precipitation method in the presence of a complexing agent [16] of Zr(IV). The only change required is the replacement of the H_3PO_4 with the proper H_2O_3PR acid. The first M(IV) phosphonates were obtained in our laboratory by using HF as complexing agent. Owing to its volatility, it is possible to regulate the rate of precipitation of zirconium phosphate by a a slow elimination of the hydrofluoric acid (e.g. by bubbling water vapour through the solution).
A higher degree of crystallinity and larger crystals can be obtained if the fluorocomplexes are decomposed by a very slow increase in the temperature (1-10°C/day from about 60 to 90°C) [17].

2.2. Structure.

Just after the synthesis of the first zirconium phosphonates of general formula $Zr(O_3PR)_2 \cdot nS$, where R is an inorganic (e.g. -OH, -H) or organic radical (e.g. $-CH_3$, $-C_6H_5$, $-O(CH_2)_n-CH_3$, etc.), S being a polar solvent intercalated in the interlayer region, it was realized, by considerations based on their density and interlayer distance [18] that these compounds can be considered as organic derivatives of $\alpha\text{-}Zr(O_3POH)_2 \cdot H_2O$ in which the inorganic -OH groups have been replaced by R groups, leaving the inorganic structure of the α-layer essentially unchanged. From the known structures of the α-layer

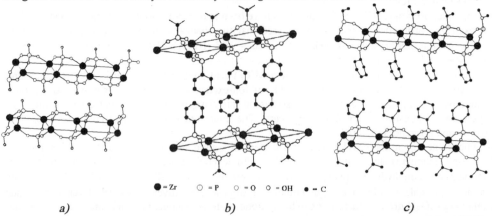

●= Zr   ○ = P   ○ = O   ○ = OH   ● = C

a)   b)   c)

Figure 1: Structural models of a) $\alpha\text{-}Zr(HPO_4)_2 \cdot H_2O$, b) $\alpha\text{-}Zr(O_3PC_6H_5)_2$ as proposed in 1978 by Alberti, c) the same compound drawn on the basis of its structural parameters (Clearfield, 1993).

and the R radical and from the experimental value of the interlayer distance it is therefore possible to foresee the most probable structural model. Waiting for the definitive structures from X-ray diffraction of single crystals (which are very difficult to obtain, owing to the extremely low solubility of this class of compounds) these model structures are temporarily very useful to understand and/or foresee the properties of the phosphonates obtained. As an example, in figure 1 the model structure of α-zirconium benzenephosphonate proposed by Alberti et al. in 1978 (fig.1b)[6], is compared with the structure of α-ZrP (fig.1a) and with the recent structure reported by Clearfield in 1993 (fig. 1c) [19]. Indeed, apart from the orientation of the benzene rings, the recent study essentially confirms the model structure proposed 15 years before.

2.3. Zirconium phosphonates with R pendant groups of the same type.

As said before, zirconium phosphonates of general formula $Zr(O_3PR)_2.nS$ can be seen as obtained from α-ZrP by the replacement of OH groups with R ones. Thus the inorganic polymeric matrix $[M(IV)(O_3P-)_2]_n$ of a single layer can be considered as a clothes -hook on which a large variety of organic and inorganic R groups can be hung. The disposition of the P-R groups on one side of the α-layer is shown in figure 2. Note that the free area around each position of the α-layer is 24 Å²; therefore two large R groups with cross sections greater than this value cannot occupy adjacent positions. The nature of the covalently attached groups depends only on the imagination and ability of the chemists to synthesize the proper phosphonate. The compositions and the interlayer distances of some zirconium phosphonates that have already been prepared are reported in table I.

Table I:Compositions and interlayer distances of some α-layered zirconium organophosphonates.

Compound	Interl. dist. (Å)	Ref.
$Zr(O_3PC_6H_5)_2$	14.7	6
$Zr(O_3PCH_2OH)_2 \cdot H_2O$	10.1	6
$Zr(O_3PCH_3)_2$	7.8	20
$Zr(O_3PC_{22}H_{45})_2$	56.2	21
$Zr(O_3PCH_2SO_3H)_2$	15.4	21
$Zr(O_3PC_2H_4COOH)_2$	13.0	22, 23
$Zr(O_3PCH_2Cl)_2$	10.1	22, 23
$Zr(O_3PCH_2COOH)_2$	11.3	23, 24
$Zr(O_3PCH_2CHOHCOOH)_2$	14.3	25
$Zr(O_3PC_5H_{10}COOH)_2$	18.5	26

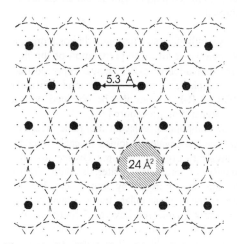

Figure 2: Idealized disposition of -PR groups on an α-layer. The free area around each site is shown.

5.3 Å

24 Å²

2.4. Zirconium phosphonates with R pendant groups of different type.

If a mixture of two different H_2O_3PR acids is used in the synthesis, zirconium phosphonates with two different R groups are obtained. These compounds can be described by the general formula $M(IV)(O_3PR)_x(O_3PR')_{2-x}.nS$. As a general rule, the x value can change almost continuously only if the two R and R' groups are very similar in size and chemical nature, or if the degree of crystallinity of the product is very low. In the greater part of the other compounds studied, discontinuous x-values are observed. Let us examine the interesting case in which the cross section of one of the two R groups is larger than the free area of an α-site (24 Å²). In this case two R groups cannot occupy two adjacent positions of the α-layer. The maximum allowed percentage of the large R group is 33.3%; note that for this percentage a regular distribution is obtained in which, as schematically shown in figure 3, six

Table II: Compositions and interlayer distances of some derivatives of α-zirconium phosphate with two different pendant groups.

Compound	Interl. dist. (Å)	Ref.
$Zr(O_3POH)_{0.66}(O_3PH)_{1.34}$	6.5	27
$Zr(O_3POH)_{1.15}(O_3PC_6H_5)_{0.85}$	12.4	28
$Zr(O_3POH)(O_3PC_2H_4COOH)$	12.9	28
$Zr(O_3PCH_2OH)(O_3PH)$	7.0	28
$Zr(O_3PC_2H_4COOH)_{1.25}(O_3PCH_2OH)_{0.75}$	13.6	28
$Zr(O_3PC_6H_5)(O_3PH)$	10.5	28
$Zr(O_3PC_6H_5)_{1.25}(O_3PH)_{0.75}$	15.0	28
$Zr(O_3PC_6H_4SO_3H)_{0.85}(O_3PC_2H_5)_{1.15}\cdot3.7H_2O$	18.5	29
$Zr(O_3PC_6H_4SO_3H)_{0.97}(O_3PCH_2OH)_{1.03}\cdot4.9H_2O$	19.6	30

small R' groups are placed around each large R group. This regular distribution of the pendant groups seems to be preferred even when the cross sections of the R and R' groups do not exceed the free area of the α-site. In this case, very probably for energetic reasons due to the interactions between adjacent layers, asymmetrical layers are formed in which one side of the layer is richer in R, the other in R' groups. The packing of the layers is that schematically shown in figure 4 in which there is an alternation of interlayer regions richer in R and R' pendant groups, respectively. Some zirconium phosphonates with two different pendant R groups are reported in table II.

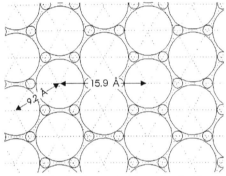

Figure 3: Idealized arrangement of a distribution of 33.3% large R groups (>24Å^2) and small R' groups on one side of the α-layer.

Although specific researches in this subject have not yet been performed, there is some evidence which seems to indicate that if the preparation is carried out in such conditions as to give compounds with a high degree of crystallinity, the $Zr(O_3PR)_2.nS$ and $Zr(O_3PR')_2.n'S$ pure phases rather than mixed phases with asymmetrical layers are formed.

2.5. Covalently pillared Zirconium diphosphonates.

Only some years after the preparation of the first organic derivative of α-ZrP, it was reported that M(IV) diphosphonates, of general formula $M(IV)(O_3P-R-PO_3)$ can also be obtained with a similar preparation procedure [31].

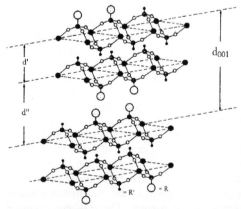

Figure 4: Schematic representation of the structure of a mixed α-zirconium phosphonate with asymmetrical layers.

In these compounds, adjacent inorganic layers of the α-type are covalently joined to each other by the divalent organic group. Owing to the possibility of preparing materials with "tailor made" interlayer porosity by simply changing the length of the organic pillar, the acronym MELS (Molecularly Engineered Layered Structures) was given to M(IV) diphosphonates to emphasize their applicative potentiality [32]. A large number of zirconium diphosphonates with divalent R groups of different length was prepared; it was however recognized that, owing to the small distance (5.3Å) between adjacent sites in the α-layer, the pillars are too crowded and insufficient free space remains in the interlayer region, independent of the length of the pillar itself.

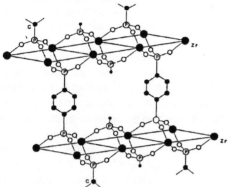

Figure 5: Schematic representation of the structure of a porous α-zirconium mixed phosphonate

2.6. Zirconium phosphite-diphosphonates with interlayer porosity.

It was initially believed that the problem of the creation of interlayer porosity could be solved simply by replacing a certain number of divalent $O_3P-R-PO_3$ groups with small monovalent O_3P-R groups (e. g. R = -H, -OH or -CH$_3$) as shown schematically in figure 5 [33].

Unfortunately it was later found in our and other laboratories that the surface area of materials having a low degree of crystallinity was due for the greater part to mesoporosity; on the other hand, when more crystalline materials were prepared, in agreement with what has been discussed above for compounds with two pendant groups, zirconium diphosphonates containing only a small percentage of phosphites and pure zirconium phosphite phases were obtained, so that negligible interlayer microporosity was found. In order to create the interlayer microporosity, a new strategy, based on the use of pillars with cross sections larger than 24Å^2, was therefore followed in our laboratory. In such a situation, two adjacent α-positions cannot be occupied by the large pillars and, according to what was discussed in section 2.4, six phosphite groups were expected to be placed around each pillar. If the cross section of the large pillar is homogeneous, no interlayer porosity is to be expected. However, if the cross section is not uniform, and in particular if the central cross-section is smaller than the terminal parts, interlayer porosity may be created. Pillars with these characteristics, i.e. pillars with bases, could be divalent R groups such as:

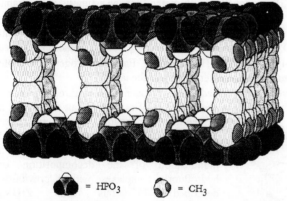

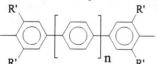
● = HPO$_3$ ◐ = CH$_3$

Figure 6: Space filling picture of the interlayer region of a microporous zirconium phosphite diphosphonate "with bases".

with n = 0, 1, 2, ... and R' = -CH$_3$, -OH, or other groups. A compound having pillars with bases (n = 0; R' = -CH$_3$) recently prepared in our laboratory indeed confirmed the above expectations. A high phosphite percentage, a good degree of crystallinity, an interlayer microporosity of 375 m^2/g (pore size of about 5Å) were indeed found [7]. A computer generated structural model of this pillared compound based in its unit cell parameters is shown in figure 6. Note that its molecular architecture is indeed particular, recalling that of a Greek temple.

3. M(IV) PHOSPHATE-PHOSPHONATES WITH γ-LAYERED STRUCTURE

The preparation of the first zirconium phosphate with a layered structure of the γ-type (1968) followed only four years after that of the α-one [34]. Nevertheless, in the following 25 years, only a small number of papers on γ-titanium and zirconium phosphates appeared in the literature, in spite of their interesting properties, such as intercalation, ionic exchange and protonic conduction.

Some organic derivatives of γ-zirconium phosphate (γ-ZrP) were obtained by Yamanaka et al. starting from 1975 [35, 36]. However, at that time γ-ZrP was believed to be an isomorphous modification of α-ZrP with only a different stacking of the layers and hydration water content so that it was not possible to give a correct interpretation of the composition and structure of the obtained organic derivatives obtained. The recent advances in the knowledge of the structure of the layers of γ -type [37] was fundamental for the development of the chemistry of γ-compounds, that nowadays seems even richer than that of α-derivatives.

In the last few years, mainly in this laboratory [11-15], a variety of organic derivatives of γ-ZrP have been prepared and the first covalently pillared γ-compounds obtained; on the basis of their structural characteristics, many properties of these compounds have been elucidated, and their potentialities for applied fields better understood. In addition, the possibilities of engineering the interlayer region by the introduction of controlled porosities and functions are presently being investigated and are really promising, offering new intriguing perspectives for the supramolecular chemistry of layered systems.

3.1. Topotactic exchange reactions on γ-ZrP: a convenient soft method for the synthesis of γ-organic derivatives.

The structure of the γ-layer consists in a rigid framework of ZrO_6 octahedra placed in two different planes and joined to each other with PO_4 tetrahedra inside and H_2PO_4 outside these planes (see figure 7). γ-ZrP must therefore be formulated as $ZrPO_4H_2PO_4 \cdot 2H_2O$. Each dihydrogen phosphate group bears two acidic protons that are situated in the external part of the layers.

It is of interest to note that γ-ZrP has never been obtained in its hydrogen form by direct synthesis. The preparation of this compound usually consists in the synthesis of one of its salt forms and a subsequent conversion in the hydrogen phase. Microcrystals with a good degree of crystallinity can be obtained with a modification of the decomposition of hydrofluoric complexes method previously described. In this case Zr(IV) fluorocomplexes are decomposed in a solution of ammonium dihydrogen phosphate [38, 15].

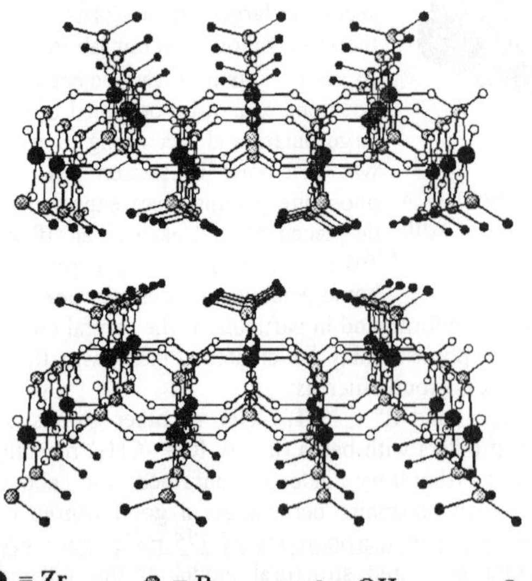

From the structure it is evident that pure γ-zirconium phosphonates or phosphites cannot exist. On the other hand it was possible to prepare mixed γ-zirconium phosphate phosphonate or phosphite compounds of general formula $ZrPO_4O_2PRR' \cdot nS$, where R and R' can be H, OH or organic group and S is the solvent intercalated.

Even if possible in principle, all our attempts to prepare mixed derivatives of γ-ZrP by direct synthesis were unsatisfactory, since it was not possible to obtain them as pure

$\bullet$ = Zr $\circledcirc$ = P $\bullet$ = OH $\circ$ = O

Figure 7: Schematic representation of the structure of γ-ZrP. The hydration water molecules are not shown.

phases.

In any case, in agreement with Yamanaka's previous work [36], it was found that the interlayer H_2PO_4 groups of γ-ZrP can be easily exchanged with other O_2PRR' groups by means of a topotactic reaction simply by contacting microcrystals of this compound with a solution of the suitable phosphonic acid. The topotactic replacement of the H_2PO_4 groups originally present in the interlayer region occurs gradually with the initial formation of a partially converted derivative:

$$\gamma\text{-ZrPO}_4O_2P(OH)_2 \rightarrow \gamma\text{-ZrPO}_4[O_2(OH)_2]_{1-x}(O_2PRR')_x \rightarrow \gamma\text{-ZrPO}_4O_2PRR'$$

It is important to note that, depending on the nature and size of the R group, the fully converted derivative may not be obtained. This fact, related to the steric hindrances of adjacent groups will be discussed later. It must also be pointed out that a partial conversion of γ-ZrP into the corresponding α-Zr phosphonate may in some cases occur parallel to the topotactic reaction. A process studied in detail in our laboratory is the full topotactic conversion of γ-ZrPO$_4$O$_2$P(OH)$_2$·2H$_2$O into γ-ZrPO$_4$O$_2$PHOH·H$_2$O. It was found that the energy of activation of the topotactic exchange process is quite low: about 4.6 Kcal mol^{-1}, indicating that the process is probably diffusion controlled. On the other hand, the process for the formation of the corresponding α-Zr phosphite needs a higher energy of activation, about 29 Kcal mol^{-1}. This value seems to indicate that the conversion to an α-compound occurs via a dissolution - reprecipitation mechanism. Very likely the dissolution is due to the formation of zirconium complexes with the phosphorous acid. In agreement with this hypothesis the formation of the α-phase was found to be negligible for concentration values < 1M.

In general the formation of the α-phase can be minimized with the choice of convenient conditions for the reaction, such as low temperatures and concentrations of phosphonic acids or short contacting times. A variety of organic derivatives was also obtained in our laboratory using aliphatic and aromatic phosphonic and phosphinic acids (see table III).

Table III: Compositions and interlayer distances of some derivative compounds of γ-zirconium phosphate

Acid employed	Composition	Interl. dist. (Å)	Ref.
H_3PO_3	$ZrPO_4\ O_2PHOH·2H_2O$	12.2	13
H_3PO_2	$ZrPO_4\ O_2PH_2·H_2O$	8.8	11
$H_2O_3PCH_3$	$ZrPO_4\ O_2POHCH_3·2H_2O$	12.8	12
$H_2O_3PC_3H_7$	$ZrPO_4\ O_2POHC_3H_7·1.2H_2O$	15.1	12
$HO_2P(CH_3)_2$	$ZrPO_4(H_2PO_4)_{0.33}(O_2P(CH_3)_2)_{0.67}·H_2O$	10.3	11
$H_2O_3PC_6H_5$	$ZrPO_4(H_2PO_4)_{0.33}(O_2POHC_6H_5)_{0.67}·2H_2O$	15.4	14
$H_2O_3P(C_6H_{11})$	$ZrPO_4(H_2PO_4)_{0.33}(O_2POHC_6H_{11})_{0.67}·H_2O$	16.9	12
$H_2O_2PC_6H_5$	$ZrPO_4O_2PHC_6H_5$	15.1	39

In all the cases the best results for the preparation of these compounds were obtained using γ-ZrP in its hydrogen form and using free phosphonic acids.

The x-ray diffraction patterns of the most crystalline γ-derivatives were indexed and the cell parameters determined. As can be seen in table IV the a and b parameters of each derivative do not change appreciably from that of γ-ZrP.

Table IV: A comparison of the unit cell dimensions for three different γ-layered compounds.

Compound	a	b	c	β
$ZrPO_4\ H_2PO_4·2H_2O$	5.386	6.636	24.806	98.70
$ZrPO_4\ O_2PHOH·2H_2O$	5.448	6.621	27.688	113.92
$ZrPO_4(H_2PO_4)_{0.33}(O_2POHC_6H_5)_{0.67}·2H_2O$	5.387	6.626	32.394	109.51
$ZrPO_4O_2POHC_3H_7·1.2H_2O$	5.364	6.621	30.710	98.86
$ZrPO_4(H_2PO_4)_{0.33}(O_2POHC_6H_{11})_{0.67}·H_2O$	5.379	6.637	34.082	98.17

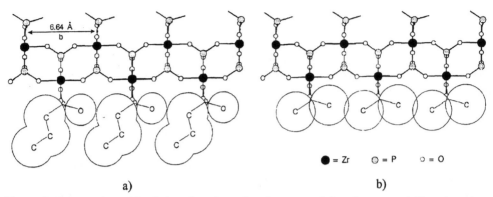

a)   b)

Figure 8: Structural model of a) γ-zirconium phosphate propylphosphonate and b) γ-zirconium phosphate dimethylphosphinate. The radii of OH and the methyl group were assumed as 1.5 and 2 Å respectively

This means that the topotactic replacement of H_2PO_4 groups with other O_2PRR' leaves the γ-layer framework essentially unaltered and that the interlayer groups effectively influence only the interlayer distance. It is therefore reasonable to try to test the accommodation of the organic moieties in the interlayer region by means of simple structural models derived by computer on the basis of the structure of the original γ-ZrP, replacing H_2PO_4 with O_2PRR' groups. These models indicate that a and b dimensions are sufficiently large to accommodate one linear alkyl chain for each O_2PRR' group (see figure 8a).

A full conversion into the organic derivative can therefore be expected for n-alkylphosphonic acids. Even though other phosphonic acids with a longer chain should be tested, our data concerning methylphosphonic and propylphosphonic acid confirm these expectations. If both R and R' are alkyl groups, the room around each γ-site is not sufficient to have a fully converted phase, as can be seen in figure 8b. A partial replacement was actually observed when using dimethylphosphinic acid. The same effect is observed when phosphonates are used with one large R group, for instance cyclohexyl phosphonic acid. A similar situation is found for phenylphosphonic acid. In this well studied case, even if the phenyl group is not as large as an alkyl chain, the rigidity of the aromatic ring causes a steric hindrance with the adjacent OH groups (see figure 9) [14].

The experimental compositions of the sterically hindered compounds mentioned seem to indicate that such constraints can be avoided if at least 33% of the original H_2PO_4 groups remain in the interlayer region, independently of the different dimensions of the hindered groups. This fact could be related to the formation of a stable configuration for the 1:2 ratio of H_2PO_4 and O_2PRR' groups in which some space, not completely filled by the organic groups, may be available.

The role of steric hindrance in the definition of the composition of the exchanged γ-derivatives

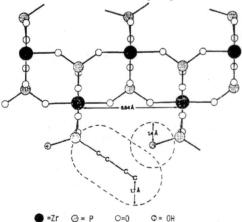

Figure 9: Schematic representation of the possible steric hindrance between adjacent phenyl and OH groups in γ-zirconium phosphate phenyl phosphonate

was also confirmed by the use of phenylphosphinic acid: $HO_2PHC_6H_5$. In this compound the second group is a hydrogen atom, which is sensibly smaller than the hydroxyl one. For such a compound a 100% exchange was observed.

3.2. Pillaring of γ-ZrP: an open way to the engineering of the γ-interlayer region.

The chemistry of γ-compounds has been further enriched by the preparation of the first covalently pillared γ-zirconium phosphate diphosphonates of general formula γ-$ZrPO_4(H_2PO_4)_{1-x}(O_2POH-R-HOPO_2)_x \cdot nS$ [15].

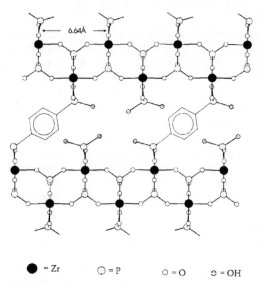

● = Zr   ◯ = P   ○ = O   ◌ = OH

Figure 10: Schematic picture of the structure of γ-zirconium phosphate phenyldiphosphonate.

On the basis of the previous experience in the preparation of γ-layered zirconium phosphate phosphonates, the synthesis of the first covalently pillared γ-zirconium phosphate diphosphonates was performed using similar procedures: topotactic exchange reactions of H_2PO_4 with diphosphonate groups, carried out at moderate temperature (70÷80°C) and concentrations (<0.1 M).

The first attempts were performed using phenyl and butyl diphosphonic acid, in which the small dimensions of the R groups could avoid any problems related to possible slow diffusion processes, high activation energies and so on. It was recently found that such problems are indeed virtually absent; at 80°C the topotactic exchange reactions proceeded at appreciable rates even if diphosphonic acids with much larger R groups were employed.

Depending on the conditions at which the reaction occurs, two different final products were found to form, corresponding to the following different pathways (where the hydration water is omitted for sake of clarity):

$$ZrPO_4H_2PO_4 + x/2 \, (O_2POH-R-HOPO_2)^{2-} \rightarrow ZrPO_4(H_2PO_4)_{1-x}(O_2POH-R-HOPO_2)_{x/2} + x \, H_2PO_4^-$$
$$ZrPO_4H_2PO_4 + x[O_2POH-R-HOP(OH)_2]^- \rightarrow ZrPO_4(H_2PO_4)_{1-x}[O_2POH-R-HOP(OH)_2]_x + x \, H_2PO_4^-$$

The first process leads to a pillared structure, and occurs with lower concentrations of diphosphonic acid, while the second leads to a layered γ-Zr phosphate diphosphonate. This latter was found when the topotactic reaction was carried out with a high concentration of diphosphonic acid (>1M). Owing to their different structure, the two possible final materials possess very different chemical properties and are well distinguishable. Even if until now our attention has been focused onto obtaining pillared compounds, we can foresee very interesting properties in γ-layered diphosphonates, too, and further investigations will be performed on them.

The γ-pillared compounds obtained with the above diphosphonic acids at the maximum percentage of exchange have the following composition respectively: $ZrPO_4(H_2PO_4)_{0.56}(O_2POH-C_6H_4-HOPO_2)_{0.22} \cdot 2.2H_2O$ and $ZrPO_4(H_2PO_4)_{0.18}(O_2POH-C_4H_8-HOPO_2)_{0.41} \cdot 1.3H_2O$. A schematic model proposed for the former compound is shown in figure 10.

In any case, owing to the high rigidity of the γ-layers, the system shows a clear tendency to the formation of solid solutions. Other than the above compounds, it was therefore possible to prepare products having a lower percentage of topotactic replacement, where only a small part of the interlayer H_2PO_4 groups are replaced by diphosphonate ones.

Even though these compounds have not the right structural characteristics to possess interlayer porosity, this behaviour, peculiar to γ-systems, is crucial for future advances in the design of new architectures based on the pillaring of γ-layers: by the proper choice of the pillaring group highly interlamellar microporous γ-zirconium phosphate diphosphonates have indeed recently been prepared.

In conclusion, the topotactic reaction is a new simple and soft way to manage the surface of the layers and/or the interlayer region to obtain materials with predictable characteristics.

4. GENERAL ASPECTS OF LAYERED M(IV) PHOSPHONATES, THEIR APPLICATIONS AND FUTURE PERSPECTIVES

Since this meeting is dedicated to the "soft chemistry routes to new materials", it could be of interest to note that M(IV) phosphates and phosphonates, different from many other solid materials, are obtained at low temperatures, often from aqueous solutions. Furthermore, the phosphate groups present on the surface of the crystals can be topotactically replaced by other phosphonates. In the case of γ-ZrP even the interlayer region can be topotactically functionalized. This fact makes it possible to prepare a large number of γ-organic derivatives in a very simple and soft manner. For the large variety of organic groups that can be covalently attached to the α- or γ-layers, we can consider the inorganic back-bone of α-ZrP and γ-ZrP, as discussed before, as two planar clothes-hooks with which two different geometrical arrangements of the covalently bound organic molecules can be obtained. These arrangements essentially depend on the structure of the layered inorganic back-bone; even two or more pendant organic groups of different size and/or chemical properties can be attached to the surface of the inorganic layers and regular arrangements with well defined ratios between these groups are usually obtained.

The field of M(IV) phosphonates has recently been enlarged with the preparation of layered M(II)- and M(III)-phosphonates which may be seen as new clothes-hook for hanging organic molecules in new geometrical arrangements [40]. The reader interested in these new metal phosphonates is referred to the recent review and references therein [5]. Here, let us only note that a large variety of clothes-hooks are already available today and new ones will be prepared in the near future. Owing to the facility with which the surface O_3PR groups can be topotactically replaced by other phosphonates, the method may be used to obtain functionalized surfaces and in many cases this method, allowing us to choose from among various geometrical arrangements, could be convenient with respect to that of silanization. We recall here, as an example, that the large surface areas obtained in special preparations of α-ZrP (e.g. α-ZrP/SiO$_2$ composites prepared from colloidal dispersions of the two components and having surface area larger than 450 m^2/g) can be topotactically functionalized with R-SO$_3$H groups. Owing to the high surface area and the presence of strong acid sites on their surface, these compounds are of interest as catalysts [41]. Furthermore, the possibility of preparing zirconium phosphonates containing -SO$_3$H groups has opened new perspectives in the field of the protonic conductors. Some materials with very good protonic conductivity even in the range 100-180°C, have already been obtained [30].

Of great interest are also the interactions in the interlayer regions between R groups belonging to adjacent layers. Furthermore, a large variety of molecules can be intercalated in these interlayer regions so that layered metal phosphonates offer new possibilities for an interlayer supramolecular chemistry.

The interlayer region of the M(IV) phosphonates can be seen as a well ordered region in which many reactions of intercalated molecules between them and/or with the R groups of the layers can be carried out. As discussed by Mallouk et al.[5] the reactivity in layered phosphonates suggest that "it should be possible to engineer solids in which guest molecules bind according to their shape and chemical properties"; e.g. some particular intralayer polymerization may be facilitated by the nature and disposition of the R groups in the interlayer region. Even chiral selectivity is possible if the R and

/or R' groups contain chiral centres. Some recent results obtained in our laboratory with α-ZrP functionalized with chiral groups, are indeed very encouraging [42].

The existence of microporous zirconium diphosphonates (zeolite-like compounds) with tailor made cavities of uniform size in their interlayer region, possibly containing catalytic sites, opens the way for their application in molecular sieving and in "shape selective catalysis" [7].

The recent experiments of Thompson et al. in photochromism via the photoreduction of zirconium phosphate-viologen compounds [43] seem to show exciting promises in the utilisation and storage of solar energy [44].

For other applications and new perspectives of metal phosphonates, such as their structural analogy with Y-type Langmuir-Blodgett films, layer-by-layer growth, topochemical reactivity and non-linear optical effects, the interested reader is referred to the previousy cited review [5].

REFERENCES

1) Clearfield, A., Stynes, J.A.: J. Inorg. Nucl. Chem., 1964, 26, 117.
2) Clearfield, A.: Chem. Rev.: 1988, 88, 125.
3) Alberti, G., Costantino U.: "Intercalation of zirconium phosphates and phosphonates", in J.L. Atwood, J.E.D. Davies , D.D. MacNicol (Eds.). Inclusion Compounds, Oxford University Press, 1991, 5, 136.
4) Alberti, G.: "Multifunctional Mesoporous Inorganic Solids", in C.A.C. Sequeira, M.J. Hudson (eds). Kluwer Academic Publishers, 1993, 179.
5) Cao G., Hong H., Mallouk, E.: Acc. Chem. Res., 1992, 25, 420.
6) Alberti, G., Costantino, U., Allulli, S., Tomassini, N.: J. Inorg. Nucl. Chem., 1978, 40, 1113.
7) Alberti, G., Costantino, U., Vivani, R., Zappelli, P.: Angew. Chem. Int. Ed. Engl., 1993, 32, 1357.
8) Alberti, G., Costantino, U., Luciani Giovagnotti, M.L.: J. Inorg Nucl. Chem., 1979, 41, 643.
9) Yamanaka, S., Tanaka, M.: J. Inorg Nucl. Chem., 1979, 41, 45.
10) Clayden, N.J.: J. Chem. Soc. Dalton Trans., 1987, 1877.
11) Alberti, G., Casciola, M., Biswas, R.K.: Inorg. Chim. Acta, 1992, 201, 207.
12) Alberti, G., Casciola, M., Vivani, R., Biswas, R.K.: Inorg Chem., in press.
13) Alberti, G., Costantino, U., Vivani, R., Biswas, R.K.: Reactive Polym., 1992, 17, 245.
14) Alberti, G., Vivani, R., Biswas, R.K., Murcia-Mascarós, S.: React. Polym., 1993, 19, 1.
15) Alberti, G., Murcia-Mascarós, S., Vivani, R.: Phys. and Chem. Mater., in press.
16) Alberti, G., Torracca, E.: J. Inorg Nucl. Chem., 1968, 30, 317.
17) Alberti, G., Costantino, U., Giulietti, R.: J. Inorg Nucl. Chem., 1980, 42, 1062.
18) Alberti, G., Costantino, U., Kornyei, J., Luciani Giovagnotti, M.L.: React. Polym., 1985, 5, 1.
19) Clearfield, A.: Acta Crystallogr. in press.
20) Dines, M. B., Griffith, P. C.: J. Phys. Chem., 1982, 86, 571.
21) Segawa, K., Sugiyama, A., Kurusu, Y.: Proc. Chemistry of Microporous Crystals. Kodansha Elsevier 1990. Tokyo.
22) Alberti, G., Costantino, U., Kornyei, J., Luciani Giovagnotti, M. L.: Chim. Ind., Milan, 1982, 64, 115.
23) Dines, M. B., Di Giacomo, P. M.: Inorg. Chem., 1981, 20, 92.
24) Alberti, G., Costantino, U., Luciani Giovagnotti, M. L.: J. Chromatogr., 1979, 180, 45.
25) Ortiz-Avila, C., Clearfield, A.: J. Chem. Soc. Dalton Trans., 1989, 1617.
26) Alberti, G., Costantino, U., Casciola, M., Vivani, R., Peraio A.: Solid State Ionics., 1991, 46, 61.
27) Alberti, G., Costantino, U., Giulietti, R.: Gazz. Chim. It., 1983, 113, 547.
28) Alberti, G., Costantino, U.: J. Molec. Catal., 1984, 27, 235.
29) Yang, C., Clearfield, A.: Reactive Polymers, 1987, 5, 13.
30) Alberti, G., Casciola, M., Palombari, R.: Solid State Ionics, 1992, 58, 339.

31) Dines, M B., Griffith, P. C.: Polyhedron, 1983, 2, 607.

32) Alper, J.: Chem. and Ind., 1986, 335.

33) Clearfield, A.: in Design of New Materials. Cocke, D. L., Clearfield, A. Eds.,1987, 121.

34) Clearfield, A., Blessing, R.H., Stynes, J.A.: J. Inorg Nucl. Chem., 1968, 30, 2249.

35) Yamanaka, S., Koizumi, M.: Clays Clay Miner., 1975, 23, 477.

36) Yamanaka, S., Sakamoto, K., Hattori, M.: J. Phys. Chem., 1984, 88, 2067.

37) Christensen, A., Andersen, E. K., Andersen, I.G.K., Alberti, G., Nielsen, N., Lehmann, M.S.: Acta Chem. Scand., 1990, 44, 865.

38) Alberti, G., Bernasconi, M.G., Casciola, M.: Reactive Polym., 1989, 11, 245.

39) Murcia-Mascarós, S.: Chem. Thesis (Dept. of Chemistry, University of Perugia, Italy, 1993).

40) Martin, K., Squattrito, P., Clearfield, A.: Inorg. Chim. Acta, 1989, 155, 7.

41) Clerici, M., Alberti, G., Malentacchi, M., Bellussi, G., Prevedello, A., Cornò, C.: European Patent., Appl n°: 90200514.9 (05-03-1990).

42) Alberti, G., Costantino, U., Dionigi, C.: unpublished data.

43) Vermeulen, L. A., Thompson, M. E.: Nature, 1992, 358, 656.

44) Dutta, P.: Nature, 1992, 358, 621.

Materials Science Forum Vols. 152 - 153 (1994) pp. 99-108
© 1994 Trans Tech Publications, Switzerland

HYDROTHERMAL SYNTHESIS OF NEW OXIDE MATERIALS USING THE TETRAMETHYL AMMONIUM ION

M.S. Whittingham, J. Li, J.D. Guo and P. Zavalij

Chemistry Department and Materials Research Center SUNY at Binghamton,
Binghamton, NY 13902-6000, USA

ABSTRACT

Mild hydrothermal synthesis leads to the formation of new metastable transition metal oxide structures which have relatively open crystal structures. The nature of the cations present in solution (the "templating ion") has a dramatic effect on the crystal structure of the phase formed, as also does the pH of the reaction medium and the particular transition metal. Thus, by appropriate choice of reaction medium new structures containing large tunnels or channels, similar to those found in aluminosilicate zeolites, can be formed that will offer unique properties for the Materials Scientist. In particular, it is expected to be possible to realize enhanced diffusion in such materials. Here the use of the spherical tetramethyl ammonium ion is described.

INTRODUCTION

Over the last two decades there has been much interest in the modification of the chemical and physical properties of materials through guest-host chemistry. In particular the intercalation reactions of two classes of host materials, graphite and the transition metal dichalcogenides, have been extensively studied. The initial interest involved enhancement of their electrical conductivity and in some cases superconductivity. Subsequently TiS_2 became the prototypical cathode material for high energy density batteries because the guest-host reaction between lithium and TiS_2 is essentially completely reversible and involves a high energy output [1]. More recently, lithium intercalated carbons are being used as the anode of secondary lithium batteries [2], where the cathode is also an intercalation compound, $LiCoO_2$ [3]. The hydrogen electrode of Ni/H_2 batteries also uses guest-host reactions, with the hydrogen being stored in variants of $LaNi_5$. It is now generally accepted that most battery electrodes operate via an intercalation mechanism. Intercalation reactions are not just found in electrochemical devices, but in many areas of chemistry and physics. Thus the 123 superconductor is formed by the intercalation of oxygen into the phase formed at high temperatures ($YBa_2Cu_3O_{7-\delta} + \delta/2O_2 \leftrightarrow YBa_2Cu_3O_7$), the use of zeolites involves the incorporation of guest molecules into the zeolitic host, and many oxidation catalysts involve the in/out diffusion of oxygen into an oxide matrix. In the case of layered structures one can envisage materials having different active properties in the host and guest lattices. As an example, luminescently active ions can be incorporated into an insulating guest host lattice, such as a silicate or zirconium phosphate. This lattice if layered can then be intercalated with organic molecules that can subsequently be polymerized into a conducting layer. By applying an electric field between two adjacent conducting layers and across the insulating host lattice the luminescent ions can be activated, leading to a molecular level electroluminescent display [4].

Zeolitic microporous materials have frameworks containing mostly the redox inactive elements aluminum and silicon. From a chemical point of view it would be highly desirable to have a framework containing, or built from, redox active species such as transition metals. In addition, for many electrochemical applications it would be preferable to work with oxides rather than sulfides because of their greater ease of handling, lower cost and for energy storage their greater electrochemical potential. Our long-term goal is to synthesize transition metal oxide materials with open structures, that might have some of the properties and applications of the aluminosilicate zeolitic materials, have novel applications and be potentially electrochemically active.

Many transition metal chalcogenides have open crystalline structures that can participate in guest-host reactions. Such host materials can have 1d, 2d or 3d framework structures as shown schematically in figure 1. Hexagonal WO_3 contains one-dimensional tunnels in which small ions can be intercalated, TiS_2 has a two-dimensional layered structure between whose layers a wide variety of electron donating species such as amines, organometallics and alkali metals can be inserted, and WO_3 has a three dimensional "ABO_3 perovskite-like" lattice with all the A sites vacant into which small cations such as H and Li can be inserted. Each of these structure types has certain advantages and disadvantages. Although the 1d structures have a very stable matrix, a single defect in the tunnel can block diffusion down it; in 2d lattices like TiS_2 there is an easy expansion perpendicular to the planes to fit almost any size specie between the sheets, but no direct diffusion of guest species between sheets; the 3d lattices have a more dimensionally stable matrix but that often places severe constraints on the size of species that can be inserted into the tunnels. Thus, only H and Li can be inserted into the perovskite-like WO_3 lattice. However, if we have larger tunnels such as in the 3d pyrochlore WO_3 structure all the alkali metals can be inserted or ion-exchanged into the tunnels [5]. Such more open structures are often thermodynamically metastable, so must be formed under conditions where the kinetic route of the reaction dominates over the thermodynamically favored product. This invariably means working at lower temperatures where kinetics can dominate (i.e. the route with the lowest activation energy will dominate product formation).

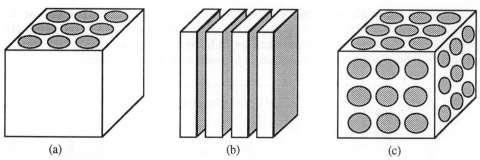

Fig. 1. Schematic of (a) 1d, (b) 2d and (c) 3d structures.

THE HYDROTHERMAL SYNTHESIS APPROACH

The Tungstates: From Na^+ to $N(CH_3)_4^+$

Key to the obtaining of metastable structures is the use of soft chemical approaches to synthesize the desired structures, and our initial work has emphasized hydrothermal methods. We previously reported that different crystalline forms of the tungsten(VI) oxides can be readily synthesized using mild hydrothermal methods [6]. Thus, if Na_2WO_4 acidified with HCl is heated in an aqueous environment at 150°C both a 1d and a 3d tunnel structure are formed depending on the initial pH of the reaction medium. These structures, shown in figure 2, contain sodium ions, oxygen ions and water in the tunnels [5,7], but they can be removed by first ion-exchanging the sodium for either H^+ or NH_4^+ ions, and then driving out these ions together with their oxygen counterions by gently heating up to about 300°C. These two materials are the same ones as first reported by Freedman [8,9], although he was unaware of their structures.

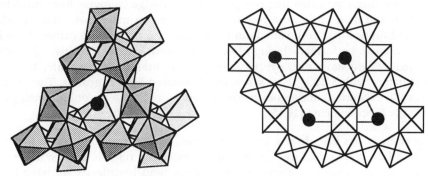

Fig. 2. Pyrochlore (left) and hexagonal (right) structures of sodium tungstates.

The ion-exchange behavior is very different between the two structures, whereas the 1d structure shows very little ion exchange the 3d, in which tunnel blockages can be easily by-passed, almost any monovalent or divalent cation can be exchanged into the structure. The ready ion exchange of the cations in the pyrochlore structure indicate that the ions should show a high diffusivity. Indeed the ionic conductivity was found to be high, and highly dependent on the contained cation, M in $M_{\approx 1}W_2O_6 \bullet xH_2O$. The H_3O^+ ion exhibits the highest conductivity [10] which is strongly dependent on the water partial pressure as indicated in figure 3 [11]. This strong dependence should permit such materials to be used as humidity sensors.

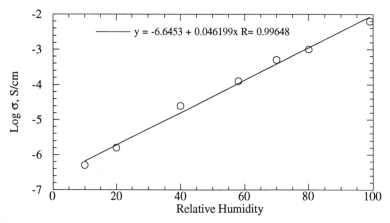

Fig. 3. Hydronium ion conductivity in pyrochlore phase as a function of the water vapor pressure.

However, these tunnels are only about 2.8Å in diameter at their narrowest point so one is restricted in the sort of reaction that can be carried out to those ions than can fit in such size tunnels. Our goal is therefore to make structures with larger tunnels, or expandable layers. Preferably we would also like to make these structures with lower mass transition metals, which is necessary if they are to be used in large-scale applications such as batteries. However, our understanding is much greater for the heavier elements such as tungsten and molybdenum, where many of the known structures are built up by the condensation of MO_6 octahedra to give rings, chains or layers. We are therefore extending our present knowledge stepwise from tungsten to molybdenum and then to the first row transition elements manganese and vanadium. There are indications that both vanadium oxides [12] and manganese oxides [13] with a layer structure can be formed through low temperature synthesis routes. In addition to these oxides, we have also synthesized a novel titanium phosphate, that has a layered structure with a TiO_2 sheet sandwiched between H_2PO_4 groups. This structure can be readily expanded, undergoes redox reactions unlike the zirconium analogs, and can be readily functionalized [14,15].

Our strategy has been to use larger templates, much along the lines of the zeolite chemist to attain layer structures or structures containing tunnels greater in size than an oxygen ion. As described above, using sodium (and hydrogen) we have formed hexagonal and pyrochlore lattices in the case of tungsten oxides. There is very little published work using templating molecules in the synthesis of transition metal oxides, but Haushalter et al [16] have shown that in the case of molybdenum phosphates a range of different structures can be formed by such an approach. Can this be done with simple oxides? It appears that the answer is yes. We have shown in the case of tungsten that using the ammonium ion in place of sodium that we can form both the hexagonal and pyrochlore tungstates [5,17]. We therefore have performed some initial experiments in which we replaced the hydrogen atoms in the ammonium ion one by one with methyl groups. In each case we formed crystalline oxide materials as indicated below:

NH_4	Pyrochlore	a=10.18Å
$NMeH_3$	Pyrochlore	a=10.34Å
NMe_2H_2	white crystalline powder, with complex x-ray powder pattern	
NMe_3H	Small crystals with unknown structure	
NMe_4	mm size single crystals with new structure (described below)	

Using the tetramethyl ammonium ion we not only obtained a totally new structure but also large single crystals, which have allowed its crystal structure to be determined [18]. In this case we did not obtain an open lattice, but one which contained an almost close-packed array of Keggin ions, $H_2W_{12}O_{40}{}^{6-}$, between which reside the $NMe_4{}^+$ ions. NEt_4 gave a similar compound, but with a larger lattice parameter.

However, when we switched from tungsten to molybdenum it appears that we have a completely different and potentially very exciting structure for the product formed at 200°C. The initial interpretation of the x-ray powder pattern, in the monoclinic system with a=11.331, b=11.949, c=11.140 Å and β=108.1°, suggests a structure formed from MoO_6 octahedra joined into sheets giving a layer structure. This structure is shown in figure 4 [19], and its layers exhibit a unique characteristic. There are tunnels about every 9Å connecting one interlayer region to the next.

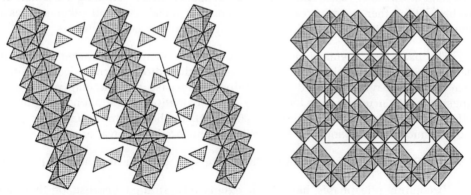

Fig. 4. Proposed structure of $NMe_4Mo_{4-8}O_{12}$, showing (left) structure perpendicular to layers, and (right) structure in layers. Tunnels between layers are along the 101 axis.

What does this mean? Unlike a true 3d structure, where atoms or molecules that one would like to incorporate into the structure are restricted to the size of the tunnels in the structure, one can spread the layers apart to incorporate species of almost any size. So we have the advantages of a layer structure, but at the same time we have some of the advantages of a 3d tunnel structure in that smaller ions and molecules can communicate from one layer to the adjoining ones through the connecting tunnels. We can now consider the possibility of placing different species in adjoining interlayer spaces (this has been shown to be feasible in the transition metal disulfides [20]), and carrying out the first step of the reaction in one layer, the product can then diffuse through the tunnels into the adjoining layer where a second step could be accomplished. These tunnels will also allow the homogenization between layers of the same reactions occurring in all layers, as for example in the discharge of a battery.

We are now trying to remove the NMe_4^+ template whilst maintaining the structure so that other cations and species can be inserted into the structure. There are a number of approaches that might be taken. These include:

> Direct thermal decomposition
> Ion exchange of the template.
> Oxidation/chemical decomposition of the template

Direct thermal decomposition is not expected to succeed without at least partial destruction of the host lattice in the case of the larger templates. However, for smaller cations we [5] and Figlarz et al [21] have made the empty WO_3 hexagonal and pyrochlore lattices by thermal decomposition of the H^+ and NH_4^+ compounds. These easily thermally decomposed cations might be substituted for the templating ions by ion exchange. In favorable cases one might also be able to directly ion-exchange to the compound of interest, e.g. to say a lithium molybdate. As an example of the use of ion exchange we have made [14] a series of new intercalation compounds from a new layered titanium phosphate, formed hydrothermally with the NMe_4^+ cation, by removing the NMe_4^+ template by ion exchanging to the H form with aqueous HCl. This solid acid readily reacts with bases such as amines thereby swelling the lattice. The swelled lattice then readily reacts with transition metal ions and other species in aqueous solution to give transition metal complexes and pillared materials.

The oxidation/chemical decomposition of the template followed by a mild thermal treatment is likely to be more successful than direct thermal decomposition. Such methods have been used successfully in the past, e.g. the empty hexagonal tungsten oxide structure was formed by H_2O_2 oxidation of the ammonium bronze $(NH_4)_xWO_3$ followed by heating to 300°C [22]. In the case of suitable templates photochemical modification may be possible to enhance its removal or reactivity.

A Second Example: Layered Titanium Phosphate Synthesized using the NMe_4^+ Cation

Zirconium phosphates have been extensively studied because of their ion exchange behavior and the possibility to functionalize the phosphate groups and thereby to add another whole dimension to the chemistry. Here the intercalate can be chemically bound to the phosphorus through a P–O– bond, permitting the incorporation of functional groups within the layer that can be subsequently used for other reactions. Much work has been carried out on the zirconium phosphates and related materials over the last decade. This work has been reviewed by Alberti, Clearfield and Mallouk [23-26] and will not be further discussed here. Our interest is in the titanium analogs, where one can also bring into play redox reactions at the metal site. Titanium can be readily reduced from the IV to the III oxidation state, whereas zirconium tends only to reduce from Zr(IV) to the metal. Very little work has been published on the titanium compounds. We have successfully prepared layered titanium phosphates hydrothermally, whose hydrogen form readily undergoes both ion exchange and acid-base reactions. These compounds are crystalline, exhibiting up to 10 basal reflections on x-ray analysis and have a repeat distance of almost 10Å, 30% greater than that of the α-$[(Ti,Zr)(HPO_4)_2].H_2O$ forms. Our proposed structure of this compound, based only on the basal plane x-ray reflections, the P MAS-NMR and the density, is indicated in figure 5. On reaction with organic bases such as $CH_3(CH_2)_nNH_2$ an expansion of the lattice is found that is linear with the number of carbons, n, in the chain; the interlayer spacing = 10.05 + 2.06n Å. We therefore expect that this compound will show all the characteristics expected of intercalation compounds, and for example lithium does readily and reversibly reduce the titanium phosphate layers. We plan on determining the effect of host on reactivity and subsequent properties.

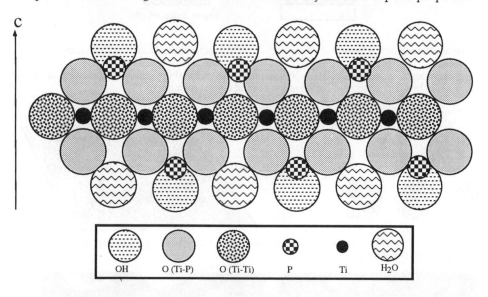

c

OH	O (Ti-P)	O (Ti-Ti)	P	Ti	H₂O

Fig. 5. Idealized structure of titanium phosphate. Actual structure has a puckered TiO_2 backbone.

The Next Step: Larger Templates

In the above examples we have shown how larger cations such as TMA can lead to very different structures than those found with the smaller alkali ions. Use of even larger templating ions are expected to lead to other structures following the experience of Beck et al [27,28] on silicates. Thus, we could similarly use long chain amines which readily form liquid crystals. We can envisage this amine approach as just the corollary of intercalating amines into 2d lattices as shown in figure 6. Normally one starts with a crystalline lattice, and then reacts it with the amine resulting in a swelled lattice. It should be feasible to do the opposite, that is start with an ordered amine array (e.g. a liquid crystal micelle) and build the oxide structure around it. Depending on the shape of the micelles (layered, columnar, or spherical) one might expect the structures formed to be layered, 1d tunnels, or 3d zeolitic following Beck's experiments [27,28]; the columnar case which could lead to one-dimensional tunnels is outlined in figure 7. Spherical micelles are expected to lead to either larger tunnel versions of the pyrochlore structure formed with the small NH_4 and $NMeH_3$ ions, or to sheet like structures as found with TMA. So we believe that the templating approach will lead to novel structures in transition metal oxides just as it has in the case of zeolites and metal phosphates. Preliminary experiments using long chain amines, such as $C_{12}(CH_3)_3amine^+$, have resulted in crystalline solids with many x-ray diffraction lines at low 2θ values and with very interesting morphologies as shown in figure 8.

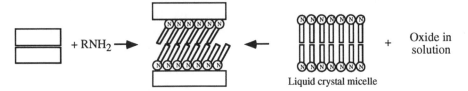

(a) Normal Intercalation Reaction   (b) Precipitation around micelle template

Fig. 6. Formation of novel intercalation compounds by (a) normal intercalatiⁿ into oxide or sulfide matrix, and (b) by condensation of lattice around liquid crystal micelle.

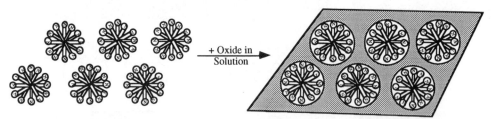

Fig. 7. Columnar micelles might lead to 1d tunnel structures, such as an enlarged version of the hexagonal bronze of figure 2.

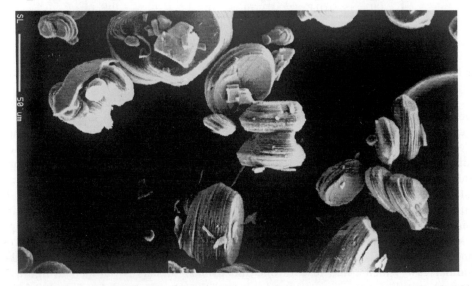

Fig. 8 Electron micrograph of the layered pancake-like crystals formed from the hydrothermal reaction of molybdate and $C_{12}H_{26}N(CH_3)_3Br$.

Redox Reactions and Control of the Oxidation State of the Transition Metal

The hydrothermal syntheses described above have tended to lead to compounds containing the transition metal in high oxidation states. However, the oxidation state can be modified either in the hydrothermal step or post synthesis. Thus, we have found that reduced oxides can be formed by using organic solvents such as ethylene glycol and acetic acid in place of water. For example, when ammonium tungstate is hydrothermally treated in a glycol/acetic acid mixture a blue ammonium tungsten bronze is formed, $(NH_4)_xWO_3$ [17]. In contrast, when water and HCl is used a white insulating solid is formed, $(NH_4)_xWO_{3+x/2} \cdot yH_2O$.

The materials described here are all expected to exhibit redox behavior. Their open crystal lattices will allow the ready intercalation of electron acceptors and donors, in addition to the ion exchange behavior observed in non redox active ion exchangers such as β-alumina, the layered silicates and zirconium phosphate. We can therefore expect redox reactions both incorporating cations and oxygen ions into the structure. Oxygen can be incorporated into the matrix and tunnels using the electrochemical method of Grenier et al. [29] so as to obtain the transition metal in its highest oxidation state. The wide range of structures found with similar chemical compositions should permit the study of the mechanism and nature of the diffusing specie(s) in this electrochemical oxygen insertion reaction, as it seems very unlikely that O^{2-} ions can diffuse sufficiently rapidly to explain the fast reaction. In the case of the hydrothermally formed tungsten oxide tunnel structures which have the general formula: $Na_xWO_{3+x/2} \cdot yH_2O$, they readily undergo chemical redox reaction either by incorporation of additional cations or by loss of oxygen:

$$Li_yNa_xWO_{3+x/2} \quad \Leftrightarrow \quad yLi + \quad Na_xWO_{3+x/2} \quad \Leftrightarrow \quad x/2 O_2 + \quad Na_xWO_3$$
Blue   White   Blue

These ionic insertion reactions occur at ambient conditions, and involve diffusion in and out of the tunnels of the structure. The redox behavior of these compounds is critically dependent on the cation in the tunnel and not well understood, even in the case of simple lithium intercalation. Thus, the sodium hexagonal tungsten oxides can react with 2 lithium atoms per tungsten, whereas the sodium pyrochlore only reacts with one lithium. On the other hand, silver pyrochlore reacts with three lithium per tungsten [5]. How the structure, and the cations contained within it, controls the degree of reduction is not clearly understood.

Comparison of the behavior of such materials should help us delineate what role crystalline structure plays in the reactivity of solids, and what are the energetics for transformation between structures. Why does one structure form under one synthesis condition, and a different one under slightly different conditions? What role does pH and the template play? Can we optimize diffusion by enlarging the diffusion path, and can we generate larger enough diffusion paths to allow the intercalation of species significantly larger than simple cations in transition metal oxide lattices. The flexibility of the hydrothermal approach to new crystalline structures may allow us to begin to generate a rational approach to the synthesis of solids with desired properties much as the organic chemist does today.

References
1) Whittingham, M. S.: J. Electrochem. Soc., 1976, 123, 315.
2) Dagani, R.: Chemical and Engineering News, 1993, January 9.
3) Mitzushima, K.; Jones, P. C.; Wiseman, P. J.; Goodenough, J. B.: Mat. Res. Bull., 1980, 17, 785.
4) Karam, R.; Whittingham, M. S.: U.S. Patent Application, 1992,
5) Reis, K. P.; Ramanan, A.; Whittingham, M. S.: J. Solid State Chem., 1992, 96, 31.
6) Reis, K. P.; Ramanan, A.; Whittingham, M. S.: Chem. of Mat., 1990, 2, 219.
7) Reis, K. P.; Prince, E.; Whittingham, M. S.: Chem. of Mat., 1992, 4, 307.
8) Freedman, M. L.: J. Am. Chem. Soc., 1959, 81, 3834.
9) Freedman, M. L.; Leber, S.: J. Less Common Metals, 1964, 7, 427.
10) Guo, J.-D.; Whittingham, M. S.: Rev. Modern Physics, 1993, September.
11) Guo, J.; Whittingham, M. S.: To be published, 1993,
12) Hinz, J.; Guo, J.; Whittingham, M. S.: NSF Summer Undergraduate Solid State Chemistry Program, 1993,
13) Ohzuku, T.; Ueda, A.; Hirai, T.: Chemistry Express, 1992, 7, 193.
14) Li, Y. J.; Whittingham, M. S.: Solid State Ionics, 1993, 63, 391.
15) Li, Y. J. Ph.D. Thesis, SUNY at Binghamton, 1993.
16) Haushalter, R. C.; Strohmaier, K. G.; Lai, F. W.: Science, 1989, 246, 1289.
17) Reis, K. P.; Whittingham, M. S.: J. Solid State Chem., 1991, 91, 394.
18) Guo, J.; Zavalij, P.; Whittingham, M. S.; Bucher, C. K.; Hwu, S.-J.: To be published, 1993,
19) Guo, J.; Zavalij, P.; Whittingham, M. S.: To be published, 1993,
20) Dines, M. B.: J. Chem. Soc. Chem. Commun., 1975, 220.
21) Figlarz, M.: Prog. Solid State Chem., 1989, 19, 1.
22) Cheng, K. H.; Jacobson, A. J.; Whittingham, M. S.: Solid State Ionics, 1981, 5, 355.
23) Alberti, G.; Costantino, U. In *Intercalation Chemistry*; M. S. Whittingham and A. J. Jacobson, Ed.; Academic Press: New York, 1982.
24) Clearfield, A.: Ann. Rev. Materials Science, 1984, 2, 219.
25) Mallouk, T. E.: Inorg. Chem., 1990, 29, 2112.
26) Mallouk, T. E.: Inorg. Chem., 1990, 29, 1531.
27) Beck, J. S.; Vartuli, J. C.; Roth, W. J.; Leonowicz, M. E.; Kresge, C. T.; Schmitt, K. D.; Chu, C. T.-W.; Olson, D. H.; Sheppard, E. W.; McCullen, S. B.; Higgins, J. B.; Schlenker, J. L.: J. Amer. Chem. Soc., 1992, 114, 10834.
28) Kresge, C. T.; Leonowicz, M. E.; Roth, W. J.; Vartuli, J. C.; Beck, J. S.: Nature, 1992, 359, 710.
29) Grenier, J. C.; Wattiaux, A.; Lagueyte, N.; Park, J. C.; Marquestaut, E.; Etourneau, J.; Pouchard, M.: Phys. C, 1991, 173, 139.

Materials Science Forum Vols. 152 - 153 (1994) pp. 109-114
© 1994 Trans Tech Publications, Switzerland

A PILLARED RECTORITE CLAY WITH HIGHLY STABLE SUPERGALLERIES

J. Guan and T.J. Pinnavaia

Department of Chemistry and Center Fundamental Materials Research,
Michigan State University, East Lansing, MI 48824, USA

ABSTRACT

A new highly stable supergallery pillared clay has been prepared by the reaction of aluminium chlorohydrate oligomers with Na^+ rectorite in the presence of polyvinyl alcohol as a pillaring precursor. This novel alumina pillared rectorite exhibits a basal spacings of 52 Å and a corresponding gallery height of 33 Å under air- dried conditions. The pillared product retains a stable gallery height of 23 Å even after treatment with 100% steam at 800°C for 17 hour . The new supergallery intercalate is charterized by a surface area, pore volume, and catalytic cracking activity superior to conventional alumina pillared rectorite and related smectites with 9 Å gallery heights.

INTRODUCTION

Most previously reported pillared lamellar structures possess gallery heights in the micropore range below 20 Å. For instance, smectite clays intelayered by organic pillaring agents, such as quaternary ammonium ions, exhibit basal spacings less than 19.0 Å, corresponding to gallery heights of less than 9 Å [1]. Smectites pillared by inorganic metal oxide aggregates normally have basal spacings of 18.0-22.0 Å and gallery heights of 8.4 - 12.4 Å[2-4]. A serious limitation of almost all these pillared clays is their relatively low hydrotherrmal stability. The agents that serve to prop open the interlayer space oftentimes are displaced from their original interlayer positions, causing the galleries to collapse at elevated temperatures, especially in the pesence of water vapor. It is apparent that the loss of gallery microporosity can seriously limit the adsorption and catalytic performance of these materials. For instance, aluminia pillared smectites treated at 730 °C for 4 hours in steam experience a 90% loss in surface area and a 80% decrease in microporous volume[5]. This loss in microporosity is accompanied by a corresponding sharp drop in petroleum cracking activity .

In the present work we describe the synthesis and characterization of a new type of pillared rectorite clay with highly stable supergalleries. Our approach utilizes poly vinyl alcohol (PVA) as a pillar precursor to complex and to stablize aluminum polycations that are larger than the Al_{13} oligomers used for forming conventional alumina pillared clays.

Experimental Section

A pillared rectorite clay with highly stable supergalleries was prepared by the reaction of aluminum chlorohydate oligomers with the sodium - exchanged form of the mineral in the presence of polyvinyl alcohol (PVA) as a pillar precursor. The pillaring agent was a commercially available aluminium chlorohydrate solution, $Al_2(OH)_5Cl$ (Reheiss Chemical Co.). The raw clay for the preparation of the stable supergallery pillared derivatives was naturally occurring rectorite (China) with a cation exchange capacity (CEC) of 60 meq / 100 g.

A Na-rectorite slurry having a solids content of 5 wt% was mixed with an aqueous solution of PVA and the slurry was then added with vigorous stirring to an aluminum chlorohydrate solution. The mixture was aged at 70-75 °C for 2-3 hours at pH 7.0. The solid product was filtered, washed with 1 liter of deionized water and air-dried. The intercalated PVA was removed by heating to 650 °C in nitrogen for 2 h, exposing the hot sample to air, and allowing it to cool slowly. This thermal treatment affords a white product with a residual carbon content of < 0.3 wt %. Altenatively, the PVA could be desorbed by exposing the pillared product to 100% steam at 800°C for 17 hours and then allowing the sample to cool slowly in air. For comparison purposes a conventional alumina pillared rectorite with a 9 Å gallery height (d_{001} = 29 Å) was prepared by the reaction of Na^+ rectorite with base hydrolyzed aluminum chloride at a pH of 4.5 , as reported in the literature [6]. Nitrogen adsorption and desorption isotherms were obtained using an Omnisorb 360 CX sorptometer. Surface areas were determined by the BET method. Pore volumes were calculated by the t - plot method.

Catalytic activity for the cracking of light diesel oil (bp = 205 - 330 °C) was determined by the microactivity test (MAT) method under the following reaction conditions: temperature, 460 °C; weight hourly space velocity (WHSV) $16hr^{-1}$;

catalyst to oil ratio (w/w) 4.0. The % conversion was obtained by chromatographic determination of the amount of unreacted diesel oil.

RESULTS AND DISCUSSION

Rectorite is a mixed layer 2:1 silicate containing regularly alternating galleries of non - swellable, high -charge density, mica - type galleries and swellable, low - charge density, smectite - type galleries. Only the smectite galleries are readily available for intercalation and ion exchange. The presence of the mica-type galleries causes the layer thickness of rectorite (19.6 Å) to be twice that of a normal smectite.

Alumina pillared clays normally are prepared by the reaction of a hydrophilic exchange form of a smectite clay, typically a Na^+- or Ca^{2+}- exchanged derivative, with Al_{13} polycations.   The present work deviates from conventional practice by utilizing   PVA as a pillaring precursor . Previous studies of related clay systems interlayered by non-ionic surfactants have shown that reducing the hydrophilic properties of the gallery surfaces greatly improves the hydrolytic stability of intercalated aluminum polycations [7].

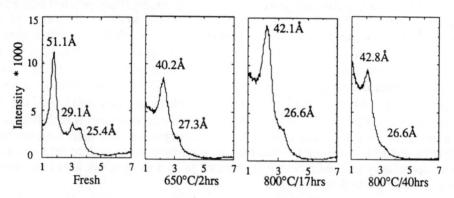

Fig.1. XRD Patterns of Calcined Supergallery Pillared Rectorite

The reaction of Na^+ rectorite with aluminum chlorohydrate polycations in the presence of PVA at pH values of 7.0 affords intercalated products with unusually large basal spacings. Figure 1 illustrates the XRD patterns obtained for a representative reaction product. The fresh, air-dried sample exhibits a first order reflection at 51.1 Å and a second order peak at 25.4 Å, indicating the presence of an ordered supergallery structure.The weak reflection at 29.1 Å most likely corresponds to a small amount of pillared rectorite with a conventional gallery height of about 9 Å. A somewhat smaller gallery height of 40.2 Å is obtained upon thermolysis in nitrogen and then in air at 650 °C. Exposing the sample to 100% steam at 800°C for up to 40 h and then to air to remove intercalated PVA results in a single supergallery reflection at 42-43 Å. This spacing corresponds to a gallery height of about 23. Å.

The term "supergallery pillared clays" has been used previously in the literature to describe microporous 2:1 layered derivatives in which the gallery height is substantially larger than the ~10Å-thickness of the clay layers [8]. Owing to its exceptional tolerance to steam, the pillared rectorite prepared in the present work not only has a supergallery structure, but it also represents a new type of *highly stable* structure. In typical structural models for alumina pillared smectites the 2:1 mica-type layers are presumed to be separated by single Al_{13} pillaring units [2-4]. Also, alumina pillared rectorites prepared by conventional methods contain "doubly - thick" 2:1 layered silicate host layers separated by single Al_{13} pillars [6]. Thus, the gallery heights of conventional alumina pillared smectites and rectorites are no more than 10Å. In contrast, the basal spacings of the pillared rectorite prepared by PVA modification coresponds to a gallery height that is twice large as conventional pillared clays. That is, our new supergallery rectorite has an unprecedented doubly - thick layer / doubly - thick pillar structure relative to conventional pillared smectites.

The surface areas and pore volumes of our supergallery pillared rectorite after calcination at 650°C for 2h and after steam treatment at 800°C for 17 h are represented in Table I. . The sample formed by thermal activation exhibits a surface area (218 m^2/g) comparable to that for the material activated in steam (200 m^2/g). Also, the supergallery pillared rectorite subjected to hydrothermal treatment at 800°C for 17 h possesses both Bronsted and Lewis acidity. The FTIR rin g stretching frequencies of chemisorbed pyridine indicate the presence of both Lewis sites (1453-1457 cm^{-1}) and Bronsted sites (1545-1547 cm^{-1}). The absorptions characteristic of both Bronsted and Lewis acid sites are more intense for the supergallery clay than for conventional pillared rectorite. Thus, on the basis of the improvement in surface area and aciditiy, our supergallery pillared rectorite is expected to be a more active acid catalysts than a conventional pillared analog.

In order to assess the catalytic properties of supergallery pillared rectorite, we have carried out microactivity tests (MAT) of gas oil cracking. Table I reports the diesel oil conversions for thermally and hydrothermally activated supergallery pillared rectorites. Both materials exhibit conversions in the range 62-64 % under standard MAT reaction conditions (see experimental section). In comparision, a conventional alumina pillared rectorite has a MAT conversion value of 54 % .It is noteworthy that air-dried supergallery rectorite gives a relatively low MAT conversion (17 %). Thus , the removal of intercalated PVA from the supergallery structure by calcination or steaming is essential for realizing high catalytic conversions.

Table I. Properties of Supergallery Pillared Rectorite

Treatment	Spacing Å	Area m^2/g	Vol. ml^3/g	MAT %
Air dried, 25º C	52	27.3	0.11	17
Calcined, 650ºC/2h	40	218	0.25	62
Steamed, 800°C/17h	42	200	0.20	64

ACKNOWLEDGMENT

The support of this research by the National Science Foundation through grants DMR-8903579 and CHE-9224102 is gratefully acknowledged.

REFERENCES

1.Barrer, R. : Pure Appl. Chem., 1989 , 61 , 1903.

2. "Pillared Clays", Catal. Today, 1988, Burch, R. (Ed.) , 2 (2,3) .

3.Figueras, F. : Catal. Rev., 1988, 30,457.

4. Schoonheydt, R. A.: Stud Surf. Sci. Catal.,1991, 58, 201.

5. Occelli, M. L.:Ind. Eng. Chem. Prod. Res. Dev., 1983, 22, 553.

6. Guan, J.; Min E.; Yu, Z. : Proc. 9th Int. Congr. Catal., Ottawa, Canada, 1988 , 104.

7. Michot, L. J.; Pinnavaia, T. J. : Clays Clay Miner., 1991, 39, 634; Michot, L.J;

   Barrès, O. ; Hegg, E. L. ; Pinnavaia,T. J.: Langmuir, 1993 , 9 , 1794.

8.Moini, A. ; Pinnavaia, T. J. :Solid State Ionics, 1988 , 26 , 119.

Materials Science Forum Vols. 152 - 153 (1994) pp. 115-124
© 1994 Trans Tech Publications, Switzerland

SYNTHESIS OF NOVEL METAL PHOSPHONATE COMPLEX STRUCTURES THROUGH SOFT CHEMISTRY

E.W. Stein, Sr.[1], C. Bhardwaj[1], C.Y. Ortiz-Avila[1], A. Clearfield[1] and M.A. Subramanian[2]

[1] Department of Chemistry, Texas A&M University, College Station, Texas 77843, USA

[2] Central Research & Development, Du Pont Co., P.O. Box 80328, Wilmington, Delaware 19880, USA

Keywords: Metal Phosphonate Synthesis, Proton Conduction of Sulfophosphonates, Ion Exchange of Metal Phosphonates, Self Assembly of Metal Phosphonates

ABSTRACT

Layered compounds containing organic pendant groups with hydrophilic functional groups may be made to exfoliate in aqueous media. These compounds behave as ion exchangers and sequestrants. Preliminary data on zirconium sulfophenylphosphonates and polyimine phosphonates will be presented. Mixing a polyimine with the sulfo-compound results in formation of insoluble interstratified complexes by self-assembly methods. Porous reactive complexes may be prepared in this way.

I. INTRODUCTION

Layered compounds are essentially two dimensional in character in the sense that the bonding forces within the layers are much stronger than those between layers. This fact allows neutral molecules and charged species to be intercalated into the interlamellar space and thus permits the manipulation of these intercalates in such a fashion as to create new structures. The method we shall emphasize in this paper is the exfoliation of the layers and use of the exfoliated layers to sequester ionic species or to combine with other exfoliates to produce new structures. The preparation of colloidal dispersions of layered compounds is discussed extensively in the paper by Allan Jacobson [1].

Recently we reported on the synthesis and properties of zirconium polyether phosphates and phosphonates [2,3]. These compounds are of two types, monophosphates (or phosphonates) of general composition $Zr[O_3PO(CH_2CH_2O)_n-H]_2$ and diphosphates $Zr[O_3PO(CH_2CH_2O)_nPO_3]$. The monophosphates swell in water and when $n \geq 9$ they exfoliate. In this condition they readily intercalate electrolytes with low lattice energies and become ionic conductors. Intercalation of the electrolyte does not result in precipitation but the dispersion may be cast into thin films or salted out by addition of water soluble organics.

The cross-linked products are totally insoluble, but can swell to a limited extent in water. We have proposed that the increased interlayer spacing on imbibing water results from a straightening of the crosslinking chains from a tilted to a more perpendicular orientation [3]. These crosslinked compounds readily incorporate transition metal species in a crown ether like manner. In this paper we now report on polyimine analogues of these polyether derivatives and how they lend themselves to the preparation of new structures. A preliminary report on these polyimines was given in ref. 3. Previously, Maya [4] had prepared $Zr(O_3PCH_2CH_2NH_2)_2\cdot 2HCl$ which he showed to exhibit anion exchange behavior. Rosenthal and Caruso [5] prepared $Zr[O_3(CH_2)_3NH_3^+]_2Cl_2$ and $Zr[O_3P(CH_2)_3NH_2]_2(O_3PCH_3)_{1.8}\cdot 2H_2O$. They found that the former compound was unreactive and apparently, even in the protonated form they were not able to exfoliate the layers. However, the mixed derivative was porous and exhibited anion exchange behavior. In addition, Cu(II) could be incorporated along with its counterion as an amine complex. We had earlier reported a similar reaction with $CuCl_2$ [3].

PREPARATION OF POLYIMINES AND PHENYLSULFONIC ACIDS

The compounds described are mainly phosphonates because of the greater stability of the P-C bond relative to the P-O bond. Both mono and diphosphonates were synthesized. The general procedure involved reaction of a polyimine, $NH_2(CH_2CH_2NH)_n-H$ with chloromethylphosphonic acid in basic solution. When an excess of the phosphonic acid is used, a diphosphonic acid is obtained. The phosphonic acids are then treated with zirconyl chloride and heated for times up to 24 h. The diphosphonates readily precipitate but the monophosphonates do not. Rather they are protonated by the HCl generated in the reaction and in this condition remain colloidally dispersed. Two methods of recovering the solids were utilized. In the first method the solid is salted out by addition of a water soluble organic (acetone, methanol). Alternatively, urea was added to the reaction mix. As refluxing is continued, the urea hydrolyses to yield ammonia which neutralizes the HCl. At a sufficiently high pH the solid zirconium polyimine precipitates.

Mixed derivatives in which a portion of the polyamine is replaced by either phosphate or phosphite were prepared by adding different amounts of H_3PO_4 or H_3PO_3 to the reactant mix. Addition of the smaller phosphate or phosphite ligands resulted in rapid precipitation of the product and in spite of longer reflux times the mixed derivatives were less crystalline. A listing of some of the polyimine derivatives and their interlayer spacings are given in Table 1.

Table 1. Representative interlayer spacings of zirconium polyimine phosphonates or phosphates of general composition $Zr(O_3PR)_x(O_3PR')_{2-x}$.

R	R'	x	Basal Spacing (Å)
$-CH_2NHCH_2CH_2NH_2$	-		17.1
$-CH_2NHCH_2CH_2NH_2$	-OH	1.7-1.8	16.3
$-CH_2NHCH_2CH_2NH_2$	-OH	0.63	14.5
$CH_2(NHCH_2CH_2)_2NH_2$	-	2	21.8
$CH_2(NHCH_2CH_2)_3NH_2$	-	2	27.6
$O-CH_2CH_2NH_2$	-	2	12.7
$O-CH_2CH_2NHCH_2CH_2NH_2$	-	2	21.0
$-CH_2NHCH_2CH_2NHCH_2PO_3$[a]	-	1	14.3
$-CH_2NHCH_2CH_2NHCH_2PO_3$[a]	-OH	0.5	Amorphous
$-CH_2NH(CH_2CH_2NH)_2CH_2PO_3$	-	1	17.7
$-CH_2NH(CH_2CH_2NH)_3CH_2PO_3$	-	1	19.2

[a]The diphosphonates conform to the general formula $Zr(O_3P-R)_x(O_3PR')_{2-2x}$.

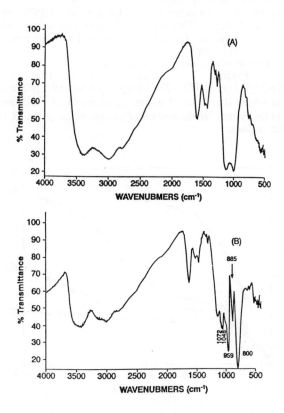

Figure 1. Infrared spectra of (A) Zr(O$_3$PCH$_2$NHCH$_2$CH$_2$NH$_2$)·2HCl·4H$_2$O and (B) the same compound containing 0.462 moles of the Keggin ion [PV$_2$W$_{10}$O$_{40}$]$^{5-}$.

   The dispersed polyimine phosphonates behave as anion exchangers. Polyvalent anions not only exchange but become encapsulated and precipitate out as tightly bound complexes between the zirconium polyimine layers. Among the species so far encapsulated are Fe(CN)$_6^{3-}$, Fe(CN)$_6^{4-}$, [PtCl$_4$]$^{2-}$ and a variety of heteropoly acid anions. The amount of anion taken up is conditioned by two factors; the anion capacity of the layered polyimine and the physical space available between the layers. The former quantity can be controlled by the amount of amine protonation. As an illustrative example of an encapsulation reaction, the polyimine derivative was protonated to the point where its formula was Zr(O$_3$PCH$_2$NHCH$_2$CH$_2$NH$_2$)$_{1.9}$(HPO$_4$)$_{0.1}$·2.3HCl. A dispersion of this compound was added to a 0.13M solution of K$_5$PV$_2$W$_{10}$O$_{40}$ with the Keggin ion being present in excess. A golden-yellow precipitate formed immediately. The mixture was stirred for 2 h and the solid collected, washed and dried at 70°C. Figure 1 shows the infrared spectra of both the original zirconium polyimine phosphonate and the complex containing the Keggin ion. The very intense bands at 3388, 3000 and 2799 cm^{-1} in spectrum A arise from combinations of O-H and N-H stretching vibrations (the latter both from protonated and free amine) as well as C-H stretching contributions and extensive H-bonding. The band at 1597 cm^{-1} is a result of the NH$_2$ and OH$_2$ scissoring motion and the very intense bands at 1000 and 1120 cm^{-1} are due to P-O$_3$ stretching vibrations. These same features are present in the IR spectrum of the Keggin ion containing solid (Figure 1B) together with the signature bands of the Keggin ion at 800, 885, 959, 1048 and 1072 cm^{-1}. Elemental analysis corresponded to the formula $Zr(O_3PCH_2NHCH_2CH_2NH_2)_{1.9}(H-PO_4)_{0.1}(H_5PVW_{10}O_{40})_{0.462}\cdot1.5H_2O$. This compound also contained a small amount (0.57%) of chloride ion which was not displaced. The protons associated with the Keggin ion in the formula are actually located on the amino groups.

The crosslinked polyimines also behave as complexing agents as shown by the potentiometric titration curves in Figure 2. Curve (A) is that of the compound $Zr(O_3PNHCH_2CH_2NHCH_2PO_3)_{0.5}(HPO_4)\cdot2HCl\cdot2H_2O$. Curves (B), (C) and (D) are the titration curves for solutions in which 0.5, 1 and 2 moles, respectively of Cu^{2+} per mole of the above polyimine was added to the system. In the process of neutralization of the HCl more and more Cu^{2+} was taken up along with co-ions.

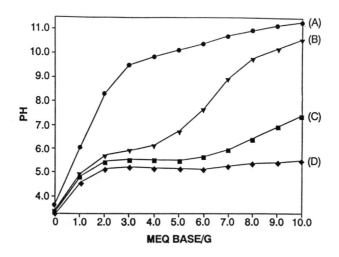

Figure 2. Potentiometric titration curves for (A) $Zr(O_3PCH_2NHCH_2CH_2NHCH_2-PO_3)_{0.5}(HPO_4)\cdot HCl$, and the same compound with (B) 0.5M Cu^{2+} added (C) 1 mole of Cu^{2+} added and (D) 2 moles of Cu^{2+} added per mole of polyimine.

Figure 3 is a computer generated schematic drawing of the fully crosslinked compound $Zr(O_3PCH_2(NHCH_2CH_2)_2NHCH_2PO_3)$. The R-groups in the drawing are understood to be additional polyimine chains. We have drawn the chains in such a way as to resemble a crystal-like structure. The layers are based on the α-zirconium phosphate structure [6] in which the polyimine chains replace -OH groups. In three dimensions the chains would be arranged in parallelogram arrays with a distance of 5.3Å between the chains [6]. Thus Cu^{2+} ions would fit comfortably between the polyimine chains. A second group of compounds that spontaneously exfoliate are the zirconium sulfophenylphosphonates [7]. They have the general composition $Zr(O_3PC_6H_4SO_3H)_x(HPO_4)_{2-x}$, where x=0.3-2, and are prepared by treatment of $Zr(O_3PC_6H_5)_x(HPO_4)_{2-x}$ with fuming sulfuric acid. In the exfoliated condition these compounds interact with large positively charged species to form insoluble complexes [7,8]. Initially we were able to prepare a series of these sulfophosphonates in which x=1 or less. The reason for this low value of x is the difficulty of isolating the products with too high a level of sulfonic acid groups. Alberti et al. [9] prepared compounds of the type $Zr(O_3PC_6H_4SO_3H)_{0.73}(O_3PCH_2OH)_{1.27}\cdot nH_2O$. By inserting the hydroxyphosphonate groups between the phenylsulfonate groups they were able to drastically reduce the swelling of the layers and thus readily recover the compound from solution. This compound was shown to be an excellent proton conductor.

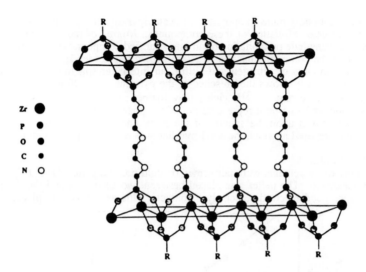

Figure 3. Schematic depiction of Zr(O$_3$PCH$_2$NH(CH$_2$CH$_2$NH)$_2$CH$_2$PO$_3$].

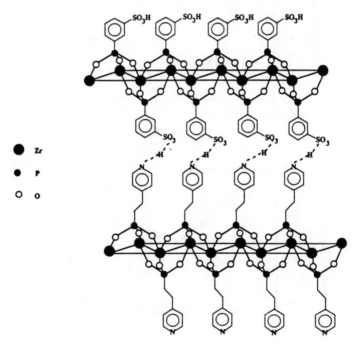

Figure 4. Schematic drawing of the interstratified compound [Zr(O$_3$PC$_6$H$_4$SO$_3$)$_2$]$^{2-}$ [Zr(O$_3$PCH$_2$CH$_2$C$_5$H$_4$NH)$_2$]$^{2+}$.

Recently we have been able to prepare the 100% sulfonated product Zr(O$_3$PC$_6$H$_4$SO$_3$H)$_2$. The term 100% needs to be qualified as a few phenyl groups (~5%) may remain unsulfonated or some hydrolysis may occur upon addition of water. A full description of the synthesis will be provided in a subsequent publication. This sulfonated compound readily exfoliates in aqueous media. Similarly protonation of the zirconium polyimine compounds or the ethylpyridinephosphonate, $Zr(O_3PC_2H_4\text{-}C_5H_4N)_2$, results in complete exfoliation of these compounds. Mixing of the sulfonate and amine exfoliates results in the immediate precipitation of an interstratified compound of the type shown in Figure 4. Evidence that the compound has the structure shown is afforded by elemental analysis which gave equimolar amounts of nitrogen and sulfur, an interlayer spacing of 18.4Å and the infrared spectrum which clearly shows the presence of the pyridine and sulfonic acid groups. In addition the MAS NMR spectrum for ^{31}P yields two distinct resonances at +6.4 ppm for the ethylpyridine phosphorus and -5.5 ppm representative of the phenylsulfonate phosphorus. These assignments were made based upon the spectra of the pure, uncombined compounds. Several similar derivatives were prepared in which the ethyl-pyridine was replaced by a polyimine chain.

CONDUCTIVITY STUDIES
    Proton conductivity measurements were performed on cold pressed pellets (150,000 psi) using gold as a blocking electrode. The pellet was placed between two gold foils and the sample was spring loaded to ensure good electrode/electrolyte contact. This assembly was placed inside a

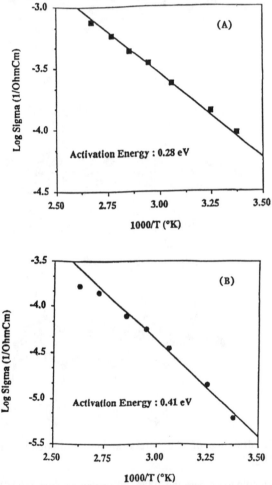

Figure 5.   Arrhenius plots of log σ versus $^1/_T \times 10^3$ for (A) $Zr(O_3PC_6H_4SO_3H)_2 \cdot nH_2O$ and (B) $[Zr(HPO_4)_{0.24}(O_3PC_6H_4SO_3H)_{1.76}]$ $[Zr(O_3PCH_2CH_2C_5H_4N)_2] \cdot nH_2O$. Data obtained at 30% relative humidity.

constant humidity cell that was maintained at a relative humidity of 30%. Impedance measurements were performed in the frequency range 5 Hz to 13 MHz using an HP 4192A impedance analyzer. A four terminal pair measurement principle was used to estimate impedance (Z) and phase angle (θ) at each frequency. The temperature range of the measurements was 293-375 K at a humidity of 30%.

Plots of Log σ versus 1000/T are shown in Figure 5 for two compounds, $Zr(O_3PC_6H_4SO_3H)_2$, and the interstratified compound depicted in Figure 4. Above 375 K sufficient water is lost to alter the conductivity from Arrhenius behavior. Actually the water content of the solids may vary slightly with temperature since they were maintained at 30% relative humidity rather than constant water content.

Alberti et al. [9] measured the conductivity of his mixed sulfonic acid alcoholate derivative from 250 K to 293 K and his results are shown as Figure 6. For comparison purposes we have added the conductivities of our two derivatives at the only common temperature of measurement. It is seen that the conductivity for our pure phenyl sulfonic acid is just about twice that calculated for the sulfonic acid-alcoholate as determined by interpolation between the values at 22 and 45% RH. This value is in accord with the fact that $Zr(O_3PC_6H_4SO_3H)_2$ contains somewhat more than twice the number of protons as does the Alberti derivative. In addition, the activation energy for our phenylsulfonic acid compound at 30% RH is much lower (0.28 eV versus 0.50 eV) than for the mixed sulfonic acid-alcoholate compound.

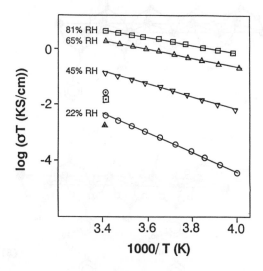

Figure 6. Plots of log (σT) versus 1/T x 10^3 for $Zr(O_3PC_6H_4SO_3H)_{0.73}(O_3PCH_2OH)_{1.27}·nH_2O$ at different relative humidities (data with solid lines drawn through the points, from ref. 9). Single points refer to values obtained for ⊙ compound A in Figure 5 at 30% RH, ▲ compound B in Figure 5 at 30% RH and ▣ to the interpolated value of $Zr(O_3PC_6H_4SO_3H)_{0.73}(O_3PCH_2OH)_{1.27}·nH_2O$ at 30% relative humidity.

The construction of interstratified compounds by the self assembly methods described here opens up a number of interesting opportunities. For example, it is possible to build in porosity to the compounds by synthesizing mixed derivatives in which phosphate or phosphite groups alternate with the larger phenyl based groups. Such compounds are schematically illustrated in Figure 7. The distance between phenyl rings interspersed by smaller groups is 10.6Å while that between the -OH groups in a direction perpendicular to the layers is 11.8Å. Even larger cavities may be built in by choice of the pendant organic groups. Such compounds may prove useful as catalysts, sorbents and ion exchangers. We have also prepared derivatives which were assembled with different tetravalent metals forming the layers, for example, titanium with amine pendant group and a zirconium sulfophenylphosphonate.

Since layered phosphonates can also be prepared from divalent [10] and trivalent [11] metals as well as the tetravalent ones it should be possible to prepare a wide range of these self-assembled compounds.

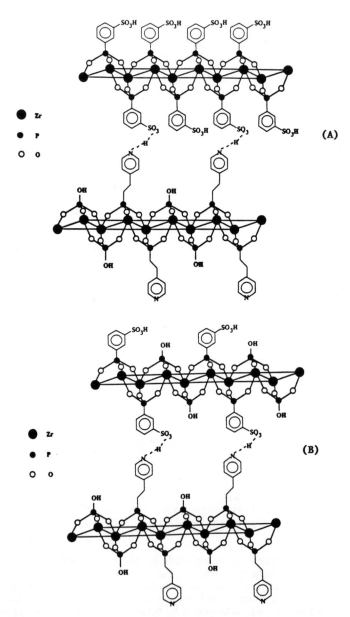

Figure 7.   Schematic drawings for self assembled molecules from (A) Zr(O$_3$PC$_6$H$_4$SO$_3$H)$_2$ and Zr(O$_3$PC$_5$H$_4$N)(HPO$_4$) and (B) Zr(O$_3$PC$_6$H$_4$SO$_3$)(HPO$_4$) with Zr(O$_3$PC$_5$H$_4$N)(HPO$_4$).

ACKNOWLEDGMENT: This study was supported by the Robert A. Welch Foundation under Grant No. A-673 for which grateful acknowledgment is made.

REFERENCES

1) Jacobson, A.: Proc. 1st Soft Chem. Symp., Kluwer Academic Publishers, The Netherlands, 1993.
2) Ortiz-Avila, C.Y. and Clearfield, A.: Inorg. Chem., 1985, 24, 1773.
3) Clearfield and Ortiz-Avila, C.Y.: In Supramolecular Architecture, T. Bein, Ed. ACS Symp. Ser. 499, Am. Chem. Soc., Wash., D.C., 1992, pp. 178-193.
4) Maya, L.: J. Inorg. Nucl. Chem., 1981, 43, 400.
5) Rosenthal, G. L. and Caruso, J.: Inorg. Chem., 1992, 31, 3104.
6) Clearfield, A. and Smith, G. D.: Inorg. Chem., 1969, 8, 431; Troup, J. M. and Clearfield, A.: Inorg. Chem., 1977, 16, 3311.
7) Yang, C.-Y. and Clearfield: Reactive Polymers, 1987, 5, 13.
8) Kullberg, L. H. and Clearfield, A.: Solvent Extr. Ion Exch., 1989, 3, 7.
9) Alberti, G.; Casciola, M.; Costantino, U.; Peraio, A. and Montoneri, E.: Solid State Ionics, 1992, 50, 315.
10) Cao, G.; Lee, H.; Lynch, V.M. and Mallouk, T.E.: Inorg. Chem., 1988, 27, 2781; Martin, K.; Squattrito, P.J. and Clearfield, A.: Inorg. Chim. Acta 1989, 155, 7; Zhang, Y.P. and Clearfield, A.: Inorg. Chem. 1992, 31, 2821.
11) Wang, R.C.; Zhang, Y.P.; Hu, H.; Frausto, R.R. and Clearfield, A.: Chem. Mater. 1992, 4, 864.; Cao, G.; Lynch, V.M.; Swinnea, J.S. and Mallouk, T.E.: Inorg. Chem., 1990 29, 2112.

Materials Science Forum Vols. 152 - 153 (1994) pp. 125-130
© *1994 Trans Tech Publications, Switzerland*

STRUCTURAL CONSEQUENCES OF SYNTHESIS PARAMETERS FOR OXYFLUORINATED MICROPOROUS COMPOUNDS

G. Férey, T. Loiseau and D. Riou

Laboratoire des Fluorures, URA CNRS 449, Université du Maine,
F-72017 Le Mans Cédex, France

Keywords: Microporous Compounds, Oxides Fluorides, Relations Synthesis-Structure, pH Dependant Reactions, Crystal Structures, Bonded and Encapsulated Fluorine, Geometry of Templates

ABSTRACT: The phases which appear in the systems M_2O_3 (M=Al,Ga) - P_2O_5 - AF (A=H^+, NH_4^+) - amine - H_2O under hydrothermal conditions (453K, autogeneous pressure, 24h) show that the pertinent parameters which influence the nature of the oxyfluorinated microporous phases are mainly the nature of the fluorinating agent (HF or NH_4F), the concentration of the amine, the acidity of the solution and the shape of the amine. NH_4^+ or a large concentration of amine suppresses the templating role of the latter. pH influences not only the nature of the microporous compound and the chemical nature of the template, but also the coordination polyhedra around M, their type of connection, and the kind of secondary building units which lead to the three dimensional structure. Finally the shape of the amines, spherical or linear, determines the type of cavity in which the template is located.

INTRODUCTION:

Hydrothermal synthesis has always been the best tool for obtaining microporous compounds. Zeolithic silicates required a basic medium, the templates being mineral or organic bases; the aluminophosphates, first synthesized by E. Flanigen *et al.* (1) needed neutral or weakly acidic conditions. The introduction of fluoride ions, due to Kessler *et al.* ((2) and references in the previous paper), strongly modifies the pH of the solutions and leads to an acidic medium. F⁻ acts as a mineralizer, but can also participate in the reaction and become trapped in the smallest cages of the microporous framework, as shown for instance for cloverite (3). This curious property contrasts with the usual behaviour of inorganic fluorides in which the F⁻ ions belong to the framework.

To introduce the fluoride ion in the skeleton of the inorganic framework, we have investigated the systems M_2O_3 (M=Al,Ga) - P_2O_5 - AF (A=H^+, NH_4^+) - amine - H_2O from the chemical and structural point of view. The parameters which influence the existence and the nature of the microporous phases were adjusted: temperature, pressure, time, concentration of reactants, pH and nature of the template. Structural information was provided by single crystal X-ray diffractometry and solid state NMR techniques on all resonating nuclei. These two approaches demonstrated a close relation between some chemical parameters and the structure of the oxyfluorinated phase.

The pertinent parameters are essentially the nature of the fluorinating agent, the concentration of the amine, the acidity of the solution, the nature and the shape of the amine. Parameters such as temperature and pressure variations, which were expected to strongly influence the formation of the phases, are not really pertinent, owing to the lability of the organic compounds at moderate (T < 250°C) temperatures, and to the conditions of hydrothermal synthesis under autogeneous pressure. The time of reaction is a second order parameter and eventually plays a role either to destroy the amine or to transform any metastable compound into the stable one by increasing reaction time. These three parameters were rapidly fixed at average values (453K, autogeneous pressure, 24h) in order to study in detail the influence of the decisive parameters noted at the beginning of this paragraph.

NATURE OF THE FLUORINATING AGENT

The cation associated with F^- in the starting AF material strongly influences the nature of the resulting phase. While H^+ easily leads to microporous compounds with the amines encapsulated in the cages of the structure, NH_4^+ exhibits a very strong templating power which masks the corresponding property of the amines introduced in the medium. This is particularly significant with the compounds $NH_4MPO_4(OH_xF_{1-x})$ [x=0.33 for M=Al and x=0.5 for M=Ga](4,5) previously ascribed as $AlPO_4$-CJ2 (6). Whatever the amine used during the synthesis, for instance cyclohexylamine, piperazine, piperidine, DABCO (1,4 diazabicyclo[2.2.2] octane), they do not enter the structure, rather NH_4^+ fills the eight-membered cavities of the structure whose three dimensional framework of corner linked tetrameric entities (Fig.1). These secondary building units (hereafter noted SBU) are formed by two PO_4 tetrahedra, one MX_6 octahedron and one MX_5 trigonal bipyramid. In the octahedron, one fluorine, which is terminal, is strongly hydrogen bonded to NH_4^+ groups; the other fluorine, in statistical distribution with OH groups, is shared between the octahedron and the bipyramid. In addition, NMR shows that the NH_4^+ groups are partially substituted by H_3O^+ in the structure (12% for Al and 7% for Ga).

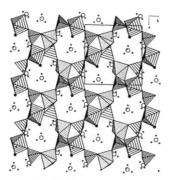

Fig.1. *[001] projection of $NH_4MPO_4(OH_xF_{1-x})$ showing the tetrameric SBU. Terminal fluorine are represented as black circles; large and small open circles correspond to N and H atoms respectively.*

The other examples described in this paper used HF as the fluorinating agent to study the role of the amine in the crystal chemistry of the oxyfluorinated microporous compounds.

CONCENTRATION OF THE AMINE

In most of the experiments, the concentration of the amine wasis of the same order of magnitude as M_2O_3 and P_2O_5. However, when a large excess of amine was used, its templating behaviour disappeared because they act as ligands for the metallic species. This was shown by $AlF(HPO_4)$, $H_2N(CH_2)_2NH_2$ synthesized from a mixture 1 Al_2O_3-1 P_2O_5-2 HF-18 amine-3O H_2O (7). In this two dimensional structure (Fig.2), the planes are built up from chains of Al octahedra corner shared via F^-, two chains being connected by HPO_4 tetrahedra. One of the vertices of the aluminium octahedron is occupied by one of the nitrogen atoms of the amine, creating a very original coordination environment AlO_3F_2N for aluminium. The stability of the structure is ensured by hydrogen bonds between the inorganic sheet and the second NH_2 group of the amine.

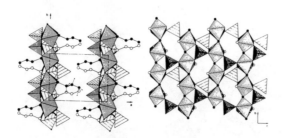

Fig.2.*[001] projection of AlF(HPO₄), H₂N(CH₂)₂NH₂ showing the ligand role of the amine (left); [100] projection showing the topology of the planes with black circles for F⁻ and open circles for the nitrogen fixed to Al (right)*

pH MODIFICATIONS AND STRUCTURE:

The influence of pH was exemplified by the study of the system Ga_2O_3 - P_2O_5 - HF - 1,3 diaminopropane (DAP) - H_2O with the ratio Ga_2O_3 - HF - H_2O fixed at 1:2:80. pH variations, obtained by modifying the content of P_2O_5 or of DAP ($1 \leq P_2O_5/Ga_2O_3 \leq 2.5$ and $0.8 \leq DAP/Ga_2O_3 \leq 2$), covered the pH range 1-10. Three phases, labelled ULM-n (for University of Le Mans) appear in this system. Orthorhombic ULM-3, which corresponds to the formula $Ga_3(PO_4)_3F_2,H_3N(CH_2)_3NH_3,H_2O$ (8) is stable in the range $5 \leq pH \leq 10$. At higher pH, it is mixed with Ga_2O_3. Monoclinic ULM-4 or $Ga_3(PO_4)_3F_2,H_3N(CH_2)_3NH_3$ (9) exists between pH 2 and 4. It could be considered as the anhydrous form of ULM-3. At pH≤1, triclinic ULM-6 $Ga_4(PO_4)_4F_2(H_2O),H_3N(CH_2)_3NH_3$, (10) is formed.

The structures of ULM-3 and ULM-4 have in common tthe same hexameric SBU. They are formed of three PO_4 tetrahedra, two GaO_4F trigonal bipyramids and one GaO_4F_2 octahedron (Fig.3). The two types of gallium polyhedra are linked together via the fluorine atoms. Within the SBU, two of the three tetrahedra are linked by corners to one octahedron and one bipyramid, whereas the third connects the three Ga polyhedra.

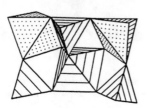

Fig.3. *Hexameric secondary building unit in ULM-3 and ULM-4. Trigonal bipyramids are dotted.*

These corner sharing SBU are arranged in two different ways in ULM-3 and ULM-4, but create 10-membered tunnels in both cases. Whereas tunnels are 'fer de lance' shaped ULM-3, they are rectangular in ULM-4 (Fig.4). The reason of the difference may be understood when looking at the inserted species in the tunnels. In ULM-3, both the water and the 1,3 DAP molecules are inserted, but examination of the hydrogen bond scheme shows that the strongest hydrogen bond exists within the tunnel between one hydrogen of the amino group and the water molecule. This indicates that the monohydrate of DAP is the templating agent, while anhydrous DAP is the template in ULM-4. The different steric occupancies can explain the different structures.

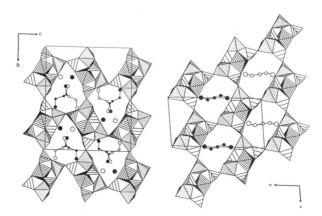

Fig.4. *[010] projections of ULM-3 (left) and ULM-4 (right), showing the different arrangement of the same secondary building units. Water molecules in ULM-3 are represented by the isolated circles.*

At very low pH, ULM-6 is formed; its complicated structure is built up from octameric units in which the gallium polyhedra are preferentially octahedra sharing this time edges instead of corners. This indicates that the pH not only influences the nature of the phases, but also the hydration of the amine (at high pH), the nature of the polyhedra around the metallic atoms (tetrahedra in basic medium, then bipyramids and octahedra when pH decreases), the size of the SBU which increases with lowering pH), and the mode of connection of the polyhedra within the SBU (favouring edge sharing at very low pH).

SHAPE, SIZE OF THE AMINE AND STRUCTURAL EVOLUTION

If the templating role of the amine is accepted, then the spherical or linear shape of the amine must play a role on the structure of the corresponding phases.

Fig.5. *Disposition of the diamino-propane (left), butane (centre) and pentane (right) in the ULM-3 topology. For n = 4 and 5, the nitrogen of one amino group is delocalized on the two positions showed in the figure.*

For the linear amines, we have begun a systematic study of the system Ga_2O_3 - P_2O_5 - HF - 1,n diamino(n)ane - H_2O with n≤6 in the pH range corresponding to that of ULM-3 described above. For n = 4-5 the topology of ULM-3 is preserved (Fig.5). For n=4, the fourth carbon implies the replacement of H_2O by one NH_3 group of the amine in the cavity; the other is delocalized on two positions. For n=5, the torsion of the aliphatic chain is important (11). This explains why a new structure appears from n=6: ULM-5 or $Ga_{16}(PO_4)_{14}(HPO_4)_{14}F_7(OH)_2$,$[H_3N(CH_2)_3NH_3]_4$, $6H_2O$. The same structure type exists also for n=7 and 8.

The three-dimensional network (12) is built up from three types of basic building units; the first, close to that already encountered for ULM-3 and 4, consists in corner linked $[Ga_3(PO_4)_2(HPO_4)F_2]$ hexameric units composed of two PO_4 tetrahedra, one HPO_4 tetrahedron, two GaO_4F trigonal bipyramids and one GaO_4F_2 octahedron , with fluorine atoms shared between the gallium polyhedra; the second is very similar to the first except that one of the trigonal bipyramids is replaced by a tetrahedron $GaO_3(OH)$; the third is octameric $Ga_4(PO_4)_4$, and can be considered as formed by a cube of corner linked GaO_4 and PO_4 tetrahedra, which encapsulate a fluorine atom, like in cloverite. This fluorine is bonded to two of the four gallium of this cube. This is the first example of both bonding and encapsulated F^- in the same structure. The framework delimits 16-and 6-membered ring channels along [100] and 8-membered ones along [010] . The diprotonated amines are inserted in the 16-membered oval channels, whose free aperture is 12.20 x 8.34Å. The water molecules are in the 6-membered tunnels.

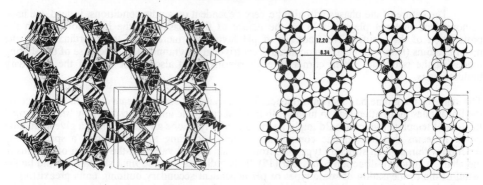

Fig.6. *Perspective view (left) and space filling representation along [100] (right) of ULM-5 showing the free aperture (12.20 x 8.34Å) of the 16-membered tunnels.*

The NMR characteristics of the encapsulated fluorine atom (δ = - 63.5 ppm/CFCl$_3$) strongly differ from that of the bonding fluorine atoms ($\delta \approx$ - 100 ppm), suggesting a less ionic character for the encapsulated species.

The use of spherical amines like DABCO (1,4 diazabicyclo[2.2.2] octane) lead to different structure types whose main feature is to lead to cages rather than tunnels for the pores. The corresponding examples that support this assumption are ULM-1 ($Ga_3(PO_4)(HPO_4)_2F_3(OH),C_6N_2H_{14}$),   0.5 H_2O)   (13)   (Fig.7)   and   ULM-2 ($Ga_4(PO_4)_2F_2(OH)_2(H_2O)_2,0.5\ C_6N_2H_{14}$)) (14).

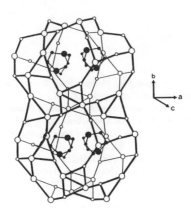

Fig.7.*Superposition of two spherical cavities in ULM-1. Only the Ga-P subnetwork is represented.*

CONCLUSION: GENERAL TRENDS

The results of the present study show that the structures of oxyfluorinated microporous compounds are dependent on the conditions of synthesis. The elaboration of new phases with significant dimensions for the pores must first prohibit the use of NH_4F as fluorinating agent, ammonium ions being better templates than the amines. In addition, the introduction of very large amounts of amines, which are then fixed on the coordination polyhedra of the metallic species, leads to the loss, at least partially, of their templating property.

The nature of the phases will also be very dependent on acidic conditions. Indeed, we have demonstrated that pH plays a role on the nature of the phases, on the hydration and on the protonation of the amines. The decrease of pH favors the increase of the coordination of the metallic atoms (from tetrahedral to octahedral via trigonal bipyramidal), the increase of the sharing modes of the polyhedra (from corner to edge sharing), and also of the size of the secondary building units.

The shape of the amine has also an influence. While spherical amines create cages in the structure, the linear ones more often lead to tunnels. In this last case, the amine does not play '*stricto sensu*' the role of template (which would imply a strict specificity of one amine to obtain a given structure type). It seems that the organic compounds exhibit, according to their length, a space filling function within a given structure type and, from certain critical values of the number of carbons in the chain, a structure-direction behaviour (15). This indicates that the final structure would be the result of the condensation of some types of pH dependent secondary building units preexisting in the solution . With this hypothesis, the resulting structure would then depend on the ratio of the sizes of both the SBU and the amine which confer to the product the maximum thermodynamic stability. *In situ* studies are currently in progress to verify this assumption.

For fluorine, we have shown that, in contrast to most of the compounds of Kessler i which exhibit encapsulated fluorine, this element can also play a ligand role in oxyfluorinated phases. This is attributed to different pressure conditions. Indeed, it is well known by solid state chemists that the increase of pressure favors the increase of the coordination of the metallic species. The Kessler phases (in which aluminium or gallium are in tetrahedral coordination and fluorine is encapsulated) would correspond to low pressure varieties, our phases being the high pressure ones. From this point of view, ULM-5, which exhibits in the same structure the two pecularities, is very interesting and pressure experiments are in progress to see eventual phase transitions.

REFERENCES

(1) Wilson, S.T., Lok B.M., Massina C.A., Cannan T.R., Flanigen E.M.: J. Am. Chem. Soc., 1982, 104, 1146.
(2) Guth J.L., Kessler H., Wey R.: Stud. Surf. Sci. Catal., 1986, 28, 121.
(3) Esterman M., McCusker L.B., Baerlocher C., Merrouche A., Kessler H.: Nature 1991, 352, 320.
(4) Férey G., Loiseau T., Lacorre P., Taulelle F.: J. Solid State Chem., 1993, 105, 179.
(5) Taulelle F.,Loiseau T., Maquet J., Livage J., Férey G.: J. Solid State Chem., 1993, 105, 191.
(6) Yu L., Pang W., Li L: J. Solid State Chem., 1990, 87, 241.
(7) Riou D., Loiseau T., Férey G.: J. Solid State Chem., 1993, 102, 4.
(8) Loiseau T., Retoux R., Lacorre P.,Férey G.: J. Solid State Chem.,in press.
(9) Loiseau T., Riou D., Férey G.: J. Chem. Soc., Chem. Comm., in press.
(10) Loiseau T., Férey G.: J. Chem. Soc., Chem. Comm., submitted.
(11) Loiseau T., Férey G.: Eur. J. Solid State Inorg.Chem., submitted.
(12) Loiseau T., Férey G.: J. Solid State Chem., submitted.
(13) Loiseau T., Férey G.: J. Chem. Soc., Chem. Comm., 1992, 1197.
(14) Loiseau T., Férey G.: Eur. J. Solid State Inorg.Chem., 1993, 30,369.
(15) Gies H., Marler B.: Zeolites, 1992, 12, 42.

Materials Science Forum Vols. 152 - 153 (1994) pp. 131-136
© 1994 Trans Tech Publications, Switzerland

THE NiO$_2$ SLAB:
A VERY CONVENIENT STRUCTURAL UNIT FOR CHIMIE DOUCE REACTIONS

C. Delmas

Laboratoire de Chimie du Solide du CNRS and Ecole Nationale Supérieure de Chimie et Physique de Bordeaux, Université Bordeaux I, F-33405 Talence Cédex, France

Keywords: Chimie Douce, Hydroxides, Oxyhydroxides, Ni-Cd Batteries, LiNiO$_2$, NaNiO$_2$, Lithium Batteries

ABSTRACT

The NiO$_2$ slab built up of edge-sharing NiO$_6$ octahedra is the structural unit of a very large number of materials that are generally obtained by chimie douce reactions from four parent phases : (α and β) Ni(OH)$_2$, LiNiO$_2$ and NaNiO$_2$. Depending on the experimental conditions, cations (H$^+$, Li$^+$, Na$^+$, K$^+$), anions (CO$_3^{2-}$, SO$_4^{2-}$, NO$_3^-$) and water molecules can be inserted. Several types of packings are observed for well crystallized materials. Moreover, in peculiar conditions, turbostratic or interstratified structures can be obtained.

Some of these materials are among the best intercalation materials used in positive electrode of lithium, Ni-Cd and Ni-MH batteries.

INTRODUCTION

The NiO$_2$ slab made of edge-sharing NiO$_6$ octahedra is the structural unit of a very large number of materials that are generally obtained by chimie douce reactions from four parent phases : (α and β) Ni(OH)$_2$, LiNiO$_2$ and NaNiO$_2$. The pristine materials are obtained either by precipitation or by high temperature solid state reaction [1]. The NiO$_2$ slabs are built up in these reactions while their relative packing and the interslab composition is monitored by the chimie douce reactions. All these materials can be classified in two main families : alkali metal intercalation compounds and hydroxides/oxyhydroxides. The second family of materials can be obtained either directly by precipitation or by exchange from the alkali metal nickelates. The chimie douce reactions allow to obtain metastable phases from the structural and/or composition point of view. Moreover, as the particle size of the pristine material is maintained during the chimie douce reaction, well crystallized hydroxides and oxyhydroxides can be obtained. These materials, which in some cases exhibit the same composition and structure than those obtained by precipitation but with very different particle sizes can be considered as model materials to explain their physical and electrochemical behavior in rechargeable batteries.

Within the NiO_2 slab, the nickel ion can be divalent ($t_2^6e^2$), trivalent ($LS-t_2^6e^1$) or tetravalent ($LS-t_2^6e^0$). Trivalent nickel ions in the low spin state exhibit a Jahn-Teller effect which leads to a distortion of the NiO_6 octahedra and in the case of $NaNiO_2$ to a macroscopic lattice distortion (monoclinic symmetry) [2]. For $LiNiO_2$, there is no distortion of the rhombohedral unit cell; nevertheless, recent EXAFS investigations have shown a local distortion of the NiO_6 octahedra [3]. The oxygen displacement, associated with the electronic transfer between a trivalent nickel ion and a divalent or a tetravalent one, may assist the cation diffusion in the interslab space resulting from the intercalation process.

In all the starting materials, cobalt or iron can be partially substituted for nickel. In all cases, the oxidation state distribution results from a compromise between the intrinsic oxidation state stabilities and the effect on the substituting cation of the ligand field imposed by the prevailing cation within the $Ni_{1-y}M_yO_2$ slab (M = Co, Fe). This constraint on the cationic oxidation state requires to modify the interslab composition in order to preserve the overall charge balance. By this new chemical route, it is possible to obtain new materials intercalated with anionic species. Moreover, such substitutions allow the monitoring of the voltage of the electrochemical cells using these materials as positive electrode.

HYDROXIDES AND OXYHYDROXIDES

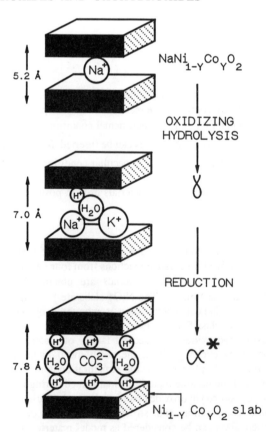

Fig. 1 - Chimie douce reaction from $NaNiO_2$.

Nickel hydroxides exist under two main varieties β and α ; while the former exhibits the classical brucite type structure, the latter, in which one layer of water molecules is intercalated between the $Ni(OH)_2$ slabs, has a turbostratic structure. Chimie douce reactions from $NaNiO_2$ allow to obtain γ-type oxyhydroxides in a first step and, after reduction, well ordered hydrated nickel hydroxides designated as α^*. The substitution of cobalt and iron ions for nickel in the starting phase leads to materials closely related to the hydrotalcite or more generally to the Layer Double Hydroxide family (LDH). Within the hydroxide, cobalt and iron ions are at the trivalent state and anions are inserted in order to compensate the excess of positive charge as schematically illustrated in Fig. 1 [4-5]. If the experiment is realized in air, carbonate anions are generally inserted, but other anions can be inserted under special conditions. The amount of inserted anions is directly related to their charge and to the amount of substituting trivalent cation. For materials obtained by direct precipitation the M^{2+}/M^{3+} ratio is mainly imposed by the stability of the formed LDH. On the contrary, when these materials are obtained by chimie douce reaction the formation of the slabs and the preparation of the LDH are disconnected. As a result, it is possible to fix the M^{2+}/M^{3+} ratio independently to the LDH stability. This remark is very important to obtain open pillar structures. In the case of materials with low amount of anions, segregation phenomena can occur and interstratified structures are therefore observed. These structures can be considered as a statistical packing of β and α^* units with Daumas-Herold domains as in graphite intercalation compounds. A schematic representation of the structure of an interstratified nickel hydroxide is given in Fig. 2 [6].

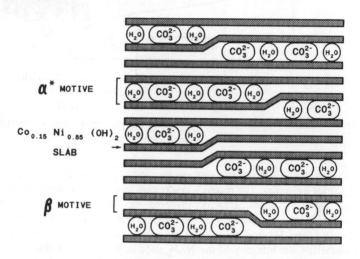

Fig. 2 - Schematic representation of the interstratified structure.

Oxidation reactions, in alkaline medium, of nickel hydroxides yield several types of nickel oxyhydroxides $\beta(III)$-NiOOH, γ-$H_xK_yNiO_2,zH_2O$ and γ-$H_xK_yNiO_2,2zH_2O$. As a result of the topotactic character of the reactions, the packing type of the precursor phases (well ordered or turbostratic) is maintained in the final materials. In the γ-type phases, the very large value of the interslab distance suggests the presence of two layers of intercalated water molecules [7].

In the γ-type substituted oxyhydroxides, which exhibit an average oxidation level close to 3.5, the magnetic and electric properties studies show that the cobalt ions stay at the trivalent state while the nickel ones are oxidized at the tetravalent state. On the contrary, in the case of iron substituted phases, both 3d cations are simultaneously oxidized at the tetravalent state.

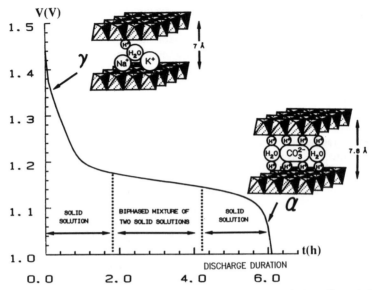

Fig. 3 - Structure modifications involved by the γ/α electrochemical cycling of nickel oxyhydroxides.

These hydroxides and oxyhydroxides can be used as positive electrode materials of alkaline batteries (Ni-Cd for example). The reaction that occurs in this case is certainly the oldest example of a Chimie Douce reaction actually involved in electrochemical energy storage. The electrochemical reaction between the α^*−hydroxide and γ-oxyhydroxide phases is extremely complicated as K^+ and H^+ cations as well as anions are simultaneously involved during the process ; nevertheless, the integrity of the NiO_2 slabs is preserved all along the cell cycling [8]. **In-situ X-Ray diffraction and Mössbauer spectroscopy** experiments have been realized for the first time during electrochemical cycling of a nickel-cadmiun battery. The X-Ray study shows the existence of a two phase domain situated between two solid solutions ; surprisingly, as illustrated in Fig. 3, there is no voltage plateau in the discharge curve and the compositions of the two materials involved in the biphased domain vary during the intercalation process as suggested by the small modifications of the X-ray diffraction patterns. This behavior can result from the simultaneous presence of three ionic species, involved during the charge/discharge process. Moreover, a reversible broadening/narrowing of the diffraction lines is observed during the cell cycling, it characterizes a modification during the redox process of the constraints within the particles resulting from an inhomogeous distribution of cationic charges in the γ-type oxyhydroxide. The Mössbauer study realized in collaboration with L. Fournès and L. Demourgues-Guerlou shows unambiguously the formation of low spin tetravalent iron during the material oxidation. The relative amount of low spin tetravalent, high spin tetravalent and high spin trivalent ions varies vs temperature and the overall amount of iron in the material. Moreover, the Mössbauer study realized on very slightly iron doped phases shows the formation of clusters of iron ions within the $Ni_{1-y}Fe_yO_2$ slabs [9].

ALKALI METAL INTERCALATION NICKELATES

The case of lithium nickelate is very interesting for both basic and application points of view [10-11]. Whatever the preparation conditions, this material presents always a non-stoechiometry with an excess of nickel in the lithium site of the layer structure. The true formula being $Li_{1-z}Ni_{1+z}O_2$, the structure exhibits a 3D character [12]. The partial substitution of cobalt for nickel leads to strictly 2D materials ($LiNi_{1-y}Co_yO_2$) for y > 0.20. For intermediate compositions, the amount of extra nickel ions in the lithium site decreases continously when the cobalt amount increases. This effect of cobalt substitution on the cationic distribution has been clearly demonstrated by Rieveld refinement of the X-ray patterns realized in collaboration with P. Gravereau and A. Rougier and by the effect on the magnetic properties studied by I. Saadoune.

The deintercalation/intercalation properties of these materials are strongly related to the amount of extra-nickel ions in the lithium sites [13-14]. As illustrated in Fig. 4, the presence of extra-nickel ions in the lithium layer restricts considerably the ability to reintercalate large amounts

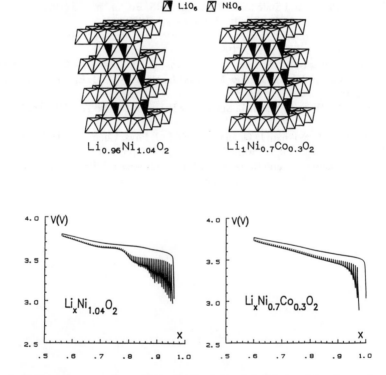

Fig. 4 - Relation between the cationic distribution and the shape of the electrochemical curve in $Li_xNi_{1-y}Co_yO_2$ systems.

of lithium. Moreover, according to the cell voltage deintercalation limit, the electrochemical cycling starting from strictly 2D materials can be reversible or can induce a small slab modification with an irreversible migration of the 3d cations from the slab to the interslab space. When the amount of deintercalated lithium is too high, the layered structure becomes metastable and the 3d cation displacement leads to a relative stabilization of the material. This behavior, that can be considered at the border of a chimie douce reaction, leads to the **formation of hazy structures** that are very convenient for lithium intercalation [15].

In the case of the homologous sodium phases, which exhibit an expanded interslab space, the weakening of the alkali ion-oxygen bonds allows, when sodium is deintercalated or intercalated, the modification of the surrounding of the Na^+ ions, leading to a reversible slab gliding.

REFERENCES

1) Delmas C., Faure C. and Borthomieu Y. : Materials Science and Engineering, 1992, B13, 89.
2) Dyer L.D., Borie B.S. and SMITH G.P. : J. Amer. Chem. Soc., 1954, 76, 1499.
3) Rougier A., Chadwick A. and Delmas C. : To be published.
4) Faure C., Delmas C., Fouassier C. and Willmann P. : J. Power Sources, 1991, 35, 249.
5) Faure C., Delmas C. and Willmann P. : J. Power Sources, 1991, 35, 263.
6) Delmas C. : Proc. Solid State Ionics Meeting, MRS, Boston, p. 335 (1990).
7) Borthomieu Y. : Thesis, University of Bordeaux I (1990).
8) Delmas C., Borthomieu Y. and Faure C. : Proc. Symp. on Nickel Hydroxide Electrodes, Electrochemical Society, p. 119 (1990)
9) Demourgues-Guerlou L., Fournès L. and Delmas C. : J. Solid State Chem. (submitted).
10) Dahn J.R., Van Sacken U., Juzhow M.W.and Janaby H.A.P. : J. Electrochem. Soc., 1991, 138(8), 2207.
11) Broussely M., Perton F., Labat J., Staniewicz R.J. and Romero A. : J. Power Sources, 1993, 43-44, 209.
12) Goodenough J.B., Wikham D.G. and Croft W.J.: J. Phys. Chem. Solids, 1958, 5, 107.
13) Delmas C. and Saadoune I. : Solid State Ionics, 1992, 53, 370.
14) Delmas C., Saadoune I. and Rougier A. : J. Power Sources, 1993, 43-44, 595.
15) Delmas C. : Solid State Ionics (submitted).

Materials Science Forum Vols. 152 - 153 (1994) pp. 137-142
© 1994 Trans Tech Publications, Switzerland

METAL CHALCOGENIDE CLUSTERS, $M_X E_Y$:
GENERATION AND STRUCTURE

I. Dance and K. Fisher

School of Chemistry, University of NSW, Kensington, NSW 2033, Australia

Keywords: Cluster, Sulfide, Metal, Laser Ablation, Mass Spectrometry, Theory, Structure, Mn, Fe, Co, Ni, Cu

ABSTRACT

Laser ablation of solid metal chalcogenides, or of mixtures of the elements, yields many cluster anions of the general formula $[M_x E_y]^-$, where E is mainly S and Se, and M is Mn, Fe, Co, Ni and Cu. The largest anion clusters detected so far for these metals are $[Mn_{22}S_{23}]^-$, $[Fe_{11}S_{10}]^-$, $[Co_{38}S_{24}]^-$, $[Ni_{15}S_{10}]^-$, and $[Cu_{45}S_{23}]^-$. The observed clusters occur in continuous series with very well defined patterns of composition: the ion maps of x vs y are narrow but continuous distributions, almost linear. This is interpreted in terms of restricted compositions for stability of metal chalcogenide clusters when devoid of terminal heteroligands. The cluster compositions correlate with electron population: for Mn x = y or y–1, while for Cu x =2y–1 or 2y–2. Structures are postulated for key members of the Mn and Cu series, and the building principles which extend the series are discussed. Postulated structures are optimised using local density functional calculations.

INTRODUCTION

We have recently reviewed the few studies of metal chalcogenide clusters in the gas phase [1]. Relatively little is known about the compositions, structures, and reactions of gaseous metal chalcogenides, which are   much better known in the condensed phases. There are two main reasons why gas phase metal chalcogenide chemistry is important in the context of new solid state materials. One is that gas phase methods are increasingly being used for controlled deposition of films, and knowledge of the composition and chemistry of the gas phase species allows better control of the technologies. The second is that clusters in the gas phase are pristine, devoid of heteroligands and devoid of the multiple external influences that occur in the condensed phases, and thus reveal the fundamental chemistry of these compounds.

CLUSTER FORMATION BY LASER ABLATION

A high power pulse (up to 1000 MW cm^{-2}) from a Nd/YAG laser at 1064 nm, directed onto the surface of an absorbing solid metal chalcogenide (or in some cases mixtures of the elements) generates a considerable number of gaseous species. We do this experiment in the cell of a Fourier Transform ion cylcotron resonance mass spectrometer, allowing the trapping and monitoring of the ions. The positive ion spectra generally do not contain large clusters, and are dominated by metal polysulfide ions. The negative ion spectra are much richer, and more interesting. Cluster ions $[M_x E_y]^-$ up to mass 4000, and up to the composition $[Cu_{45}S_{23}]^-$, are detected [2-5]. The smaller clusters are more abundant, and can be subjected to collisional dissociation and ion-molecule reactions.

The laser impact process generates very high temperatures, and cavities, at the solid surface. There is an associated pressure surge, presumably due to a considerable number of neutral species which are rapidly pumped back to the cell pressure of 10^{-7} mbar. The negative ions are believed to be formed by attachment of low energy electrons to neutrals, a process which occurs in the gas phase as short period after the laser ablation. The resulting ions are trapped by the potential on the end plates of the cell, excited to a coherent ion cyclotron motion by an RF pulse, and monitored at some later time (msec to sec) by following the FID of a subsequent measurement pulse.

All indications are that the negative ions formed possess considerable stability. They can be retained in the cell, the smaller ones with sufficient intensity are shown to not readily dissociate on collision with inert gases, and to be unreactive with background reagent gases, except H_2S. A fundamental question in these experiments is whether the negative ions observed are revealing the compositions of all neutral clusters that survive the laser ablation plasma (that is all neutrals have similarly favourable electron affinities), or whether the negative ions indicate simply the neutrals with the most favourable electron affinities. Although this is an oversimplification of complex gas phase chemistry, we can proceed with the assumption that the species observed are the more stable combinations of the elements.

Figure 1 shows the spectrum of negative ions formed by laser ablation of MnS (green, stable, rock salt form). There is a clear sequence of ions $[Mn_xS_x]^-$ and $[Mn_xS_{x+1}]^-$ for x = 4 to 22 (except for $[Mn_5S_5]^-$) and also a sequence $[Mn_xS_xO]^-$ with the oxygen derived from residual water.

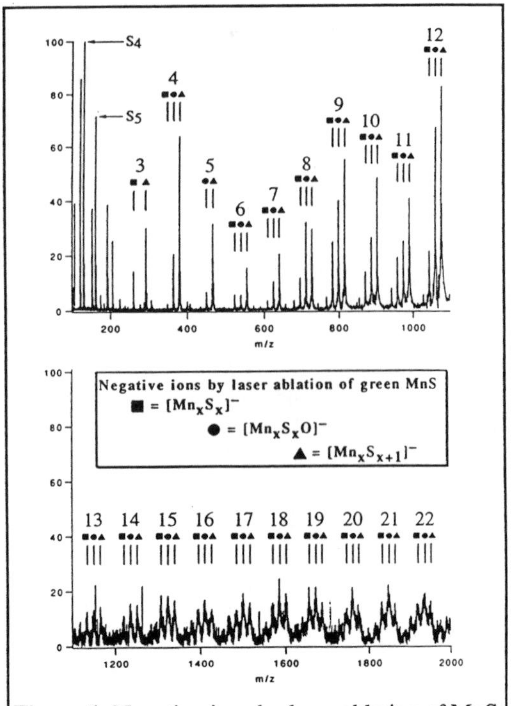

Spectra with similar continuous series of ions are obtained for FeS and $KFeS_2$, CoS, NiS and Ni_3S_2, and CuS and KCu_4S_3. The series of ions for Fe, Co, Ni and Cu are presented in the ion maps contained in Figure 2.

Figure 1 Negative ions by laser ablation of MnS

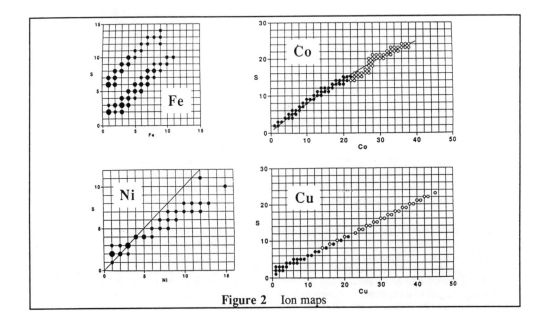

Figure 2   Ion maps

DISTRIBUTIONS OF ANIONIC CLUSTERS $[M_XE_Y]^-$

The ion maps reveal: (1) there are very well-defined regions of stability for $[M_xS_y]^-$, and many compositions for which no ions are observed; (2) the ion distribution lines are generally straight and continuous, with some curvature and additional structure for Co; (3) the slopes of the distributions are dependent on the identity of the metal, and related to the metal electron population. This latter point is demonstrated in the Table, which lists a selection of values for N_M, the number of valence electrons per metal, and calculated as the sum of the metal 3d and 4s electrons and six electrons per S, all divided by the number of metal atoms in the cluster. It is clear that there is a preferred electron population for these clusters, 14 to 16e for the smaller clusters, and approaching 12e for the larger clusters, and that this is independent of the composition and the number of bonding connections [5].

Table   Valence electron populations per metal atom (N_M, see text) in $[M_xS_y]^-$ clusters.

Mn		Fe		Co		Ni		Cu	
Mn_4S_5	14.8	Fe_4S_4	14.3	Co_4S_4	15.3	Ni_4S_4	16.3	Cu_4S_3	15.8
		Fe_6S_7	15.2	Co_6S_6	15.2	Ni_6S_4	14.0		
		Fe_9S_9	14.1	Co_9S_8	14.4	Ni_9S_6	14.1	Cu_9S_5	14.4
$Mn_{10}S_{11}$	13.7	$Fe_{10}S_9$	13.5	$Co_{10}S_8$	13.9	$Ni_{10}S_7$	14.3	$Cu_{10}S_6$	14.7
$Mn_{12}S_{13}$	13.8			$Co_{12}S_9$	13.8	$Ni_{12}S_8$	14.0	$Cu_{12}S_7$	14.6
				$Co_{15}S_{12}$	13.8	$Ni_{15}S_{10}$	14.1	$Cu_{15}S_8$	14.3
				$Co_{23}S_{16}$	13.2			$Cu_{23}S_{12}$	14.2
				$Co_{37}S_{24}$	12.9			$Cu_{37}S_{19}$	12.1

QUESTIONS OF STRUCTURE

The occurrence of the continuous series of clusters, for all metals except Ni, suggests that there is a single building principle that extends each series. In the case of Mn there is a regular increment of MnS to extend the series, while the copper chalcogenide clusters increment by Cu_2S. A fundamental question about their structures is whether each member of the series is constructed by addition to the previous, or whether there are many structural rearrangements with growth. As the clusters grow there is increasing opportunity for structural isomerism, and a relatively flat energy surface is expected.

There are no data on geometrical structure. Therefore local density functional calculations of electronic structure have been undertaken, with energy gradient calculations allowing geometry optimisation. The density functional methodology embodied in the program DMol [6] is used. The exchange correlation terms are based on those of a uniform electron gas; double numerical basis functions are used, with d polarisation functions; core orbitals are frozen, and the calculations are spin restricted for even electron populations. We have tested DMol on a variety of known metal chalcogenide clusters, and find that correct cluster topologies are obtained, albeit with contracted geometry and bond distances shortened by ca 0.1Å with the local density method. In particular, the geometry calculated for $Cu_{29}Se_{15}(PH_3)_{12}$ is in very good agreement with that of the crystal structure of $Cu_{29}Se_{15}(PPr^i_3)_{12}$ [7]. For the purposes of establishing first the general picture of the structures of metal chalcogenide clusters, the DMol method is valuable and powerful. We discuss here the structural principles for manganese and for copper, the two extremes of the transition metals considered in this article.

POSTULATED STRUCTURES FOR $[Mn_XS_Y]^-$

The most stable form of solid MnS has the rock salt structure, and with approximately equal numbers of Mn and S atoms appearing in the gaseous clusters, it is reasonable to propose that they are fragments of the solid state structure. This hypothesis has frequently been made for comparable clusters of metal nitrides, oxides and halides [8]. For instance, the ion $[Mn_{13}S_{14}]^-$ could be postulated to have the face-centered cubic unit cell structure of Figure 3 (a), and $[Mn_{13}S_{13}O]^-$ could be the same. Optimisation of this structure for $[Mn_{13}S_{14}]^-$ by density functional methods reveals however that it is less stable than the modification shown in Figure 3 (b), in which the metal atoms have moved inwards to generate Mn-Mn bonds, and the S atoms outwards to cap all of the faces of the resulting Mn-centered Mn_{12} cuboctahedron: the relative binding energies are -2545 kcal mol^{-1} and -3051 kcal mol^{-1} respectively. All calculations for other cluster structures demonstrate this principle of maximisation of the M–M bonding, with the chalcogen atoms capping faces of metal polyhedra.

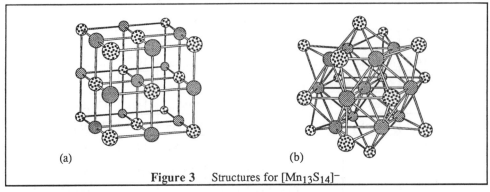

(a)   (b)

Figure 3   Structures for $[Mn_{13}S_{14}]^-$

For $[Mn_{12}S_{12}]^-$ two of the structural postulates are (1) a partially capped Mn_{12} icosahedron, with T_h symmetry, and (2) four Mn_3S_3 triangles stacked in abab mode. The principles in this latter structure, together with end capping by S or MnS, provide reasonable postulates for other members of the $[Mn_xS_x]^-$, $[Mn_xS_{x+1}]^-$ and $[Mn_xS_xO]^-$ series. Structures well known in iron sulfide cluster chemistry can account for $[Mn_4S_4]^-$ and $[Mn_6S_6]^-$, but we are puzzled by the apparent absence of $[Mn_5S_5]^-$ for which a favourable pentacapped square pyramid can be suggested.

POSTULATED STRUCTURES FOR $[Cu_XS_Y]^-$

Professor D. Fenske of the University of Karlsruhre in Germany has reported the syntheses and crystal structures of some amazing megaclusters with copper selenide cores and terminal phosphine ligands. Representative compounds are $Cu_{20}Se_{13}(PEt_3)_{12}$ [9], $Cu_{29}Se_{15}(PPr^i_3)_{12}$ [7], $Cu_{36}Se_{18}(PBu^t_3)_{12}$ [7], $Cu_{70}Se_{35}(PEt_3)_{22}$ [9] and $Cu_{146}Se_{73}(PPh_3)_{20}$ [10], and it is significant that all except the first have the ratio x = 2y or x = 2y–1. These terminated clusters, like the naked $[Cu_{2y-1}S_y]^-$ clusters we observe, are electron precise in the sense that the Cu is formally Cu(I). The Fenske clusters provide pointers to the structures of specific members of our series, and guide us in the formulation of pertinent structural principles.

Using the core framework of $Cu_{29}Se_{15}(PPr^i_3)_{12}$ as starting point, we have optimised the structure of $[Cu_{29}S_{15}]^-$, and find that the Cu atoms on the three-fold axis and at the ends of $Cu_{29}Se_{15}(PPr^i_3)_{12}$ move inside the cluster by one hexagonal layer, yielding the more compact structure shown in Figure 4: the solid symbols indicate the two Cu atoms that move inside.

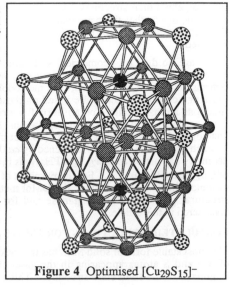

Figure 4 Optimised $[Cu_{29}S_{15}]^-$

For $[Cu_{25}S_{13}]^-$ we propose the optimised structure shown in Figure 5 (a). This is very similar to the structure of $[Cu_{29}S_{15}]^-$, and contains within it a Cu_{13} centered cuboctahedron. There is a straightforward sequence of structures between $Cu_{25}S_{13}$ and $Cu_{13}S_7$ (see Figure 6) which accounts for the ions observed. This sequence involves detachment of Cu_2S units from the base of the $Cu_{25}S_{13}$ structure, as shown in Figure 5 (b): first three Cu_2S units are successively removed to yield $Cu_{23}S_{12}$, $Cu_{21}S_{11}$ and $Cu_{19}S_{10}$, and then three more Cu_2S units using Cu_2 pairs in the layer above are removed to generate the structures of $Cu_{17}S_9$, $Cu_{25}S_8$ and $Cu_{13}S_7$. The Cu atoms of the groups removed are indicated as filled symbols in Figure 5 (b).

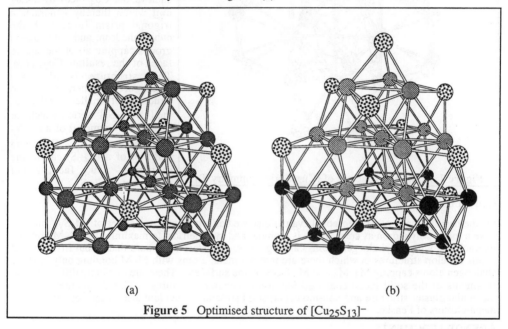

(a)   (b)

Figure 5 Optimised structure of $[Cu_{25}S_{13}]^-$

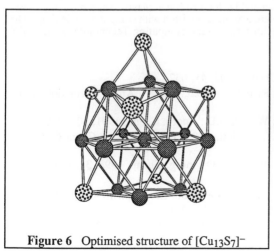

Figure 6 Optimised structure of $[Cu_{13}S_7]^-$

We have optimised three structural isomers for $[Cu_{13}S_7]^-$, namely the Cu-centered cuboctahedron with $(\mu_4-S)_6(\mu_3-S)$ capping (binding energy -1869 kcal mol^{-1}), the Cu-centered anti-cuboctahedron with $(\mu_4-S)_6(\mu_3-S)$ capping (binding energy -1823 kcal mol^{-1}), and the Cu-centered cuboctahedron with $(\mu_3-S)_7$ capping (binding energy -1801 kcal mol^{-1}). Figure 6 shows the most stable isomer, which is also part of the structure of $[Cu_{25}S_{13}]^-$ and the sequence from $[Cu_{25}S_{13}]^-$ to $[Cu_{13}S_7]^-$.

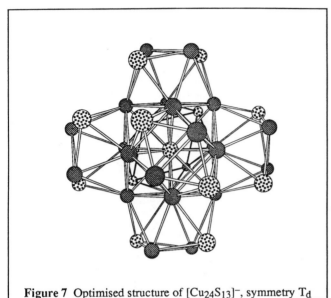

Figure 7 Optimised structure of $[Cu_{24}S_{13}]^-$, symmetry T_d

In Figure 7 we present our optimised structure for $[Cu_{24}S_{13}]^-$, a member of the $[Cu_{2x-2}S_x]^-$ series.

This structure, with T_d symmetry, can be conceived as an S-centered Cu_{12} cuboctahedron, capped on each of the Cu_4 faces by a Cu_2 unit which thereby generates a trigonal prism fused with the cuboctahedron, and with (μ_4–S) groups capping all of the square faces of the resulting Cu_{24} core. This structure is also the paradigm for generation of the structures of smaller members of the $[Cu_{2x-2}S_x]^-$ series, which are obtained by removal of a Cu_2S_2 protuberance of $[Cu_{24}S_{13}]^-$ and addition of a (μ_4–S) cap to the resulting Cu_4 face, a net decrement of Cu_2S.

CONCLUSIONS

On the basis of many other theoretical evaluations, we believe that the structures of $[M_xS_y]^-$ clusters have a metal polyhedron as core, which is globular and symmetrical to maximise the M–M bonding interactions, with the chalcogen atoms capping faces. Clusters with more than about fifteen metal atoms develop structures in which there are internal metal atoms with M–M bonding only, and the chalcogen atoms capping M_3, M_4 and M_5 faces on the surfaces. These are distinctly different from fragments of the solid metal chalcogenide non-molecular structures. This distinction between molecular cluster structure and non-molecular solid structure is evident also in the copper selenide mega-clusters of Fenske.

ACKNOWLEDGMENTS

This research is supported by the Australian Research Council. KF acknowledges award of a National Research Fellowship. Dr Garry Willett and Dr John El Nakat played major roles in the laser ablation mass spectrometry program, and Dr Marcia Scudder assisted with computer modelling. The generous allocation of computer resources by Australian Supercomputing Technology is gratefully acknowledged

REFERENCES

1) Dance, I.G.: Fisher, K.J: Prog. Inorg. Chem., 1993, 41, 637.

2) El Nakat, J.H.: Dance, I.G.: Fisher, K.J.: Rice, D.: Willett, G.D.: J. Amer. Chem. Soc., 1991, 113, 5141.

3) El Nakat, J.H.: Dance, I.G.: Fisher, K.J.: Willett, G.D.: Inorg. Chem., 1991, 30, 2958.

4) El Nakat, J.H.: Dance, I.G.: Fisher, K.J.: Willett, G.D.: J. Chem. Soc., Chem Commun., 1991, 746.

5) El Nakat, J.H.: Fisher, K.J.: Dance, I.G.: Willett, G.D.: Inorg. Chem., 1993, 32, 1931.

6) DMol v 2.2, Biosym Technologies, Inc.

7) Fenske, D.; Krautscheid, H.; Balter, S.: Angew. Chem., Int. Ed. Engl., 1990, 29, 796.

8) Chen, Z.Y.; Castleman, A.W.; J. Chem. Phys., 1993, 98, 231.

9) Fenske, D.; Krautscheid, H.: Angew. Chem., Int. Ed. Engl., 1990, 29, 1452.

10) Fenske, D; private communication 1993.

Materials Science Forum Vols. 152 - 153 (1994) pp. 143-148
© 1994 Trans Tech Publications, Switzerland

SOME CHALCOGENIDES SYNTHESES VIA SOFT CHEMISTRY

G. Ouvrard, E. Prouzet, R. Brec and J. Rouxel

Institut des Matériaux de Nantes, CNRS UMR 110, 2 rue de la Houssinière,
F-44072 Nantes Cédex 03, France

Keywords: Chalcogenides, Deintercalation, Ion Condensation, XAS

ABSTRACT

Most of the examples of chalcogenides syntheses via soft chemistry are related to deintercalation or ion condensation. The first process is illustrated by lithium deintercalation from the lamellar Li_2FeS_2 compound and the investigation of the Li_xFeS_2 ($0 \leq x \leq 2$). For $1 \leq x \leq 2$, lithium is removed from tetrahedra and iron is oxidized from its pristine oxidation state II to III. For lower lithium content, lithium is extracted from octahedra and the redox process involves sulfur oxidation by sulfur pairing. For $1 \leq x \leq 1.5$, a structural rearrangement occurs that stabilizes the partially deintercalated slab through a new distribution of iron atoms in tetrahedral sites. The ion condensation process is illustrated by the synthesis of the amorphous nickel thiophosphate a-$NiPS_3$. An XAS study at the nickel K edge allows to define precisely the condensation process and to confirm the formation of very small particles, in agreement with XRD, HREM and magnetic susceptibility measurements.

INTRODUCTION

Soft chemistry has been used to prepare chalcogenides either by deintercalation reaction [1] or ion condensation [2] in solution. In this paper we will use a few examples that have been studied in the depth (i) to emphasize the facts that, due to anion-cation redox competition or preferential sites effects, a given structure is not necessarily a rigid framework during deintercalation/intercalation processes (ii) to understand condensation processes and their structural aspects. The deintercalation reaction will be illustrated by the Li_xFeS_2 system obtained by removing lithium atoms at room temperature from the lamellar compound Li_2FeS_2. In the second part of the paper, the synthesis of the amorphous nickel thiophosphate a-$NiPS_3$ will be described as an example of condensation reaction.

I. THE Li_xFeS_2 SYSTEM.

Li_2FeS_2 is a high temperature (800°C) synthesized ternary sulfide whose structure [3] is composed of hexagonal close packing layers of sulfur atoms with iron in tetrahedral sites in every other slab (Figure 1). Lithium atoms are found in equal proportions in octahedra between the [S-Fe-S] layers and in tetrahedra inside these layers. This phase can be considered as the intercalated form of an hypothetical lamellar iron disulfide. Because of its well known mobility, lithium can be removed from the host lattice, either chemically by stirring the Li_2FeS_2 powder suspension in an iodine/acetonitrile solution or electrochemically by recharging a lithium battery whose positive is made of the ternary sulfide. It has been shown [4-6] that lithium can be completely extracted leading to a poorly crystallized new iron disulfide readily differenciated from the known high temperature modifications, pyrite or marcassite. EXAFS [7] has shown that, in this new phase, iron keeps the tetrahedral coordination found in the pristine material. Infrared [6] and Mössbauer [3] studies showed that initially lithium is removed from tetrahedral sites with a corresponding oxidation of iron from the oxidation state II to III. Further deintercalation follows a biphased process with the occurrence of the new disulfide where iron keeps its tetrahedral coordination and its oxidation state III, the redox process involving sulfur oxidation by sulfur pairs formation. The resulting phase can be written as $Fe^{III}S^{-II}(S_2)^{-II}_{1/2}$. Concerning the structural modifications induced by intercalation or deintercalation, a break observed at x=1.5 in the equilibrium recharge or discharge curve for a Li_xFeS_2/Li battery [5] can be compared with the drastic modification of the Mössbauer spectrum around this composition.

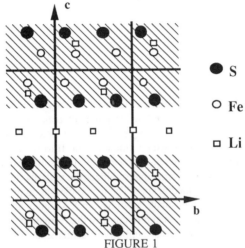

FIGURE 1

Projection along the *a* axis of the Li_2FeS_2 structure. The shaded area underlines the (2D) character of the FeS_2 framework.

Table I gathers the Mössbauer characteristics of Li_xFeS_2 phases with respect to x. For the Li_2FeS_2 compound, a doublet is recorded corresponding to two structurally inequivalent tetrahedrally coordinated iron atoms. The isomer shift (IS) is very close to 0.50 mm s^{-1} and consistent with previously recorded values [8] for high spin Fe^{2+} in tetrahedral sulfur coordination. The difference in quadrupole splitting (QS) values may reflect a more or less distorted sulfur environment. For $Li_{1.78}FeS_2$ a decrease of the IS and QS is recorded in agreement with a change in iron oxidation state. Site filling is noticeably altered and reveals an unexpected easy site change of iron. The change of both characteristics indicates a single-phased process consistent with the OCV curve where a continuous potential variation is observed. Between $Li_{1.78}FeS_2$ and $Li_{1.31}FeS_2$ compositions, a drastic change takes place. The two original iron sites give rise to two new sites labelled Fe1-2 and Fe3, whose characteristics remain roughly the same during further lithium deintercalation, their relative proportions being only changed at the composition Li_1FeS_2. The new Fe3 isomer shift, lower than the previous one, is the undisputed signature of tetrahedrally sulfur coordinated Fe^{III}. The Fe1-2 characteristics are not so different from those observed for Fe1 and Fe2. This could indicate that the oxidation process is biphased between the $Li_{1.5}FeS_2$ and Li_1FeS_2 compositions. Nevertheless we know that for x=1, all iron atoms have been oxidized to the oxidation state III. More over, the OCV curve indicates that the process is always single-phased (a biphased process corresponds to a voltage plateau). We must then consider that iron atoms are continuously oxidized during the lithium deintercalation process and that the changes in the Mössbauer characteristics reflect a structural modification rather than an electrochemical process.

Table I. Mössbauer isomer shift (IS), quadrupole splitting (QS) and relative intensities of iron sites in the Li_xFeS_2 compounds ($0.21 \leq x \leq 2$).

x in Li_xFeS_2	IS(mm s^{-1})			QS(mm s^{-1})			Relative Intensities		
	1	2	3	1	2	3	1	2	3
2.00	0.50	0.49		0.90	1.61		77	23	
1.78	0.45	0.38		0.84	1.39		64	36	
1.31		0.35	0.18	1.24		0.37	73		27
1.10		0.36	0.20	1.18		0.35	55		45
0.85		0.32	0.16	0.93		0.49	60		40
0.57		0.32	0.23	0.94		0.53	65		35
0.21		0.32	0.26	0.85		0.56	50		50

In a first publication [3], it was proposed that the structural modification could correspond to a partial iron shift from tetrahedra to octahedra. Iron K edge EXAFS spectroscopy [7] indicates that the iron coordination is not markedly changed, either in the coordination number or in the iron-sulfur distances. We must then consider an iron shift to tetrahedral sites. Infrared spectroscopy experiments indicate that the tetrahedral lithium in the slab are removed first, octahedral lithium in the so-called van der Waals gap in a fully deintercalated phase being removed only for x≤1. The tetrahedrally coordinated lithium atoms probably stabilize the slab. When 50% of these tetrahedral lithium are deintercalated, iron atoms must move to stabilize the new lacunar slab structure. These moved iron atoms correspond to the new Fe3 Mössbauer peak, whose relative intensity increases markedly from x=1.31 to x=1.1. The structural rearrangement occurs at a threshold value, to which corresponds a limit for the stability of the deintercalated phase and the process is perfectly reversible. This probably originates in the fact that the atomic displacement is made inside the slab, without noticeable modification in the general structural framework or the iron-sulfur bonds. These changes explain why a regular variation can be observed in the X-ray diffraction patterns in terms of parameter variations and peak intensities.

Many examples are now known in which intercalation/deintercalation reactions induce important transition metal displacements. Such displacements can be related either to the stabilization of the whole structure in the so-called Crystal Energy Minimization (CEM) process or to the stabilization of the new electronic configuration of transition metal itself (Ligand Field Stabilization (LFS)) [9].

II. SYNTHESIS OF THE AMORPHOUS THIOPHOSPHATE a-NiPS3.

The lamellar thiophosphates MPS3 (M=V, Mn, Fe, Co, Ni, Zn, Cd) have been widely studied for their physical properties and their ability to intercalate various species [10]. They are classically prepared by high temperature reactions between the elements. It has been shown that a soft chemistry procedure allows to obtain them as amorphous phases a-MPS3 [11], by using a metathesis reaction between M^{2+} cations and $(P_2S_6)^{4-}$ anions. The reaction is performed in aqueous media, at room temperature. In the case of NiPS$_3$, the cations are brought in as nitrate and the anions from a solution of Li$_4$P$_2$S$_6$, a very ionic salt which readily dissolves in water to yield Li$^+$ and (P$_2$S$_6$)$^{4-}$. The thiophosphate is formed instantaneously as a sol or a gel, depending on the concentrations of the starting solutions. After several washings then dryings under mild conditions (100°C under vacuum), a very fine powder of a-NiPS$_3$ is obtained.

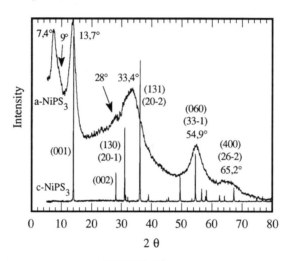

FIGURE 2
X-ray diffraction patterns of a-NiPS$_3$ and c-NiPS$_3$.

Long exposure time X-ray diffraction experiments, performed on an INEL CPS 120 diffractometer, allow to observe some broad peaks situated at angles corresponding to the main c-NiPS$_3$ lines (Figure 2). A thorough structural study of this diagram has shown that the phase contains small crystallized domains of size approximately 25 x 25 x 40 Å^3 formed by a core phase having the c-NiPS$_3$ structure and limited by sheets with stacking faults [12].

Transmission electron microscopy experiments confirm these results by showing, on bright field and dark field images, that the phase is made of small crystallites with a mean diameter of 30±10 Å, embedded into an amorphous matrix [13]. The simultaneous occurrence in a-NiPS$_3$ of such small crystallites and an amorphous phase allows to explain the magnetic susceptibility behaviour versus temperature (Figure 3).

It is known that the crystallized NiPS$_3$ antiferrromagnetically orders below 155K [10]. This temperature corresponds to a slope change in the χ^{-1} versus T variation for a-NiPS$_3$. The high temperature branch linear variation corresponds to a Curie-Weiss paramagnetism. From the slope of the high temperature straight line, a Curie constant of 12.6 x 10^{-6} m^3 K mol^{-1} is calculated in perfect agreement with the spin-only value calculated for Ni^{II} ions. For the low temperature part, the Curie constant drops to 7.2 x 10^{-6} m^3 K mol^{-1}, half that of the high temperature one. This behaviour can be explained by superimposed crystalline and amorphous phases magnetisms [13]. At high temperature all the spins contribute to the Curie-Weiss behaviour with a calculated number of spin bearers of 6.021 x 10^{23} per mole. Below 155K, the small crystallites undergo an antiferromagnetic ordering as in c-NiPS$_3$ and mainly the amorphous matrix contributes to the low temperature Curie-Weiss law.

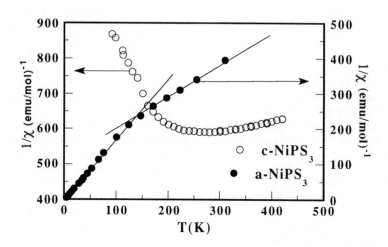

FIGURE 3
Reciprocal magnetic susceptibilities versus temperature for a-NiPS$_3$.and c-NiPS$_3$

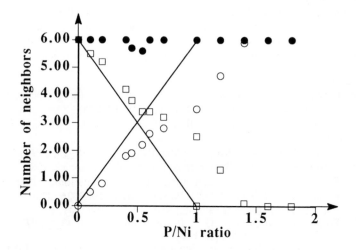

FIGURE 4
Evolution of the fitted numbers of oxygen (squares) and sulfur (open circles) atoms surrounding
nickel versus P/Ni ratio. Black points represent the sum of sulfur and oxygen atoms and the straight
lines the expected variations for the gradual formation of a pure sulfur environment.

In this temperature range the number of observed spin bearers is 3.4 x 10^{23} per mole. This
corresponds to the loss of 40% of the magnetic moments, which would be the fraction of small
crystallized particles formed during the synthesis.

In order to better understand the process of formation of a-NiPS$_3$, an XAS study has been undertaken, at the nickel K edge, on solutions made of mixtures of nickel nitrate and lithium thiophosphate aqueous solutions in various proportions. Various volumes of the thiophosphate solution have been added to the same quantity of the nitrate solution, in order to vary the P/Ni ratio from 0 to 1.8. For P/Ni=0, the nickel is coordinated only by oxygen atoms and it can be supposed that its environment is purely sulfur for P/Ni=1 corresponding to the complete formation of NiPS$_3$ and the end of the reaction. In effect the radial distribution fonction of nickel extracted from EXAFS spectra is gradually modified from a pure oxygen to a pure sulfur, going through mixed environment but a noticeable proportion of oxygen atoms is found for P/Ni=1. Figure 4 represents the fitted numbers of oxygen and sulfur atoms surrounding nickel, versus P/Ni.

The variation of this surrounding is linear, sulfur replacing gradually oxygen, the sum of sulfur and oxygen atom being always very close to six. Nevertheless, as compared to a variation corresponding to a pure sulfur environment for P/Ni=1 as in c-NiPS$_3$, an excess of oxygen and a lack of sulfur is systematically observed. This could correspond either to a mixed environment for nickel atoms with intermediate formation of thiophosphate or to a particle size effect. In this last case either nickel octahedra are peripheral, hence a mixed coordination for nickel, sulfur inside the particle and oxygen outside, or P$_2$S$_6$ octahedra are peripheral, corresponding to a P/Ni ratio higher than 1. EXAFS is not able to distinguish between a mixed nickel environment and a mixture of pure oxygen and pure sulfur surroundings. The edge shape can be perfectly interpreted as a linear combination of these two last surroundings, with the oxygen and sulfur proportions calculated from EXAFS experiments [13]. This fact proves undoubtedly that the condensation is performed in such a way that P$_2$S$_6$ groups are external, nickel atoms inside the particles being surrounded only by sulfur. The fact that a pure sulfur environment is observed in EXAFS only for P/Ni=1.4 is due to the small size of the particles. From a condensation model, its diameter has been estimated to 30 Å, in very good agreement with the particle size determined by X-ray diffraction and HREM experiments on powders. Such a size corresponds to a limit of stability of the entities in water solution. We may then consider that small one slab particles are formed in solution or gel and the ordered or disordered stacking condensation of these species upon drying led to the crystalline or amorphous part in the powder.

REFERENCES

1) Murphy, D.W., Cros, C., Di Salvo, F.J. and Waszczak, J.V.: Inorg. Chem., 1977, 16, 3027
2) Chianelli, R.R. and Dines, M.B.: Inorg. Chem., 1978, 17, 2758-
3) Blandeau, L., Ouvrard, G., Calage, Y., Brec, R. and Rouxel, J.: J. Phys. C: Solid State Physics, 1987, 20, 4271
4) Brec, R., Dugast, A. and Le Mehaute, A.: Mat. Res. Bull., 1980, 15, 619
5) Dugast, A., Brec, R., Ouvrard, G. and Rouxel, J.: Solid State Ionics, 1981, 5, 375
6) Gard, P., Sourisseau, C., Ouvrard, G. and Brec, R.: Solid State Ionics, 1986, 20, 231
7) Brec, R., Prouzet, E. and Ouvrard, G.: J. Power Sources, 1989, 26, 325
8) Goodenough, J.B. and Fatseas, G.: J. Solid State Chem., 1982, 41, 1
9) Brec, R., Prouzet, E. and Ouvrard, G.: J. Power Sources, 1993, 43-44, 277
10) Brec, R.: Solid State Ionics, 1986, 22, 3
11) Prouzet, E., Ouvrard, G., Brec, R. and Seguineau, P.: Solid State Ionics, 1988, 31, 79
12) Fragnaud, P., Prouzet, E. and Brec, R.: J. Mater. Res., 1992, 7, 1839
13) Fragnaud, P., Prouzet, E., Ouvrard, G., Mansot, J.L., Payen, C., Brec, R. and Dexpert, H.: J. Non Cryst. Solids, 1993, 160, 1

Materials Science Forum Vols. 152 - 153 (1994) pp. 149-154
© *1994 Trans Tech Publications, Switzerland*

NANOPHASE CERAMICS BY THE SOL-GEL PROCESS

C. Guizard, C. Mouchet, J.C. Achddou, S. Durand, J. Rouvière and L. Cot

Laboratoire des Matériaux et des Procédés Membranaires CNRS EP 50, ENSC, 8, rue de l'Ecole Normale, F-34053 Montpellier Cédex, France

Keywords: Nanophase Ceramics, Sol-Gel Process, Zirconia, Titania

The control of sol-to-gel transition with transition metal alkoxides allows to prepare nanophase ceramic layers exhibiting individual particle size of less than 10 nm with a residual microporosity. Two ways are described, one in inverse micelle reaction media the other with acacH modified alkoxides. Basic principles of these two methods are discussed with titanium and zirconium alkoxides as molecular precursors. Titania and zirconia coatings on various substrates have been obtained with the aim of different applications. Examples of optical materials, coloured protective coatings or microporous membranes are given.

INTRODUCTION

The sol-gel process has been proved to be a very effective and a very versatile method to prepare glasses and ceramics from molecular precursors (metal salts or metal organic compounds) [1]. Concerning metal organic precursors, by far silicon alkoxides have been the most investigated ones. They can be easily handled in ambiant atmosphere and they exhibit slow hydrolysis and condensation rates at room temperature allowing to keep under control chemical reactions involved in sol-to-gel transition. However glasses and ceramics are not limited to silica derived materials and other molecular precursors such as transition metal alkoxides are also of interest in ceramic preparation. The main limitation in using these precursors is that they are highly sensitive to hydrolysis and cannot be simply handled as silica alkoxides. Recent advances in sol-gel chemistry have shown new methods for the control of hydrolysis and condensation of metal alkoxides. These methods have been largely described in literature and can be classified in two main topics. The first one deals with the chemical modification of metal alkoxides using hydroxylated ligand additives such as polyols [2], organic acids [3,4], β-diketones [5] and allied derivatives. The second one consists in a novel process in which colloidal particles and gels are formed in amphiphilic media by the control of hydrolysis and condensation reactions [6,7].

In other respects a new field of interest in advanced materials has arised recently with nanophase ceramics. These materials exhibit a fine grained structure, the size of the individual atomic clusters or particles being of the order of one to few nanometers. Because the individual grains are extremely small, on the order of 50% or more of the atoms are found in or very near the interface between grains [8]. Another important feature is the residual micropores existing in these materials. Consequently quite specific properties (optical, electrical, magnetic, mechanical and catalytic) must be expected for these nanophase ceramics.

The aim of this paper is to stress the sol-gel process as a very convenient method to prepare nanophase ceramics starting from metal alkoxides precursors. Two transition metal alkoxides (Ti and Zr) and derived ceramic coatings have been investigated more systematically in our laboratory. The two chemical approachs above mentionned have been used and recent results are discussed hereafter with special emphazis on the structural characterization of these materials.

OXIDE GEL AND CERAMIC FORMATION IN INVERSE MICELLE SYSTEMS

Experimental procedure and characterization methods applied to gel formation

Titania and zirconia gels have been prepared by controlling the hydrolysis and condensation reactions in a ternary sytem (water - Triton X surfactants - alkane) used as reaction medium. Inverse micelles act as many water reservoirs providing water molecules in a controlled manner during alkoxide hydrolysis and cluster formation. From a kinetic point of view, three parameters have been defined (n, h and m) which directly influence sol-to-gel transition, in particular cluster growing and cluster size distribution as well as gelation time t_g [9]:

(i) h is the hydrolysis level represented by the water/alkoxide mole ratio,

(ii) m is defined as the molality of the alkoxide in the final mixture expressed in mol/kg,

(iii) n is the hydration level which corresponds to the water/surfactant mole ratio.

$$t_g = A\, e^{kn} \tag{1}$$

For $Ti(O^tBu)_4$ and $Ti(O^iPr)_4$ precursors, the variation of t_g as a funtion of n, has been laid down to be an exponential law (equation 1) in which the exponent k only depends on the hydrophilic chain length of the Triton X used to prepare the inverse micelle system. The local and long range structures of these titania gels have been investigated by small angle X-ray (SAXS) and neutron diffraction (SANS). A fractal dimension of 2.08 for titania gels obtained with $m = 1$, $n = 1$ and $h = 2$, suggests a reaction limited cluster-cluster aggregation during gel formation [10]. We have shown recently that this power law, as a function of n, is not valid with low values of m, typically $m < 0.5$. Very tenuous titania gels with low volume fraction of oxide ($m = 0.1$) have been synthesized with a fractal dimension of 1.4. An aggregation model which can be related to this low fractal dimension is the tip-to-tip model which considers cluster polarizability [11]. The structure of a zirconia gel obtained with $m = 1$ and $n = 2$ has also been investigated by SANS. In this case the mass fractal dimension is 2.3, suggesting that the aggregation mechanism is diffusion limited [12].

The mesoporous texture of titania gels has been characterized using thermoporometry. This method allows to measure pore size distribution in wet gels. The influence of n and m on the porous texture evolution of fresh and aged gels has been investigated showing that very homogeneous gels can be obtained depending on preparation conditions [13]. The water/surfactant mole ratio n has been evidenced as the main parameter allowing to control gelation time and gel nanoscale homogeneity; in other words sol-gel processing of nanophase ceramics from metal alkoxide precursors can be easily managed using inverse micelle systems as reaction media. In a recent work [14], this concept has been extended to lamellar phase systems proving that anisotropic oxide gels can be obtained thank to the anisotopic surfactant bilayer structure. This is of peculiar interest for the preparation of oriented ceramic coatings exhibiting an anisotropic structure.

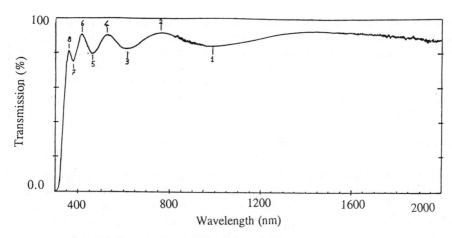

Figure 1. Transmission spectrum of a coated titania multilayer system (5 successive layers).

Application to titania coatings with optical properties
The versatility of this process has been exploited to prepare pure or doped titania layers as well as oxide patterns on glass substrates. According to rheological properties of the sols prepared from inverse micelle systems [15], spin or dip coating technique have been used to prepare titania multilayer coatings doped with Eu^{3+} [16]. Transparent crystallized coatings have been obtained at 500°C (Figure 1). As shown in figure 2, they exhibit fluorescent properties when irradiated at 393 nm. Another possible application is transparent oxide patterning which can be carried out on fresh coated layers with no deformation of the pattern during heat treatment [17].

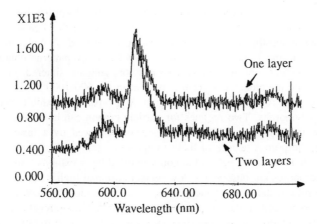

Figure 2. Emission spectrum of a coated titania multilayer system doped with Eu^{3+}.

NANOPARTICLE SOLS AND DERIVED CERAMIC FROM MODIFIED ALKOXIDES

Experimental procedure and characterization methods
Zirconia and titania sols were prepared by reacting corresponding alkoxides with acetylacetone (acacH). More emphasis has been put on zirconium isopropoxyde with the aim of producing supported zirconia layers exhibiting nanophase ceramic properties. Alkoxide groups on zirconium atoms are substituted by acacH groups according to equation 2, where *r* is the

acacH/alkoxide mole ratio. Different r values have been used to prepare zirconia sols by adding under stirring acetylacetone to the zirconium alkoxide solution (10% weight in isopropanol). A clear sol is obtained which can be handled under air atmosphere. A binder (polyethylene glycol 3% weight) can be added to the sol to adjust rheological properties before coating. Non-supported layers were prepared using the same conditions by pouring the sol in a plate shaped Teflon vessel, let the solvent to evaporate at 80°C and put the dried film in a crucible for sintering. Different sintering temperature T from 400°C to 1000°C have been used.

$$Zr(O^iPr)_4 + r\,acacH \rightarrow Zr(O^iPr)_{4-r}(acac)_r + r\,ROH \qquad (2)$$

Three characterization methods have been used for structural investigation. X-ray diffraction provides information on the crystalline structure. Moreover, it is possible to determine the mean dimension L of the individual crystallites forming the ceramic using equation 3 proposed by Scherrer. The pure X-ray diffraction broadening β can be related to the experimentally observed breadth B by using a correcting factor b to account for instrumental broadening. This factor can be measured by using a standard specimen with crystallite size in excess of 300 nm. The shape factor K has been set equal to 1 in this case. Following this way, T and r dependences of the mean particle size have been measured. Size and aggregated or non aggregated state of individual grains have been checked using transmission electron microscopy [18]. Residual micro and mesoporosity have been measured using N_2 adsorption/desorption. Specific suface area has been calculated according to the BET equation and micropore size has been determined using the Horwath-Kawazoe method.

$$L = K\,\lambda \,/\, \beta \cos \theta \qquad (3)$$

Discussion on zirconia ceramic microstructure and related applications

The main advantage of using acacH as chelating agent is that this ligand acts as a blocking functionnal group when substoichiometric hydrolysis ratios are used. A mole ratio r greater than 1 prevents precipitation and leads to nanoparticle or gels. X-ray patterns obtained on sintered samples show the existence of tetragonal zirconia for samples sintered in the 400-600°C temperature range while the monoclinic form is present in the 600-1000°C range. The transition occurs between 600 and 700°C and in this region X-ray patterns exhibit characteristic lines related to the two polymorphs. Two regimes for particle growth can be observed for L versus T in figure 3, a slow increase in the 400 - 600°C region followed by a faster increase beyond 600°C. Concerning the porous texture of these samples one can see in figure 4 the evidence of two domains. At high T and low r values the non-supported layer exhibits a mesoporous texture, on the contrary micropores are observed at low temperature and high r value. These results are consistent with the role of the ligand which limits the condensation between alkoxide molecules allowing nanoparticles to form at the sol stage and to pack in a non-aggregated manner during the coating process. Transmission microscopy images showing the microstructure of these samples have been reported elsewhere [18]. Another remarkable result is the sudden increase in pore diameter which occurs at 600°C. We can see that this phenomenon corresponds to the allotropic transformation from tetragonal to monoclinic zirconia which occurs at this temperature with a significant increase in grain growth rate. When calculating the ratio between particle size and pore diameter, this ratio varies between 1 and 2 in the mesoporous range and between 2 and 3 in the microporous range. These values show that a more compact arrangement of non-aggregated particles is obtained in microporous sample. Beyond 600°C, the mesoporous structure results from sintering of partially aggregated particles.

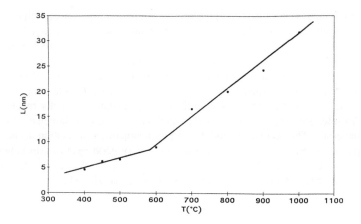

Figure 3. Evolution of zirconia particle growth versus sintering temperature.

From these results, it appears that tetragonal zirconia exhibits basic properties of nanophase ceramics with particle size of less than 10 nm and a residual microporosity. Two kinds of application have been investigated for this material. The first one consists in thin coloured zirconia films coated on stainless steel substrates [19]. The protective properties of these films have been studied using an electrochemical method to determine the corrosion resistance of the coated samples. The almost dense structure of zirconia layers obtained with modified alkoxide showed better corrosion resistance than samples obtained from conventional sols prepared with pure alkoxides. Also a good plasticity has been evidenced for these coatings under strain deformation. The second application deals with membrane preparation. Microporous ceramic membranes are expected for new applications in nanofiltration of liquids, gas separation or membrane reactors. In that sort of membranes, the presence of mesopores has to be avoided in order to preserve high selectivity. The properties of tetragonal zirconia investigated here fit perfectly with the expected properties of these membranes [18].

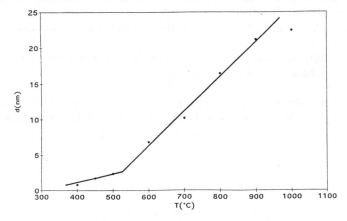

Figure 4. Evolution of residual porosity in zirconia ceramics versus sintering temperature.

CONCLUSION

The two methods discussed here allow to prepare homogeneous titania and zirconia ceramics at the nanometer scale. Basic properties of nanophase ceramics have been evidenced for these materials. A common feature of these methods is their ability to produce stable sols under ambient atmosphere using alkoxide precursors generally very sensitive to hydrolysis and precipitation. However quite different mechanisms are involved in gel and ceramic formation. The first one is based on interpenetrated oxide clusters which form the ceramic material under drying and firing. In this case particle size in the ceramic cannot be directly related to cluster size in the sol. An interesting development of this method is the use of lamellar phases to produce anisotropic materials. The second method consists in nanoparticle formation at the sol stage from which nanophase ceramics have been obtained with a very good relationship between individual particles in the sol and individual grains in the sintered ceramic. Work is in progress to confirm on cubic stabilized zirconia the results obtained with tetragonal zirconia.

References

1.   Brinker C. J. and Scherer G.W., Sol-Gel Science, Academic Press, New York, 1990.
2.   Guizard C., Cygankiewicz N., Larbot A. and Cot L., J. Non-Cryst. Solids, 1986, 82, 86.
3.   Doeuff S., Henry M., Sanchez C. and Livage J., J. Non-Cryst. Solids, 1987, 89, 206.
4.   Laaziz I., Larbot A., Julbe A., Guizard C. and Cot L., J. Non-Cryst. Solids, 1992, 98, 393.
5.   Sanchez C. and Livage J., New J. Chem, 1990, 14, 513.
6.   Guizard C., Stitou M., Larbot A., Cot L. and Rouviere J., in *Better Ceramics through Chemistry III* (Brinker C.J. Clark D.E. and Ulrich D.R. Edrs.), MRS Pittsburg, 1988, 115.
7.   Osseo-Assare K. and Arriagada F.J., Colloids and Surfaces, 1990, 50, 321.
8.   Siegle R.W., MRS Bull., October 1990, 60.
9.   Guizard C., Larbot A., Cot L., Perez S. and Rouvière J., J. Chim. Phys., 1990, 87, 1901.
10.   Marignan J., Guizard C. and Larbot A., Europhys. Lett., 1989, 8(7), 691.
11.   Barthez J.M., Ayral A., Hovnanian N., Marignan J., Guizard C. and Cot L., to be published in J. of Sol-Gel Science and Technology (*Special issue on the 7th Internat. Workshop on Glasses and Ceramic from Gels*).
12.   Auvray L., Ayral A., Cot. L., Dabadie T., Guizard C. and Ramsay J.,to be published in J. of Sol-Gel Science and Technology (*Special issue on the 7th Internat. Workshop on Glasses and Ceramic from Gels*).
13.   Quinson J.F., Chatelut M., Guizard C., Larbot A. and Cot L., J. Non-Cryst. Solids, 1990, 121, 72.
14.   Dabadie T., Ayral A., Guizard C. and Cot L., to be published in these proceedings.
15.   Guizard C., Achddou J.C., Larbot A. and Cot L., J. Non-Cryst. Solids, 1992, 147&148, 681.
16.   Guizard C., Achddou J.C., Larbot A., Cot L., Le Flem G., Parent C. and Lurin C., in *Sol-Gel Optics* (Mackenzie J.D. and Ulrich D.R. Eds), SPIE, 1990, 1328, 208.
17.   Achddou J.C., PhD thesis, Université des Sciences et Techniques du Languedoc, Montpellier, 1990.
18.   Julbe A. Guizard C., Larbot A., Cot L. and Giroir-Fendler A., J. of Memb. Science, 1993, 77, 137.
19.   Chino C., Charbonnier M., de Becdelievre A.M., Guizard C., Pauthe M. and Quinson J.F. in *Prog. in Research and Development of Processes and Products from Sols and Gels,* North Holland, Elsevier Sci. Pub. B.V. (Vilminot S., Nass R. & Schmidt H. Eds)1992, 327.

Materials Science Forum Vols. 152 - 153 (1994) pp. 155-162
© 1994 Trans Tech Publications, Switzerland

PHYSICO-CHEMICAL PROPERTIES OF SOLID ONE-DIMENSIONAL INCLUSION COMPOUNDS

K.D.M. Harris

Department of Chemistry, University College London, London WC1H OAJ, England

Keywords: Solid Inclusion Compounds, Incommensurate Solids, Molecular Dynamics

ABSTRACT

In recent years, much attention has been devoted to the study of those solid inclusion compounds in which the "host" solid contains uni-directional tunnels; materials in this category include certain zeolites as well as several crystalline organic inclusion compounds such as those containing urea or thiourea as the host. In this paper we focus upon urea and thiourea inclusion compounds containing a wide range of different types of organic guest molecule. Aspects of the preparation, structure and dynamics of these inclusion compounds are highlighted by discussing selected results obtained from solid state NMR spectroscopy, X-ray diffraction, incoherent quasielastic neutron scattering, computer simulation and mathematical modelling. Aspects of the commensurate *versus* incommensurate structural relationship between the host and guest substructures within these inclusion compounds are discussed in detail.

INTRODUCTION

There is currently considerable interest in the chemistry and physics of solid inclusion compounds as a consequence of the wide range of important physico-chemical properties associated with them. In part, scientific interest in these solids is motivated by the desire to compare the structural, dynamic, and chemical properties of organic "guest" molecules embedded within different crystalline environments, and to investigate how the properties of the guest molecule may be influenced by the structural characteristics of its environment. In this paper, we focus upon those solid inclusion compounds in which the "host" structure contains uni-directional tunnels, with particular attention devoted to the urea and thiourea inclusion compounds [1-3].

In the urea inclusion compounds [4,5], the urea molecules form an extensively hydrogen-bonded arrangement (figure 1) that contains linear, parallel tunnels with an effective "diameter" ranging between *ca.* 5.1 Å and 5.9 Å. The guest molecules are densely packed along these tunnels. Urea crystallizes in this tunnel-containing host structure only in the presence of appropriate guest

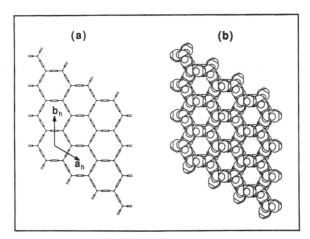

Figure 1   Two representations (showing nine complete tunnels) of the host substructure in
conventional urea inclusion compounds, viewed along the tunnel axis (c_h) [note: $|a_h|$ =
$|b_h| \approx 8.23$ Å; $|c_h| \approx 11.02$ Å]. The atomic radii are zero in (a) whereas conventional van
der Waals radii are used in (b). Note that in the real inclusion compounds, the tunnels of
this host structure are filled with a dense packing of guest molecules (not shown).

molecules, and the guest therefore serves as an essential template for construction of the tunnel
structure. Structural compatibility between host and guest components is a fundamental requirement
for the formation of most inclusion systems, and, as a consequence, urea will only form inclusion
compounds with guest molecules that are based on a sufficiently long *n*-alkane chain, with a further
requirement that the degree of substitution of this chain must be small. Appropriate guest molecules
include *n*-alkanes and derivatives such as α,ω-dihalogenoalkanes, diacyl peroxides, carboxylic acids
and carboxylic acid anhydrides. The urea tunnel structure is unstable [6-8] if the guest molecules are
removed from the inclusion compound (the urea then recrystallizes in its "pure" crystalline phase,
which does *not* contain empty tunnels). Thus, appropriate guest molecules must be present as an
essential template for the formation of the host framework *and* as an essential buttress in maintaining
the existence of this host framework. The urea host tunnel is a chiral structure, and the degree to
which the properties of the guest molecules are affected by this chiral environment is an issue with
important chemical implications regarding the prospects for chiral recognition and asymmetric
synthesis.

Thiourea also crystallizes in a tunnel structure in the presence of appropriate guest molecules, with
the tunnels having a larger cross-sectional area than those in urea inclusion compounds. As a
consequence, the urea and thiourea host structures preferentially incorporate different types of guest
molecule. The thiourea tunnel, for example, can include guest molecules such as cyclohexane and
some of its derivatives, ferrocene and other organometallics, and certain compounds containing a
benzene ring; such guest molecules do not generally form inclusion compounds with urea.

Detailed discussions of the molecular features in the guest molecules that form inclusion compounds
with urea and thiourea have been elucidated previously [9-12].

PREPARATION OF UREA INCLUSION COMPOUNDS
Urea inclusion compounds are generally prepared by spontaneous co-crystallization of urea and a suitable guest species from solution. The possibility that solvent molecules can be co-included together with the intended guest molecule is an important experimental problem that can be overcome either by using a very small solvent molecule (such as methanol) that will not form an energetically favourable inclusion compound (in competition with the intended guest molecule) or by using a solvent (such as t-amyl alcohol) that is too bulky to fit within the urea tunnel structure. An illustrative example of the importance of selecting an appropriate solvent has been discussed [13] in connection with the preparation of urea inclusion compounds containing carboxylic acid anhydride guests.

ESSENTIAL STRUCTURAL PROPERTIES OF UREA INCLUSION COMPOUNDS
In the conventional urea inclusion compounds, the host substructure has space group $P6_122$ at room temperature [4]. The guest molecules are densely packed within the tunnels, and generally exhibit sufficient positional ordering to allow an average three-dimensional lattice to be defined, and measured, despite substantial dynamic disorder of the guest molecules at room temperature. The periodicity (denoted c_h) of the host substructure along the tunnel axis and the periodicity (denoted c_g) of the guest substructure along the tunnel axis are usually incommensurate [5] – in classical terms, this means that it is *not* possible to find sufficiently small integers p and q for which $pc_g \approx qc_h$ (a detailed discussion of commensurate and incommensurate behaviour in one-dimensional inclusion compounds, typified by the urea inclusion compounds, is given in ref. 14). Although the host and guest substructures are chemically distinguishable from each other, and possess different structural periodicities, these two substructures are *not* independent, since each substructure will exert an incommensurate modulation upon the other. The host substructure is best considered in terms of a "basic structure" which is subjected to an incommensurate modulation mediated by the guest substructure; the basic structure can be described *via* conventional crystallographic principles (e.g. three-dimensional space group symmetry). In a similar way, the guest substructure can be considered in terms of an incommensurately modulated "basic structure". The incommensurate modulations describe perturbations to the basic structures which arise as a result of host-guest interaction. A full discussion of these structural issues for the urea inclusion compounds is given in ref. 5.

X-RAY DIFFRACTION FROM UREA INCLUSION COMPOUNDS
Upon transforming this structural description to reciprocal space, it is clear that the urea inclusion compound will give two distinguishable diffraction patterns: the "h" diffraction pattern, which arises from diffraction by the basic host structure (and by the incommensurate modulation within the guest substructure), and the "g" diffraction pattern, which arises from diffraction by the basic guest structure (and by the incommensurate modulation within the host substructure). The "h" diffraction patterns from urea inclusion compounds generally comprise discrete diffraction maxima, the positions of which can be rationalized on the basis of a unique three-dimensionally periodic reciprocal lattice. The "g" diffraction pattern usually comprises *both* diffuse scattering and discrete scattering. The diffuse scattering is usually in the form of two-dimensional sheets perpendicular to the tunnel axis (in direct space), indicating that there are regions of the inclusion compound crystal in which the guest molecules are ordered *only* in one dimension (along the tunnel axis). The presence of discrete diffraction maxima within the "g" diffraction pattern indicates that there are other regions of the crystal in which the guest molecules are ordered in three dimensions (*vide infra*).

THE INCOMMENSURATE STRUCTURAL NATURE OF UREA INCLUSION COMPOUNDS

As discussed above, the conventional urea inclusion compounds generally contain a regular packing (repeat distance c_g) of guest molecules along their tunnels. In general, c_g is *incommensurate* with the repeat distance (c_h) of the urea molecules along the tunnel ($c_h \approx 11.0$ Å for the conventional urea inclusion compounds). A new mathematical approach that has been developed [14,15] for understanding the fundamental distinction between commensurate and incommensurate behaviour in one-dimensional inclusion compounds is discussed below.

THREE-DIMENSIONAL ORDERING OF THE GUEST MOLECULES IN UREA INCLUSION COMPOUNDS

It has been discovered recently that well-defined positional correlations can exist between guest molecules in *different* tunnels (separated by more than 8 Å) in urea inclusion compounds, despite the fact that there is no well-defined positioning of guest molecules relative to the host (a consequence of the incommensurate relationship between the host and guest substructures). The mode of three-dimensional ordering of the guest molecules in urea inclusion compounds is conveniently described in terms of two parameters (figure 2) – the periodic repeat distance (c_g) of the guest molecules along

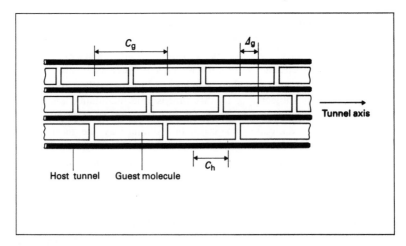

Figure 2   Schematic two-dimensional representation of a urea inclusion compound, viewed perpendicular to the tunnel axis, indicating the definitions of c_g and Δ_g.

the tunnel axis (corresponding to the distance between the centres of mass of adjacent guest molecules in the tunnel), and the offset (Δ_g), along the tunnel axis, between the centres of mass of guest molecules in adjacent tunnels. The exact nature of the three-dimensional ordering of the guest molecules depends on the functional groups present on the guest molecule, with different families of guest molecule giving rise to different characteristic modes of inter-tunnel ordering. Results for selected families of guest molecule at room temperature are summarized as follows:

(1)   For *n*-alkane/urea inclusion compounds [16], $\Delta_g = 0$ (independent of the value of c_g); c_g increases linearly with the number of CH_2 groups in the *n*-alkane molecule.

(2)   For diacyl peroxide/urea inclusion compounds [17], $\Delta_g = 4.6$ Å (independent of the value of c_g).

(3)   For α,ω-dibromoalkane/urea inclusion compounds [18], Δ_g depends on the value of c_g, with Δ_g and c_g related by the exact relationship: $\Delta_g = c_g/3$.

(4)   For carboxylic acid anhydride/urea inclusion compounds [13], $\Delta_g = 0$, with the exception of heptanoic anhydride/urea, for which $\Delta_g = 2.3$ Å.

The development of a detailed understanding of the fundamental factors controlling the three-dimensional ordering of guest molecules in urea inclusion compounds is an issue of particular importance at present.

MATHEMATICAL MODEL OF ONE-DIMENSIONAL INCLUSION COMPOUNDS

To a good approximation, inclusion compounds in which the host substructure comprises uni-directional tunnels can be considered as one-dimensional. A mathematical model of one-dimensional inclusion compounds has been developed recently [14,15] to allow structural properties of such inclusion compounds to be predicted and rationalized on the basis of potential energy functions (describing host-guest interaction, guest-guest interaction and the intramolecular potential energy of the guest molecules) computed for the inclusion compound. Within this mathematical model, the one-dimensional inclusion compound is considered as a linear, infinite host tunnel (with periodic repeat distance c_h) containing a finite number (n) of identical guest molecules in a periodic arrangement (with periodic repeat distance c_g). The characteristic energy function $\hat{E}(\alpha,n)$ for the inclusion compound is defined (taking $\alpha = c_g/c_h$) as:

$$\hat{E}(\alpha,n) = \frac{1}{\alpha}\left(\inf_{\lambda} \left(\frac{1}{n} \sum_{k=0}^{n-1} E_h(k\alpha+\lambda) \right) + \hat{E}_{guest}(\alpha) + \hat{E}_{intra} \right) \tag{1}$$

with the entailment that the optimum guest structure in the inclusion compound corresponds to the minimum characteristic energy. In this expression, $E_h(t)$ represents the energy of an individual guest molecule, due to host-guest interaction, when the guest molecule is located at position t along the host tunnel, $\hat{E}_{guest}(\alpha)$ is the guest-guest interaction energy, per guest molecule, when the periodicity of the guest structure is α, and $\hat{E}_{intra}$ is the intramolecular potential energy of the guest molecule. A graphical method (construction of the "characteristic energy diagram") has been developed to allow the structural properties of the one-dimensional inclusion compound to be assessed from the potential energy functions $E_h(t)$, $\hat{E}_{guest}(\alpha)$, and $\hat{E}_{intra}$ computed for the inclusion compound of interest.

Using this approach, the optimum guest periodicities (c_g) have been determined [19] for the n-alkane/urea inclusion compounds, and the question of whether these inclusion compounds exhibit commensurate or incommensurate behaviour has been assessed. The values of c_g predicted theoretically [19] for these inclusion compounds are in good agreement with the values of c_g determined experimentally [19,20]. For all the n-alkane/urea inclusion compounds studied, the optimum c_g is *ca.* 0.5 Å shorter than the value of c_g corresponding to the minimum guest-guest interaction energy; this fact substantiates the claim that, at the optimum guest periodicity in n-alkane/urea inclusion compounds, the interaction between adjacent guest molecules is repulsive.

A further application of the mathematical model of one-dimensional inclusion compounds, to predict the optimum conformational properties of the guest molecules in a one-dimensional inclusion compound, is discussed below.

CONFORMATIONAL PROPERTIES OF GUEST MOLECULES IN THIOUREA
INCLUSION COMPOUNDS

The conformational properties of monosubstituted cyclohexane guest molecules in their thiourea
inclusion compounds represent an interesting example of the way in which the behaviour of a
molecule can be altered markedly by incarcerating it within a solid host material. From high-
resolution solid state ^{13}C NMR spectra of thiourea inclusion compounds containing monosubstituted
cyclohexane guest molecules $C_6H_{11}X$ (with X = CH$_3$, NH$_2$, OH, Cl, Br, I), the relative populations
of the axial and equatorial conformers of the $C_6H_{11}X$ guest molecules have been determined [21].
On this basis, the guest molecules studied can be divided into two classes: those with X = Cl, Br, and
I have a predominance of the axial conformer (relative populations of equatorial conformer ≈ 0.05 –
0.15), whereas those with X = CH$_3$, NH$_2$, and OH have a predominance of the equatorial conformer
(relative populations of equatorial conformer ≈ 0.82 – 0.97). There is a marked contrast between the
conformational behaviour of the $C_6H_{11}X$ molecules with X = Cl, Br, and I as guest molecules in the
thiourea tunnel structure in comparison with the behaviour of the same molecules in solution (for
which there is a predominance of the equatorial conformer).

The dramatic dependence of these properties on the identity of the substituent X reflects the fine and
subtle energetic balances that exist for these inclusion compounds, and important insights into the
reasons underlying the preference for the axial conformation for $C_6H_{11}Cl$ within the thiourea tunnel
structure have been obtained by applying the mathematical model, described above, to the
$C_6H_{11}Cl$/thiourea inclusion compound [22]. As a consequence of the factor $1/\alpha$ in equation 1, the
characteristic energy refers to an energy *per unit length of host tunnel* rather than an energy per guest
molecule, and this emerges as a critical factor underlying the preference for the axial conformation of
the $C_6H_{11}Cl$ guest molecules in the thiourea tunnel structure. Thus, while the equatorial
conformation corresponds to a more negative $\hat{E}_{intra}$ than the axial conformation, the $C_6H_{11}Cl$ guest
molecules in the axial conformation can be packed more efficiently (smaller α) within the thiourea
tunnel, giving a more favourable host-guest interaction energy *per unit length of tunnel*. The analysis
provides theoretical justification for the observation that the axial conformation of $C_6H_{11}Cl$ is
preferred within the thiourea tunnel structure, and confirms that the optimum periodicity of the guest
molecules along the tunnel is commensurate with the thiourea host structure ($c_g = c_h/2$).

DYNAMIC PROPERTIES OF UREA INCLUSION COMPOUNDS

Incoherent quasielastic neutron scattering (IQNS) has been used to investigate the dynamic properties
of *n*-alkane [23,24] and α,ω-dibromoalkane [25] guest molecules constrained within the urea tunnel
structure. These investigations were carried out on samples of urea inclusion compounds containing
urea-d$_4$ to ensure that the incoherent scattering is dominated by the guest molecules, and two
experimental geometries of semi-oriented polycrystalline samples were considered (specifically, with
the tunnel axes of all crystals either parallel or perpendicular to the neutron momentum transfer
vector), allowing translational motions of the guest molecules along the tunnel to be investigated
separately from reorientational motions of the guest molecules about the tunnel axis. In the low-
temperature phase of these inclusion compounds there is an oscillatory motion along the tunnel axis
which becomes overdamped above the phase transition temperature. Above the phase transition
temperature, the dynamic properties of the guest molecules can be understood in terms of
translational motions along the tunnel (modelled as continuous diffusion between rigid impermeable
boundaries) and reorientational motions about the tunnel axis (modelled as uni-axial rotational
diffusion in a one-fold cosine potential). Quantitative dynamic information relating to these motional
models has been derived for the *n*-alkane/urea and α,ω-dibromoalkane/urea inclusion compounds.

Most investigations of the dynamic properties of urea inclusion compounds have focused upon the motion of the guest molecules, with little attention given to the possibility that the urea molecules may also undergo interesting dynamic processes. Recent ^{2}H NMR studies of *n*-alkane/urea-d$_4$ [26,27] and α,ω-dibromoalkane/urea-d$_4$ [28] inclusion compounds have led to the proposal that, at ambient temperature, the urea molecules undergo 180° jumps about their C=O axes. These ^{2}H NMR investigations provide no evidence for reorientation of the NH$_2$ groups about the C–N bond (on the ^{2}H NMR timescale). The timescale for the 180° jump motion of the urea molecules is substantially longer than the known timescale for motion of the guest molecules at the same temperature (*vide supra*), confirming that the reorientational motions of the host molecules and the reorientational and translational motions of the guest molecules are uncorrelated (at least at ambient temperature).

REFERENCES
1) Takemoto, K.; Sonoda, N.: "Inclusion compounds" (Ed.: Atwood, J.L.; Davies, J.E.D.; MacNicol, D.D.), Academic Press, New York, 1984, vol. 2, p. 47
2) Harris, K.D.M.: Chemistry in Britain, 1993, 29, 132
3) Harris, K.D.M.: J. Solid State Chem., 1993, 106, 83
4) Smith, A.E.: Acta Crystallogr., 1952, 5, 224
5) Harris, K.D.M.; Thomas, J.M.: J. Chem. Soc., Faraday Trans., 1990, 86, 2985
6) McAdie, H.G.; Frost, G.B.: Can. J. Chem., 1958, 36, 635
7) McAdie, H.G.: Can. J. Chem., 1962, 40, 2195
8) Harris, K.D.M.: J. Phys. Chem. Solids, 1992, 53, 529
9) Schiessler, R.W.; Flitter, D.: J. Amer. Chem. Soc., 1950, 74, 1720
10) Fetterly, L.C.: "Non-stoichiometric compounds" (Ed.: Mandelcorn, L.), Academic Press, New York, 1964, p. 491
11) Schlenk, W.: Liebigs Ann. Chem., 1949, 565, 204
12) Schlenk, W.: Liebigs Ann. Chem., 1951, 573, 142
13) Shannon, I.J.; Stainton, N.M.; Harris, K.D.M.: J. Mater. Chem., manuscript in press
14) Rennie, A.J.O.; Harris, K.D.M.: Proc. Royal Soc. A, 1990, 430, 615
15) Rennie, A.J.O.; Harris, K.D.M.: J. Chem. Phys., 1992, 96, 7117
16) Fukao, K.; Miyaji, H.; Asai, K.: J. Chem. Phys., 1986, 84, 6360
17) Harris, K.D.M.; Hollingsworth, M.D.: Proc. Royal Soc. A, 1990, 431, 245
18) Harris, K.D.M.; Smart, S.P.; Hollingsworth, M.D.: J. Chem. Soc., Faraday Trans., 1991, 87, 3423
19) Shannon, I.J.; Harris, K.D.M.; Rennie, A.J.O.; Webster, M.B.: J. Chem. Soc., Faraday Trans., 1993, 89, 2023
20) Laves, F.; Nicolaides, N.; Peng, K.C.: Z. Kristallogr., 1965, 121, 258
21) Aliev, A.E.; Harris, K.D.M.: J. Amer. Chem. Soc., 1993, 115, 6369
22) Schofield, P.A.; Harris, K.D.M.; Shannon, I.J.; Rennie, A.J.O.: J. Chem. Soc., Chem. Commun., 1993, 1293
23) Guillaume, F.; Sourisseau, C.; Dianoux, A.J.: J. Chem. Phys., 1990, 93, 3536
24) Guillaume, F.; Sourisseau, C.; Dianoux, A.J.: J. Chim. Phys. (Paris), 1991, 88, 1721
25) Smart, S.P.; Guillaume, F.; Harris, K.D.M.; Sourisseau, C.; Dianoux, A.J.: Physica B, 1992, 180&181, 687
26) Heaton, N.J.; Vold, R.L.; Vold, R.R.: J. Amer. Chem. Soc., 1989, 111, 3211
27) Heaton, N.J.; Vold, R.L.; Vold, R.R.: J. Magn. Reson., 1989, 84, 333
28) Aliev, A.E.; Smart, S.P.; Harris, K.D.M.: J. Mater. Chem., manuscript in press

Materials Science Forum Vols. 152 - 153 (1994) pp. 163-168

SOFT CHEMISTRY ROUTES TO OXIDE CATALYSTS

K.R. Poeppelmeier and D.C. Tomczak

Department of Chemistry and Ipatieff Catalytic Laboratory,
Northwestern University Evanston, IL 60208 USA

Keywords: Acid-Base Catalysis, High Surface Area Oxides, Lithium Aluminum Oxides

ABSTRACT

The difference in reactivity of solid α-LiAlO$_2$ versus γ-LiAlO$_2$, the former when in contact with molten benzoic acid will exchange Li$^+$ for H$^+$ and the latter will not, demonstrates that the Li$^+$ cation in α-LiAlO$_2$ is less acidic yielding the oxide more basic. Similarly when the surfaces of α-LiAlO$_2$ and γ-LiAlO$_2$ are contacted with the organic molecule isopropyl alcohol at similar temperature and conversion, the selectivity to base-catalyzed aldol condensation products are greatest with α-LiAlO$_2$. Both examples demonstrate that the surface basicity depends on the environment of the O^{2-} anion and the cation(s) to which it is bound.

1. INTRODUCTION

In the last several years low temperature "chimie douce" techniques have been applied in the synthesis of new materials. Soft chemistry routes can be used to prepare new metastable solids, which are not obtainable by high temperature techniques, with useful chemical, physical and morphological properties. Soft chemistry techniques have been subclassified [1] in three categories 1) intercalation-deintercalation associated with redox chemistry 2) acid-base chemistry based on selective protonation of oxygen anions of varying basicity and 3) metathesis reactions on internal surfaces or van der Waals gaps inside solids. These materials have potential application in the field of heterogeneous catalysis because solids with significant surface area or intracrystalline surfaces are required in order to design a useful catalytic reactor which must transform some 10^{-6} moles per second per cubic centimeter of space filled with catalyst [2]. In this paper we compare three syntheses of α-LiAlO$_2$ and describe the surface area and porosity of the chimie douce and conventionally prepared materials.

2. EXPERIMENTAL

2.1 α-LiAlO$_2$

Three procedures were used to synthesize α-LiAlO$_2$ a) Li$_2$CO$_3$ and boehmite (248 m^2/g) AlO(OH) were reacted at 600°C in air for 30 hours b) Li$_2$CO$_3$ and gelatinous boehmite AlO(OH)•nH$_2$O (208 m^2/g Kaiser chemical) were reacted at 600°C in air for three days and c) LiAl$_2$(OH)$_7$•2H$_2$O and LiOH•H$_2$O were ground together and kept at room temperature under flowing nitrogen gas (CO$_2$ free) saturated with water vapor for 3 days [3]. The sample was calcined in air at 400°C for 30 hours (C1), 400°C for 100 hours (C2) and a third sample was calcined at 400°C for 30 hours followed by 500°C for one hour (C3).

2.2. Surface Area and Porosity Measurements

Adsorption studies were performed on an Omicron Omnisorp 360 instrument using nitrogen as the adsorbate gas. Both the surface areas and porosities of the α-LiAlO$_2$ samples prepared by various routes were studied. A typical adsorption isotherm for a type IV material was observed. Data from zero to 0.3 P/P$_0$ were used in the BET calculation of surface area. Desorption data from one to approximately 0.5 P/P$_0$ were used to calculate the pore size distribution (mesopores) and pore volume.

3. RESULTS AND DISCUSSION

The thermal history of polycrystalline oxide materials plays a significant role in determining their surface area. Conventional solid state synthetic methods require higher temperature to reach complete reaction, by which time the products are desurfaced through sintering. For example, α-LiAlO$_2$ produced by the reaction of Li$_2$CO$_3$ with boehmite (AlO(OH)) has a plate-like morphology with a particle size in the range of 1-25 μm and a surface area of less than 5 m^2/g with no measurable porosity. In contrast, decomposition between 400-500°C of the mixture of LiAl$_2$(OH)$_7$•2H$_2$O and LiOH•H$_2$O resulted in α-LiAlO$_2$ with surface areas greater than 50 m^2/g and porosities in the range 0.2-0.3 cc/g. Similarly reaction of pseudoboehmite and Li$_2$CO$_3$ resulted in a material similar to starting with lithium dialuminate [4], presumably because of the hydrated nature of pseudoboehmite, which after decomposition at 600°C resulted in α-LiAlO$_2$ with a surface area of 42 m^2/g and substantial mesoporosity of 0.45 cc/g.

Catalytic studies with α-LiAlO$_2$, γ-LiAlO$_2$ and calcined 'HAlO$_2$' were carried out with the isopropyl alcohol (IPA) probe reaction [5,6,7]. 'HAlO$_2$' (Li$_{1-x}$H$_x$AlO$_2$; x ≈ 0.9), a proton-stuffed γ-alumina, occurs when the dense 3-dimensional mixed metal oxide α-LiAlO$_2$ is reacted with molten lauric or benzoic acid at 220 °C [8]. 'HAlO$_2$' retains the morphological characteristics (SA ≈ 40 m^2/g) of α-LiAlO$_2$ but undergoes on irreversible framework rearrangement from the rock-salt to spinel structure. After calcination the dehydroxylated "HAlO$_2$" maintains its γ-alumina-like spinel structure (SA ≈ 70 m^2/g). IPA is a useful molecule to study the surface chemistry of a mixed metal oxide because it provides information about acid and base sites. All three solids investigated in this study α-LiAlO$_2$, γ-LiAlO$_2$ and the calcined 'HAlO$_2$' catalyzed the formation of propylene (dehydration) and acetone (dehydrogenation). The formation of propylene indicates acid-base pair sites, while acetone is indicative of basic sites. In addition to the formation of propylene and acetone,

Table 1. Surface Area and Porosity Data for α-LiAlO$_2$

	SA (m^2/g)	Porosity (cc/g)
Conventional Method		
(A) Boehmite	<5	NA
Soft Chemistry Preparations		
(B) Pseudoboehmite	42	0.45
(C1) 400°C/30 hrs.	68	0.25
(C2) 400°C/100 hrs.	58	0.22
(C3) 500°C/1 hr. after 400°C/30 hrs.	69	0.28

Table 2. Pore Volume vs. Pore Radius

Pore Radius (Å)	Pore Volume (cc/g)			
	B	C1	C2	C3
>300	0.036	0.036	0	0.056
300-200	0.105	0.018	0.012	0.017
200-100	0.232	0.023	0.026	0.025
100-50	0.067	0.023	0.050	0.022
50-40	0.003	0.018	0.041	0.034
40-30	0.002	0.047	0.052	0.057
30-20	0.000	0.060	0.027	0.054
20-10	0	0.024	0.012	0.019

the most interesting result was that base catalyzed acetone aldol condensation products were formed on calcined 'HAlO$_2$' and α-LiAlO$_2$. Both materials produced the C$_6$-diene 2-methyl-1,3 pentadiene or 4-methyl-1,3-pentadiene, while only α-LiAlO$_2$ produced the additional aldol condensation products 4-methyl-2-pentanone and 4-methyl-2-pentanol. The reaction pathways that describe the formation of the condensation products can be used to infer the nature of the basic (anionotropic) and acidic (cationotropic) features of the oxide's surface. The difference in reactivity of solid α-LiAlO$_2$ (octahedral cations, Figure 1(A)) versus γ-LiAlO$_2$ (tetrahedral cations, Figure 1(B)), the former when in contact with molten benzoic acid will exchange Li$^+$ for H$^+$ and the latter will not, demonstrates that the Li$^+$ cation in α-LiAlO$_2$ is less acidic yielding the oxide more basic. Similarly when the surfaces of α-LiAlO$_2$ and γ-LiAlO$_2$ are contacted with isopropyl alcohol at similar temperature and conversion, the selectivity to base-catalyzed aldol condensation products are greatest with α-LiAlO$_2$.

(A)

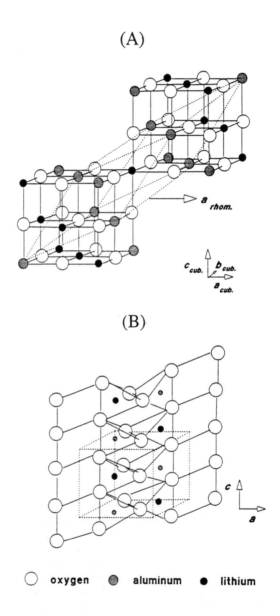

(B)

○ oxygen   ● aluminum   ● lithium

Figure 1. (A) Rhombohedral α-LiAlO₂ with all cations in octahedral coordination. (B) Tetragonal γ-LiAlO₂ with all cations in tetrahedral coordination.

These experimental results are in agreement with the theoretical findings of Dronskowski [9] that showed it is the less acidic Li atom in α-LiAlO$_2$ versus γ-LiAlO$_2$ which is exchanged against a proton.

However, it is not sufficient to only detect catalytic activity, but the surface should produce useful quantities of product. Because solids provide their catalytic function on their surface, considerable surface area is required. Amorphous or highly subdivided solids with properties that were difficult to correlate with the bulk structure have found many uses in heterogeneous catalysis. In the last twenty years crystalline aluminosilicates have found widespread applications owing to the surface area associated with the intrazeolite channels ($\leq$20 Å). In contrast to highly subdivided solids, the uniform size and distribution of the channels in zeolites has allowed many detailed studies of catalytic reactions on their interior surfaces. Efforts to increase the pore sizes of these and other materials to the mesoporous (~20 to 500 Å) region is at the present time an active area of research which has met with some success [10].

Soft chemistry provides an interesting synthetic alternative and opportunity for the chemist to prepare mesoporous solids of fundamental and practical interest. In Tables I and II the surface area (SA m^2/g), total porosity (cc/g) and pore size distributions (pore volume in cc/g versus pore radius in Å) are given for four samples of α-LiAlO$_2$ prepared from the reaction of a lithium salt with pseudoboehmite (B) or bayerite (C1, C2, and C3), i.e. allowing the salt to diffuse followed by calcination in the temperature range 400-600°C. In each case significant surface area and porosity was maintained with crystallites in the submicron range. The crystallites form a mesoporous secondary structure. Those derived from bayerite at 400°C (C1 and C2) develop porosity in the 20-50 Å (radius) range and with heating (C3) larger pores develop. The α-LiAlO$_2$ derived from pseudoboehmite (sample B) at 600°C does not contain a significant number of smaller pores but has the highest pore volume from pores greater than 50 Å radius.

4. CONCLUSION

Soft chemistry techniques have the advantage that new and more complex solids with surface area and porosity can be prepared which possess surface chemical features that can be related to the bulk structure. These solids can advance our understanding of many fundamental chemical reactions and catalytic phenomena.

5. ACKNOWLEDGMENTS

The authors gratefully acknowledge support from the Department of Energy, contract DE-F605-86ER75295 for the surface area and porosity measurement equipment, and from the National Science Foundation, Solid State Chemistry, contract DMR-8915897 for support on compound precursors to mixed metal oxides.

REFERENCES

1) "Chemical Reactivity of Low-Dimensional Solids", J. Rouxel, *Chemica Scripta,* <u>61</u>, 33 (1988).

2) "Zeolites-New Horizons in Catalysis", P. B. Weiss, Chemtech, August 1973, p. 498.

3) "Synthesis of High Surface Area α-LiAlO$_2$", K. R. Poeppelmeier, C. K. Chiang, and D. O. Kipp, *Inorg. Chem.*, <u>27</u>, 4523 (1988).

4) "Structure of LiAl$_2$(OH)$_7$•2H$_2$O", J. P. Thiel, C. K. Chiang, and K. R. Poeppelmeier, *Chem. Mater.*, <u>5</u>, 297, 1993.

5) "Characterization of the Li$_{1-x}$H$_x$AlO$_2$ System; $0.00 \leq x \leq 0.90$", D. C. Tomczak, S. H. Thong, and K. R. Poeppelmeier, *Catal. Lett.*, <u>12</u>, 139 (1992).

6) "Cation Replacement in αLiAlO$_2$ and Catalytic properties of αLiAlO$_2$ and Calcined "HAlO$_2$"", D. C. Tomczak and K. R. Poeppelmeier, *ACS Symposium Series*, Division of Petroleum Chemistry, San Francisco, CA, <u>37</u>(3), 996 (1992).

7) "Characterization of the Acidic and Basic Properties of αLiAlO$_2$, γLiAlO$_2$, and Calcined "HAlO$_2$" Using Isopropyl Alcohol", D. C. Tomczak and K. R. Poeppelmeier, *J. Catal.*, accepted.

8) "Cation Replacement in α-LiAlO$_2$", K. R. Poeppelmeier and D. O. Kipp, *Inorg. Chem.*, <u>27</u>, 766 (1988).

9) "Reactivity and Acidity of Li in LiAlO$_2$ Phases", R. Dronskowski, *Inorg. Chem.*, <u>32</u>, 2 (1993).

10) "A New Family of Mesoporous Molecular Sieves Prepared with Liquid Crystal Templates", J. S. Beck, J. C. Vartali, W. J. Roth, M. E. Leonowicz, C. T. Kresge, K. D. Schmitt, C. T.-W. Chu, D. H. Olson, E. W. Sheppard, S. B. McCullen, J. B. Higgins and J. L. Schlenker, *J. Am. Chem. Soc.* <u>114</u>, 10834 (1992); *Nature*, <u>359</u>, 710 (1992).

Materials Science Forum Vols. 152 - 153 (1994) pp. 169-174
© 1994 Trans Tech Publications, Switzerland

ACID-BASE ROUTES IN SOFT CHEMISTRY INVOLVING
PREFABRICATED OXIDES

M. Tournoux

Institut des Matériaux de Nantes, UMR 110 CNRS, Université de Nantes,
2 rue de la Houssinière, F-44072 Nantes Cédex 03, France

Keywords: Acid-Base, Soft Chemistry, $TiO_2(B)$, Lamellar Perovskites, Phosphatoantimonic Acids

ABSTRACT. - Acid-base routes involving prefabricated oxides lead to new, frequently metastable materials closely related to the starting materials. The first step of the process is a high temperature synthesis and the second consists of an ion exchange in acidic medium near room temperature. It may be followed by a condensation reaction as the temperature is increased. The thumb rule which governs the design of the precursors is rather simple : the higher the covalency in the anionic framework the more mobile the cations in the interlayer region or eventually in the channels. Layered compounds such as $K_2Ti_4O_9$, lamellar perovskites or phosphatoantimonates can be used as starting materials for this type of chemistry.

I - INTRODUCTION

Over the last twenty years there have been exciting developments in the synthesis at relatively moderate temperature of new, frequently metastable materials, in connection with their applications in the areas of energy storage and catalysis. In opposite to classical methods, this type of chemistry leads to a close structural relationship between the starting material and the final product.

Several routes exist for the preparation of such materials. We intend to focus here on acid-base routes involving prefabricated oxides. The first step of the process is the synthesis at high temperature of an alkali compound, more often a potassium compound having a 2D-anionic framework. The second step is an ion-exchange reaction in an acidic medium leading to a 2D-solid acid. This acid form may be used as a starting material for a lot of reactions, some of them allowing exfoliation.

Sometimes it is possible from the protonated form to carry out a topotactic condensation reaction as the temperature is increased.

II - HYDROXYLATION-CONDENSATION PROCESS APPLIED TO $K_2Ti_4O_9$

A classical illustration of the hydroxylation-condensation process is the preparation of the octatitanate $K_2Ti_8O_{17}$ and of the TiO_2 (B) form of titanium dioxide, from the potassium tetratitanate $K_2Ti_4O_9$ which exhibits a corrugated layer structure [1,2]. The 2D anionic framework is built up from TiO_6 octahedra sharing edges and vertices. The potassium cations are located between the layers and can be easily exchanged. We recall here this classical example because it has been recently evidenced that TiO_2 (B), new form of titanium dioxide, synthesized for the first time in Nantes, occurs in the nature which performs all sorts of chemistry including soft chemistry.

The ion-exchange in an acidic medium preserves the sheet structure of the tetratitanate and leads to $K_{2-x}(H_2O)_nTi_4O_{9-x}(OH)_x$. According to the experimental conditions, all the x values between 0 and 2 may be obtained. The nature of the products obtained after heating at about 500° C depends on the x value in the precursor. For x = 2, TiO_2 (B) is obtained. Its structure is that of the framework of the Wadsley's bronze Na_xTiO_2. For x = 1 in the precursor the final product is the potassium octatitanate which exhibits the same iono-covalent framework as the Watt's bronze $K_3Ti_8O_{17}$ previously prepared by high temperature methods. For x values between one and two, a mixture of the two adjacent phases, TiO_2 (B) and potassium octatitanate, is obtained. For x between zero and one the final product is an intergrowth between 2D-tetra and 3D-octatitanate.

Considering the positions and the nature of their neighbours it is possible to assess the basicity of framework oxygens qualitatively. In the tetratitanate the most basic oxygen is that of the unshared vertices of octahedra within the layer. This point is confirmed by two different quantitative approaches recently proposed by R. Dronskowski [3,4] and M. Henry [5,6] respectively and discussed elsewhere in this symposium. After hydroxylation leading to the oxohydroxo-tetratitanate the unshared vertices are occupied by OH groups. Upon heating a topotactic dehydration reaction occurs. Tetratitanate and octatitanate contain identical corrugated sheets. These sheets are joined by sharing corners of octahedra in the octatitanate.

The structural relationship between the precursor and the final product is not so clear in the case of TiO_2 (B). Investigations on this reaction by electron diffraction and microscopy are presently in progress. It must be noted that A. Jacobson, P. Davies et al. have recently prepared TiO_2 (B) from other layered titanates [7,8].

It must be emphasized that TiO_2 (B) and the octatitanate have the same framework as the Wadsley and Watt's bronzes respectively. This point suggests a formal relationship between the two processes leading to metastable materials. TiO_2 (B) and the octatitanate prepared by the hydroxylation condensation process could indeed be considered as formally resulting from the topotactic oxidation of the parent bronze.

The transformation of TiO_2 (B) to anatase either by heating above 550° C or under high pressure can be explained by the close structural relationship between these two forms of titanium dioxide. In anatase there is a cubic close packing of oxygen anions. Cubic close packing is also present but with vacancies in TiO_2 (B). In this cubic packing one out of nine oxygen positions is unoccupied. The structure of both TiO_2 (B) and anatase may be described in terms of rock salt structure as $(Ti_4\square_5)(O_8\square)$ and $(Ti_4\square_4)(O_8)$ respectively. The transformation of TiO_2 (B) to anatase results from a crystallographic shear [9] and reminds the formation of the Magneli phases which implies the elimination of anionic vacancies along a shear plane of edge sharing octahedra previously linked by vertices. The difference between the formation of the Magneli phases and the TiO_2 (B) to anatase transformation is that, in the latter, there is elimination of both anionic and cationic vacancies.

Rutile, anatase and brookite are well-known minerals, weathering products of Ti-bearing rocks. After describing the transformation of TiO_2 (B) to anatase one could speculate about the possible occurence of TiO_2 (B) in the nature. Two years ago in American Mineralogist, Jillian Banfield et al. published the identification of naturally occuring TiO_2 (B) by structure determination using high-resolution electron microscopy, image simulation and distance-least-squares refinement [10]. Coherently intergrown lamellae of TiO_2 (B) has been identified in natural anatase crystals coming from Binntal in Switzerland. The bulk samples analysed by electron microprobe were very pure and the TiO_2 (B) was, in term of composition, indistinguishable from the anatase. High resolution heating experiments carried out within the electron microscope showed that the polymorph TiO_2 (B) converts to anatase with increasing temperature according to the mechanism that we have previously described.

III - PROTONATED LAMELLAR PEROVSKITES

As far as I know the first reaction of acid-base soft chemistry, with lamellar oxides having a structure related to perovskite or K_2NiF_4 type, has been carried out in Nantes [11,12,13].

Two types of layered oxides derived from the perovskite structure are known. One is the Ruddlesden-Popper series [14] with this general formula $(M_y[A_{n-1}B_nO_{3n+1}])$ where the slab composition is indicated between the brackets. The slab thickness is n octahedra. The perovskite sheets are interleaved by M cations. Most of these previously reported phases correspond to $y = 2$ and contain two divalent M cations by formula unit. They do not exhibit any exchange property.

In order to perform chemistry in the interlayer space we decided to increase the covalency of the bonds within the anionic network. We have synthesized a lot of perovskite layered oxides having this generic formula $(M^I(A_{n-1}B_nO_{3n+1}))$ where M stands for an alkali cation. In these phases the charge of the layer is four times weaker than in the Ruddlesden-Popper series. This lower charge decreases the ionic interaction between layer and interlayer cations thus allowing ion exchange. These lamellar perovskites have also been studied by Jacobson et al [15,16]. The relative position of the layers depends on the nature of the M cations. The way in which adjacent layers are stacked up tends to optimize the coordination of the monovalent interlayer cation.

The loss of c axis doubling when a protonated phase is formed by exchange from the potassium parent phase indicates that a relative translation of adjacent layers occurs to adopt the same position as in the cesium phases. The separation is of course contracted. The stoichiometry corresponds to one proton for every pair of terminal oxygens from adjacent layers. N-alkylamines and other organic bases can be intercalated. Jacobson has shown that such intercalated phases can be used to cause spontaneous exfoliation of the structure into thin sheets that form a stable dispersion in a polar solvent [17].

J. Gopalakrishnan [18] and J.C. Joubert [19] have simultaneously and independently shown that it is also possible to carry out ion-exchange reactions in lamellar perovskites exhibiting a layer charge only two times weaker than in the Ruddlesden-Popper series. In the protonated phases $H_2[Ln_2Ti_3O_{10}]$ there is the required number of OH groups to allow a condensation reaction :

$$H_2Ln_2Ti_3O_{10} \longrightarrow 3Ln_{2/3}TiO_3 + H_2O$$

which is observed around 400° C. According to Gopalakrishnan the resulting product crystallizes around 1000° C to give a cation defective perovskite for the lanthanum compound and a pyrochlore in the case of Sm, Gd and Dy. M. Richard and L. Brohan have recently studied in Nantes, the thermal behavior of the neodymium phase. Their most stricking result concerns the existence between 600° C and 900° C of a lamellar oxygen defective perovskite $Nd_2Ti_3O_9$. In other words dehydroxylation and condensation does not occur simultaneously. However a part of the Nd^{3+} cations of the intermediate phase is present in the interlayer space. Above 900° C, a 3D cation defective perovskite is obtained. These results are discussed in more details in a poster.

The comparison between the Ruddlesden-Popper series and the alkali layered perovskite clearly shows that an increase in the covalency within the layers allows to carry out chemical reactions which preserve the structure of the layers.

IV - PHOSPHATOANTIMONIC ACIDS AND RELATED COMPOUNDS

To increase the covalency in lamellar oxides layers it may be of interest to build them not only from octahedra but also with XO_4 tetrahedra. Such materials have motivated considerable research as it is the case for the α-ZrP type compounds $(Zr(HPO_4)_2 , H_2O)$ which are layered compounds. The covalent layers are built up from ZrO_6 or other MO_6 octahedra and phosphate groups sharing vertices. These compounds, intensively studied in particular by A. Clearfield and G. Alberti [20], are remarkable for their ion-exchange properties.

An other way to increase the covalency of the layer is to choose a higher oxydation state for M in the MO_6 octahedra. We have studied the possibility of using Sb(V) instead of tetravalent cations. Several compounds have been prepared in the $K_2O-P_2O_5-Sb_2O_5$ system. We focus here on two bidimensional compounds [21,22,23].

$KSb(PO_4)_2$ has a layered structure. The layers are built up from SbO_6 octahedra and PO_4 tetrahedra sharing corners. Three oxygen atoms of each phosphate group are linked to three antimony atoms. The fourth is unshared and points into the interlayer space. The potassium atoms are situated within the interlayer space. In fact this compound contains the same type of layer as α-ZrP but since Zr^{IV} is replaced by Sb^V the layer charge is twice as less in the antimony compound.

$K_3Sb_3O_6(PO_4)_2$, xH_2O has also a layered structure. The covalent layers are built up from SbO_6 octahedra and PO_4 tetrahedra sharing vertices. The SbO_6 octahedra are linked together in the same way as the WO_6 octahedra in the (001) plane of the hexagonal tungsten bronze. The phosphate groups are linked to these layers of octahedra via three of their vertices. The fourth, which is unshared, points into the interlayer space wherein potassium atoms are situated. This compound which is prepared by solid state reaction at 1000° C is in fact hydrated at room temperature and its water content corresponds to x = 6 for almost the whole range of relative humidities.

The ion-exchange reaction from these 2D-phosphatoantimonates is performed in acidic medium near room temperature. It is rather fast, complete without any hydrolysis. The phosphatoantimonic acids are obtained under an hydrated form ($HSb(PO_4)_2,xH_2O$ and $H_3Sb_3O_6(PO_4)_2$, xH_2O). The dehydration of each layered acid is reversible, up to 560° C for the first and 300° C for the second.

These layered solid acids are very strong. When they are dispersed in water they give colloidal solutions which can be titrated by alkali hydroxydes. The titration curves of the two layered phosphatoantimonic acids are similar to that of chlorhydric acid. So strong solid acids that can exchange all sorts of cations may be of interest in heterogeneous catalysis. Unfortunately their surface areas are rather low, about 5 m^2 g^{-1}. In an attempt to increase them the intercalation of amines such as 1,4-diazabicyclo [2,2,2] octane (DABCO) has been carried out [24]. It must be noted that the titration of $H_3Sb_3O_6$ $(PO_4)_2$ by DABCO leads to the neutralization of only two out of three of the acid sites. The solid resulting from this titration exhibits a surface area of 60 m^2/g. It is not possible to remove the DABCO from the interlayer region by leaching with an acidic solution, however such a leaching removes the amine molecules from the surface leading to acid sites at the solid surface. The DABCO intercalated phosphatoantimonic acid can also be quantitatively obtained by heating at 180° C for six days this mixture (Sb_2O_5, $3H_2O$; H_3PO_4 ; DABCO ; H_2O in a molecular ratio : 1, 2, 1, 3600) in an autoclave lined with teflon.

From characterization studies including X-ray diffraction, I.R. and MAS NMR spectroscopies, it is possible to propose a structural model. The interlayer distance is very close to that of the hydrated potassium parent phase. The protonated DABCO is perpendicular to the layer. The remaining protons are still acidic as indicated by their chemical shift and by the fact that they can be titrated by a KOH solution leading to $K(H_2DABCO)Sb_3O_6(PO_4)_2$, xH_2O.

While continuing to work on these phosphatoantimonic acids, mainly in collaboration with industry, we have extended our investigations to systems including Si or Ge as the X tetrahedral cation. All the compounds obtained in A_2O-Sb_2O_5-XO_2 (A = alkali) systems exhibit 3D frameworks [25,26]. One of these frameworks, that of $Cs_3Sb_3O_6(Si_2O_7)$, is built up from SbO_6 octahedra and XO_4 tetrahedra, can be described as resulting from a condensation of $M_3O_6(XO_4)_2$ covalent layers similar to those which occur in $K_3Sb_3P_2O_{14}$. This condensation takes place via the outwardly pointing vertex, as yet unshared, of each XO_4 group. They join to give X_2O_7 groups. This framework sounds like a pyrochlore structure in which the bridging octahedra, i.e. those octahedra which are linking HTB-like planes of octahedra would be replaced by a X_2O_7 group. This framework is more open than the pyrochlore one and exhibit a different stacking of the HTB-like planes.

In a first attempt to obtain a structure of this type, the ion-exchange properties of $H_3Sb_3O_6(PO_4)_2$ have been used. After a titration with CsOH up to 1/3 of the ion-exchange capacity, a thermolysis of the resulting product has been performed with the hope that the removal of the last water molecule would imply the departure of one of the unshared oxygen atoms of a PO_4 group thus leading to 3D-$CsSb_3O_6(P_2O_7)$. Unfortunately, this reaction does not work, probably because there is not enough cesium to stabilize such a structure that contains three possible Cs-sites per formula unit. This means that this structure might be stabilized with a disilicate group and/or a digermanate group.

By high temperature methods we have effectively prepared these disilicate and digermanate related to the pyrochlore type although with a different stacking of the HTB-like layers [27]. From $Cs_3Sb_3O_6(Si_2O_7)$ the protonated form may be obtained by exchange in acidic medium near room temperature.

V - CONCLUSION

This few results illustrate acid-base soft chemistry involving prefabricated oxides. In opposite to the sol-gel process, the prefabrication technique is a hybrid method that combines the use of high temperature precursors and soft chemistry. The general thumb rule which governs the design of the precursor is rather simple : the higher the covalency in the anionic framework the more mobile the cations in the interlayer region or eventually in the channel. The choice of the oxidation states and of the coordination polyedra of the cations in the framework, allows to adjust the covalency and the number of protons in the protonated form. According to this number further condensation are possible or not as illustrated by the thermal behavior of the lamellar perovskites.

REFERENCES

1) Marchand, R., Brohan, L. and Tournoux, M.: Mat. Res. Bull., 1980, 15, 201
2) Tournoux, M., Marchand, R. and Brohan, L.: Prog. Sol. St. Chem. , 1986, 17, 33
3) Dronskowski, R.: J. Am. Chem. Soc., 1992, 114, 7230
4) Dronskowski, R.: Inorg. Chem., 1993, 32, 1
5) Henry, M.: Mat. Res. Soc. Symp. Proc., 1992, 271, 243
6) Henry, M.: Poster Nantes, Sept. 1993
7) Feist, T.P., Mocarski, S.J., Davies, P.K., Jacobson, A.J., Lewandowski, J.T.: Solid State Ionics, 1988, 28-30, 1338
8) Feist, T.P. and Davies, P.: J. Sol. St. Chem., 1992, 101, 275
9) Brohan, L., Verbaere, A., Tournoux, M. and Demazeau, G.: Mat. Res. Bull., 1982, 17, 355
10) Banfield, L. and Veblen, D.R.: Am. Min., 1991, 76, 343
11) Dion, M., Ganne, M. and Tournoux, M.: Mat. Res. Bull., 1981, 16, 1429
12) Dion, M.: Thesis Univ. of Nantes (1984)
13) Dion, M., Ganne, M., Tournoux, M. and Ravez, J.: Rev. Chim. Min., 1984, 21, 92
14) Ruddlesden, S.N. and Popper, P.: Acta Cryst., 1957, 10, 538, 1958, 11, 54
15) Jacobson, A.J., Johnson, J.W. and Lewandowski, J.T.: Inorg. Chem., 1985, 24, 3727
16) Jacobson, A.J., Lewandowski, J.T. and Johnson, J.W.: J. Less Common Met., 1986, 116, 1137
17) Treacy, M.M.J., Rice, S.B., Jacobson, A.J. and Lewandowski, J.T.: Chem. Mat., 1990, 22, 273
18) Gopalakrishnan, J. and V. Bhat, V.: Inorg. Chem., 1987, 26, 4301
19) Gondrand, M. and Joubert, J.C.: Rev. Chim. Min., 1987, 24, 33
20) Clearfield, A.: "Inorganic Ion Exchange Materials" C.R.C. Press Inc., (1982)
21) Piffard, Y., Verbaere, A., Lachgar, A., Deniard-Courant, S. and Tournoux, M.: Rev. Chim. Min., 1986, 23, 766
22) Piffard, Y., Verbaere, A., Oyetola, S., Deniard-Courant, S. and Tournoux, M.: Eur. J. Solid State Inorg. Chem., 1989, 26, 113
23) Oyetola, S., Verbaere, A., Guyomard, D., Piffard, Y. and Tournoux, M.: Eur. J. Solid State Inorg. Chem., 1989, 26, 175
24) Galarneau, A., Bujoli, B., Taulelle, F., Piffard, Y. and Tournoux, M.: Chem. Mat., (in press)
25) Pagnoux, C., Verbaere, A., Piffard, Y. and Tournoux, M.: Acta Cryst. C, 1991, 47, 2297
26) Pagnoux, C., Verbaere, A., Kanno, Y., Piffard, Y. and Tournoux, M.: J. Sol. St. Chem., 1992, 99, 173
27) Pagnoux, C., Verbaere, A., Piffard, Y. and Tournoux, M.: Eur. J. Solid State Inorg. Chem., 1993, 30, 111

Materials Science Forum Vols. 152 - 153 (1994) pp. 175-182
© 1994 Trans Tech Publications, Switzerland

SYNTHESIS OF NOVEL METAL OXIDES BY SOFT-CHEMISTRY ROUTES*

J. Gopalakrishnan, S. Uma, K. Kasthuri Rangan and N.S.P. Bhuvanesh

Solid State and Structural Chemistry Unit, Indian Institute of Science,
Bangalore-560 012, India

Keywords: Layered Perovskite Oxides, Bronsted Acidity, Intercalation/Deintercalation, NASICON and $KTiOPO_4$-Type Phosphates, H-V-W-O Oxides

ABSTRACT.-Protonated layered perovskites, $H_yA_2B_3O_{10}$ (A=La/Ca; B=Ti/Nb), framework phosphates of NASICON and $KTiOPO_4$ (KTP) structures and oxide hydrates of the formula, $H_xV_xW_{1-x}O_3 \cdot yH_2O$ for x = 0.125 and 0.33, have been synthesized by soft-chemical routes involving respectively ion-exchange, redox deintercalation and acid-leaching from appropriate parent oxides. $K_{0.5}Nb_{0.5}M_{0.5}OPO_4$ (M=Ti,V) prepared by redox deintercalation are new nonlinear optical materials exhibiting second harmonic generation. $H_xV_xW_{1-x}O_3$ oxides possessing α-MoO_3 and hexagonal WO_3 structures exhibit acid-base and redox intercalation /insertion of chemical species, characteristic of layered and tunnel structures.

INTRODUCTION

There is a growing interest in oxide materials especially with the advent of 'soft-chemistry' routes [1] which have enabled synthesis of several metastable metal oxides [2] that are inaccessible through conventional high-temperature methods. We have been pursuing soft-chemical/low-temperature methods for the synthesis of metastable phases of ReO_3, perovskite,rutile and related structures for some time [3]. In this paper, we describe three types of metal oxides synthesized by us in recent times. They are (i) protonated layered perovskites,(ii) framework phosphates belonging to NASICON and $KTiOPO_4$ (KTP) structure types,and (iii) layered and three-dimensional oxides in the H-V-W-O system - all synthesized by 'soft-chemistry' routes involving respectively ion-exchange, redox deintercalation and acid-leaching of appropriate parent oxides.

EXPERIMENTAL

Protonated layered perovskites, $H_yA_2B_3O_{10}$ (A=La/Ca; B=Ti/Nb), were prepared from the parent alkali metal analogues, $(K/Rb)_yA_2B_3O_{10}$, by ion-exchange in aqueous HNO_3. Intercalation of organic bases was investigated by refluxing the host materials with a 10% solution of the base in n-heptane around 90 °C. Deintercalation of alkali metal from $Na_3V_2(PO_4)_3$, $A_xTiV(PO_4)_3$ (A=Na,K) and $KM_{0.5}M'_{0.5}OPO_4$ (M=Nb, Ta; M'=Ti,V) was investigated by passing chlorine through a suspension of the solid phosphate in $CHCl_3$. Acid-leaching of

---

*Contribution No. 934 from the Solid State and Structural Chemistry Unit.

$LiVWO_6$ was carried out by treating the solid with varying concentrations of HNO_3/HCl both over water bath and under reflux. Oxidation state of transition metal, where essential, was determined by potentiometric titration using Ce(IV) as oxidizing agent. Solid products were characterized by EDX analysis, X-ray powder diffraction and thermogravimetry. Second harmonic generation (SHG) intensities were measured using a pulsed Q-switched Nd:YAG laser.

RESULTS AND DISCUSSION

Protonated layered perovskites

There are two series of layered perovskites, one is the Ruddlesden-Popper series [4] of the general formula, $A'_2[A_{n-1}B_nO_{3n+1}]$ of which $Sr_4Ti_3O_{10}$ is a typical n=3 member, and the other is the Dion-Jacobson series [5,6] of the general formula, $A'[A_{n-1}B_nO_{3n+1}]$, of which $CsCa_2Nb_3O_{10}$ [7] and $KLaNb_2O_7$ [8] are typical n=2 and n=3 members. A few years back, we prepared a new series of titanates, $A'_2[Ln_2Ti_3O_{10}]$, where A'=K,Rb and Ln=La or rare earth [9], which are isostructural with $Sr_4Ti_3O_{10}$. Members of both the series of oxides, for instance $K_2La_2Ti_3O_{10}$ and $CsCa_2Nb_3O_{10}$, undergo facile ion-exchange in aqueous acids to yield protonated derivatives, $H_2La_2Ti_3O_{10}$ and $HCa_2Nb_3O_{10}$, that retain the parent structures. Of these, the latter is a strong Bronsted acid intercalating a wide variety of organic bases [10], including pyridine ($pK_a = 5.3$) and aniline ($pK_a = 4.6$), while members of the former series do not show such an obvious acidic property.

To understand the relation between the acidic property and the structure of the layered perovskite oxides, we prepared protonated oxides of the general formula $H_yA_2B_3O_{10}$ - $H_{1-x}La_xCa_{2-x}Nb_3O_{10}$ (0 < x < 1), $HCa_{2-x}La_xNb_{3-x}Ti_xO_{10}$ (0 < x ≤ 2) and $H_{2-x}La_2Ti_{3-x}Nb_xO_{10}$ (0 ≤ x ≤ 1), starting from the corresponding K/Rb compounds by ion-exchange. In table 1, we list the compositions and lattice parameters and in figure 1, X-ray powder diffraction (XRD) patterns of a few representative members. The data reveal that $H_yA_2B_3O_{10}$ oxides adopt two distinct structures: for y ≤ 1, the structure is primitive tetragonal, being similar to $CsCa_2Nb_3O_{10}$ [7] and $HCa_2Nb_3O_{10}$ [11]; for (1 < y ≤ 2), on the other hand, the structure is body-centered tetragonal (I4/mmm) similar to that of $Sr_4Ti_3O_{10}$ [4] and $K_2La_2Ti_3O_{10}$ [9]. The difference between the two structures arises from the stacking of the triple-perovskite $[A_2B_3O_{10}]$ layers in the c-direction. While the adjacent perovskite slabs are displaced by $1/2(a+b)$ giving a body-centered tetragonal cell for $Sr_4Ti_3O_{10}$, there is no such displacement with the $CsCa_2Nb_3O_{10}$ structure (figure 2).

This difference in the structures of $H_yA_2B_3O_{10}$ oxides seems to have a profound influence on the Bronsted acidity, as revealed by intercalation of n-alkylamines and other bases. While the members of $H_{1-x}La_xCa_{2-x}Nb_3O_{10}$ exhibit Bronsted acidity just as the parent $HCa_2Nb_3O_{10}$ [10], intercalating several organic bases including pyridine, $H_2La_2Ti_3O_{10}$ and $H_{2-x}La_2Ti_{3-x}Nb_xO_{10}$ for 0 < x ≤ 0.75 do not intercalate even very strong bases such as piperidine ($pK_a = 11.2$) and guanidine ($pK_a = 12.5$) [12]. The lack of Bronsted acidity for the protons of $H_{2-x}La_2Ti_{3-x}Nb_xO_{10}$ seems to be related to the special interlayer structure arising from the displacement of the perovskite slabs in [110] direction (figure 2). It is significant that even the isostructural $H_2Ca_2Nb_2TiO_{10}$ (table 1) is exactly similar to $H_2La_2Ti_3O_{10}$ in its intercalation behaviour, revealing that the lack of Bronsted acidity of these solids has its origin in the structure rather than in the intrinsic acidity of the protons attached to TiO_6/NbO_6 octahedra.

Recently we prepared anion-deficient layered perovskites, $HCa_2Nb_{3-x}M_xO_{10-x}$ (M = Fe, Al) for 0 < x ≤ 1.0), possessing structure and properties similar to the parent $HCa_2Nb_3O_{10}$.

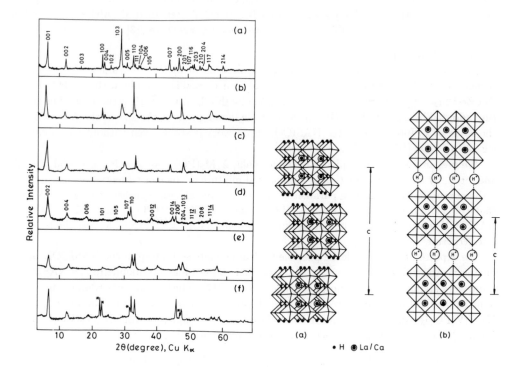

Fig.1 - X-ray powder diffraction patterns of
(a) $H_{0.5}La_{0.5}Ca_{1.5}Nb_3O_{10}$, (b) $HCaLaNb_2TiO_{10}$,
(c) $HLa_2Ti_2NbO_{10}$, (d) $H_{1.5}La_2Ti_{2.5}Nb_{0.5}O_{10}$,
(e) $H_2La_2Ti_3O_{10}$ and (f) $H_2Ca_2Nb_2TiO_{10}$.
In (f), asterisks denote impurity phase.

Fig.2 - Idealized structures of
(a) $H_2La_2Ti_3O_{10}$ and (b) $HCa_2Nb_3O_{10}$.

Table 1- Composition and lattice parameters of representative protonated layered
perovskites, $H_yA_2B_3O_{10}$

Composition	Lattice parameters (Å)		Structure type
	a	c	
$HCa_2Nb_3O_{10}{}^a$	3.850(6)	14.379(3)	$CsCa_2Nb_3O_{10}$
$H_{0.5}La_{0.5}Ca_{1.5}Nb_3O_{10}$	3.881(5)	14.39 (1)	$CsCa_2Nb_3O_{10}$
$HCaLaNb_2TiO_{10}$	3.855(6)	14.41 (6)	$CsCa_2Nb_3O_{10}$
$HCa_{0.5}La_{1.5}Nb_{1.5}Ti_{1.5}O_{10}$	3.835(4)	14.39 (6)	$CsCa_2Nb_3O_{10}$
$HLa_2Ti_2NbO_{10}{}^b$	3.832(5)	14.52 (5)	$CsCa_2Nb_3O_{10}$
$H_{1.5}La_2Ti_{2.5}Nb_{0.5}O_{10}$	3.832(6)	27.51 (6)	$Sr_4Ti_3O_{10}$
$H_2La_2Ti_3O_{10}$	3.824(5)	27.40 (6)	$Sr_4Ti_3O_{10}$
$H_2Ca_2Nb_2TiO_{10}$	3.844(6)	28.67 (7)	$Sr_4Ti_3O_{10}$

^{a} values taken from ref. [11] and ^{b} anhydrous phase is unstable

Intercalation experiments suggest that the oxygen vacancies in the perovskite slabs are probably ordered in $HCa_2Nb_2AlO_9$ giving octahedral (NbO_6) - tetrahedral (AlO_4) - octahedral (NbO_6) layer sequence reminiscent of the brownmillerite structure [13].

Framework phosphates

Phosphates of the formula, $A_xM_2(PO_4)_3$, consisting of $M_2(PO_4)_3$ framework formed by corner-sharing MO_6 octahedra and PO_4 tetrahedra crystallize mainly with the NASICON $[Na_3Zr_2PSi_2O_{12}]$ [14] (figure 3a) and the langbeinite $[K_2Mg_2(SO_4)_3]$ [15] structures. The structure adopted by a given $A_xM_2(PO_4)_3$ depends on, among others, the size of the A cation and the value of x. Interestingly, the NASICON framework is stable even without the A cation, e.g., $Nb_2(PO_4)_3$ [16].

We prepared several phosphates of the formula, $A_xM_2(PO_4)_3$ for A = Na or K and M = Ti and/or V, by hydrogen-reduction of a preheated mixture of the constituents. We list in table 2 the composition and lattice parameters of the new phosphates synthesized by us. We see that the sodium compounds, $Na_3V^{III}{}_2(PO_4)_3$ and $Na_3Ti^{III}V^{III}(PO_4)_3$, crystallize with the NASICON structure, while the potassium compound, $K_2Ti^{IV}V^{III}(PO_4)_3$, adopts the langbeinite structure.

We expected that it would be possible to deintercalate sodium from the NASICON phosphates for the following reasons. It is known that sodium ions are mobile in the NASICON framework giving rise to fast sodium ion conductivity [17]. The presence of titanium and vanadium in the III state should favour oxidative deintercalation of sodium from $Na_3V_2(PO_4)_3$ and $Na_3TiV(PO_4)_3$. Indeed we could deintercalate sodium completely from $Na_3V_2(PO_4)_3$ using Cl_2 in $CHCl_3$, giving $V_2(PO_4)_3$ according to the reaction [18]

$$Na_3V_2(PO_4)_3 + 3/2\ Cl_2 \rightarrow V^VV^{IV}(PO_4)_3 + 3\ NaCl.$$

$V_2(PO_4)_3$ retains the NASICON framework showing that the deintercalation is a topochemical reaction. It should be mentioned that, unlike $Nb_2(PO_4)_3$ which could be prepared by a direct solid state reaction [16], $V_2(PO_4)_3$ could not be prepared by a direct method. The presence of vanadium in V and IV oxidation states together with the 'empty' NASICON framework of $V_2(PO_4)_3$ renders it an excellent host material for reductive insertion of electropositive species such as hydrogen and lithium. In table 2, we have listed the characteristics of $Li_3V_2(PO_4)_3$ and $H_3V_2(PO_4)_3$ prepared by reductive insertion of Li/H into $V_2(PO_4)_3$.

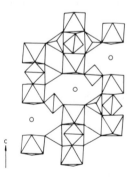

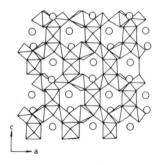

Fig.3-Framework structures of (a) NASICON and (b) $KTiOPO_4$.

(a)   (b)

Table 2 -Composition, colour and lattice parameters of $A_xM_2(PO_4)_3$

Composition	Colour	Reducing Power of the Sample		Lattice Parameters ($\mathring{A}$)		Structure type
		Found	Calcd.	a	c	
$Na_3V^{III}_2(PO_4)_3$	Green	3.99	4.00	8.68(2)	21.71(2)	NASICON
$V^{IV}V^V(PO_4)_3$	Brown	0.96	1.00	8.52(3)	22.02(4)	NASICON
$Li_3V^{III}_2(PO_4)_3$	Green	3.95	4.00	8.31(2)	22.50(2)	NASICON
$H_3V^{III}_2(PO_4)_3$	Greenish black	3.96	4.00	8.57(3)	22.48(3)	NASICON
$Na_3Ti^{III}V^{III}(PO_4)_3$	Brown	3.00	3.00	8.759(3)	21.699(4)	NASICON
$NaTi^{IV}V^{IV}(PO_4)_3$	Green	0.96	1.00	8.460(5)	21.619(8)	NASICON
$K_2Ti^{IV}V^{III}(PO_4)_3$	Green	2.10	2.00	9.855(3)	-	Langbeinite
$Ba_{1.5}V^{III}_2(PO_4)_3$	Gray	3.96	4.00	9.884(3)	-	Langbeinite
$BaKV^{III}_2(PO_4)_3$	Greenish yellow	3.94	4.00	9.873(2)	-	Langbeinite

A similar deintercalation of sodium from $Na_3TiV(PO_4)_3$ proceeds to the extent of removal of two sodium atoms,

$$Na_3TiV(PO_4)_3 + Cl_2 \rightarrow NaTiV(PO_4)_3 + 2\,NaCl.$$

Chemical analysis and magnetic susceptibility establish the formula of the deintercalation product to be $NaTi^{IV}V^{IV}(PO_4)_3$.

Interestingly, we could not deintercalate potassium from the langbeinite-$K_2TiV(PO_4)_3$ under similar conditions. The lack of deintercalation reactivity of this phase in contrast to the facile deintercalation of sodium from $Na_3V_2(PO_4)_3$ and $Na_3TiV(PO_4)_3$ is presumably related to the structure. Langbeinite being a true cage structure with small windows connecting the cages does not permit mobility of potassium ions in $K_2TiV(PO_4)_3$, while NASICON being a skeletal structure with an interconnected interstitial space permits facile mobility of sodium ions through the framework resulting in deintercalation, provided appropriate transition metal ions such as Ti(III) and V(III), which can undergo oxidation, is incorporated in the framework.

$KTiOPO_4$ (KTP) is another framework phosphate containing one dimensional channels parallel to [001] where potassium ions reside [19] (figure 3b). Since it is known that this

Table 3 -Composition, colour, lattice parameters and SHG intensity of KTP-like phosphates.

Composition	Colour	Reducing Power of the Sample		Lattice Parameters($\mathring{A}$)			SHG inten sity*
		Found	Calcd.	a	b	c	
$KNb_{0.5}Ti_{0.5}OPO_4$	Dark blue	0.45	0.50	12.976(5)	6.488(4)	10.773(6)	-
$K_{0.5}Nb_{0.5}Ti_{0.5}OPO_4$	Light blue	-	-	12.879(9)	6.402(7)	10.659(4)	0.8
$KNb_{0.5}V_{0.5}OPO_4$	Brown	0.98	1.00	12.949(6)	6.431(8)	10.686(4)	-
$K_{0.5}Nb_{0.5}V_{0.5}OPO_4$	Green	0.52	0.50	12.801(6)	6.357(4)	10.569(5)	0.4
$KTa_{0.5}V_{0.5}OPO_4$	Grey	0.98	1.00	12.985(4)	6.442(3)	10.696(4)	-
$K_{0.5}Ta_{0.5}V_{0.5}OPO_4$	Yellow	0.53	0.50	12.819(5)	6.367(4)	10.615(5)	0.5

*Normalized with respect to that of $KTiOPO_4$ which is taken as unity.

structure also allows mobility of potassium ions through the channels resulting in ion-exchange and ionic conductivity [20], we anticipated that it would be possible to oxidatively deinter-calate potassium from this structure, by incorporating appropriate transition metal atoms in the framework. To realize this possibility, we prepared new KTP-analogues, $KM_{0.5}M'_{0.5}OPO_4$ ($M = Nb^V$, Ta^V; $M' = Ti^{III}$, V^{III}) and investigated oxidative deintercalation of potassium using Cl_2. Typically, deintercalation proceeds according to the following reaction

$$KNb^V_{0.5}Ti^{III}_{0.5}OPO_4 + 1/4\ Cl_2 \rightarrow K_{0.5}Nb^V_{0.5}Ti^{IV}_{0.5}OPO_4 + 1/2\ KCl$$

giving new KTP analogues, $K_{0.5}M_{0.5}M'_{0.5}OPO_4$ (table 3). Significantly, while the parent $KM_{0.5}M'_{0.5}OPO_4$ do not show second harmonic generation (SHG) of 1064 nm radiation, the deintercalated products, $K_{0.5}M_{0.5}M'_{0.5}OPO_4$, do show SHG activity. The SHG response from these powder materials is comparable to that of KTP.

Layered and three-dimensional vanadium-tungsten oxide hydrates related to α-MoO$_3$ and WO$_3$.1/3H$_2$O

We prepared two new vanadium-tungsten oxide hydrates of the general formula, $H_xV_xW_{1-x}O_3 \cdot yH_2O$ for x=0.125; y=1.5 and x=0.33; y=0.33 by acid-leaching of LiVWO$_6$ in aqueous HNO$_3$/HCl (table 4). $H_{0.125}V_{0.125}W_{0.875}O_3 \cdot 1.5H_2O$ (I), obtained by leaching of LiVWO$_6$ in dilute HNO$_3$/HCl, crystallizes in a layered, α-MoO$_3$-like, structure. I dehydrates around 130 °C to give a hexagonal phase similar to the one reported by Feist and Davies [21]. $H_{0.33}V_{0.33}W_{0.67}O_3 \cdot 1/3H_2O$ (II) is another new hydrate obtained by refluxing LiVWO$_6$ in concentrated HNO$_3$.II is isostructural with WO$_3$.1/3H$_2$O [22] topochemically dehydrating around 330 °C to give hexagonal WO$_3$-like [23] $H_{0.33}V_{0.33}W_{0.67}O_3$ (III). A schematic represen-tation of the structures of the hydrates and their dehydration products is shown in figure 4.

Both I and III could be regarded functionalized derivatives of α-MoO$_3$ and hexagonal WO$_3$, exhibiting redox and acid-base intercalation/insertion reactivity characteristic of lay-ered and tunnel structures. Thus, I readily intercalates n-alkylamines resulting in large layer expansions (table 4). Especially significant is the insertion of ammonia molecules into the hexagonal tunnels of III through an acid-base reaction arising from the Bronsted acidity of the protons.

In summary, we have shown that soft-chemical methods based on ion-exchange, acid-leaching and redox deintercalation afford inexpensive routes for the synthesis of valuable inorganic materials possessing novel structures and properties.

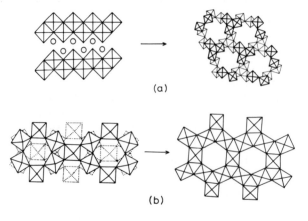

(a)

(b)

Fig.4-Schematic representation of the structural changes accom-panying dehydration of (a) $H_{0.125}V_{0.125}W_{0.875}O_3 \cdot 1.5H_2O$ and (b) $H_{0.33}V_{0.33}W_{0.67}O_3 \cdot 1/3H_2O$.

Table 4- Composition and lattice parameters of vanadium-tungsten oxide hydrates and their derivatives.

Composition	Lattice Parameters (Å)		
	a	b	c
$H_{0.125}V_{0.125}W_{0.875}O_3 \cdot 1.5H_2O$ (I)	7.77(3)	13.87(6)	7.44(3)
$H_{0.33}V_{0.33}W_{0.67}O_3 \cdot 1/3H_2O$ (II)	7.22(3)	12.54(7)	7.66(4)
$H_{0.33}V_{0.33}W_{0.67}O_3$(III)	7.25(4)	-	3.87(2)
$K_{0.18}[H_{0.125}V_{0.125}W_{0.875}O_3 \cdot 1.5H_2O]$	7.63(3)	13.74(7)	7.41(3)
$(NH_4)_{0.125}V_{0.125}W_{0.875}O_3 \cdot 1.5H_2O$	7.88(3)	24.70(9)	7.17(2)
$(n\text{-}C_6H_{13}NH_2)_{0.58}H_{0.125}V_{0.125}W_{0.875}O_3 \cdot 1.5H_2O$	7.73(4)	38.4 (2)	7.43(4)
$(n\text{-}C_{10}H_{21}NH_2)_{0.61}H_{0.125}V_{0.125}W_{0.875}O_3 \cdot 1.5H_2O$	7.78(9)	58.0(6)	7.55(9)
$H_{0.90}[H_{0.33}V_{0.33}W_{0.67}O_3]$	7.26(2)	-	3.87(2)
$K_{0.33}[H_{0.33}V_{0.33}W_{0.67}O_3]$	7.25(4)	-	3.87(3)
$(NH_4)_{0.30}H_{0.03}V_{0.33}W_{0.67}O_3$	7.24(2)	-	3.86(1)

Acknowledgement. We express our grateful thanks to Professors J. ROUXEL and M. TOURNOUX for providing generous financial support to present this work in the International Symposium on Soft-Chemistry Routes to New Materials. We also thank the Department of Science and Technology, Government of India, for support.

REFERENCES

1) Rev. Chim. Miner. 1984, 21(4) devoted to Soft-Chemistry

2) Stein, A., Keller, S.W., and Mallouk, T.E.: Science, 1993, 259, 1558

3) Rao, C.N.R., and Gopalakrishnan, J.: Acc.Chem.Res.1987,20, 228; Gopalakrishnan, J.: in*Chemistry of Advanced Materials* (Ed. Rao, C.N.R.) Blackwell, Oxford (1993)

4) Ruddlesden, S.N., and Popper, P.: Acta Crystallogr., 1957,10, 538; 1958, 11, 54

5) Dion, M., Ganne, M., and Tournoux, M.: Mater. Res. Bull., 1981, 16, 1429

6) Jacobson, A.J., Johnson, J.W., and Lewandowski, J.T.: Inorg. Chem., 1985, 24, 3727

7) Dion, M., Ganne, M., Tournoux, M., and Ravez, J.: Rev. Chim. Miner., 1984, 21, 92

8) Gopalakrishnan, J., Bhat, V., and Raveau, B.: Mater. Res. Bull., 1987, 22, 413

9) Gopalakrishnan, J., and Bhat, V.: Inorg. Chem., 1987, 26, 4299

10) Jacobson, A.J., Johnson, J.W., Lewandowski, J.T.: Mater.Res. Bull., 1987, 22, 45

11) Jacobson, A.J., Lewandowski, J.T., and Johnson, J.W.: J. Less Common Met., 1986, 116, 137

12) Uma, S., Raju, A.R., and Gopalakrishnan, J.: J. Mater. Chem., 1993, 3

13) Collville, A.A., and Geller, S.: Acta Crystallogr., Sect. B, 1971, 27, 2311

14) Hong, H.Y-P.: Mater. Res. Bull., 1976, 11, 173

15) Zemann, A., and Zemann, J.: Acta Crystallogr., 1957, 10, 409

16) Leclaire, A., Borel, M.M., Grandin, A., and Raveau, B.: Acta Crystallogr. Sect. C, 1989, 45, 699

17) Goodenough, J.B., Hong, H.Y-P., and Kafalas, J.A.: Mater. Res. Bull., 1976 11, 203

18) Gopalakrishnan, J., and Kasthuri Rangan, K.: Chem. Mater. 1992, 4, 745

19) Tordjman, I., Masse, R., and Guitel, J.C.: Z. Kristallogr., 1974, 139, 103

20) Jarman, R.H.: Solid State Ionics, 1989, 32/33, 45

21) Feist, T.P., and Davies, P.K.: Chem. Mater., 1991, 3, 1011

22) Gerand, B., Nowogrocki, G., and Figlarz, M.: J.Solid State Chem., 1981, 38, 312

23) Gerand, B., Nowogrocki, G., Guenot, J., and Figlarz, M.: J. Solid State Chem., 1979, 29, 429

Materials Science Forum Vols. 152 - 153 (1994) pp. 183-186

TERNARY NITRIDE SYNTHESIS: AMMONOLYSIS OF TERNARY OXIDE PRECURSORS

D.S. Bem, J.D. Houmes and H.-C. zur Loye

Department of Chemistry, Massachusetts Institute of Technology,
Cambridge, MA 02139, USA

Keywords: Ternary Nitrides, Ammonolysis of Metallates, Oxide Precursors, $FeWN_2$, $\alpha\text{-}MnWN_2$, $\beta\text{-}MnWN_2$

ABSTRACT: The ternary nitrides $FeWN_2$, $\alpha\text{--}MnWN_2$, and $\beta\text{--}MnWN_2$, have been synthesized through the ammonolysis of transition metal tungstate precursors. $FeWN_2$ is hexagonal with a unit cell of $\mathbf{a} = 2.8724(1)$ Å and $\mathbf{c} = 10.973(4)$Å. Rietveld refinements were performed in the space group $P\bar{3}1C$ with R_{wp}=9.91%, and R_p= 7.23%. Ammonolysis of $MnWO_4$ yields two products: at 700°C $\alpha\text{--}MnWN_2$ [$\mathbf{a}$=2.914(8), $\mathbf{c}$=16.562(6) space group R3] and at 800°C $\beta\text{--}MnWN_2$ [$\mathbf{a}$ =2.922(3), $\mathbf{c}$=10.897(9) space group P6$_3$/mmc]. Both $MnWN_2$ compounds have similar $\mathbf{a}$ lattice parameters, however the $\mathbf{c}$ parameter of $\beta\text{--}MnWN_2$ is shorter by 5.665 Å. This shortening of the $\mathbf{c}$ axis results from a change in the stacking sequence of the metal layers in the structure. This is the first example of a ternary nitride that can be synthesized in two structural modifications.

We are exploring the use of transition metal oxides as precursors for the low temperature synthesis of new nitrides. Previous synthetic strategies for ternary nitrides involved the stabilization of late transition metals via the inductive effect.[1,2] This approach has been very successful; however, it limits the number of potential new ternary nitrides. Consequently, there are few known and well characterized ternary nitrides that do not contain a highly electropositive element. The nitrides that do not contain electropositive elements are often not synthesized directly, but rather by the conversion of one nitride into another. For example, $CuTaN_2$ was synthesized via an ion exchange reaction between CuI_2 and $NaTaN_2$.[3]

Currently we are investigating the use of transition series metallates as precursors to ternary transition metal nitrides in single step reactions. We have previously reported the synthesis of the intermetallic nitrides Co_3Mo_3N, Fe_3Mo_3N and Ni_3Mo_3N[4] by the ammonolysis of cobalt, iron and nickel molybdate precursors. Investigation of the ammonolysis of tungstate precursors has led to the synthesis of $FeWN_2$[5], $\alpha\text{--}MnWN_2$, and $\beta\text{--}MnWN_2$. The successful preparation of these nitrides in the absence of highly electropositive elements demonstrates the utility of this synthetic approach, and promises to lead to additional ternary nitride phases. Furthermore, the preparation of $MnWN_2$ is of particular interest, because it represents the first example of a ternary nitride that can be synthesized in two structural modifications.

In this study, transition metal tungstates, $FeWO_4$ and $MnWO_4$, were used as precursors for ternary transition metal nitrides. Hydrated metal tungstates were prepared by drop-wise addition of 400 mL (0.25 M) aqueous solution of metal chloride, $FeCl_2$ (Cerac, 99.99%) or $MnCl_2$ (Cerac, 99.5%) to a 150 mL (0.55 M) solution of $Na_2WO_4 \cdot (H_2O)_2$ (Aldrich 99%). The product was isolated by vacuum filtration and rinsed with two washings of water followed by a single washing with ethanol. The solid was dried at 150°C for 24 hours. The products were brown powders and amorphous by powder X-ray diffraction. The metal tungstate precursor was placed into an alumina boat which was inserted into a quartz flow through reactor located in a hinged tube furnace. The samples were heated under flowing ammonia gas (150cc/min) at 700°C ($FeWN_2$ and α–$MnWN_2$) or 800°C (β–$MnWN_2$). The reaction temperature was held constant for 12 hours followed by quenching to room temperature by turning off and opening the furnace.

The absence of oxygen and the metal ratios of $FeWN_2$, α–$MnWN_2$ and β–$MnWN_2$ were determined from pressed pellets using energy dispersive spectroscopy on a Jeol JSM 6400 scanning electron microscope collected by a Noran Z-max windowless detector with quantification performed using virtual standards on associated Voyager software. Nitrogen content was determined by combustion C,H,N analysis (Oneida). The results of the metal and nitrogen analysis are consistent with the compositions $FeWN_{1.93}$, α–$MnWN_{1.95}$, and β–$MnWN_{1.95}$. Further characterization of the products was carried out by X-ray powder diffraction using a Rigaku RU300 diffractometer with Cu$K\alpha$ radiation ($\lambda = 1.54184$ Å). NBS silicon was used as a standard for accurate peak positions.

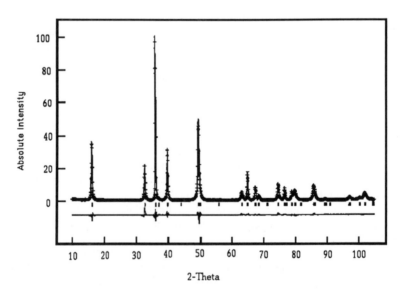

Fig 1: Data fit and residuals for $FeWN_2$ (space group $P\bar{3}1C$, $a=2.8724$ $c=10.973$, $R_{wp}=9.91\%$, $R_p=7.23\%$)

The powder X-ray diffraction data of FeWN$_2$ was previously reported as hexagonal with **a** = 2.867(2) Å and **c** = 16.458(9) Å [5]. However when Rietveld refinements were attempted no adequate fit could be achieved. Reindexing the X-ray diffraction data using a unit cell of **a** = 2.8724(1) Å and **c** =10.973(4)Å gave a satisfactory refinement in the space group P$\bar{3}$1C [W (0,0,1/4), Fe (0,0,0) and N (1/3,2/3,0.128)] R$_{wp}$=9.91% , R$_p$= 7.23%) (Figure 1) The proposed structure has similar metal-nitrogen coordination to LiMoN$_2$ [6] and consist of layers of trigonal prismatically coordinated tungsten and octahedrally coordinated iron.

The powder pattern of α–MnWN$_2$ is isostructural with that of LiMoN$_2$ [6] and was indexed hexagonal (R3; **a**=2.914(8) and **c**=16.562(6) Table 2). Powder X-ray diffraction of β–MnWN$_2$ was similarly indexed hexagonal, (**a** =2.922(3) and **c**=10.897(9) space group P6$_3$/mmc, Table 3), and appears to be structurally related to Ta$_3$MnN$_4$ [7] with manganese on one-third of the tantalum sites.

h k l	d$_{obs}$	d$_{calc}$	I$_{obs}$	I$_{calc}$	h k l	d$_{obs}$	d$_{calc}$	I$_{obs}$	I$_{calc}$
0 0 3	5.5362	5.521	100	100	0 0 2	5.487	5.489	78	84
0 0 6	2.7625	2.7603	59	51	0 0 4	2.733	2.737	44	42
1 0 1	2.4957	2.4956	97	100	1 0 1	2.465	2.467	100	100
1 0 2	2.4167	2.4147	54	55	1 0 2	2.296	2.296	51	72
1 0 4	2.1558	2.1554	33	40	1 0 3	2.078	2.079	68	55
1 0 5	2.0083	2.0078	42	62	1 0 4	–	1.857	–	5
0 0 9	1.8402	1.8402	6	4	0 0 6	1.823	1.825	7	3
1 0 7	1.7256	1.7263	24	33	1 0 5	1.655	1.655	32	22
1 0 8	1.6004	1.6008	10	12	1 0 6	1.478	1.479	12	11
1 1 0	1.4562	1.4575	31	19	1 1 0	1.461	1.462	24	17
1 1 3	1.4080	1.4092	15	8	1 1 2	1.412	1.413	12	7
1 0 10	1.3867	1.3848	7	11	0 0 8	1.368	1.368	8	4
0 0 12	1.3803	1.3802	11	5	1 0 7	1.329	1.330	12	8
1 0 11	1.2937	1.2931	10	6	1 1 4	1.289	1.288	22	12
1 1 6	1.2886	1.2889	21	14	2 0 1	1.258	1.256	2	6
2 0 1	1.2584	1.2586	8	6	2 0 2	1.233	1.232	5	4
2 0 2	1.2479	1.2478	7	3	2 0 4	–	1.148	–	0
2 0 4	1.2081	1.2074	4	3	1 1 6	1.141	1.140	5	2
2 0 5	1.1802	1.1795	5	5	1 0 9	1.096	1.096	6	3
1 1 9	1.1420	1.1425	5	2	2 0 5	1.093	1.094	4	3
1 0 13	1.1373	1.1374	5	3	0 0 10	–	1.094	0	0

Table 2: Observed and calculated d-spacings and peak intensities of α–MnWN$_2$[8]

Table 3: Observed and calculated d-spacings and peak intensities of β–MnWN$_2$[8]

The structures of α–MnWN$_2$ and β–MnWN$_2$ appear to be related by a change in the stacking of the metal layers along the **c** axis. Both phases have layers of octahedrally coordinated manganese and trigonal prismatically coordinated tungsten and similar **a** lattice parameters, however there is a difference of 5.665 Å in the **c** parameter. Defining each metal layer as A, B, or C, the metal layers of α–MnWN$_2$ have a stacking sequence of AA'BB'CC' similar to the metal layer stacking of LiMoN$_2$.[6] In β–MnWN$_2$, however, the stacking sequence is ABAC similar to that found in FeTa$_3$N$_4$.[7] This stacking rearrangement results in a reduction of the **c** lattice parameter by one tungsten-nitrogen and one manganese-nitrogen layer or 5.665 Å. Based on the structures of LiMoN$_2$ and Ta$_3$MnN$_4$ the change in the **c** parameter is ≅ 5.52 Å, which is in reasonable agreement with the observed difference of 5.665 Å. Similar structural rearrangements have been observed in the transition metal sulfide NbS$_2$. Niobium sulfide exists as rhombohedral NbS$_2$ below 800°C (space group R̄3m; **a**= 3.33, c=17.81) and hexagonal NbS$_2$ above 850°C (space group P6$_3$/mmc; **a**=3.31, **c**=11.89).[9] Heating of α–MnWN$_2$ in a atmosphere of argon to 800°C results in decomposition of the product, however heating under ammonia results in β–MnWN$_2$. Further structural studies of these transitions are currently under investigation, as well as studies of the physical properties of these nitrides.

References:

1.   F. J. DiSalvo, Science,1990, 247, 649-655.

2.   J. Etourneau, J. Portier, F. Ménil, J. Alloys Compd.,1992, 188, 1-7.

3.   U. Zachwieja, H. Jacobs, Eur. J. Solid State Inorg. Chem.,1991, 28, 1055-1062.

4.   D. S. Bem, C. P. Gibson, H.-C. zur Loye, Chem. Mater.,1993, 5, 397-399.

5.   D. S. Bem, H.-C. zur Loye, J. Solid State Chem.,1993, 104, 467.

6.   S. H. Elder, F. J. DiSalvo, L. H. Doerrer, Chem. Mater.,1992, 4, 928-937.

7.   N. Schönberg, Acta Chem. Scand.,1954, 8, 213-220.

8.   A. C. Larson, et al., . P. S. White, Eds. Chemistry Division, NRC, Ottawa, Canada, K1A 0R6,

9.   W. G. Fisher, M. J., Sienko, Inorg. Chem.,1980, 19, 39-43.

Acknowledgment is made to the donors of The Petroleum Research Fund, administered by the ACS, for partial support. Additional financial support from DuPont and Hoechst-Celanese is acknowledged. DSB would like to acknowledge 3M for support.

Materials Science Forum Vols. 152 - 153 (1994) pp. 187-192
© *1994 Trans Tech Publications, Switzerland*

ROOM TEMPERATURE TAILORING OF ELECTRICAL PROPERTIES OF SEMI- AND SUPERCONDUCTORS VIA CONTROLLED ION MIGRATION

D. Cahen [1], L. Chernyak [1], K. Gartsman [1], I. Lyubomirsky [1], Y. Scolnik [1], O. Stafsudd [2] and R. Triboulet [3]

[1] Weizmann Institute of Science, Rehovot 76100, Israel

[2] Engn. & Appl. Science, UCLA, USA

[3] Lab. Phys. Solides, CNRS, F-92195 Meudon, France

Keywords: Electronics, Semiconductor, Superconductor, Diffusion, Electromigration, Junction, Device, Chalcopyrites, Chalcogenides, Cuprates

ABSTRACT   In semi- or superconducting compounds, with a relatively mobile component, this component's local concentration can be varied significantly by ion electromigration that occurs while the bulk of the material remains at or below room temperature. Because in ABX_2 semiconductors (with A=Cu or Ag, B= In or Ga and X= S, Se or Te) or in (Hg,Cd)Te changing the concentration of the mobile Cu, Ag or Hg(?) affects their electrical properties (via control of the electronic carrier density), this allows for actual room temperature device creation. In high T_c cuprate superconductors oxygen is mobile and its local concentration can be changed electrochemically at room temperature to control T_c and other electronic properties.

INTRODUCTION   By viewing electronic device structures from a solid state ionic, and mixed conductors from a solid state electronic point of view, and by studying the behaviour of mixed ionic/electronic conductors far from equilibrium, a new direction in ionics*, can be explored. Among the first results are new, gentle (douce), ways for creating electronic (p/n/p and S/N/S) junctions and the emergence of the concept of self-stabilizing junctions.

Recall that junctions represent non-equilibrium (thermodynamically unstable) distributions of chemical species, be they dopants or major constituents. Contrary to, for example, semiconductors, mixed conductors are normally considered using equilibrium thermodynamics, which teaches that non-equilibrium distributions cannot be maintained without applying an external potential and, therefore, no internal electric field can exist in them. In semiconductors such distributions are commonly obtained by inducing dopant diffusion at high temperatures or by direct, high energy ion implantation of dopants.   Similar high energy methods have been employed to create superconducting device structures.

"SEMIONIC" SEMICONDUCTORS   In such   materials the concentration of at least one dopant can be changed in a soft chemical manner (e.g. by low temperature dopant migration), while such changes in composition do not lead to degeneracy or metallic behavior, i.e. the material remains a semiconductor [1]. The electrical properties of IB-III-VI$_2$ type chalcopyrites, such as $CuInS(e)_2$, ambipolar semiconductors, are dominated by *native* point defects, including Cu-related ones, whose concentration is strongly dependent on the sample stoichiometry [2]. Especially the Ag compounds show ionic conductivity, but also in $CuInS(e)_2$ the chemical diffusion coefficient of Cu was found to be appreciable at low temperatures, depending on the concentration of V_{Cu} [3]. Because of doping action of Cu-related defects, gentle creation of doping profiles becomes feasible (cf. Fig. 1).

Indeed, by applying electric (E) - fields *at ambient temperatures* non-equilibrium doping profiles can form, that are stable after removal of the E-field. This was shown by formation of µm-sized single and multiple diode structures in initially homogeneous and uniformly doped single crystals of $CuInSe_2$ [3,4] & (Hg,Cd)Te [5]. These effects are seen by electron (light) beam-induced current, EBIC and by I-V (current-voltage) measurements, after the external E-field was removed. For such measurements the sample is inside a scanning electron microscope (SEM).

The principle of EBIC is best illustrated by first explaining the optical equivalent OBIC or LBIC (Optical- or Light-beam induced current). When a photon with > bandgap energy hits a semi-conductor sample and is absorbed, an electron is promoted to the conduction band and a positively charged hole is left in the valence band, i.e. an electron-hole pair is formed. If this process occurs in a homogeneous semiconductor, then the electron and hole will recombine and heat will be produced. If, however, the process occurs in a region where there is an electric field, the electron and hole will be separated, flowing into opposite directions. If suitable electrical contacts are provided, an external electric current can be measured. By scanning the photon beam across a sample and measuring the current as a function of position, a line profile or map can be obtained, that reflects electrical inho-mogeneities in the material. Thus, if a p/n junction exists in the sample an OBIC signal will be ob-tained in that region. This is so because the p/n junction is characterized by an internal electric field. Instead of photons, electrons can be used. In SEM electrons with energies of 10's of keV are often used. One such electron can produce many electron hole pairs in the material (theoretically some 10,000 e^--h^+ pairs can form after a30 keV electron hits $CuInSe_2$). Thus EBIC can provide informa-tion on the electrical nature of a sample and especially on the existence of internal E-fields in it, with high sensitivity and with the spatial resolution of the SEM. The information comes from a layer that is several μm's deep, the penetration depth of the electrons. The above is a highly simplified descrip-tion of EBIC, but one that suffices for the present purpose. More information can be found in ref.6.

Because the internal E-fields result from gradients in doping density, due to sub-ppm charge imbalances, EBIC serves essentially as an ultra-sensitive analytical tool to follow compositional changes in a spatially resolved manner. Similar results are obtained with OBIC, when the sample is illuminated in a micromanipulator.

Figure 1: Schematic of an experimental set-up, for applying voltage to (p-)$CuInSe_2$, between small area point and large area back contact. Electron flow and Cu ion migration are shown. The latter results in changing the electronic properties of the area close to the point contact. To create the doping profiles, reverse bias voltages of 10 to 100's of Volts, depending on the materials bulk resistivity, are applied intermittently to the small top contact, as described in ref. 4.

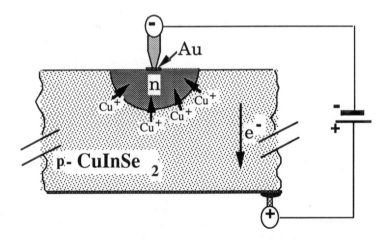

Because a (junction) transistor is a structure of two diodes, we looked for transistor action. Two-terminal phototransistor action and amplification was obtained, i.e. clear evidence for sharp junctions, resulting from non-equilibrium doping profiles [7] (Fig. 2). Before applying the E-Field, the top contact behaved as a Schottky barrier. Such *E-field-induced* structures are stable for > 8 months, after removal of the E-field, also in terms of device action. No longer term stability mea-surements have been done, as yet. We note that in the case of *thermally created* doping profiles we have in some cases noted changes on a scale of ~10's of μm, after 100 days.

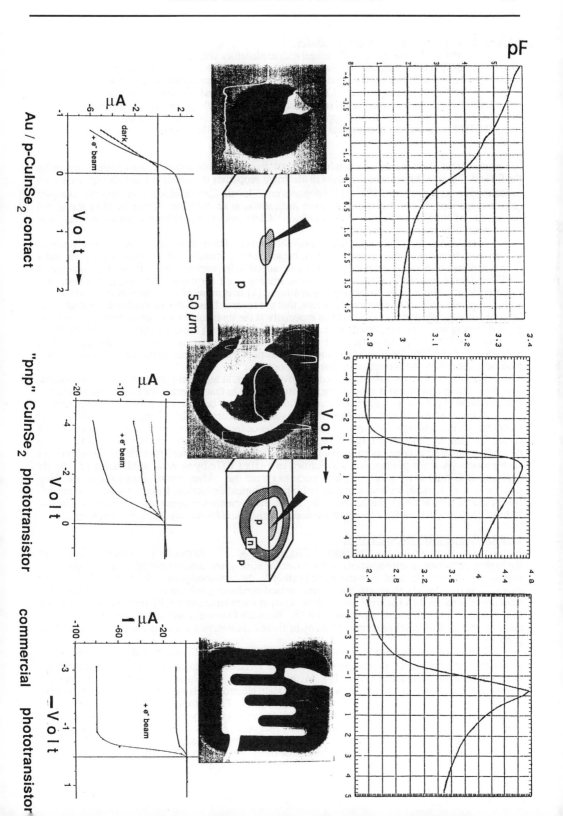

Figure 2: TOP: change in C-V characteristics,
MIDDLE: change in EBIC image(with explanatory sketches);
BOTTOM: change in I-V characteristics with and without e-beam "illumination",
before and after applying external E-field, at room temperature, in a sample of
originally homogeneous p-CuInSe$_2$ (left & middle) and of commercial transistor (right)
LEFT: Characteristics of 50 μm diameter Au/CuInSe$_2$/large area Au back contact.
MIDDLE: Same after E-field application between top and bottom Au contacts.
RIGHT: Same for commercial p/n/p transistor (only EBIC, without sketch, is shown).

We interpret this phenomenon as a result of internal dopant redistribution/electromigration, driven by the E-field. To give the observed changes in electrical properties it suffices to have changes in dopant concentration on the order of 10^{17} - 10^{18} dopants/cm-3 level. Careful electron microprobe analyses (electron-induced, X-ray fluorescence) gave indications at an 80% confidence level of the actual occurrence of changes in elemental composition. Clear evidence will have to await the results of experiments with radioactive tracers.

Simulations show that under the non-equilibrium conditions that apply here dopant electromigration leads to compositional inhomogeneity, because of the formation of a region enriched in, and a region depleted of dopants, embedded in the unchanged bulk [unpublished]. Further application of the E-field leads apparently to self-amplification, leading to the formation of a double structure, such as a transistor. This can be understood by realizing [4] that more of the potential drop will occur across regions depleted of electronic carriers, that are reverse biased during further E-field application. Because the phenomenon is observed especially at the threshold of sudden current increases, localized heating will occur, especially in the regions of non-uniform dopant distribution. Although ΔT is found to be < 100°C at the surface and < 200°C inside the material [unpublished], the temperature dependence of the diffusion coefficient, together with such localized heating can also provide positive feedback, needed for the formation of sharp junctions.

Elsewhere [4,7] we have discussed the involvement, at high field strengths, of electron wind, to explain that in several cases we observe polarity opposite to that expected for Coulombic migration. Electron wind requires a high electronic carrier flux, which may be achieved at avalanche breakdown.

While such mechanism can explain the stability of structures, EBIC and Seebeck effect measurements on the Ag/(Hg,Cd)Te system provide evidence for a novel self-stabilization process, based on the fact that Ag diffuses in (Hg,Cd)Te by a dissociative diffusion mechanism, i.e. it diffuses as a donor (interstitial), but dopes as an acceptor (substitutional). Recent results provide vivid evidence for such process. A sharp p/n junction was formed in n-(Hg,Cd)Te by thermal in-diffusion of Ag. By applying an E-field to the junction it was made to spread out. After several days the original sharp junction was re-formed, without any external input. Logical deduction, based on a dissociative diffusion mechanism leads to sharp, stable doping profiles, created in apparently counter-Coulombic fashion. *Such profiles could be thermodynamically, rather than kinetically stable p/n junctions.*

CUPRATE SUPERCONDUCTORS:
Formation of superconducting junctions in YBa$_2$Cu$_3$O$_{7-x}$ Room temperature modification of the high temperature superconductors is important for semi- *cum* super-conducting technologies, due to possible adverse effects of even moderate heating on the semiconductor and interface quality. The cuprate superconductors show low temperature mixed conducting behavior, due to the relatively high mobility of some of the O atoms / ions in them, even at room temperature [8] (in some cases it is due to H insertion and extraction, alone, or with O). Because O-content and order affect the superconducting (and normal state) properties of many of these cuprates, this suggests a way to control these properties.

Earlier we reported that bulk polycrystalline YBa$_2$Cu$_3$O$_7$(123) can be reduced quantitatively, in a controlled manner, at room temperature, using an electrochemical set-up with 123 as the cathode in a non-aqueous liquid electrolyte [9]. In thin films not only quantitative reduction (via wet electrochemistry), but also re-oxidation (via solid state electrochemistry) was achieved *at room temperature,* by way of extraction or insertion of oxygen [10]. The homogeneity of the products allows now spatial control over the films' O-content. Both epitaxial and polycrystalline thin films were used. After reduction the films were re-characterized by magnetic, electrical and structural (X-ray diffraction; XRD) measurements. Depending on the amount of charge passed, complete or partial loss of superconductivity was found. XRD showed loss of oxygen (x~6.4 to 6.7). Upon re-oxidation, after passing a sufficient amount of charge, XRD showed a phase with increased oxygen content (x~6.8), as well as near-complete restoration of the electrical and magnetic superconducting properties. Moreover, the diffraction patterns showed, in some cases, a decrease in peak width (FWHM) (up to ~15%) which we interpret as an increase in the homogeneity of the samples after the treatment.

By exposing only very thin regions of the film to the solution we can prepare Superconducting - Non Superconducting -Superconducting (SNS)-like structures. We did so by using lithographic methods to delineate narrow bridges on the film surface [11]. The structures that are obtained by such locally selective reduction process showed low temperature behavior consistent with the formation, not only of SNS junctions, but of that expected for multiple Josephson junctions [12]. This presents a reversible, soft chemical way of patterning 123 thin films. Junctions prepared by this method give I-V characteristics and $I_c \cdot R_n$ results, typical for S/N/S junctions. Preliminary results of the dependence of the critical current, I_c, on the magnitude of an externally applied magnetic field show a dependence with a periodicity of some 50 - 75 Gauss. This dependence is not clean enough to provide proof for a single Josephson junction, because I_c does not approach zero, but rather drops by 30 - 50 %. Still, it strongly indicates that multiple Josephson junctions have been created, as indeed is to be expected for this process. Such junctions may present the first case of monolithic Josephson junctions.

We note that, for medium quality films, the superconducting behavior, as measured resistively, can actually be improved after slight reduction of 123. Even larger improvements are then observed after subsequent re-oxygenation, as is to be expected on the basis of the known dependence of T_c on oxygen content We interpret the improvement upon reduction tentatively in terms of reduction-induced changes in the concentration and order of oxygen in the outermost layers of individual grains [13]. The same effect is probably involved in the observation that, even for high quality films, reduction, as well as subsequent re-oxygenation lead to a decrease in FWHM of XRD peaks.

Acknowledgements: We thank the US-Israel Binational Science Foundation, Jerusalem, ISRAEL, the Israel Science Foundation, the Israeli Ministry of Science and Technology and the Levine Fund for Applied Research for financial support.

References and Footnotes

*   "Semionics", in the case of semiconductors. The name "superionics" is used for another purpose.

1   Cahen,D., in *Ternary and Multinary Compounds,*(M.R.S.,Pittsburgh,1987), vol.TMC, p.433 ff.
2.   Neumann,H.in*Verbindungshalbleiter* (eds.Unger,K.&Schneider,H.;Geist &Portig,Leipzig,1986)
3   Dagan, G., Ciszek, T.F., Cahen, D., J.Phys. Chem., 1992, 96,11009
4.   Jakubowicz, A., Dagan,G., Schmitz, C. and Cahen,D., Adv. Mater., 1992,4, 741
5.   Gartsman,K., Chernyak,L.,Gilet,J.M.,Triboulet,R.,Cahen,D.,Appl. Phys.Lett., 1992, 61, 2428
6.   Holt, D.B. in *SEM Microcharacterization of Semiconductors* (Holt, D.B. and Joy, D.C. eds., Academic, London, 1989), Ch. 6
7.   Cahen, D., Gilet, J.M., Schmitz, C., Chernyak, L., Gartsman, K.,Science, 1992, 258, 271
8   Scolnik, Y., Sabatani, E., Cahen,D., Physica C, 1991,174, 273
9   Schwartz, M., Scolnik, Y., Rappaport, M., Hodes, G., Cahen,D., Physica C, 1988,153-155, 1467; Mater. Lett.,1989 ,7, 411; J. Mater. Chem., 1991,1, 339
10   Scolnik, Y., Cahen, D., J. Electron. Mater., 1991, 20, 34
11   D. Cahen, D. and Scolnik, Y., U.S.A.. Pat. Appl. 1991
12   Tolpygo, S., et al., Physica C , 1993, 209 , 211
13   Rudolf, P.,Paulus, W., and Schöllhorn, R., Adv. Mat., 1991, 9, 438

Materials Science Forum Vols. 152 - 153 (1994) pp. 193-196
© 1994 Trans Tech Publications, Switzerland

ELECTROCHEMICAL OXIDATION OF La$_2$CuO$_4$.
PHASE EQUILIBRIA AMONG SUPERCONDUCTORS

S. Ondoño-Castillo, P. Gomez-Romero, A. Fuertes and N. Casañ-Pastor

Institut de Ciencia de Materials de Barcelona, C.S.I.C., Campus U.A.B.,
E-08193 Bellaterra, Barcelona, Spain

Keywords: Electrochemical Oxidation, Oxygen Doping, Copper Oxides, Superconductivity, Charge Reorganization, Phase Equilibria

ABSTRACT

Electrochemical oxidation of La$_2$CuO$_4$ in aqueous OH$^-$ yield superconducting materials with Tc values that depend on the degree of oxidation and the temperature. The chemical nature of the charge carriers changes with temperature, yielding notable changes in Tc values. For the maximum oxidation of 0.15 holes per copper, an internal redistribution of holes between Cu and O is proposed from RT to 100°C. Partial derivatives of magnetic susceptibility vs susceptibility show the existence of several coexisting phases with distinct Tc's, possibly within each grain.

INTRODUCTION

High-Tc superconducting oxides are obtained by cation or oxygen doping of the corresponding stoichiometric parent compounds. In many cases of oxygen doping, the use of high pressures is needed to ensure insertion of excess oxygen that may in turn create the appropriate number of charge carriers to make the oxide superconducting. That is the case for La$_2$CuO$_4$ [1]. Electrochemical methods constitute an alternative to the use of high pressures and high temperatures, in which a soft route to these materials is followed by applying an externally imposed potential. The redox reaction forces the insertion of ions of the appropriate charge to maintain charge neutrality.

Electrochemical oxidation and reduction of several cuprates have been reported [2,3,4]. Some of those electrochemical reactions have been carried out at room temperature using an aqueous basic medium or a non-aqueous system. In the case of La$_2$CuO$_4$ oxidation at room temperature in presence of OH$^-$ ions yields a superconducting material [3,5] with Tc values higher than those reported for the high pressure phases and that depend on the number of carriers introduced [5a,b] and on the thermal treatment that follows the oxidation [5c]. The maximum Tc value observed corresponds to a sample with a number of holes around 0.15 and that has not been treated above room temperature.

We have previously observed that the orthorhombic distortion of the cell found in electrochemically oxidized samples [3,5] decreases as the doped La$_2$CuO$_{4+\delta}$ is treated at 100°C [5c], going even below the orthorhombic distortion of the parent phase La$_2$CuO$_4$. At the same time the Tc values at which superconductivity appears get reduced by about 10K, going from 45 to 35K, approaching those found in high pressure samples. That occurs without change in the total number of carriers, as TGA analysis demonstrate [5c]. This paper reports evidence indicating the existence of an equilibrium of phases near room temperature that we interpret as an internal charge redistribution among copper and oxygen ions. Also, the paper reports a study of the derivatives of magnetic susceptibility with temperature that are indicative of the existence of distinct phases coexisting within each grain, and that we believe are inherent to the method of preparation. Therefore, the phases obtained electrochemically at room temperature cannot be considered the same as the ones obtained with high pressure, but the thermal studies performed demonstrate that they are related to each other [5c].

EXPERIMENTAL

La_2CuO_4 was prepared by solid state reaction of stoichiometric amounts of La_2O_3, and CuO (Aldrich 99.99%) at 1050°C for 48 hours and subsequent annealing at 1050°C until no impurities were observed in the X-ray diffraction pattern. Pellets of La_2CuO_4 were obtained by grinding this material, pressing it into a 13 mm diameter pellet (10 tons), and sintering it at 1050°C for 24 hours. The density was around 80%. The pellets were semiconducting before the oxidation. Electrochemical oxidation experiments were carried out in 1M KOH or NaOH media (bidistilled water), using a double compartment cell. The counterelectrode was Pt wire (0.5 mm diameter), and the reference electrode was Ag/AgCl doubly-bridged with 3M KCl. Electrical contact with the oxide pellet was made through platinum wire and silver paint. The contact and back side of the pellet were protected from the solution by a watertight teflon holder and a Kelrez O-ring. All experiments were carried out at constant potential (0.55 and 0.7 V vs Ag/AgCl for La_2CuO_4 using a EG&G PAR 273A Potentiostat interfaced to a PC. Electrolysis times ranged from 10 to 70 hours. Cyclic voltammograms were run between 1 and 10 mV/sec. Current efficiencies are close to one for potentials below 0.65 V, but above this value they get reduced significantly. Oxygen evolution is always observed in large or small extent. Pellets were washed with water and ethanol after oxidation and dried either at room temperature under vacuum for at least 12 hours, or at 100°C for one to two hours. All experiments were carried out in argon atmosphere to eliminate carbon dioxide and oxygen. For each oxidation a blank pellet of La_2CuO_4 was dipped in the electrolyte for the same period of time than the pellet being oxidized. Further analyses were carried out both on blank and oxidized pellets.

Oxygen contents and formal oxidation states were determined in homogeneized powdered samples by TGA in Ar/H_2(5%) using a Perking Elmer TGA7, and by iodometry according to the method proposed by Nazzal and Torrance [6]. TGA experiments in Ar/H_2 were performed with an initial weight equilibration of 60 minutes at room temperature followed by a 2 °C/min ramp to 550 °C, and an isotherm of five hours at this temperature. TGA data reported in Table 1 are based on calculations that assume dinegative inserted species, O^{-2}. X-ray powder diffraction data were obtained on a Rigaku D/max-RC rotating anode diffractometer (RU 200-B generator) equipped with a germanium (111) monochromator (Cu Kα1). Magnetic properties were measured on a Quantum Design SQUID Magnetometer at fields below 10 Oe, in powder form. Resistance measurements were performed with the help of a EG&G PAR lock-in (model 5209) by the four probe method.

DISCUSSION

Applied potentials above 0.5 V vs Ag/AgCl yield samples oxidized up to a limiting value of 0.15 holes per Cu. According to the analytical data, values above that number are never obtained. All facts discussed below correspond to samples with a number of carriers corresponding to that limiting value. As Table 1 and Figure 1 show, it is possible to obtain samples with an overall number of charge carriers equal, but with distinct susceptibility responses, by changing the applied potential. Thus, at 0.55 V we obtain a very small fraction of a phase with a Tc of 43K, while at 0.7V, that phase is the main component.

Table 1. Oxidation states and Tc values obtained in electrochemically oxidized samples.

$E_{applied}$ vs Ag/AgCl	Drying Temperature	Tc	Formal Oxid. State for Cu (TGA)	Formal Oxid. State for Cu (Iodom.)
0.7 V	RT	45 K	2.13	2.07
0.7 V	100°C	33 K	2.12	2.14
0.55 V	RT	43 (+30) K	2.12	2.06
0.55 V	100°C	32 (+20) K	2.12	2.12

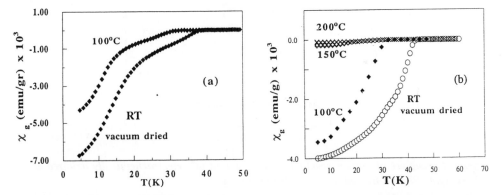

Figure 1. Mass susceptibility vs T for La₂CuO₄ oxidized electrochemically and treated at several temperatures. a) Oxidation at 0.55 V, b) oxidation at 0.7 V vs Ag/AgCl.

TGA analyses show that in all cases studied the number of charge carriers is the same. However, iodometry yields lower number of carriers for non-heated samples. Such inconsistency may only be explained if the chemical nature of the charge carriers varies from room temperature to 100°C treated samples, the oxidant species being less reactive vs iodine in the low temperature samples. It is known that iodine reacts very slowly with peroxides [7], and therefore, iodometry may not be giving the total oxidative power if oxygens in intermediate oxidation states are present. A possible interpretation of these facts, that would also explain the change in Tc values observed, is the existence of an internal redistribution of charge carriers among Cu and O, that would keep the total number of holes constant between room temperature and 100°C. (Further support of this interpretation is given by the fact that water or hydroxide oxidation on perovskites yield peroxo groups [8]). This would correspond to an equation of the type:

$$La_2Cu^{+2}_{1-x}Cu^{+3}_xO^{-2}_4(O^-)_\delta \xrightarrow{\;\;100°C\;\;} La_2Cu^{+2}_{1-x-\delta}Cu^{+3}_{x+\delta}O^{-2}_{4+\delta}$$

in which, oxygen in low oxidation states oxidizes Cu^{+2} to Cu^{+3}.

As the temperature is raised above 100°C, the Meissner fraction lowers considerably, although the superconducting signal is still large at 200°C (corresponding to about 10% superconducting fraction). TGA data in oxygen shows weight loss at this temperature [5c], implying a reduction of the oxide that destroys the superconductivity.

The derivatives of mass susceptibility vs T show clearly the existence of several transitions in all samples, with the exception of the 200°C material. It is very probable that these correspond to an equilibrium of phases with a different arrangement of excess oxygen or of charge carriers.

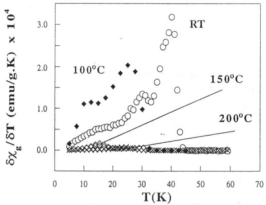

Figure 2. Derivatives of mass susceptibility vs T for La$_2$CuO$_4$ electrochemically oxidized at 0.7 V vs Ag/AgCl, and treated at several temperatures.

ACKNOWLEDGEMENTS. The authors thank Spanish CICYT (Ministry of Education) and MIDAS Program for financial help.

REFERENCES
[1] e.g. Schirber, J.E. et al; Physica C, (1988), 152, 121
[2] a) Schwartz, M., et al.; Physica C, (1988), 153-155, 1457. b) Schwartz, M. et al.; Solid State Ionics, (1989), 32/33, 1137. c) Rochani, S. et al.; J. Electroanal. Chem., (1988), 248, 461
[3] a) Wattiaux, A., at al.; Comp. R. Ac. Sci. (Paris), (1990), 310, 1047. b) Grenier, J.C. et al.; Physica C ,(1991), 173, 139. c) Grenier, J.C. et al.; J. Solid State Chem., (1992), 96, 20.
[4] Koshkin, V.M. et al.; Superconductivity, (1991), 4, 158
[5] a) Chou, F.C. et al.; Physica C (1992), 197, 303. b) Grenier, J.C. et al.; Physica C, (1992), 202, 209. c) Casañ-Pastor, N. et al.; Solid State Ionics (1992) in press. Casañ-Pastor, N. et al.; Physica C, accepted
[6] Nazzal, A.I. et al.; Physica C (1988), 153-155, 1367
[7] Kolthoff, I.M. "Analisis Químico Cuantitativo" Ed. Nigar, Buenos Aires 1965, p. 883
[8] Kinoshita, K. "Electrochemical Oxygen Technology" John Wiley & Sons, New York 1992

Materials Science Forum Vols. 152 - 153 (1994) pp. 197-200
© 1994 Trans Tech Publications, Switzerland

OBTENTION OF SUPERCONDUCTING YBa$_2$Cu$_3$O$_{7-\delta}$ DEPOSITS BY ELECTRODEPOSITION AND ELECTROPHORESIS

S. Ondoño-Castillo, J. Bassas, P. Gomez-Romero, A. Fuertes and
N. Casañ-Pastor

Institut de Ciencia de Materials de Barcelona, C.S.I.C., Campus U.A.B.,
E-08193 Bellaterra, Barcelona, Spain

Keywords: Superconductivity, Electrodeposition, Electrophoresis, Deposits

ABSTRACT

Simultaneous electrodeposition of Y, Ba and Cu from solution is achieved in presence of complexing agents, using metal wires as substrates. The resulting wire is superconducting after thermal treatment at 900°C. The main problem is the presence of excess copper in the deposit that may be easily overcome using cyanide as a complexing agent which inhibits copper deposition in DMSO. On the other hand, deposition of YBa$_2$Cu$_3$O$_{7-\delta}$ particles by electrophoresis yields superconducting wires with preferential orientation and high critical currents.

INTRODUCTION

The use of electrochemical methods to make superconducting deposits over metallic substrates constitutes a soft chemistry route with important advantages over more established methods. The poor physical properties of superconducting cuprates, related to their ceramic character, constitute a considerable problem in terms of their applicability. Using electrochemical methods, substrates of practically any shape can act as physical supports, avoiding the need of any further mechanical processing. On the other hand, scaling up the electrochemical methods is a problem largely solved, given the extent of application of these methods in industry. That, together with their relative low cost makes these methods very attractive from an industrial point of view. Two possible methods that make use of an electric field to obtain deposits over a metallic substrate may be used. First, the simultaneous electrodeposition of the metals present in the final oxide, by electrolysis of solutions containing salts of those metals, constitutes a way of obtaining the superconducting phase in situ by reaction of highly reactive precursors at high temperatures in the appropriate stoichiometric ratio. Since superconducting oxides contain very electropositive elements aside from copper, aqueous solutions must be avoided and the simultaneous deposition of all metals is not straightforward. Thus, in the case of YBa$_2$Cu$_3$O$_{7-\delta}$, for a given reducing overpotential, copper will deposit preferentially over Y and Ba. The optimum approach to solve this problem involves inhibition of copper deposition. This could be done by complexing Cu^{+2} or Cu$^+$, which destabilizes metallic copper with respect to its oxidized forms causing a considerable decrease in its reduction potential. This paper reports the first attempts that have been made, as far as we know, to deposit copper at potentials near those of Y and Ba, using a variety of complexing agents.

Second, the deposition of solid particles of superconducting material, previously formed, by electrophoresis of their suspensions, constitutes a simple physical method that only requires a further thermal treatment to eliminate the solvent from the deposit and to connect the grains. Both methods have been widely used in industry, but their application to superconductors is still object of much study [1,2,3,4]. This paper reports the results obtained in a variety of conditions and substrates.

EXPERIMENTAL

Electrodeposition of precursors . Cyclic voltammetries were studied for each element prior to the deposition, as well as for binary mixtures and ternary mixtures of the elements. Depositions have been performed in dimethylsulfoxide (DMSO), or DMSO-H_2O, at constant potential. The reference electrode was a Ag/AgCl electrode doubly bridged with a 1M tetrabutylammonium chloride (TBACl) solution in DMSO. The working electrode is a wire of the metallic substrate (Ag, Ni, Pt, Ti) 0.5 mm in diameter. The counterelectrode is a platinum wire 0.5 mm diameter. The solutions were deoxygenated with bubbling argon prior to their study. All depositions were carried out in argon atmosphere. All experiments were run using a EG&G PAR 273A Potentiostat. The deposits were treated at 900°C in air and annealed at 450°C. XRay powder diffraction was performed on a Siemens D-500 diffractometer. Magnetic measurements were done on a Lakeshore susceptometer.

Electrodeposition of particles by electrophoresis. Suspensions of $YBa_2Cu_3O_{7-\delta}$ crystals (size 2 to 6 μm were prepared in alcohols and ketones by agitation with mechanical and ultrasonic methods. In most cases, iodine was added as an electrolyte generator. A cylindrical cell was used with concentric electrodes. The anode was a stainless steel cilinder and the cathode was the metallic substrate (Ag, Ni, Fe, Ti). Deposition occurred always at the cathode. Potentials applied ranged from 500 to 800V/cm. The current dropped exponentially with time during the electrophoresis experiment which typically lasted several minutes. The deposits were treated at 900°C in air and annealed at 450°C. Magnetic measurements were made on a Lakeshore AC susceptometer. Resistivity measurements were done by the four point method with a high current power supply. Scanning Electron Microscopy pictures were obtained from a Hitachi S-570 microscope.

DISCUSSION

Electrodeposition of precursors . When deposition from solutions containing copper, ytrium and barium are carried out in absence of any additive, copper is deposited preferentially, as expected, yielding an excess of copper oxide in the final deposit even for low copper concentrations in the starting solution. This occurs even in organic solvents as DMSO where reported reduction potentials for Cu^{+2}, Y^{+3} and Ba^{+2} are around 0.2-0, -2.2, and -2.0 Volts respectively [5]. We observe reduction waves for copper and barium with maxima at -0.2 and -2.3 V vs Ag/AgCl, in absence of additives or complexing agents. The wave for yttrium reduction, around -2V, does not have a maximum.

Table 1. Reduction waves observed in cyclic voltammetry experiments of copper nitrate solutions in DMSO. (Values correspond to wave's maxima)

Electrolyte	Complexing agent	E (Volts vs Ag/AgCl)
-	-	-0.2/-0.67
TBACl	-	-0.4/-1.25
-	Glycine	-0.45/-0.96/-1.61
TBACl	Glycine	-0.32/-0.81/-1.44
TBACl	Ethylendiamine	-0.70/-0.91/-1.42
-	Cyanide	-2.33/-2.77
TBACl	Cyanide	-2.12/-2.73

When complexing agents are used, the yttrium and barium waves shift only slightly, while copper shows little change in the case of amines, and a considerably large shift in presence of cyanide. Table 1 shows the reduction waves observed for copper nitrate solution in DMSO in presence of some additives and complexing agents. The first wave reported in each case corresponds always to copper deposition. The existence of several waves is ascribed to the presence of an equilibrium among different complexes within the solution, and to the reduction of the complexing agent. It is known that several cyanide complexes of copper exist [6] in presence of excess CN⁻, all with very large stability constants and negatively charged. The stability of these complexes is responsible for the large shift in the copper reduction potential and their negative charge may account for the strong inhibition of copper deposition observed.

Table 2 shows the conditions and deposits obtained from simultaneous depositions of Cu, Y and Ba in various media. Even in absence of any additive, we can observe the existence of superconducting $YBa_2Cu_3O_{7-\delta}$, after a thermal treatment at 900°C. The kinetics of deposition are very favorable when copper is deposited in excess. However, when the stoichiometric ratios are achieved, the growth of dendrites constitutes an important problem, which is bypassed in part using complexing agents. All wires are weakly superconducting with a weak Meissner fraction unless platinum is used as substrate, case in which the wire does not superconduct.

Table 2. Deposits obtained over Ag wire after thermal treatment at 900°C

Y:Ba:Cu (mM) starting conc.	Applied E (V) vs. Ag/AgCl	Solvent	Additives	123:211:CuO: $BaCO_3$ from Xray
40:70:95	-4	DMSO	Thiourea	100:-:50:-
40:70:95	-4	DMSO	Ethylendiamine	10:-:100:-
16:57:83	-2.5	DMSO	Cyanide, TBACl	100:-:-:66
16:57:83	-2.75	DMSO	Cyanide, TBACl	100:34:-:59
20:40:60	-2	DMSO-10%H₂O		100:-:32:-
20:40:60	-2	DMSO-10%H₂O	Ethylendiamine	80:-:64:-
20:40:30	-2	DMSO-10%H₂O	Ethylendiamine	60:-:15:-
20:40:60	-2	DMSO-10%H₂O	Thiourea	10:-:100:-

Rigorous absence of water was not needed. In fact, when DMSO was dried with activated alumina, a smaller proportion of $YBa_2Cu_3O_{7-\delta}$ was observed in the deposit. This fact induced us to study the effect of controlled additions of water to the DMSO medium. For 10% water, a dramatic change may be observed in Table 2. In presence of ethylendiamine the amount of $YBa_2Cu_3O_{7-\delta}$ phase after thermal treatment increases notably according to diffraction data. It is very probable that, at the applied potentials, the reduction of water that occurs simultaneously with copper deposition, creates a local excess of OH⁻ ions near the cathode that precipitate Y and Ba (as their hydroxides) on the surface of it.

Thiourea has been used widely to improve the quality of copper deposits [7]. In our case the addition of thiourea, other variables being constant, increases the content of copper in the deposit, and therefore the amount of CuO in the final deposit. Curiously, the same happens when complexing amines are present.

Electrodeposition of particles by electrophoresis

Deposition of $YBa_2Cu_3O_{7-\delta}$ particles over metallic wires yields high quality, well adhered deposits in alcohols and ketones when long chain alcohols (butanol) and complex ketones (eg. methyl-isobutylketone) are used. In absence of any added salt, electrophoresis allows the obtention of a deposit in case of butanol, while iodine needs to be added in the case of methyl-isobutylketone. We believe that a side reaction occurs in butanol at the electrodes that generates solvating ions. Figure 1 shows Scanning electron microscopy pictures of a $YBa_2Cu_3O_{7-\delta}$ wire over Ag, obtained at 800V in methyl-isobutylketone, and treated at 900°C. No fractures are evident in these conditions. When no stirring is used the plates show a preferential orientation with the c axis perpendicular to the main axis of the wire. This orientation is not observed for samples obtained with stirring. When Ag is the substrate, the thermal treatment only reaches 900°C and the oxide plates are well connected only in the surface. The critical currents obtained vary in function of the thermal treatment. The maximum values observed (1000 A/cm^2 at N_2(l) and zero magnetic field) correspond to deposits obtained over nickel with thermal treatments made at 960°C and annealing at 450°C in oxygen. (Figure 2 shows the observed resistance for a deposit over nickel). Considering that these samples are not textured, the values obtained are high as compared to other literature values [1].

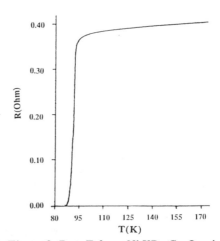

Figure 1. Scanning electron micros-
copy of Ag-YBa₂Cu₃O₇ wire (from methyl-
isobutylketone. Long. view. Top=Bottomx5

Figure 2. R vs T for a Ni-YBa₂Cu₃O₇ wire
by electrophoresis in butanol

ACKNOWLEDGEMENTS. The authors thank the CICYT (Spanish Ministry of Education) and the MIDAS program for Grants to help carry out this work.

REFERENCES
[1] eg. Bhattacharya, R.N. et al.; J. Mater. Res., (1991), 6, 1389.
[2] a) Maxfield, M. et al.; Appl. Phys. Lett., (1989), 54, 1932. b) Weston, A. et al.; J. Mater. Sci: Mater. Electronics, (1991), 2, 129.
[3] Pawar, S.J. et al.; Mater. Chem. Phys., (1991), 28, 259. Bhattacharya, R.N. et al.; Electrochem. Soc., (1991), 138, 1643.
[4] Minoura, H. et al.; Chem. Lett., (1991), 3, 379.
[5] Koch, T.R. and Purdy, W.C.: Talanta, 1972, 19, 989
[6] Massey, A.G.: "Comprehensive Inorganic Chemistry" Bailar, J.B. ed, Pergamon Press 1973, Vol 3, chapter 27
[7] Spiro, P.: "Electroforming" R. Draper LTD, Teddington 1968

Materials Science Forum Vols. 152 - 153 (1994) pp. 201-204
© 1994 Trans Tech Publications, Switzerland

INTERSTRATIFICATION IN THE SUBSTITUTED NICKEL HYDROXIDES

Y. Borthomieu, L. Demourgues-Guerlou and C. Delmas

Laboratoire de Chimie du Solide du CNRS and Ecole Nationale Supérieure de Chimie et Physique de Bordeaux, Université de Bordeaux I, F-33405 Talence Cédex, France

Keywords: Cobalt and Iron Substituted Nickel Hydroxides, Interstratification, Infrared

ABSTRACT

In the cobalt or iron substituted nickel hydroxides with small substitution amounts, segregation in the distribution of the anions (CO_3^{2-}, OH^-...) within the interslab space leads to interstratified materials. X-ray diffraction and IR studies have shown that their structure is constituted of a random distribution of α^* (intercalated with anions) and $\beta(II)$ (unintercalated) interslab spaces.

INTRODUCTION

Chimie douce reactions (hydrolysis and reduction) from $NaNi_{1-y} M_y O_2$ (M = Co, Fe) layered oxides lead to a new family of substituted nickel hydroxides [1,2] which are strongly apparented to the "Layer Double Hydroxides" (LDH's)[3] and to the mineral reevesite [4]· In these hydroxides, the substituting ions (Co, Fe) within the $Ni_{1-y} M_y (OH)_2$ slabs are trivalent while the nickel ions are divalent, so that anions (CO_3^{2-}, SO_4^{2-}, NO_3^-, OH^- ...etc) are inserted between the slabs in order to compensate the excess of positive charge. Water molecules are also inserted in the interslab space; the general formula of such materials is $Ni_{1-y} M_y (OH)_2 (CO_3)_{y/2}$, z H_2O in the case of inserted carbonate ions. The structural evolution vs the amount of substituting cation is discussed in the present paper.

RESULTS AND DISCUSSION

Study of the carbonate inserted hydroxides

Since the iron and cobalt systems exhibit a similar behavior, only the results concerning the iron-substituted nickel hydroxides are detailed in this paragraph.

X-ray diffraction study

Depending on the substitution amount (y), two types of materials are obtained.

▶ Substitution amounts (y) higher than 0.2 lead to α^* type phases. The X-ray diffraction pattern corresponding to the y = 0.3 composition is reported in Fig.1. The α^*

phase crystallizes in the rhombohedral system (SG : R3m) and the interreticular distance assigned to the (003) diffraction line (indexed on the hexagonal cell) corresponds to the interslab distance ($\approx$ 7.8 Å).

▶ The X-ray diffraction pattern of the 10% iron substituted hydroxide (y = 0.1) is also reported in Fig.1. The (003) and (006) diffraction lines observed for the y = 0.3 composition are replaced, for the y = 0.1 composition, by two broader asymmetric lines corresponding to distances that are no more multiple of one another. Such a behaviour may result from a disorder in the periodicity of the (00l) planes. Indeed, the re-oxidation of the hydroxide leads to a well-crystallized phase, which suggests that the (Fe,Ni)O$_2$ slab is maintained in the reduced phase.

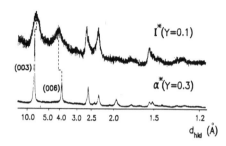

Fig.1 : X-ray diffraction patterns of the iron substituted nickel hydroxides with y = 0.1 (I*) and y = 0.3 (α^*).

Infrared study

The IR spectra of the hydroxides with increasing amount of iron can be compared to one another in Fig.2. The IR spectrum of the α^* phase exhibits, in addition to the bands characteristic of the intercalated CO$_3^{2-}$ ions and H$_2$O molecules, a broadening of the band assigned to OH groups of the slab (3400cm^{-1}) due to hydrogen bonding with the intercalated water molecules. The species intercalated within the interslab space in the α^* phase are schematically represented in Fig.3.

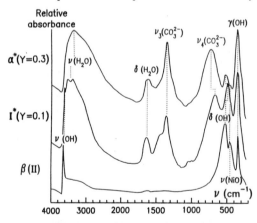

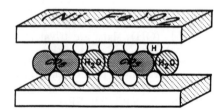

Fig.3 : Schematic representation of the interslab space for an iron substituted α^* hydroxide (y $\geq$ 0.20).

Fig.2 : Comparison of the infrared spectra of the α^* (y$_{Fe}$ = 0.3), I* (y$_{Fe}$ = 0.1) and Ni(OH)$_2$ hydroxides.

Besides, in the case of the ß(II) phase (unhydrated variety), a narrow band characteristic of "free" i.e non bonded OH groups is observed at 3650 cm^{-1}, as shown in Fig.2. The IR spectrum of the material obtained for y < 0.2 shows the existence of two types of interslab spaces : hydrated α^* type motives and unhydrated ß(II) motives. This behaviour suggests an interstratified structure which can be considered as a packing of α^* and ß type motives randomly distributed along the c_{hex} axis, as schematized in Fig.4.

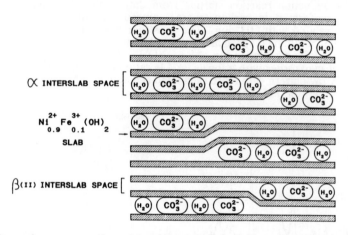

Fig.4 : Schematic representation of the slab packing in the interstratified structure.

Simulation of the X-ray diffraction patterns of the interstratified materials

The profile of the (001) X-ray diffraction lines for the interstratified materials has been simulated according to the Hendricks and Teller method, starting from the values of both interslab distances of the α^* and ß(ıı) motives, as well as from the molar ratio of the number of ß(ıı) type interlab spaces over the overall number of α and ß(ıı) ones ($r_ß$)[5]. Fig.5 gives the variation of the diffracted intensity vs diffraction angle for increasing $r_ß$ values. The spectra calculated for $r_ß = 0.99$ and 0.01 correspond to the ß(ıı) phase and the α^* phase respectively, while interstratified materials are obtained for intermediate values. For y = 0.1, comparison of the experimental and calculated data leads to $r_ß = 0.75$ in the case of cobalt substitution and 0.55 in the case of iron substitution.

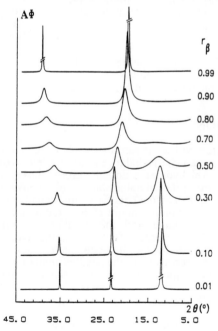

Fig.5 : Variations of the calculated diffracted intensity vs diffraction angle, for various values of the $r_ß$ ratio

Study of the OH⁻ inserted hydroxide with y = 0.2

The reduction reaction starting from the cobalt substituted γ (y = 0.2) oxyhydroxide, performed in carbonate free atmosphere, allows one to obtain an hydroxide intercalated with OH^- ions. The corresponding X-ray diffraction spectrum, reported in Fig.6, cannot be indexed and is characteristic of an interstratified material. It is similar to that of the coalingite : $Mg_{5/6}\ Fe_{1/6}\ (OH)_2\ (CO_3)_{1/12}\ (\ O)_{1/3}$, which belongs to the family of pyroaurites with a lower Fe/Mg ratio (1/5 vs 1/3). Comparison of the experimental X-ray pattern with those calculated according to the Hendricks and Teller method leads to $r_\beta = 0.57$ (Fig.7).

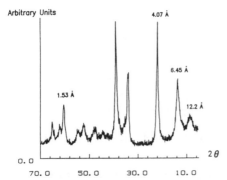

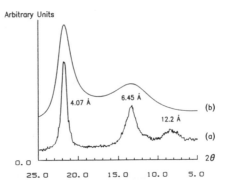

Fig.6 : X-ray diffraction pattern of the hydroxide with $y_{Co} = 0.2$, intercalated with OH^- ions.

Fig.7 : Comparison of the experimental X-ray diffraction pattern of the I*(OH⁻) phase (a) and of the Aφ function calculated for $r_\beta = 0.57$ (b).

The additive line observed for 12.2 Å is related to the sum of the interslab distances of ß(ɪɪ) and α^* motives (4.6 Å and 7.6 Å respectively) and gives evidence of a not fully random distribution with a trend to form a 1-1 order since the r_β ratio is close to 0.50.

CONCLUSION

Interstratified materials are obtained when the amount of anions required to perform the overall charge balance is unsufficient to occupy all the interlab spaces in the substituted nickel hydroxide and therefore causes an unhomogeneous distribution of these anions within the interslab space. This behaviour leads to the coexistence of both α^* (intercalated with anions) and ß(ɪɪ) (unintercalated) domains.

The minimum amount of inserted anions required to stabilize an α^* phase depends on the charge and the number of oxygen atoms of the anion (steric contribution). As an example for y = 0.2, the hydroxide inserted with OH^- ions is an interstratified material while the homologous phase inserted with CO_3^{2-} is an α^* phase. Such a behaviour results from the smaller number of oxygen atoms per negative charge brought by the OH^- ions, which enhances the trend to segregation.

REFERENCES

[1] C. Delmas, J.J. Braconnier, Y. Borthomieu and P. Hagenmuller, Mat. Res. Bull., 1987, 22, 741.
[2] L. Demourgues-Guerlou, J.J. Braconnier and C. Delmas, J. Solid State Chem., in press.
[3] W.T. Reichle, Solid State Ionics, 1986, 22, 135.
[4] H.F.W. Taylor, Mineralog. Mag., 1973, 39, 377
[5] S.B. Hendricks and E. Teller, J. Chem. Physics, 1942, 10, 147

Materials Science Forum Vols. 152 - 153 (1994) pp. 205-208
© 1994 Trans Tech Publications, Switzerland

LOW TEMPERATURE SYNTHESIS OF PbMo$_6$S$_8$ SUPERCONDUCTING MATERIAL

S. Even-Boudjada [1], L. Burel [1], V. Bouquet [1], R. Chevrel [1], M. Sergent [1], J.C. Jegaden [1], J. Cors [2] and M. Decroux [2]

[1] Laboratoire de Chimie du Solide et Inorganique Moléculaire, URA CNRS n° 1495, Université de Rennes I, Avenue du Général Leclerc, F-35042 Rennes Cédex, France

[2] Département de Physique de la Matière Condensée, Université de Genève, 24, quai Ernest Ansermet, CH-1211 Genève 4, Switzerland

Keywords: Chevrel Phase, PbMo$_6$S$_8$, Superconductor

ABSTRACT :
A new synthesis process of PbMo$_6$S$_8$ Chevrel phase has been carried out at low temperature (500°-600° C). This new method, using Mo$_6$S$_8$ compound and PbS instead of Pb under hydrogen gas flow, provides PbMo$_6$S$_8$ (PMS) by intercalation with good superconducting properties and allows to get very fine particles. PbMo$_6$S$_8$ formation at low temperature is very interesting for technological development of Chevrel phase wires.

INTRODUCTION :
  PbMo$_6$S$_8$ has quasi-isotropic superconducting properties and a very high critical field Bc$_2$ ~ 50 Tesla [1]. This material can be an excellent candidate for production of intense magnetic fields.
  With the exception of Nb-Ti alloy, a ductile superconductor, PbMo$_6$S$_8$, Nb$_3$Sn and the high Tc ceramic oxides are brittle materials. So, the technological process already used to make filamentary wires with Nb$_3$Sn is named "wind and react" or "in situ" method. We have mainly choiced this method using the cold powder metallurgy process and as matrix the niobium-copper billet drawn at room temperature.
  The proposed challenge is to make wires carrying high current densities.

LOW TEMPERATURE PbMo$_6$S$_8$:
  With this "in situ" method, chemical reaction for PbMo$_6$S$_8$ formation occurs directly inside the wire. The purpose is to prepare ultra-fine precursors (50-100 nm) of Chevrel phases in order to get a very reactive and "homogeneous" mixture of pure powders at nanoscale. The interest of powders with very small grain size is to increase the pinning forces at the grain boundaries for improving critical current densities, under conditions to densify and to sinter the superconducting powders.
  The purity is obtained by handling all the precursors in a glove-box under controlled atmosphere and the homogeneity by synthesizing powders with similar morphology and grain size; most of these precursors are prepared by liquid route.
  A new low temperature synthesis of PbMo$_6$S$_8$ Chevrel phase has been carried out below 470° C for three days [2]. This method used Mo$_6$S$_8$ compound and lead sulfide PbS instead of soft Pb large grains. Effectively, PbS can be synthesized very easily by precipitation as fine particles (100 nm). Mo$_6$S$_8$ cannot be prepared directly by high temperature solid state reaction, so it is obtained from Li$_x$Mo$_6$S$_8$ by the HCl leaching method [3], and its grain size is about 200 nm. Excess of sulfur coming from the decomposition of PbS under hydrogen gas flow is evacuated as H$_2$S. This decomposition of PbS provides extremely small droplets of reactive lead which inserts very quickly into Mo$_6$S$_8$ host lattice.
  PbMo$_6$S$_8$, synthesized by this insertion process, has a very small grain size (200-300 nm)- figure 1, compared to 5-10 μm for PbMo$_6$S$_8$ grains formed by high temperature synthesis

(1000° C) from classical precursors prepared in sealed quartz tubes - figure 2. The grain size of this new $PbMo_6S_8$ can be compared to that of lead Chevrel phase obtained from precursors synthesized by liquid route [4]-figure 3.

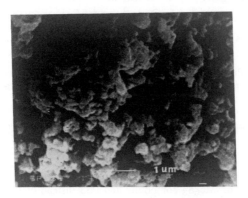

Fig.1 : SEM photograph of low temperature PMS

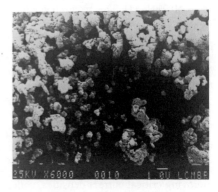

Fig.2 : SEM photograph of high temperature   Fig.3 : SEM photograph of PMS from
        PMS from sealed tubes precursors   liquid route precursors

     Previously, Tarascon et al. [5] prepared $PbMo_6S_8$ using large grains of lead, but the duration required for complete insertion into Mo_6S_8 was long (~ 3 weeks).
     In order to observe the homogeneity of the lead distribution in Mo_6S_8, we compare our new process (1) with another one deriving from Tarascon's method (2) :
     - PbS + Mo_6S_8 - reaction 1
     - Pb + Mo_6S_8 - reaction 2
     Both reactions take place under hydrogen gas flow at 440° C for 58 hours. The X-ray powder diffraction pattern results are Pb, Mo_6S_8 and $PbMo_6S_8$ peaks for reaction1 and Pb, Mo_6S_8 peaks for reaction 2. This comparison is made by line scan profil analysis and on X-ray map - figure 4 for reaction 1 and figure 5 for reaction 2:

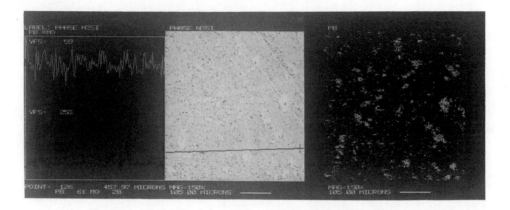

Fig.4 : Line scan profil analysis and X-ray map of lead distribution into Mo_6S_8 for PbS + Mo_6S_8 /H_2

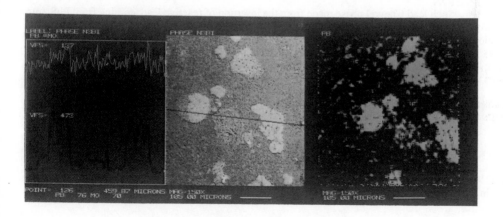

Fig.5 : Line scan profil analysis and X-ray map of lead distribution into Mo_6S_8 for Pb + Mo_6S_8 / H_2

SUPERCONDUCTING PROPERTIES :

$PbMo_6S_8$ grains made from PbS and Mo_6S_8 under hydrogen gas flow, have very good superconducting properties. Specific heat measurements on this $PbMo_6S_8$ after further annealing treatment at 600° C show a very sharp jump at Tc onset of 14 K [6]- figure 6. This result has never been observed before for such a low temperature preparation but only at very high temperature, 1600°C, by ceramic method- figure 7. For samples prepared at 1000°C, the specific heat anomaly at

Tc is very smooth which shows the inhomogeneity of the different PMS grains or the internal inhomogeneity of the grains.

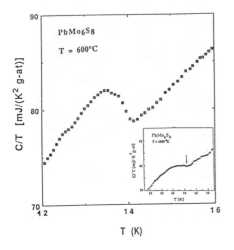

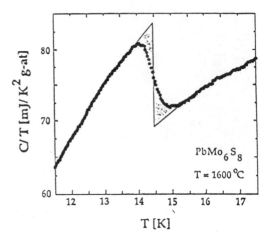

Fig.6 : Specific heat data of PMS prepared at 600°C- in insert, PMS prepared at 1000°C

Fig.7 : Specific heat data of PMS prepared at 1600°C

CONCLUSION

Such lead Chevrel phase made from the insertion of reactive lead, coming from PbS, into Mo_6S_8 compound under hydrogen gas flow at low temperature is very interesting for technological development of Chevrel phase wires. Effectively, a such "homogeneous" mixture of Pb and Mo_6S_8 could be used as powder for wire fabrication and the heat treatment temperatures needed on wires using "in-situ" method will considerably decrease. So, the technical process to make superconducting wires will be compatible with known insulation and winding method.

The challenge to have high current densities with good homogeneity on large lengths of filamentary wires, with lowest heat treatment temperatures will be succeed if a good grain connection exists with this new synthesis process of $PbMo_6S_8$.

ACKNOWLEDGEMENT :

This work is financially supported by Ministère français de la Recherche et de l'Espace, GEC Alsthom Company and Agence française De l'Environnement et de la Maîtrise de l'Energie, within the EUREKA EU 96 project.

REFERENCES :
1) O.Fischer, Appl. Phys. 16 (1978) 28.
2) M.Rabiller-Baudry, P.Rabiller, S.Even-Boudjada, L.Burel, R.Chevrel, M.Sergent, M.Decroux, J.Cors, J.L.Maufras to be published.
3) R.Chevrel, M.Sergent, J.Prigent, Mat. Res. Bull., 9, 487 (1974).
4) S.Even-Boudjada, V.Bouquet, L.Burel, R.Chevrel, M.Sergent, M.Decroux, H.Massat, P.Genevey to be published.
5) J.M.Tarascon, F.J.Disalvo, D.W.Murphy, G.W.Hull, E.A.Rietman and J.V.Waszczak, J. Sol. State Chem. 54, 204-212 (1984).
6) M.Decroux, P.Selvam, J.Cors, B.Seeber and O.Fischer, R.Chevrel, P.Rabiller and M.Sergent, Proceedings of Applied Superconductivity Conference, 1992, Chicago, Illinois, march 1993- Vol 3- N°1 p. 1502.

Materials Science Forum Vols. 152 - 153 (1994) pp. 209-212
© 1994 Trans Tech Publications, Switzerland

TERT-BUTYLLITHIUM, A SELECTIVE AND CONVENIENT REAGENT FOR THE REDUCTIVE INTERCALATION OF INTERSTITIAL SOLIDS: A NOVEL ROOM TEMPERATURE SYNTHESIS OF $(LiMo_3Se_3)_n$, A SOLUBLE POLYMERIC CHEVREL CLUSTER COMPOUND

J.H. Golden, F.J. DiSalvo and J.M.J. Fréchet

Cornell University, Dept. of Chemistry, Baker Labs, Ithaca, N.Y. 14853, USA

Keywords: Soft Chemistry, Intercalation, *tert*-Butyllithium, Inorganic Polymer, Chevrel Phase, $(InMo_3Se_3)_n$, $(LiMo_3Se_3)_n$

ABSTRACT : As part of our ongoing study of the polar solvent soluble inorganic polymer $(LiMo_3Se_3)_n$, we have investigated its synthesis via the heterogenous reductive intercalation of $(InMo_3Se_3)_n$ with a variety of organolithium reagents at room temperature. Herein we describe the results of our investigation of the soft chemical synthesis of $(LiMo_3Se_3)_n$, and the determination of the reducing potential of *tert*-butyllithium vs. the intercalable rutile solid WO_2.

Introduction

Interest in organic and inorganic low dimensional solids has increased steadily over the past decade, as part of a search for new materials displaying novel physical properties including: anisotropic conductivity, superconductivity, non - linear optical phenomena, and piezoelectric behavior [1]. Parallel to this search for new materials is the impetus to develop low temperature synthetic methods, popularly known as "chimie douce", or "soft chemistry" methods. Soft chemical strategies typically encompass a range of low temperature solution techniques [2,3,4], which bypass traditional high temperature synthetic strategies (>400 °C), and can yield metastable phases and/or compounds with elements in unusual oxidation states. To this end, we have examined soft chemical routes to synthesize the metastable linear chain compound, $(LiMo_3Se_3)_n$.

In 1980, Potel et al. [5] described a series of metallic linear chain compounds, $(MMo_3X_3)_n$ (M = a monovalent main group metal, X = Se, S). The structure of MMo_3X_3 is based on the condensation of octahedral clusters of molybdenum face capped by chalcogen atoms, to form linear $(Mo_3X_3^-)_n$ chains. The structure can also be viewed as antiprismatically stacked triangles of Mo, with edge bridging chalcogen atoms, as shown in figure 1. These compounds are one member of the family of materials generally known as the Chevrel phases [6] , with the general chemical composition of $M_xMo_{3n}Se_{3n+2}$, for $n \geq 2$.

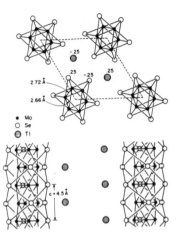

Figure 1. Projection of $(TlMo_3Se_3)_n$ structure onto (001) and $(11\bar{2}0)$ planes. Note that Mo-Mo bondlengths are close to Mo metal lengths of 2.72 Å.

These linear chain compounds display a unique variety of physical properties [7,8]. One of the most interesting properties arises when lithium is the M interstitial cation. If $(LiMo_3Se_3)_n$ is treated with highly polar solvents such as DMSO or water, the compound dissolves to form highly absorbing burgundy solutions consisting of rigid rod polyanions and solvated lithium cations [8]. The strong Mo-Mo bonding in the $(Mo_3Se_3^-)_n$ "inner core" imparts considerable stiffness to the chains, producing chain alignment under shear and flow of the solutions [8].

As shown in eq. 1, metastable $(LiMo_3Se_3)_n$ is traditionally produced by ion exchange:

$$(InMo_3Se_3)_n + LiI \xrightarrow[\text{3-8 weeks}]{456\,^\circ C} (LiMo_3Se_3)_n + InI_{(sublimes)} \qquad \text{eq. 1}$$

In an attempt to develop a direct low temperature soft chemical route to $(LiMo_3Se_3)_n$, we investigated the heterogeneous solution reaction of $(InMo_3Se_3)_n$ with several commonly available or easily prepared organolithium reagents. Specifically, three lithium radical anion reagents and two alkyl lithium reagents were investigated, both types having been used previously with success in a variety of intercalation reactions involving layered and interstitial solids [9].These organolithium reagents differ in their reduction potentials [9], and as we shall see, in their chemical reactivity towards $(InMo_3Se_3)_n$. Possible reaction pathways are illustrated in eqs. 2-4 :

$$(InMo_3Se_3)_n + LiR \longrightarrow (LiMo_3Se_3)_n + InR \qquad \text{eq. 2}$$

$$(InMo_3Se_3)_n + xLiR \longrightarrow Li_xInMo_3Se_3 + xR \qquad \text{eq. 3}$$

$$(InMo_3Se_3)_n + 6LiR \longrightarrow In + 3Mo + 3Li_2Se + 6R \qquad \text{eq. 4}$$

The reaction of the lithium radical anion and alkyl lithium reagents is easily monitored. Specifically, the radical anions become colorless as they undergo oxidation, while the alkyl lithium reagents form irreversible radical disproportionation products easily monitored by GC/MS or ^{1}H NMR.

Results and Discussion $(LiMo_3Se_3)_n$ was prepared by treating $(InMo_3Se_3)_n$ with a ten-fold excess of *tert*-butyllithium in pentane/toluene at 23 °C for seven to ten days. While there is some reductive attack of the chain structure, as exemplified by eq. 4, reductive intercalation of lithium competed successfully to form $(Li_xInMo_3Se_3)_n$, as illustrated by eq. 3. X-ray powder diffraction, TEM, SEM microprobe were amoung the techniques used to analyze the reaction products. Pure $(LiMo_3Se_3)_n$ was isolated by solubilization of crude $(Li_xInMo_3Se_3)_n$ in DMSO/water, followed by centrifuging off insoluble byproducts (eq.4) and unreacted $(InMo_3Se_3)_n$. After stripping of solvent, yields of $(LiMo_3Se_3)_n$ were in excess of 50%.

Proton NMR was used to confirm that *tert*-butyllithium reacts by a radical disproportionation process, providing lithium atoms that diffuse into the interstitial spaces between the chains. Specifically, in an NMR tube reaction, the course of the intercalation was monitored for seven days at 23 °C. The spectra displayed the disappearence of *tert*-butyllithium and the simultaneous evolution of the radical disproportionation products isobutane, isobutylene, and 2,2,3,3-tetramethylbutane, suggesting that intercalation occurs by a radical pathway.

Other organolithium reagents investigated proved to be non-selective towards $(InMo_3Se_3)_n$. *n*-Butyllithium, a commonly used intercalation reagent, displayed no reductive behavior towards $(InMo_3Se_3)_n$, while the radical anion reagents lithium naphthalide, lithium 4,4'-di-tert-butylbiphenylide, and the dianion of benzophenone caused decomposition, as confirmed by x-ray powder diffraction. Attempts to dissolve this material in DMSO/water did not produce the dark burgandy solutions characteristic of $(LiMo_3Se_3)_n$.

Estimation of the non-reversible reduction potential of *tert*-butyllithium was achieved by preparation of the reference compound $Li_{0.37}WO_2$. Specifically, pure WO_2 was prepared [10], and stirred with one equivalant of *tert*-butyllithium in hexane for ten days with exclusion of light. Titration of the unreacted organolithium reagent [11] was found to be consistent with the stoichiometric product $Li_{0.37}WO_2$. From the Li=0.37 stoichiometry, we then estimated the relative reducing potential of *tert*-butyllithium by comparison to known charge-discharge curves for lithium intercalated rutile solids [12].

The obtained value of 0.83 V above Li/Li$^+$ was consistent with the observed reactivity of *tert*-butyllithium towards $(InMo_3Se_3)_n$, and has allowed us to

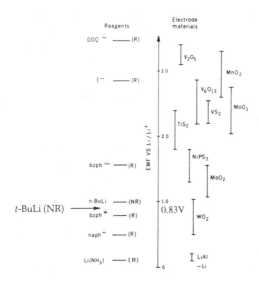

Figure 2. Adapted from reference 9b. Potentials of a wide variety of electrode materials and reagents for modeling cell reactions relative to Li/Li⁺. The reduced form of the reagent is shown. (R) denotes reversible, (NR) denotes nonreversible.

To our knowledge, this is the first example of the use of *tert* - buyllithium in the intercalation of interstitial solids, and the use of proton NMR to monitor the course of such a reaction. Determination of the reduction potential of *tert* - butyllithium, and thus it's relative reactivity towards intercalable solids such as WO_2, now gives the synthetic chemist greater flexibility in his or her choice in reagents for the reductive intercalation of layered and interstitial solids.

REFERENCES

[1] (a) DiSalvo, F.J. *Solid State Chemistry : A Rediscovered Chemical Frontier; Science* , **1990**, 247, 649 - 655. (b) Rouxel, J., *Acc. Chem. Res.* **1992**, 25, 328 - 336. [2]Pujalt - Paz, G.R. *Physica C* . **1990**, 166, 177 - 184. [3] Rabenau, A. *Angew. Chem. Intl. Ed. Engl* . **1985**, 24, 1026 - 1040. [4] Wood, P.T.; Pennington, W.T.; Kolis, J.W. *J. Am. Chem. Soc..*,**1992**, 114, 9233 - 9235. [5] Potel et al. *J. Solid State Chem* . **1980**, 35, 286 - 290. [6] Chevrel, R. et al. *J. Solid State Chem.* **1971**, 3, 515. [7] Tarascon, J.M.; DiSalvo, F.J.; Waszczak J.V.*Solid State Comm* . **1984** , 52, 227 - 231. [8] Tarascon, J.M.; DiSalvo, F.J.; Chen, C.H.; Carroll, P.J.; Walsh,M.; Rupp, L. *Solid State Comm.* **1985**, 58, 227 - 231. [9] (a) Wardell, J.L. *Comprehensive Organometallic Chemistry* ; Pergamon : New York, 1982, 43 - 120. (b) Murphy, D.W.; Christian, P.A. *Science* , **1979**, 205, #4407, 651 - 656. [10] Brauer, G. , Ed. ; *Handbook of Preparative Inorganic Chemistry* ; Academic Press : New York, 1963, 1421 - 1422. [11] Watson, S.C., Eastham, J.F. *J. Organometal. Chem* . , **1967**, 9, 165 - 168. [12] Murphy, D.W., DiSalvo, F.J., Carides, J.N., Waszczak, J.V. *Mat. Res. Bull.* ,**1978**, 13, 1395 - 1402.

Materials Science Forum Vols. 152 - 153 (1994) pp. 213-216
© *1994 Trans Tech Publications, Switzerland*

LITHIUM DEINTERCALATION IN THE SPINEL LiMn$_2$O$_4$

F. Coowar[1], J.M. Tarascon[1], W.R. McKinnon[2] and D. Guyomard[3]

[1] Bellcore, Red Bank, NJ, USA

[2] N.R.C., Ottawa, Canada, K1AOR9

[3] I.M.N., Université de Nantes, F-44072 Nantes, France

Keywords: Lithium Intercalation, Manganese Oxide, Lithium Batteries, Cation Mixing, Oxygen Deficiency

ABSTRACT: Using a new electrolyte composition stable against oxidation up to 5V vs. Li, the full electrochemical deintercalation of lithium from the spinel LiMn$_2$O$_4$ has been studied. The origin of two new reversible oxidation-reduction peaks centered around 4.5 V and 4.9 V vs. Li is examined. The capacity associated with these peaks is shown to depend on both the nominal composition x in Li$_x$Mn$_2$O$_4$ and the synthesis conditions, namely annealing temperatures and cooling rates. We discuss the evidence that these peaks are related to local structural defects and propose a reasonable explanation for their origin.

INTRODUCTION

The spinel LiMn$_2$O$_4$ is a material of choice for the positive electrode of rocking-chair batteries based on a carbon negative electrode [1]. It has been shown that most of the lithium is deintercalated reversibly from LiMn$_2$O$_4$ at a voltage close to 4.1 V vs. Li [2,3]. Using an electrolyte resistant to oxidation up to 5V vs. Li, we recently demonstrated [4] that the remaining lithium in the structure (about 0.1 Li) can be reversibly deintercalated in 2 steps at about 4.5 and 4.9V vs. Li.

We report here a study of the influence of both the nominal lithium composition (x) of the Li$_x$Mn$_2$O$_4$ powders and the preparation conditions (annealing temperatures and cooling rates) on the electrochemical behavior at high voltage. We show that the 4.5V and 4.9V peaks can be used as fingerprints to tune the synthesis of electrochemically optimized Li$_x$Mn$_2$O$_4$ powders for rocking-chair lithium batteries.

EXPERIMENTAL

Li$_x$Mn$_2$O$_4$ powders were first prepared at 800°C in air by reacting a mixture of x/2 Li$_2$CO$_3$ (or x LiNO$_3$) and 2 MnO$_2$ for 1 day, followed by 2 successive grinding and identical annealing sequences. For several x lithium compositions in the range 0.75 to 1.2, the powders were annealed during 48 hours at the temperature T (varying from 700°C to 1100°C) and then either slowly cooled at 5 to 10°C per hour (these powders will be denoted SC) or quenched (denoted Q) to RT between 2 stainless steel plates.

The experimental conditions, including preparation of composite electrodes, preparation of the electrolyte and a description of swagelock laboratory cells, have been described in detail elsewhere [1,5]. The cyclic voltammetry experiments were performed by means of a "Mac-Pile" system using a two-electrodes configuration in a potentiostatic mode.

RESULTS

1) Unit cell parameter and oxygen non-stoichiometry

The variation of the cubic lattice parameter **a** as a function of the nominal composition x for Q and SC samples is shown in figure 1. The powders are single phases within the range $0.9 < x < 1.15$. Q samples have a larger **a** parameter than SC samples. For both series the **a** parameter decreases with increasing x. For a given x composition, the increasing of the quenching temperature lead to an increase of the **a** parameter, while the slow cooling from increasing temperatures did not affect the **a** parameter. Note that quenching temperatures above 900°C lead to mixed phases. The X-ray pattern of a sample quenched from 1000°C allowed to identify $LiMnO_2$ as the major phase. These results indicate a strong dependence of the spinel unit cell parameter on the synthesis conditions of the materials.

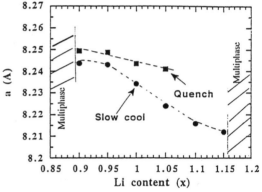

Figure 1: Dependence of the lattice cubic parameter on the cooling rate after annealing at 800°C.

TGA experiments have been performed in order to get information on the oxygen content of the samples. During the first heating in air, the weight begins to decrease for temperatures above 780°C and reaches a loss of about 5% at 1000°C. The faster the cooling rate, the smaller the weight uptake on subsequent cooling. To study the effect of quenching on the oxygen content, some powders quenched from various temperatures have been reheated at constant rate. The results, reported in figure 2, show that all powders quenched from above 800°C present a weight uptake, starting at 400°C, that increases with increasing previous quenching temperature, while sample quenched at 700°C did not gain weight. It can then be concluded that oxygen deficiency within the spinel structure is drastically enhanced when the quenching temperatures and the cooling rates are increased.

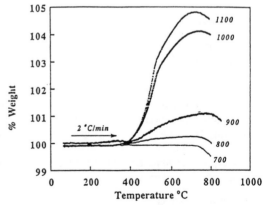

Figure 2: Oxygen weight uptake for samples quenched from various temperatures.

2) Electrochemical behavior

The influence of the synthesis conditions on the electrochemical deintercalation behavior of the $Li_xMn_2O_4$ spinel materials has been systematically investigated in the 4.3V-5.1V range.

For the same cooling rate and the same quenching temperature, the capacity of the 4.5V process decreased with increasing nominal composition x, while the capacity of 4.9V process increased, see figure 3. Increasing the quenching temperature resulted in a decrease of the 4.9V

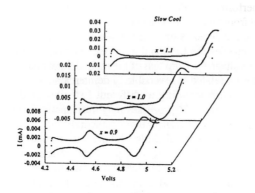

Figure 3: Influence of the nominal composition x in $Li_xMn_2O_4$ on the cyclovoltammogram for SC samples.

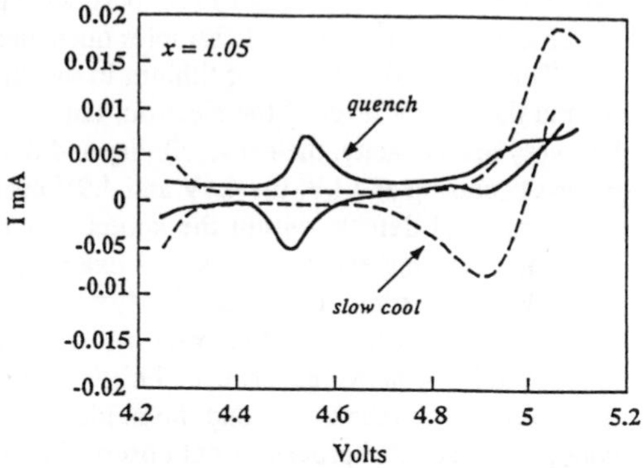

Figure 4: Influence of the cooling rate on the cyclovoltammogram for $Li_{1.05}Mn_2O_4$ samples.

capacity while the 4.5V capacity increased. The effect of the cooling rate on the electrochemical behavior is depicted in figure 4. Decreasing the cooling rate decreased the capacity at 4.5V and increased the capacity at 4.9V.

Most samples showed electrochemical processes at 4.5V and 4.9V vs. Li. Slowly cooled $Li_xMn_2O_4$ (x>1.0) samples were the only one showing no 4.5V process. The capacities measured at 4.5V and 4.9V were correlated so that one capacity decreased when the other one increased. We did not succeed in preparing samples with no capacity at 4.9V, however quenched samples with low nominal x composition (x<1.0) showed the lowest capacities. These capacities correspond to very low values, their sum being always lower than 10% of the total capacity obtained between 3.5V and 5V vs. Li.

DISCUSSION

The strong dependence of the relative amplitude of 4.5V and 4.9V peaks with nominal composition and treatment history rules out the possibility of these peaks being associated with an impurity phase. The spinel $LiMn_2O_4$ is a cubic structure where lithium is present in the 8a tetrahedral sites and Mn is present in the 16d octahedral sites. Partial inversion in the spinel structure at high temperature, corresponding to cation mixing between tetrahedral and octahedral sites, has to be considered, as well as oxygen deficiency, corresponding to a decrease of the manganese average oxidation state.

Consequently we will represent the defect spinel $LiMn_2O_4$ by the following general formula: $[Li_{xt} Mn_{yt} V_{zt}]_{Te} [Mn_{yo} Li_{xo} V_{zo}]_{Oc} O_{4-\delta}$ to account for cation mixing and vacancies that may occur at high temperature. There are several possibilities of oxidation state for Mn: 2+, 3+, 4+. For crystal field reasons, Mn^{3+} and Mn^{4+} prefer octahedral sites, while for thermodynamic reasons (due to Madelung energy) Mn^{2+} is stabilized in tetrahedral sites [6]. However there have been convincing experimental data for the presence of Mn^{3+} in tetrahedral sites within quenched $NiMn_2O_4$ samples for example [7]. Quenching should enhance cation mixing between octahedral and tetrahedral sites compared to slow cooling. Increasing the quenching temperature should increase cation mixing and favor the presence of Mn^{2+} and/or Mn^{3+} in tetrahedral sites,

The unit cell parameter of Q samples larger than that of SC samples, and its increase with quenching temperature is compatible with an increase of the degree of inversion. The intensity ratio of (311) to (111) X-ray diffraction lines is larger in Q samples than in SC samples, consistent with inversion as well. Preliminary EPR experiments performed by Ph. Molinie in Nantes demonstrated the presence of Mn^{2+} cations in samples quenched from above 800°C and not in SC samples.

The chemical potential of lithium in the structure depends on both the site energy of the Li^+ ion and the energy level of the electron, that accompanies the Li^+ ion during intercalation. Due to their very low capacity, their reversibility and their voltage far from that of the main deintercalation process occurring at 4.1V, the 4.5V and 4.9V peaks are related to reversible Li deintercalation from local structural defects within the spinel structure, corresponding to different Li^+ sites and/or different electronic energy levels. For example, 4.5V and 4.9V processes may correspond to the removal of tetrahedral or octahedral Li^+, with the oxidation of tetrahedral or octahedral Mn.

It is difficult to prove the exact mechanism, however the following explanations account for and reconcile the above experimental observations. As a reminder, the 4.1V process corresponds to the removal of tetrahedral Li and the oxidation of octahedral Mn^{3+} to Mn^{4+}, as shown in previous work [2,3]. The 4.5V process is not observed in SC samples with x>1.0, in which no tetrahedral Mn is expected, and has a large capacity in Q samples. It corresponds then to the oxidation of tetrahedral Mn with probably the removal of tetrahedral Li. The capacities at 4.5V and 4.9V being correlated, the 4.9V peak should correspond to the removal of Li from the same sites, namely tetrahedral sites.

CONCLUSION

We have shown through X-ray diffraction, thermogravimetric analyses and potentiostatic studies that the electrochemical lithium deintercalation in the spinel $Li_xMn_2O_4$ is highly sensible to changes in synthesis conditions and revealed for the first time the importance of oxygen non-stoichiometry.

The present study has a direct impact on the application of $LiMn_2O_4$ spinel material in lithium rechargeable batteries. As a matter of fact, samples that were slowly cooled have led to optimal electrochemical performance, leading to the use of 4.5V peak as an indicator to synthesize electrochemically optimized $LiMn_2O_4$ powders.

REFERENCES

1) Guyomard, D. and Tarascon, J.M.: J. Electrochem. Soc., 1992, 139, 937.

2) Rossow, M.H., De Kock, A., De Picciotto, L.A., Thackeray, M.M., David, W.I.F. and Iberson, R.M.: Mat. Res. Bull., 1990, 25, 173.

3) Ohzuku, T., Kitagawa, M. and Hirai, H.: J. Electrochem. Soc., 1990, 137, 769.

4) Guyomard, D. and Tarascon, J.M.: "The Electrochemical Society Extended Abstracts", abstract 37, Toronto, Ontario, Canada, October 11-16, 1992, Vol 92-2, p.50.

5) Guyomard, D. and Tarascon, J.M.: J. Electrochem. Soc., accepted for publication.

6) Navrostky, A and Kleppa, O.J.: J. Inorg. Nucl. Chem., 1967, 29, 2701.

7) Baudour, J.L., Bouree, F., Fremy, M.A., Legros, R., Rousset, A., Gillot, B.: Physica B, 1992, 180 & 181, 97.

Materials Science Forum Vols. 152 - 153 (1994) pp. 217-220

MECHANISMS OF THE REVERSIBLE ELECTROCHEMICAL INSERTION OF LITHIUM OCCURING WITH NCIM$_s$ (NANO-CRYSTALLITE-INSERTION-MATERIALS)

S.D. Han[1], N. Treuil[1], G. Campet[1], J. Portier[1], C. Delmas[1], J.C. Lassègues[2] and A. Pierre[3]

[1] Laboratoire de Chimie du Solide du CNRS, 351 cours de la Libération, F-33405 Talence, France

[2] Laboratoire de Spectroscopie Moléculaire et Cristalline du CNRS, 351 cours de la Libération, F-33405 Talence, France

[3] Department of Mining, Metallurgical and Petroleum Engineering, University of Alberta, Edmonton, Alberta, T6G-2G6, Canada

Keywords: Nanocrystallite, Dangling Bond, Lithium Insertion

ABSTRACT

A new family of insertion-compound electrodes so called NCIM$_s$ (Nano-Crystallite-Insertion-Materials) has been proposed : the major requirement is that the electrode materials have to be polycristalline with a crystallite and particle size as small as possible (the accepted definition being that many crystallites make a particle). Indeed by minimizing the size of the crystallites the formation of defects is favoured, particularly at the crystallite surface, acting as reversible (de)grafting sites of Li$^+$. Also related to that the cation-anion bonding is weakened not only in the grain boundary region but also within the crystallite close to its surface : then the electrochemical insertion of Li$^+$ takes place through easy bonding rearrangements.

I. INTRODUCTION

In the last 20 years much attention has been focussed on A_xMO_2 - type intercalation compounds (A=Li, Na and M=Co, Ni, Mn...) which are used as positive electrodes in reversible alkali electrochemical cells (see for example refs. 1).

However, a very long-term cyclability (i.e. over 10^3 cycles) might be hardly achievable, particularly for corresponding electrodes having a large grain size, probably because the Li^+ (de)intercalation process slightly perturbs the host lattice.

Some of us have patented, a few years ago, a new strategy and related experiments that have enabled us to put forward a rather new family of insertion-compound electrodes able to sustain long-term Li^+ electrochemical cyclability [2]. The major requirement is that the electrode materials are polycristalline with a crystallite and particule sizes as small as possible. Therefore we later called the polycristalline electrode materials $NCIM_s$ (for nano-crystallite-insertion material).

Table 1 gives important examples related to mixed-valency metal oxides.

For clarity the examples listed in Table 1 have been divided into two classes, I and D, according to whether the resistivity tends to increase (class I) or to decrease (class D) upon the electrochemical Li^+ insertion process.

Table 1 - Some nanocrystallite insertion materials

Sample	Class	Average grain size (Å)	Insertion rate x (measured in $LiClO_4$ (p.c.), $1.5V<V(Li)<3.5V$)
Li_xSrTiO_3	D	80	$0 \leqslant x \leqslant 0.3$
Li_xCrO_2	I	30	$0 \leqslant x \leqslant 1$
$Li_xMn_2O_3$	I	50	$0 \leqslant x \lesssim 2$
$Li_xFe_2O_3$	D	150	$0 \leqslant x \lesssim 0.5$
Li_xNiO_2	I	60	$1 \leqslant x \leqslant 2$
Li_xCuO_2	I	50	$1 \leqslant x \lesssim 2$
Li_xWO_3	D	40	$0 \leqslant x \lesssim 2$

Rather similar considerations were reported by Barloux et al. and concern the spinel $LiMn_2O_4$ [1]. Also apparently related to that, Kumagai et al. [1] have reported that the positive electrode $MnO_2.yV_2O_5$ was formed by incorporation of V_2O_5 into MnO_2 matrices and the crystallinity of the oxide decreased with increase in V_2O_5 content incorporated : they have shown that the amount of Li^+ ions which can be reversibly electrochemically (de)insered increased with increasing y value, i.e. with decrease in the crystallinity ; it reached about 1 Li^+ per mole of transition metal with y = 0.6 [1].

In this paper the framework of the model accounting for the reversible electrochemical Li^+ insertion occuring in the $NCIM_s$, is presented.

We also show, for the first time, that the model accounts for the evolution of the open circuit voltage or the electrodes, vs the fraction, x, of the alcali.

II. MECHANISMS OF THE REVERSIBLE ELECTROCHEMICAL INSERTION OF LITHIUM OCCURING WITH NCIM$_s$

First of all, by minimizing the size of the crystallites we tend to :

(i) favour the formation of defect bonds, particularly at the crystallite surface (of its vicinity), such as anions adjacent to cation vacancies : *these defects may act as reversible (de)grafting sites for Li+.*

(ii) weaken the cation-anion bonding not only on the grain boundary region but also within the grain close to its surface : *then the electrochemical insertion of Li+ may be expected to occur through easier bonding rearrangements.*

That is depicted here-below for $SrTiO_3$-NCIM, *taken as a non-limiting but illustrative example.*
First of all fig. 1 illustrates the electron conduction via $[Ti:3d]_{sub}$ or $[Ti:3d]_{bw}$ energy states :

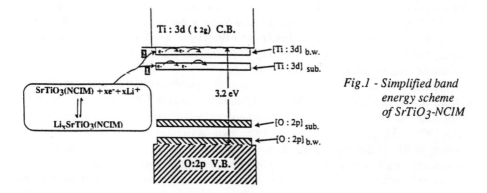

Fig.1 - Simplified band energy scheme of $SrTiO_3$-NCIM

¤ $[Ti:3d]_{sub}$ represents deep subband-gap energy states arising from cation defects adjacent to an anion vacancy. They are lowered below the Π^*conduction band of Ti^{4+} : $3d^0(t_{2g})$ parentage. Conversely anion defects adjacent to cation vacancies occur. They introduce acceptor states $[O:2p]_{sub}$ arising from the O^{2-} : $2p^6$ valence band. According to the model the latter defects act as reversible (de)grafting sites for Li+, (see (i)).

¤ the $[Ti:3d]_{bw}$ and $[O:2p]_{bw}$ energy states originate from Ti-O bond weakening. This bond weakening induces Li+ (de)insertion as mentioned above (see (ii)).

We will see, now, that the model accounts for the differences observed between the open-circuit voltage (OCV) vs x (the fraction of the alkali) curves related to polycrystalline electrodes having different sizes of crystallites. For sake of simplicity, such a behavior is illustrated only for two n-type electrodes Li_xSnO_2 and Li_xWO_3 (fig. 2a, b).

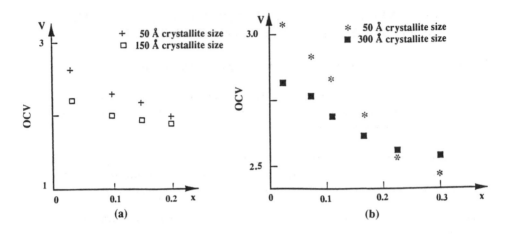

Fig.2 - Equilibrium OCVs vs x for some Li / LiCF$_3$SO$_3$ / NCIMs (a : Li$_x$SnO$_2$; b : Li$_x$WO$_3$) .

The concentration of the "sub" and b.w." states increases as the crystallite size is reduced. This obviously causes, *only for the lower x values*, a decrease of the Fermi-energy (E$_F$) (i.e. higher electron affinity) and thereby of OCV towards anodic values. Indeed, for the lower x values, the OCV are higher for the electrodes having the smallest crystallite size (fig. 2a and 2b for x≲0.15).

For higher x values (x >> 0.15) and when the inside-crystallite structure is well adapted for the reversible intercalation of lithium as occurs for Li$_x$WO$_3$, an inversion of the OCV is observed (fig. 2c) : indeed for x >> 0.15 ail the subband gap energy states [W^{6+} : 5d^0]$_{sub}$ and [W^{6+} 5d^0]$_{bw}$ (the "twin states" of [Ti^{4+} : 3d^0]$_{sub}$ and b.w. reported in fig. 1) are filled with electrons : therefore the lithium intercalation within the nanocrystallites can now take place ; it is accompanied with a "delocalization" of the injected electrons in the conduction band. On the other hand it is well established that the band-energy width increases as the crystallite size decreases; therefore the WO$_3$ electrodes having the smallest crystallite size have their conduction-band edge shifted towards cathodic values : this causes a pushing of E$_F$, and thereby of OCV, towards cathodic values (as it is illustrated on fig. 2b for x >> 0.15).

References

1.   J. Rouxel, in F. Levy (ed.), Physics and Chemistry of Layered Materials, vol. VI, Reidel, Dordrecht, 1979.

    P. Barloux, J.M. Tarascon and F.K. Shokoohi, J. Solid State Chem. 94 (1991) 185.

    N. Kumagai, S. Tanifuji, T. Fujiwara and K. Tanno, Electrochim. Acta, 37(6)(1992) 1039.

2.   J.P. Couput, G. Campet, J.M. Chabagno, M. Bourrel, D. Muller, R. Garrié, C. Delmas, B. Morel, J. Portier and J. Salardenne, Int. Appl. Publ. under PCT. Int. Pat. Class GO2F 1701, FO1 G9/00, C 23C 14/34, WO 91/01510, 1989.

Materials Science Forum Vols. 152 - 153 (1994) pp. 221-236
© 1994 Trans Tech Publications, Switzerland

ELABORATION BY REACTION IN MOLTEN NITRATES AND CHARACTERIZATION OF PURE TITANIUM OXIDE WITH LARGE SURFACE AREA

V. Harlé[1,2], J.P. Deloume[1], L. Mosoni[1], B. Durand[1], M. Vrinat[2] and M. Breysse[2]

[1] Laboratoire de Chimie Minérale 3, URA 116, Université Claude Bernard Lyon1, 43 Boulevard du 11 Novembre, F-69622 Villeurbanne Cédex, France

[2] Institut de Recherches sur la Catalyse, 2 Avenue Einstein, F-69626 Villeurbanne Cédex, France

ABSTRACT.- The reaction of titanium oxysulfate with a molten bath made of an equimolar mixture of KNO_3-$NaNO_3$, in the range 523-723K, leads to a pure crystallized TiO_2 with large specific surface area.

By acting on the temperature of preparation, it is very easy to control the surface area and the porosity of the oxide. With adjusted synthesis parameters, a 100 m^2/g TiO_2 with average pore radii in the range of 50-70 Å is obtained after annealing at 873K.

INTRODUCTION

TiO_2 with large specific surface area attracts more and more attention recently and especially in catalysis. This oxide has been used as a support for many catalysts, such as V_2O_5/TiO_2 , Mo/TiO_2 , Pt/TiO_2 and, Rh/TiO_2, for selective oxidation reaction and NOx selective catalytic reduction [1].

For hydrotreating reactions, commercial catalysts commonly consist of CoMo or NiMo supported on a transition alumina. Nevertheless, in a search for more active catalysts, a number of studies have been made using other oxides as support. Even if the sulfide-support interaction is still not well understood, recent reviews of support effect on hydrotreating catalysis [2,3] clearly indicated that TiO_2 seems to be promising.

Nevertheless, in order to be used as a support for hydrotreating catalysts, TiO_2 has to present a large surface area and large pore radii. Moreover, as reactions are carried out between 573 and 673K and regeneration of the catalyst at about 773K, the textural properties of the oxide have to be stable under heating at 773-873K. TiO_2 has traditionally been obtained using a low temperature synthesis method, such as hydrolysis of $TiCl_4$ [4] or $Ti(SO_4)_2$ [5]. Oxides with large surface area in the range 40-60 square meters per gram but with weak thermal stability are generally obtained. Higher TiO_2 surface areas have been reported by Motoya et al [6] for samples prepared by hydrolysis of alkoxides (sol-gel method), but this method appears expensive to be employed industrially.

Recently, Hamon et al [7] and Jebrouni [8] showed that highly divided ZrO_2 could be easily obtained by synthesis in molten salts. For TiO_2, Kerridge [9] reported that the reactivity of potassium hexafluorotitanate with molten alkaline nitrates leads to a mixture of anatase and LiF. However, no attention was paid to the morphological characteristics of the powder. The objective of this work is therefore to prepare TiO_2 by reaction of the titanium oxysulfate in molten nitrates. Study is focused on the effect of the synthesis parameters on the textural properties of the titanium oxide obtained, before and after annealing at 873K, in order to check its stability under thermal treatment.

EXPERIMENTAL PROCEDURE

Materials

Titanium oxysulfate $TiOSO_4$, 2 H_2O from Riedel de Häen was used after dehydration at 573 K for 10 hours. Control of the weight loss by thermogravimetric analysis (TGA) and XRD analysis agreed with the formation of anhydrous $TiOSO_4$.
The molten bath was an equimolar mixture of sodium and potassium nitrates (Prolabo) previously dehydrated at 383 K for 10 hours (m.p.: 498 K). Some further experiments were carried out in a KNO_3-$LiNO_3$ molten

bath (34% molar in LiNO3) exhibiting a lower melting point than KNO3-NaNO3 (402 K instead of 498 K).

Procedure

Reactions were carried out in a pyrex batch reactor. The TiOSO4 was introduced as a mixture with the nitrates (which were in large stoichiometric excess : 2, 4.3, 8.5 and 17). The reaction was performed between 603 and 703K (heating rate of 400 K/h), the reactivity of TiOSO4 with the nitrates being evidenced by nitrogen dioxide release according to

$$TiOSO_4 + 2\,NO_3^- \rightarrow TiO_2 + SO_4^{2-} + 2\,NO_2 + 1/2\,O_2 \quad (1)$$

In molten salts synthesis, the temperature of preparation and the time of reaction are important parameters. Therefore, in order to determine the shortest time of reaction required for total conversion, thermogravimetric analysis of the reaction between TiOSO4 and the nitrates have been carried out at 603, 673 and 703K. Results presented in figure 1 indicate that, in the three cases, the reaction begins at about 510K (i.e. as soon as the bath is molten) and is mainly performed during the heating of the bath. Reaction was evidenced by NO_2 release, confirmed by infrared spectroscopy analysis of the gases given off. The weight loss determined during these experiments agreed with the calculated one due to NO_2 and O_2 release according to reaction (1).

Moreover, the higher the final temperature of the reaction, the shorter the time required for total conversion. In the following, the duration of the experiment at each temperature was choosen according to the time of reaction determined from these TGA experiments.

After reaction and air quenching, the solid phase was extracted by water-washing for 24 hours in a soxhlet device. In the last washing, no sulfate ions were detected by addition of $Ba(NO_3)_2$ to the filtrate. The oxide was finally dried at 373 K for 15 hours. Heat treatments were performed under dried air for 2 hours at 873K (heating rate : $2K.min^{-1}$) in order to study the textural stability of the solids.

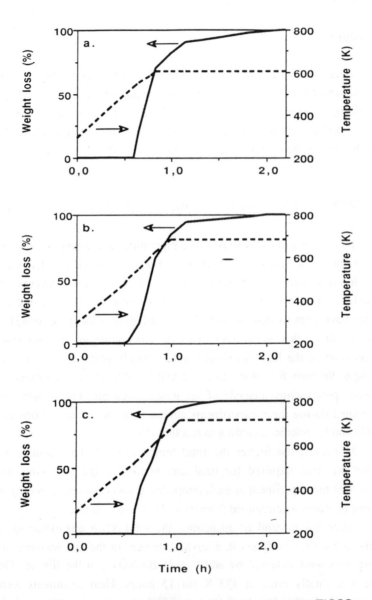

Fig. 1 - Thermogravimetric analysis of the reaction between $TiOSO_4$ and the nitrates : reaction carried out at 603 K (a), 673 K (b) and 703 K (c).

A previous experiment performed at 673K had indicated that no significant variation of the TiO_2 surface area before and after annealing was observed when titanium oxysulfate was ball milled for 30 minutes (milling performed in an alumina container). Moreover, a decrease in the heating rate from 400K/h to 130K/h had no influence on the surface area before and after annealing. Therefore, all the experiments have been carried out without milling of the titanium oxysulfate and with a 400 K/h heating rate.

The specific conditions of synthesis of each sample are listed in Table I.

Characterization

The X-Ray diffractograms (before and after annealing at 873K) were recorded on a Siemens D500 diffractometer using Ni-filtered Cu Kα radiation. Regardless of the preparation temperature, the X-Ray patterns agreed with TiO_2 anatase (JCPDS 21-1272) as shown in figure 2. However, peak widths varied with the preparation conditions. The size of the crystallites was estimated according to the Scherrer method after correcting the experimental effect by a parabolic law [10].

Specific surface area, specific pore volume (Vp) and mean pore radius (Rp) were determined by nitrogen adsorption after outgasing the powders under vacuum for 1h at 673K. For all samples, the nitrogen adsorption-desorption isotherms were similar. An example given in figure 3 indicates an E type isotherm, according to the nomenclature proposed by De Boer [11]. Such a type of hysteresis is generally associated with "ink bottle" pores or voids between close-packed spherical-like particles.

Chemical analyses were performed by the Service Central d'Analyse - CNRS, in order to evaluate the impurities content of the TiO_2 samples. Sodium (or lithium), potassium and nitrate ions contents were always lower than 50 ppm, although sulfate ion contents were found between 0.1 and 0.5% (wt).

Experiment	Stoichiometric excess of nitrates	Temperature (K)	Duration at final temperature (h)
E 1	8.5	603	1.4
E 2	8.5	673	1
E 3	8.5	703	0.4
E 4 *	8.5	673	1
E 5 **	8.5	523	2
E 6	8.5	673	2
E 7	8.5	673	4
E 8	2	673	1
E 9	4.3	673	1
E 10	17	673	1
E 11	4.3	673	2

* : $TiOSO_4$ ball milled for 30 minutes in a alumina container.

** : molten bath made of KNO_3-$LiNO_3$ instead of KNO_3-$NaNO_3$

TABLE I - Conditions of synthesis of the titanium dioxides.
(molten bath : KNO_3-$NaNO_3$; heating rate : 400K/h)

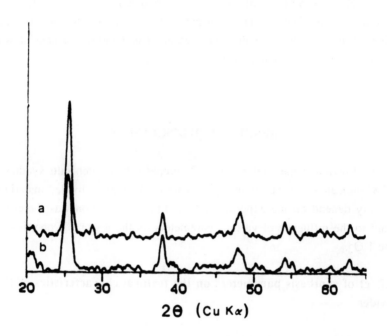

Fig. 2 - X-Ray diffractograms of E1 (a) and E3 (b) before annealing.

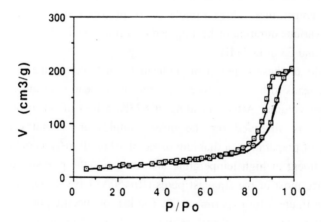

Fig. 3 - Isotherms of N₂ adsorption-desorption on E2 annealed at 873K.

High resolution microscopy was realized with a Jeol 100 CX instrument fitted with a VHP polar piece. Its resolution power was 0.2 nm. The solids were ultrasonically dispersed in ethanol and the suspension was collected on carbon-coated grids.

RESULTS AND DISCUSSION

Previous papers related to ZrO_2 prepared by molten salts synthesis [7,8] indicated that the textural properties of the highly divided materials greatly depend on the experimental conditions. Attention was therefore paid to the influence of various parameters on the textural properties of the TiO_2.

Effect of synthesis parameters on the textural characteristics of the oxides

Effect of the preparation temperature:

Experiments performed in a microbalance (Fig.1) have indicated that the time for the reaction to go to completion decreased with increasing temperature. Three samples were prepared at 603, 673 and 703K with various duration of heating deduced from the TGA curves (see exp. E1, E2 and E3 of table II).

Results of these experiments indicated that lowering the reaction temperature leads to higher surface area and lower mean pore radius and specific pore volume. After annealing at 873K, a loss of 50 to 60% in surface area is observed for the three samples and, whatever the temperature of preparation, a titanium oxide of 70 to 80 m^2/g is obtained. Such a treatment at high temperature also modifies the porosity of the material since most of the smallest pores (Rp<50Å) have disappeared as illustrated in figure 4 for experiment E2. The loss of specific pore volume observed after annealing is therefore due to a decrease of the number of the small pores.

Experiment	Before annealing		After annealing	
	S_{BET} (m^2/g)	$< Rp (\text{Å}) <$ $Rm(\text{Å}) / Vp(cm^3/g)$	S_{BET} (m^2/g)	$< Rp (\text{Å}) <$ $Rm(\text{Å}) / Vp(cm^3/g)$
E 1	216	20< Rp <70 30 / 0.28	83	20< Rp < 120 55 / 0.21
E 2	159	20< Rp <120 65 / 0.38	82	20< Rp < 120 80 / 0.32
E 3	144	20< Rp <120 85 / 0.40	73	20< Rp < 120 80 / 0.30
E 4	180	/	81	/
E 5	305	/	105	20< Rp < 50 30 / 0.21
E 6	144	/	100	20< Rp < 120 65 / 0.28
E 7	137	/	89	20< Rp < 120 60 / 0.28
E 8	190	/	105	20< Rp < 120 45 / 0.30
E 9	182	/	97	/
E 10	138	/	74	/
E 11	/	/	100	20< Rp <120 70 / 0.28

TABLE II - Effect of synthesis parameters
on the textural properties of the oxides.

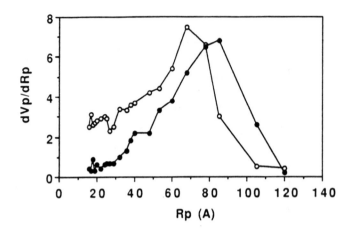

Fig. 4 - Pore radii distribution of E2
before (o) and after (•) annealing at 873K.

A temperature of 673 K seems to be a good compromise between large pore radii and large surface area ; the effect of other parameters of the preparation were then studied for reaction carried out at 673 K for one hour.

Influence of the nature of the molten bath
For ZrO_2 preparation, Jebrouni [8] showed that the nature of the cation associated with the nitrates could influence the textural properties of the oxides. Therefore, synthesis of TiO_2 was made with a KNO_3-$LiNO_3$ molten bath.

The reaction of $TiOSO_4$ with either KNO_3-$NaNO_3$ or KNO_3-$LiNO_3$, when performed at the same temperature, leads to powders exhibiting the same surface area, before and after annealing. However, the low melting point of KNO_3-$LiNO_3$ (402K instead of 498K for KNO_3-$NaNO_3$) allows to prepare titanium dioxide at lower temperature and therefore to increase the surface area ; powders reaching 305 m^2/g are obtained at 523K (see E5 in tables I and II). Annealing at 873 K, decreases the surface area down to 105 m^2/g but the average pore radii of the sample is much lower, 25 Å instead of 80 Å for the sample E2 prepared at 673K in a KNO_3-$NaNO_3$ molten bath.

Effect of the duration of heating

It is well known that prolonged thermal treatments of divided powders in the presence of a molten bath involve a growth of particles and modifications in the textural properties of the samples. This effect, investigated at 673 K in experiments E2, E6 and E7, evidences a moderate decrease of the surface area of the resulting oxide by lengthening the time of heating. This decrease of surface area before annealing can be explained by a phenomenon of maturation of the crystallites leading to an increase of their size as evidenced by XRD. Such a modification could also explain the lower loss of surface area during annealing.

Reaction at 673K for 2 hours, leading to the largest surface area and maintaining a large porosity, appears to be the optimum conditions.

Influence of the ratio nitrates-TiOSO4

Experiments E8, E9, E2 and E10, performed with various ratios nitrates-$TiOSO_4$, indicate large variations in the textural properties of the resulting TiO_2 (table II). The higher the $TiOSO_4$ concentration, the higher the surface area before and after annealing. Nevertheless, this increase in surface area resulted in a decrease in porosity (Fig. 5).

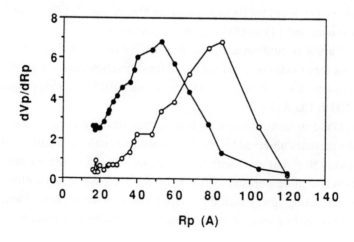

Fig. 5 - Pore radii distribution of E2 (o) and E6 (•) annealed at 873K.

Similar experiments followed by TGA show that the reaction rate is lowered by a decrease in the ratio nitrate/$TiOSO_4$. Such a behaviour could be associated with an increase in the viscosity of the molten mixture, limiting the growth of the crystallites by diffusionnal effects.

Comparison between the surface area and the pore size distribution of the various samples suggests that an intermediate nitrate/oxysulfate ratio (i.e. 4.3 corresponding to 5g of nitrates per gram of $TiOSO_4$) and reaction for 2 hours at 673K would give a TiO_2 powder with appropriate textural properties for catalysis. This was done for experiment E11, which gives after annealing, a TiO_2 sample with a surface area of 100 m^2/g and a pore size centered at 70 Å.

Study of the variation of texture during annealing

In order to understand the loss of surface area observed during the annealing process, High Resolution Electron Microscopy (HREM) was performed on the E2 oxide prepared at 673K. Before annealing, TiO_2 appears as both square particles, 75 Å in size, and diamond shaped particles, 115 Å in size (Fig 6a). After annealing, the shapes of the particles are retained (Fig 6b) but an increase in the size is noticed (75 to 130 Å for the first ones and 145 to 185 Å for the second ones).

This was confirmed by the broadening of the X-Ray diffraction lines. As the broadening is similar for all the diffraction lines, the crystallites size was determined from the most intense line (101) ; the width increased from 100 to 165 Å (Fig 7).

Good agreement between both techniques allows us to conclude that the particles observed by HREM are monocrystals. Comparison of the pore radii (50 Å and over) with the crystallite size (75 to 185 Å) indicates an intercrystalline porosity. Otherwise, the increase of crystallite size during annealing agrees with the disappearance of small pores. Thus, the decrease of surface area can be attributed to a beginning of sintering.

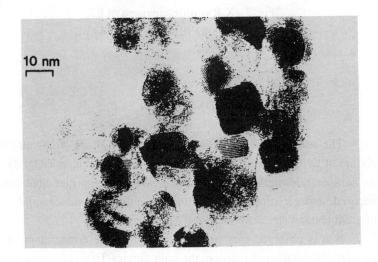

Fig. 6 - HREM micrograph of E2
before (a) and after (b) annealing at 873K

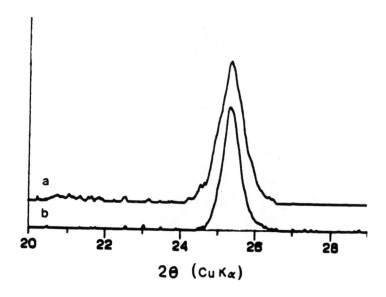

Fig. 7 - X-Ray diffractograms of E2
before (a) and after (b) annealing at 873 K.

CONCLUSION

The reactivity of $TiOSO_4$ in a molten bath made of an equimolar mixture of KNO_3-$NaNO_3$ leads to a pure crystallised anatase TiO_2. By this low temperature method, porous oxides with large surface area can be prepared, and variations in the synthesis conditions produce solids with different textural characteristics.

It was observed that parameters such as the temperature of the reaction, the duration of reaction, the ratio nitrates-$TiOSO_4$, influence the surface area and the porosity of the powder obtained. Optimization of the experimental conditions results in TiO_2 with a large surface area of 100 m^2/g and a mean pore size radius close to 60 Å after annealing at 873 K. This result confirms that the molten salts synthesis is a promising way to prepare new supports for catalysts which maintain a large surface area after annealing at 873 K.

Acknowledgments

This work was carried out in the framework of the contract "Hydrogénation des polyaromatiques". Support was received from ELF,IFP,TOTAL and the CNRS-PIERSEM.

REFERENCES

[1] S. MATSUDA and A. KATO, *Appl. Catal.*, 1983, **8**, p.149.

[2] M. BREYSSE, J.L. PORTEFAIX and M. VRINAT, *Catalysis Today*, 1991, **10**, p.489.

[3] F. LUCK, *Bull. Soc. Chim. Belg.*, 1991, **100**, p.781.

[4] H. SHIMADA, T. SATO, Y. YOSHIMURA, J.HIRAISH and A. NISHIJIMA, *J. Catal.* , 1981, **69**, p.434.

[5] W. ZHAOBIN, X. QUIN, G. XIEMAN, E.L. SHAM, P. GRANGE and B. DELMON, *Appl. Catal.*, 1990, **63**, p.305.

[6] J.A. MONTOYA, T. VIVEROS, J.M. DOMINGUEZ, L.A. CANALES and I. SCHIFTER, *Catalysis Letters* , 1992, **15**, p. 207.

[7] D. HAMON, M. VRINAT, M. BREYSSE, B. DURAND, F. BEAUCHESNE, T. des COURIERES, *Bull. Soc. Chim. Belg.*, 1991, **100,** p.933.

[8] M. JEBROUNI, thesis, Lyon, 1990, 169-90.

[9] D.H. KERRIDGE and J. CANCELA REY, *J. Inorg. Nucl. Chem.*, 1975, **37**, p. 2257.

[10] H.P. KLUG and L.E. ALESSANDER, in *"X Ray diffraction procedure"* edited by John Wiley and Sons.(Inc. New-York, 1954) p.509.

[11] J.H. De BOER in *"The Structure and Properties of Porous Materials"* D.H. EVERETT and F.S. STONE (eds) , Butterworths, London, 1958.

Materials Science Forum Vols. 152 - 153 (1994) pp. 237-240

BINARY AND TERNARY TELLURIDE SOLIDS FROM THE OXIDATION OF ZINTL ANIONS IN SOLUTION

C.J. O'Connor, B. Wu and Y.S. Lee

Department of Chemistry, University of New Orleans, New Orleans, Louisiana 70148, USA

Keywords: Intermetallics, Amorphous Semiconductor, Spin Glass

ABSTRACT: The series of solids of composition MTe_2, $M_5(InTe_4)_2$ and $M_3(SbTe_3)_2$ (where M = Cr, Mn, Fe, Co, and Ni) are obtained from the rapid precipitation metathesis reaction of the binary or ternary Zintl phase materials with transition metal halides in solution. The intermetallic solids formed are amorphous and metastable. We report on the magnetic and electrical properties of these materials and a variety of phenomena have been observed. Several materials (*eg.*, $Fe_5(InTe_4)_2$, $CrTe_2$, $Ni_3(SbTe_3)_2$, *etc.*) exhibit spin glass behavior with freezing temperatures in the range 2-25K. An interesting photomagnetic effect has also been observed in several spin glass materials that results in the formation of magnetic bubbles. Electrical resistivity of these materials ranges from metallic to amorphous semiconductor.

INTRODUCTION

Solid state chemists have for some time been looking for low-temperature techniques to prepare solid materials.[1] Recent reports from our laboratory have described the synthesis of novel molecular solids from the oxidation of Zintl anions in solution.[2] The synthetic protocol consists of the preparation of so-called Zintl phases at high temperature followed by an examination of their chemical reactivity with metal ions in solution at room temperature to produce intermetallic solids. For example, the series of solids of composition MTe_2, $M_5(InTe_4)_2$ and $M_3(SbTe_3)_2$ (where M = Cr, Mn, Fe, Co, and Ni) are obtained from the reaction of the binary or ternary Zintl phase materials with transition metal halides in solution. These intermetallic materials are of amorphous nature due to the very rapid metathesis reaction between the Zintl polyanion and the metal cation and the insolubility of the metathesis product that is formed.

As part of our effort in synthesis and characterization of novel solid materials, we have been investigating the magnetic and electrical properties of these materials. The magnetic and electric properties are examined as a function of temperature and a variety of phenomena have been observed.

EXPERIMENTAL

Synthesis - The Zintl phase materials to be considered here are generally obtained by fusing the stoichiometric elements together under high temperature, for example,[3]

$$3\,K + Sb + 3\,Te \xrightarrow{\ 550^\circ\ } K_3SbTe_3$$

Owing to the extreme air-sensitivity of both starting materials and products, all the manipulations are carried out under an argon atmosphere.

Methods of preparing intermetallics consist of electron transfer from precursor Zintl polyanions to transition-metal cations, which results in the rapid precipitation of the neutral solid product. For example,

$$2\,K_3SbTe_3 + 3\,MX_2 \xrightarrow{\ solvent\ } M_3(SbTe_3)_2\ (s) + 6\,KX$$

$$2\,K_5InTe_4 + 5\,MX_2 \xrightarrow{\ solvent\ } M_5(InTe_4)_2\ (s) + 10\,KX$$

$$K_2Te_2 + MX_2 \xrightarrow{\ solvent\ } MTe_2(s) + 2KX$$

where M = Cr, Mn, Fe, Co and Ni and X = Cl, Br.[4-6] The precursor method described here takes advantage of the soluble Zintl anions and the reaction takes place at room temperature rather than at high temperatures normally employed in solid state synthesis.

Elemental Analysis - The presence of the expected elements in each sample and the homogeneity of each sample were confirmed with an EDS-equipped AMRAY Model 1820 scanning electron microscope. The quantitative elemental analysis for these materials was carried out on a IL-S-12 AA Spectrometer and gives results consistent with the proposed metathesis reaction.

Resistivity Measurement - Resistivity measurements for all materials were performed on pressed pellets (16,000 psi) over the temperature range 20-300 K using the four-probe van der Pauw technique.[7] The current was supplied by a Keithley Model 224 programmable current source and the voltage drop across the sample was measured with a Keithley Model 181 digital nanovolmeter.

Magnetic Measurement - The dc magnetic susceptibility and magnetization measurements were conducted on a SHE Corp. VTS-50 superconducting susceptometer that is interfaced to an IBM XT computer system. Two types of experiments were mainly conducted during the magnetic measurement: magnetic susceptibility χ (M/H) as a function of temperature and remanent magnetization as a function of field or temperature.[8]

Photomagnetic Measurement - The quantitative measurement of photomagnetic response in the time domain was performed using the STEPS technique[9] which permits illumination of the sample with high intensity radiation while very precise magnetic data was being recorded on the SQUID susceptometer.[10]

RESULTS AND DISCUSSION

Zintl materials are molecular compounds that lie at the boundary between intermetallic alloys and ionic salts, and may have a combination of properties related to both classes of materials. Because of the difference in the electronegativity of the elements that comprise a Zintl phase, there is often a great deal of ionic character in the Zintl phase material, which may be sufficient to allow the solvation of salt like ions (cations and Zintl anions) in polar solvents. It has been observed that there is a trend that correlates the solubility properties of ternary Zintl phases with their crystal structures. The Zintl-phase materials which contain isolated Zintl anions in their solid are soluble in some polar solvents.

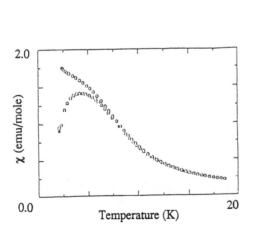

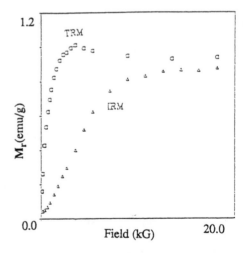

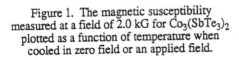

Figure 1. The magnetic susceptibility measured at a field of 2.0 kG for $Co_3(SbTe_3)_2$ plotted as a function of temperature when cooled in zero field or an applied field.

Figure 2. The field dependence of the isothermal remanent magnetization (IRM) and thermal remanent magnetization (TRM) for $Fe_5(InTe_4)_2$.

For example,

$$K_5InTe_4 \xrightarrow{H_2O} 5 K^+ + InTe_4^{5-}$$

$$K_3SbTe_3 \xrightarrow[DMF]{DMSO} 3 K^+ + SbTe_3^{3-}$$

In the presence of a transition-metal cation of sufficient electron affinity, a rapid precipitation reaction results in insoluble intermetallic products of formula M_2SnTe_4, $M_5((InTe_4)_2$ and $M_3(SbTe_3)_2$. These materials resulting from rapid precipitation from solutions are amorphous. The X-ray powder diffraction patterns exhibit a lack of diffraction peaks and are consistent with a minimal amount of crystalline order for the freshly prepared materials. The observed physical properties such as the spin glass state and amorphous semiconductivity are also consistent with those expected for an amorphous solid.

The magnetic susceptibilities of most of the metal telluride solids under investigation can be interpreted on the basis of localized magnetic moments. For example, It has been observed that the dominant magnetic exchange interaction is antiferromagnetic between the localized M^{2+} (M = Cr, Mn) moments and ferromagnetic on the Fe^{2+} ions.

Several materials (*e.g.*, Fe_2SnTe_4, $Fe_3(GaTe_3)_2$, $CrTe_2$, $Ni_3(SbTe_3)_2$, *etc.*) exhibit magnetic properties which are characterized as spin glass behavior with freezing temperature in the range 2-25 K. The spin glass phenomenon is characterized by some very unusual behavior in the bulk magnetic properties of the materials.[11] Figure 1 illustrates the effect of zero field cooling and field cooling of the spin glass material $Co_3(SbTe_3)_2$. The most diagnostic experiment for the characterization of the spin glass state is the analysis of the field dependence of the isothermal remanent magnetization (IRM) and thermal remanent magnetization (TRM), as shown in Figure 2 for $Fe_5(InTe_4)_2$. The hump in the TRM curve has been observed in many spin glasses and is characteristic of the spin glass state.

The unusual magnetic properties of spin glasses make them excellent candidates for an experiment in which radiation is used to generate the photo-induced magnetic effects on the surface of a material. A process for generating magnetic bubbles in several spin glass materials has been illustrated for Fe_2SnTe_4, $Fe_3(SbTe_3)_2$, $Ni_3(SbTe_3)_2$, and others. This phenomenon has also been observed in crystalline spin glass materials.[12]

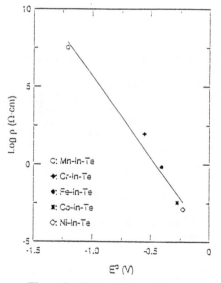

Figure 3. The logarithm of the room temperature electrical resistivity of $M_5(InTe_4)_2$ plotted as a function of the standard reduction potential of the precursor divalent transition metal ions.

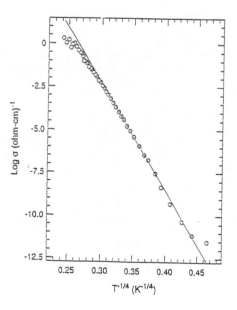

Figure 4. Temperature dependence of the specific conductivity plotted as log σ versus $1/T^{1/4}$ for $CrTe_2$.

The magnetic and especially the electrical properties of these materials are roughly correlated with the ability of the precursor transition metal cations to be reduced by the Zintl polyanion. Figure 3 illustrates the plot of the logarithm of the room temperature resistivity of $M_5(InTe_4)_2$ versus the standard reduction potential of the precursor divalent transition metal ions. This correlation likely arises from the degree of the transfer of electrons from the Zintl polyanion to cation when redistribution of charge occurs during the formation of the intermetallic materials.

Although a few of the intermetallic materials exhibit conductivity in the range expected for a metallic material, the variable range hopping mechanism proposed by Mott[13] is found to dominate the electrical conductivity over a wide temperature range in the semiconducting intermetallic solids such as $Cr_5(InTe_4)_2$, $CrTe_2$, and $Cr_3(SbTe_3)_2$, As shown in Figure 4, the specific conductivity of $CrTe_2$ exhibits $exp(-B/T^{-1/4})$ behavior over the temperature range of 180 to 50 K. The linear range verifies the variable range hopping mechanism.

CONCLUSION

Three groups of new intermetallic materials, MTe_2, $M_5(InTe_4)_2$, and $M_3(SbTe_3)_2$ where M = Cr, Mn, Fe, Co, Ni, have been prepared by reaction of the Zintl polyanions Te_2^{2-}, $SbTe_3^{3-}$, and $InTe_4^{5-}$ with divalent transition metal cations in aqueous solution. The resultant intermetallic materials exhibit varieties of properties, such as spin glass in $Fe_5(InTe_4)_2$, variable range hopping conduction in $CrTe_2$, and photo-magnetism in $Ni_3(SbTe_3)_2$.

It appears that the magnetic and conducting properties of these intermetallic materials are roughly correlated with the standard reduction potentials of M^{2+}, that is, with the ability of the precursor M^{2+} ions to be reduced by the Zintl polyanion. Although the formation of these materials is perhaps better viewed as a metathesis reaction rather than a formal redox reaction, there is some degree of electron transfer or redistribution from electron rich polyanion to the cation. For instance, Mn^{2+} is the most difficult to reduce ion in this series and thus there is little electron transfer from anion to cation in forming $Mn_5(InTe_4)_2$. The closed shell is localized on the $InTe_4^{5-}$ and the d electrons are localized on the individual Mn^{2+}. The localization of the unpaired electrons on the Mn^{2+} is responsible for the observed magnetic property and the limited degree of electronic conduction in $Mn_5(InTe_4)_2$.

REFERENCES:

1) Stein, A.; Kellor, S. W.; and Mallouk, T. E.; *Science,* **1993**, *259*, 1558.
2) Jung, J.-S.; Ren, L.; and O'Connor, C. J.; *J. Mater. Chem.*, **1992**, 2, 829; Ren, L.; Jung, J.-S.;. Ferre, J.; and O'Connor, C. J.; *J. Phys. Chem. Solids*, **1993**, *54*, 1; Wu, B.; Ren, L.; O'Connor, C. J.; and Jung, J -S.; *J. Applied Phys.*, **1993**, *73*, 5463; Zhang, J. H.; Wu, B.; O'Connor, C. J.; and Simmons, W. B.;*J. Applied Phys.*, **1993**, *73*, 5718.
3) Jung, J.-S.; Stevens, E. D.; and O'Connor, C. J.;*J. Solid State Chem.*, **1991**, *94*, 362.
4) Jung, J.-S.; Ren, L.; and O'Connor, C. J.; *J. Mater. Chem.*, **1992**, 2, 829; Wu, B.; Ren, L.; O'Connor, C. J.; Tang, J.; Jung, J.-S.; Ferré, J.; and Jamet, J.-P.; *J. Mater. Res.*, submitted.
5) Zhang, J. H.; Wu, B.; and C. O'Connor, C. J.; *Chem. Mater.*, **1993**, *5*, 17; Zhang, J. H.; vanDuyneveldt, A. J.; Mydosh, J. A.; and O'Connor, C. J.; *Chem. Mater.*, **1989**, *1*, 404.
6) Zhang, J. H.; Wu, B.; O'Connor, C. J.; and Simmons, W. B.; *J. Applied Phys.*, **1993**, *73*, 5718.
7) van der Pauw, L. J.; *Philips R   Repts*, **1958**, *13*, 1.
8) O'Connor C. J.; in *"Resear.. Frontiers in Magnetochemistry"*, O'Connor, C. J., Editor, (World Scientific Publishing, Inc., Singapore, New Jersey, London, **1993**) 109-138.
9) O'Connor, C, J,; Sinn, E.; Bucelot, T. J.; and Deaver, B. S.; *Chem. Phys. Lett.*, **1980**, *74*, 27.
10) O'Connor, C. J.; *Prog. Inorg. Chem.*, **1982**, *29*, 203.
11) Moorjani, K.; and J. Coey, M. D.; "Magnetic Glasses", Elsevier, **1985**; Mydosh, J. A.; *J. Mag. and Mag. Mat.*, **1978**, *7*, 237; Edwards, S. F.; and Anderson, P. W.; *J. Phys. F, Metal Phys.*, **1975**, *5*, 965; Fisher, K. H.; *Phys. Status Solidi B.*, *116*, **1983**, 353; **1985**, *116*, 130; Binder, K.; and Young, A. P.; *Rev. Mod. Phys.*, **1986**, *58*, 801.
12) Ayadi, M.; and Ferre, K.; *J. Magn. Magn. Mafer.*, **1986**, *91*, 54; Ayadi, M.; and Ferre, J.; *Phys. Rev. B*, **1991**, *44*, 10079.
13) Mott, N. F.; and Davis, E. A.; *"Electronic Processes in Non-Crystalline Materials"* 2nd ed., (Oxford, **1979**).

Materials Science Forum Vols. 152 - 153 (1994) pp. 241-244
© 1994 Trans Tech Publications, Switzerland

SYNTHESIS, CHARACTERIZATION AND PROPERTIES OF NEW SILICOANTIMONIC ACIDS:
$H_4Sb_4O_8(Si_4O_{12})\cdot xH_2O$ and $H_3Sb_3O_6(Si_2O_7)\cdot xH_2O$

C. Pagnoux[1], A. Verbaere[1], F. Taulelle[2], M. Suchaud[1], Y. Piffard[1]
and M. Tournoux[1]

[1] Institut des Matériaux de Nantes, UMR CNRS 110, 2 rue de la Houssinière,
F-44072 Nantes Cédex 03, France

[2] Laboratoire de RMN, Institut de Chimie, Université Louis Pasteur, 2 rue Blaise Pascal,
F-67000 Strasbourg, France

Keywords: Silicoantimonic Acids, Protonic Conductivity, Ion Exchange

ABSTRACT: - The preparation of $H_4Sb_4O_8(Si_4O_{12})\cdot xH_2O$ and $H_3Sb_3O_6(Si_2O_7)\cdot xH_2O$ consists of an ion exchange process in acidic medium starting from the corresponding potassium and cesium compounds respectively, that both exhibit a strongly covalent and rather open mixed octahedral-tetrahedral framework. These silicoantimonic acids are hydrated and their water content has been studied at 20°C as a function of relative humidity (RH). They are true lattice hydrates that dehydrate reversibly up to 400°C. Their protonic conductivity has been measured at 20°C as a function of RH and at RH $\cong$ 60% as a function of temperature. Titrations with alkali hydroxide solutions show that they behave as solid acids with ion exchange properties. ^{23}Na MAS NMR experiments on Na$^+$ exchanged derivatives of $H_3Sb_3O_6(Si_2O_7)\cdot xH_2O$ with various Na$^+$ loadings have shown that Na$^+$ fills the alkali sites selectively.

INTRODUCTION

In the area of mixed octahedral-tetrahedral framework solids it has already been shown that some alkali antimony phosphates [1,2] exhibit interesting features with respect to fast ion mobility. They can be ion exchanged in acidic medium thus leading to solid acids that behave as good ion exchangers and protonic conductors [3-6]. As part of search for similar materials we have recently extended our investigations to alkali antimony germanates and silicates. Some of these, $Cs_4Sb_4O_8(Si_4O_{12})$ [7] and $Cs_3Sb_3O_6(Si_2O_7)$[8] , exhibit rather open frameworks that enables ion exchange reactions. We report here on the preparation, characterization and properties of their proton exchanged derivatives.

PREPARATION

$H_4Sb_4O_8(Si_4O_{12})\cdot xH_2O$ and $H_3Sb_3O_6(Si_2O_7)\cdot xH_2O$ have been prepared by an ion exchange process in acidic medium starting from the corresponding potassium and cesium compounds, respectively. $K_4Sb_4O_8(Si_4O_{12})$, whose structure is isotypic with that of $Cs_4Sb_4O_8(Si_4O_{12})$, was prepared via a sol-gel route as previously described [7]. $Cs_3Sb_3O_6(Si_2O_7)$ was prepared by solid state reaction starting from $CsNO_3$, $Sb_2O_5 \cdot xH_2O$ and SiO_2 [8]. The alkali antimony silicates were ion exchanged in a 9N nitric acid solution at 50°C. The solution to solid ratio was 100 ml/g of alkali material. Three stages, with renewal of the acidic solution after each 2 hours stage, are necessary to obtain the solid acids. At the completion of the whole ion exchange process, the alkali content of the solid acids is always less than 3% of the original one in the alkali compounds [9].

CHARACTERIZATION

The silicoantimonic acids are hydrated and their water content has been studied at 20°C as a function of relative humidity (RH). The adsorption isotherms (figure 1) for both acids show that the water content remains nearly the same over almost the whole range of RH. This behavior is similar to that observed for the 3D phosphatoantimonic acid $H_5Sb_5O_{12}(PO_4)_2 \cdot xH_2O$ [5]. The very beginning of the dehydration process is observed for RH $\cong$ 0%, whereas additional water is physisorbed onto the surface of the grains when RH becomes greater than 85%. According to the shape of these isotherms it is very likely that, for both acids, the water content corresponding to the wide plateau is associated to water molecules located within the channels of the structure. This conclusion is supported by the fact that the dehydration processes, leading to the anhydrous phases $H_4Sb_4O_8(Si_4O_{12})$ and $H_3Sb_3O_6(Si_2O_7)$, are completly reversible [9]. They occur between 40 and 400°C. Upon heating at temperatures above 450°C both acids decompose and, at 700°C, the final products are mixtures of Sb_6O_{13} and SiO_2.

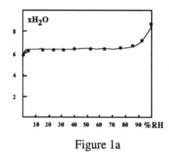 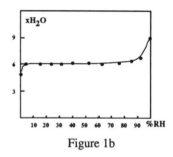

Figure 1a   Figure 1b

Figure 1. - Evolution of the water content x as a function of the relative humidity at 20°C for $H_4Sb_4O_8(Si_4O_{12}) \cdot xH_2O$ (1a) and $H_3Sb_3O_6(Si_2O_7).xH_2O$ (1b).

PROTONIC CONDUCTIVITY AND ION EXCHANGE BEHAVIOR

The existence of strong covalent bonding within the framework of the starting alkali antimony silicates [7,8] makes the bonding between alkali ions and framework of strong ionic character and enables ion exchange behavior. Accordingly, the proton exchanged derivatives should behave as Brönsted acids and exhibit protonic conductivity as well as ion exchange properties.

For both acids the protonic conductivity has been measured at 20°C as a function of RH and for a given RH as a function of temperature. The powders, equilibrated at various RH were pressed at $10^5 N/cm^2$ into about 1 mm thick pellets. Opposite sides of the disk were coated with a gold paste and leads were attached with a silver conductive resine. The pellets were mounted in a cell maintained at 20°C with a controlled RH. As it was not possible with our equipment to work at a fixed RH when varying temperature, activation energy measurements were performed on pellets equilibrated at a given RH at 20°C and embedded afterwards in an insulating resine.

As shown in figure 2, the evolution of the conductivity as a function of RH is closely related to that of the water content, for both acids. Since the small change that is observed in the water content, mainly for high RH values, does not involve any modification of the lattice parameters, this additional water can be considered as physisorbed. Then, it may be responsible for the increase of the conductivities by about one order of magnitude when RH goes from 20 to 85%. Accordingly, when RH is close to 10%, the amount of physisorbed water becomes negligible and the measured conductivities can be considered as the true "bulk-type" conductivities of $H_4Sb_4O_8(Si_4O_{12})\cdot6.5H_2O$ ($\sigma = 0,5.10^{-5}$ $\Omega^{-1}cm^{-1}$) and $H_3Sb_3O_6(Si_2O_7).6$ H_2O ($\sigma = 10^{-4}\Omega^{-1}cm^{-1}$).

Activation energy measurements were performed between -50°C and room temperature on pellets equilibrated first at RH = 60%. The straight-line slopes in figure 3 give activation energies of 0.61 eV for $H_4Sb_4O_8(Si_4O_{12})\cdot6.5H_2O$ and 0.45eV for $H_3Sb_3O_6(Si_2O_7)\cdot6H_2O$ .Such values fall in the range of those usually observed for compounds in which proton diffusion occurs within an ordered network of hydrogen bonds [10].

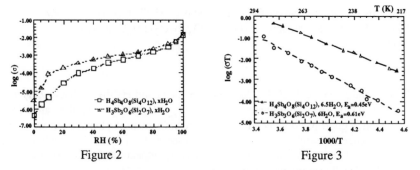

Figure 2   Figure 3

Figure 2 - Evolution of the protonic conductivity as a function of RH at 20°C
Figure 3 - Evolution of the protonic conductivity as a function of temperature at RH=60%

Preliminary investigations of the ion exchange properties have been undertaken by titrating the solid acids with 0.1N alkali hydroxide solutions and using 0.1N ACl (A = alkali) as supporting electrolyte. The titration rate was adjusted in order not to modify both the shape of the curves and the position of the endpoints. Under these conditions, for both acids, the extent of exchange increases when the size of the alkali ion decreases but the full exchange capacity is never obtained. For $H_3Sb_3O_6(Si_2O_7)\cdot6H_2O$ (figure 4) two endpoints have been observed on both the Li^+-H^+ and Na^+-H^+ curves. Such a behavior could well be related to structural features. As a matter of fact, it has been shown that in the structure of $Cs_3Sb_3O_6(Si_2O_7)$ [8], Cs atoms occupy two independent crystallographic positions with multiplicities of 4 and 2; consequently the potentials of these sites are certainly different and Li^+ as well as Na^+, that are more polarizing than the other alkali ions, could then fill such sites selectively. This hypothesis, supported by the fact that the first endpoint occurs at 2/3 of the total exchange capacity, has been verified from ^{23}Na MAS NMR experiments on samples with various Na^+ loadings. Figure 5 shows ^{23}Na MAS NMR spectra corresponding to compositions $Na_2HSb_3O_6(Si_2O_7)\cdot xH_2O$ and $Na_3Sb_3O_6(Si_2O_7)\cdot xH_2O$, i.e. a 67% and a fully exchanged phase. For the first one there is only one line in the spectrum whereas two lines are observed for the second one. Furthermore in this latter case the line intensities are in a 2:1 ratio thus giving evidence for two Na sites with multiplicities in the same ratio.

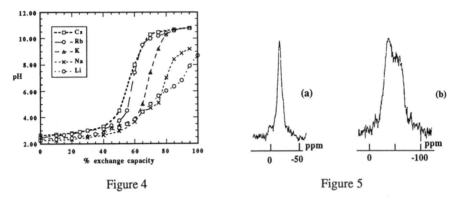

Figure 4   Figure 5

Figure 4 - Potentiometric titration curves for alkali cations on $H_3Sb_3O_6(Si_2O_7)\cdot xH_2O$
Figure 5 - ^{23}Na MAS NMR experiments for $Na_2HSb_3O_6(Si_2O_7)\cdot xH_2O$ (a) and
$Na_3Sb_3O_6(Si_2O_7)\cdot xH_2O$(b)

CONCLUSION

The 3D silicoantimonic acids $H_4Sb_4O_8(Si_4O_{12})\cdot6.5H_2O$ and $H_3Sb_3O_6(Si_2O_7)\cdot6H_2O$ behave very much like the 3D phosphatoantimonic acid $H_5Sb_5O_{12}(PO_4)_2\cdot6H_2O$ [5]: they are true lattice hydrates that exhibit a good protonic conductivity. The influence of physisorbed water on this conductivity is evidenced clearly but its importance is lesser than in particle hydrates such as $HSb(PO_4)_2\cdot xH_2O$ [5]. The ion exchange properties clearly appears as limited by the diffusion processes within the 3D silicoantimonic frameworks.

REFERENCES

1) Piffard, Y. , Lachgar , A. and Tournoux, M.: J. Solid State Chem., 1985, 58, 253.

2) Piffard, Y., Oyetola, S., Courant, S. and Lachgar, A.: J. Solid State Chem., 1985, 60, 209.

3) Tournoux, M. and Piffard, Y.: French Patent 85-01839.

4) Piffard, Y., Verbaere, A., Lachgar, A., Deniard-Courant, S. and Tournoux, M.: Rev. Chim. Min., 1986, 23, 766.

5) Deniard-Courant, S., Piffard, Y., Barboux, Ph., Livage, J.: Solid State Ionics, 1988, 27, 189.

6) Piffard, Y., Verbaere, A., Oyetola, S., Deniard-Courant, S. and Tournoux, M.: Eur. J. Solid State Inorg. Chem., 1989, 26, 113.

7) Pagnoux, C., Verbaere, A., Kanno, Y., Piffard, Y. and Tournoux, M.: J. Solid State Chem., 1992, 99, 173.

8) Pagnoux, C., Verbaere, A., Piffard, Y. and Tournoux, M.: Eur. J. Solid State Inorg. Chem., 1993, 30, 111.

9) Pagnoux, C., Verbaere, A., Piffard, Y. and Tournoux, M.: Solid State Ionics,1993, 61, 149.

10) Barboux, Ph.: Thesis, Paris VI, 1987.

Materials Science Forum Vols. 152 - 153 (1994) pp. 245-250
© 1994 Trans Tech Publications, Switzerland

INVESTIGATIONS ON LAYERED PEROVSKITES:
$Na_2Nd_2Ti_3O_{10}$, $H_2Nd_2Ti_3O_{10}$ AND $Nd_2Ti_3O_9\square$

M. Richard, L. Brohan and M. Tournoux

Institut des Matériaux de Nantes, Laboratoire de Chimie des Solides,
2 rue de la Houssinière, F-44072 Nantes Cédex 03, France

Keywords: Layered Perovskites, Structure Determination, T.E.M. Diffraction, Acid-Exchange

ABSTRACT

The structure of the layered perovskite $Na_2Nd_2Ti_3O_{10}$ was investigated by refining X-ray powder diffraction data using the Rietveld method. The Nd^{3+} cation is located in the intra-slab perovskite site and the Na^+ cation in the interlayer space. Acid exchange of the potassium homologous compound leads to the formation of $H_2Nd_2Ti_3O_{10}$, xH_2O. Prior to condensation at 900°C, the thermolysis of this solid acid yields an intermediate phase which, despite the total removal of water, retains the layered structure from 600°C to 850°C. The thermal evolution of the protonated form was studied by thermal analysis and crystallographic techniques and was found to exhibit two main steps. The first corresponds to the removal of water, which is complete at 600°C, and results in an intermediate phase containing five-coordinated titanium cations and a statistical distribution of Nd^{3+} in the available intra- and inter-slab sites. The second step can be considered as a complex condensation reaction leading to a 3D-cation defective perovskite.

INTRODUCTION

There is a growing interest in the proton exchange and subsequent dehydration of layered alkali oxides in connection with possible application of the resulting solids in catalysis, photocatalysis or electrochemistry. For example, this soft chemistry process has been applied to the layered titanates $A'_2Ti_nO_{2n+1}$ (A' =Na, K, Cs ; $3 \leq n \leq 6$) to form a metastable form of titanium oxide, $TiO_2(B)$, at temperatures below 350°.[1,2,3,4,5] Two families of lamellar perovskites containing interlayer alkali-metal cations $A'[A_{n-1}B_nO_{3n+1}]$ and $A'_2[A_{n-1}B_nO_{3n+1}]$ have been recently reported. These compounds are related to the Ruddlesden-Popper phases $M^{II}[A_{n-1}B_nO_{3n+1}]$, but exhibit layer charges four and two times weaker respectively. This lower charge allows ion exchange in an acidic medium and can yield protonated derivatives [6,7,8]. In this paper we report the Rietveld refinement of the $Na_2Nd_2Ti_3O_{10}$ structure, the thermal behaviour of $H_2Nd_2Ti_3O_{10}$, xH_2O and the characterization of the phases that result from its complete dehydration. According to J. Gopalakrishnan the complete dehydration of $H_2Ln_2Ti_3O_{10}$ leads to a defective perovskite for the lanthanium compound whereas the dehydrated products for Ln=Sm, Gd and Dy retain their layer-like features before transforming to pyrochlores at 950°C [6]. M. Gondrand et al. have also shown that the complete dehydration of $H_2Gd_2Ti_3O_{10}$ leads to a mixture of $Gd_2Ti_2O_7$ pyrochlore and TiO_2 anatase at 900°C [7].

$Na_2Nd_2Ti_3O_{10}$ - STRUCTURAL DETERMINATION

The X-ray powder diffraction (XRD) pattern for $Na_2Nd_2Ti_3O_{10}$ was recorded using a D5000 Siemens diffractometer in a Bragg-Brentano geometry.The Rietveld refinement of the X-ray powder diffraction data was performed in an I4/mmm space group beginning with the atomic positions first proposed by Ruddlesden and Popper [9]. The Rietveld refinement was conducted using the MPROF program [10]. The XRD pattern exhibits some extra peaks which can be attributed to an impurity. These unexplained peaks correspond to the most intense reflections of a cubic perovskite, most likely $Na_{0.5}Nd_{0.5}TiO_3$. This perovskite was included as a second phase in the refinement. After an initial refinement, a difference Fourier map for $Na_2Nd_2Ti_3O_{10}$ was generated from the observed and calculated structure factors. The difference Fourier map was then used to locate Nd^{3+} in the perovskite cavity and Na^+ between the layers. Table I indicates the refined lattice and atomic parameters .

Table I : Final atomic parameters from the Rietveld refinement of $Na_2Nd_2Ti_3O_{10}$

Atoms	x	y	z	$B(\text{Å}^2)$	occupancy
Ti 1	0	0	0	-0.08(7)	1
Ti 2	0	0	0.1485(1)	-0.08(7)	1
Na	0	0	0.2883(3)	2.4(2)	1
Nd	0	0	0.42531(6)	0.96(4)	1
O 1	0	0.5	0	1.0(1)	1
O 2	0	0	0.0656(4)	1.0(1)	1
O 3	0	0.5	0.1348(3)	1.0(1)	1
O 4	0	0	0.2133(5)	1.0(1)	1

Note : a=3.8182(2)Å, c=28.369(1)Å, Space group I4/mmm, Z=2
R_{exp}=0.0706, R_p=0.166, R_{wp}=0.1207, R_B=0.0725

The interatomic distances within a I4/mmm cell are listed in Table II and an idealized picture of the structure is shown in figure 2a. The Ti1-O distances show the usual spread of values expected for titanate structures whereas the Ti2 cation has a relatively long bond to the O2 oxygen (d_{Ti2-O2}=2.350(1)Å). The weakening of the Ti2-O2 bond may be due to the reinforcement of the Ti2-O4 bond that results from the high basicity of the O4 atom. The sodium atoms occupy an 8+1 coordinated site; a short bond is observed with O4, which belongs to the central layer of octahedra of the perovskite slab. The neodynium is 12-fold coordinated in the intra-layer perovskite site with a rather symmetric environment.

Table II : Metal-oxygen distances in the structure

M-O	(Å)		
Ti1-O1	1.909(1)	Na-O3	2.900(9)
Ti1-O2	1.862(1)	Na-O4	2.7003(3)
		Na-O4	2.13(2)
Ti2-O2	2.350(1)	Nd-O1	2.852(1)
Ti2-O3	1.948(2)	Nd-O2	2.712(1)
Ti2-O4	1.840(1)	Nd-O3	2.559(5)

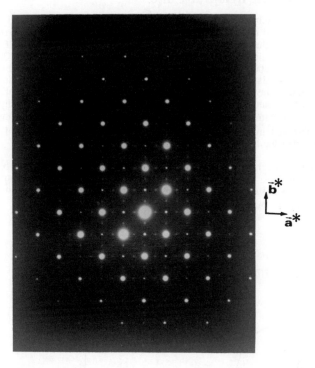

Figure 1 : Experimental [001] zone axis electron diffraction pattern exhibiting superlattice reflections along a* and b* axis.

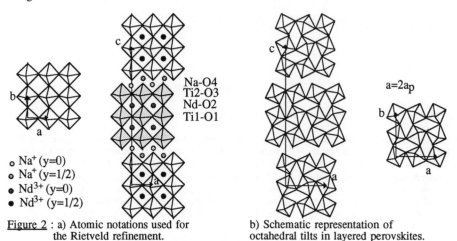

Figure 2 : a) Atomic notations used for the Rietveld refinement.

b) Schematic representation of octahedral tilts in layered perovskites.

However this description is only that of the average structure and [001] zone axis electron diffraction patterns exhibit superlattice reflections that involves a doubling of the a parameter of the previous tetragonal cell. (figure 1) This superstructure is probably due to small atomic displacements resulting from octahedral tilts around b and c axis, which have been frequently reported for many layered perovskites.[8,11] Such a situation is schematically depicted in figure 2b. A Rietveld refinement of neutron diffraction data is presently in progress.

THE PROTONATED PHASE $H_2Nd_2Ti_3O_{10}$, xH_2O

Almost complete exchange of K^+ by protons (more than 98%) in $K_2Nd_2Ti_3O_{10}$ was achieved by stirring the starting layered titanate in excess of 1M HNO_3 for two days. The X-ray powder diffraction pattern, closely related to that of $K_2Nd_2Ti_3O_{10}$, shows the presence of strong (00l) lines, indicating that the layered structure is retained. Indexing the four peaks at d=13.949Å, 6.872Å, 3.752Å and 2.675Å as (002), (004), (110) and (200) respectively, yields a unit cell which is consistent with a≈3.8 Å and c ≈28Å. The [010] zone axis electron diffraction pattern shows a k+l=2n+1 extinction explained by a (a+b)/2 shift of adjacent layers as in $Na_2Nd_2Ti_3O_{10}$. The [001] zone axis exhibits some weak extra reflections involving the same doubling of the a_p parameter observed for the anhydrous sodium parent and in good agreement with previous data due to Gondrand et al.. In addition the enlargement of the 00l reflections observed in electron diffraction experiments can be attributed to a small stacking disorder resulting from the hydroxylation process.

Thermogravimetry for the protonated phase shows a multistep weight loss behavior which can be related to the formation of discrete structural intermediates during the dehydration reaction. As shown in figure 3, the total weight loss (complete at 600°C) corresponds to 1.7 H_2O leading to the single phase $Nd_2Ti_3O_9\square$. The starting material may be written as $H_2Nd_2Ti_3O_{10}$, $0.7H_2O$.

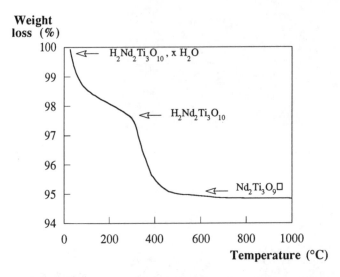

Figure 3: Thermogravimetric analysis of $H_2Nd_2Ti_3O_{10}$, xH_2O (x=0.7).

High temperature in situ X.R.D and powder diffraction measurements on samples quenched from high temperatures were used to isolate and characterize the structures of these intermediates. During the dehydration of $H_2Nd_2Ti_3O_{10}$, $0.7H_2O$ to the 3D cation defective perovskite $Nd_{2/3}TiO_3$ two distinct structural intermediates were isolated : $H_2Nd_2Ti_3O_{10}$ and $Nd_2Ti_3O_9\square$ at 350°C and 600°C respectively. Between room temperature and 300°C the materials are multiphasic. The refined lattice parameters of the tetragonal I centered unit cell are close to those of the parent layered perovskite in the 350°C to 600°C temperature range. The layer-like feature of the structure is thus maintained. However, there is a 10% decrease in the c parameter during the reaction :

$$H_2Nd_2Ti_3O_{10}, 0.7H_2O \rightarrow Nd_2Ti_3O_9\square + 1.7H_2O.$$

The most interesting aspect of this reaction seems to be the conservation of the lamellar structure between 600°C and 850°C. In this temperature range, the c parameter decreases by 3% without any significant structural change.

In order to confirm that condensation did not actually occur, a Rietveld refinement of $Nd_2Ti_3O_9\square$ (obtained at 600°C) was performed. Initial refinements using a structure in which the Nd the perovskite site and in the interlayer space were tested and a structure with 1/3 of the Nd cations in the interlayer space and 2/3 in the layer yielded the best refinement result. In this case the interlayer (0, 0, 0.2363) site is half occupied by the O4 oxygen and one third by Nd; the final atomic positions are shown in table III. Of course it is unusual to distribute an anion and a cation on the same site. This result correponds to an average structure. Certainly some local relaxations occur that are not taken into account in this refinement due to the poor quality of the available data related to this particular type of chemistry.

Table III : Final atomic parameters from the Rietveld refinement of $Nd_2Ti_3O_9\square$.

Atoms	x	y	z	B($\mathring{A}^2$)	occupancy
Ti 1	0	0	0	3.8(1)	1
Ti 2	0	0	0.1558(2)	3.8(1)	1
Nd1	0.5	0.5	0.0915(2)	3.5(1)	2/3
Nd2	0	0	0.2363(3)	2.5(2)	1/3
O 1	0	0.5	0	4.2(3)	1
O 2	0	0	0.0633(9)	4.2(3)	1
O 3	0	0.5	0.163(1)	4.2(3)	1
O 4	0	0	0.2363(3)	4.2(3)	1/2

Note : a=3.8334(6)Å, c=24.363(4)Å, Space group I4/mmm, Z=2
R_{exp}=0.076, R_p=0.291, R_{wp}=0.195, R_B=0.1675

When the neodynium atom is located on the O4 site the Ti2 titanium atom adopts a five-fold coordination with four O3 neighbours in a square planar arrangement and the fifth oxygen O2 at the apex of a square pyramid. Fivefold coordination for Ti^{4+} was first found by Anderson and Wadsley in the hygroscopic layered compound $K_2Ti_2O_5$ [12] and has subsequently been identified in Ln_2TiO_5 (Ln=Y,La) [13,14] and Tl_2TiO_3 [15]. This coordination has been also observed in titano-silicates such as $Ba_2TiOSi_2O_7$ [16] or $Na_2TiOSiO_4$ [17,18]. In these compounds the coordination is square pyramidal and in all cases the shortest distance is from the metal to the apical oxygen.

Although this model gave the best fit to the X-ray profile, in order to explain the relatively high values of the reliability factors it must be consider that : (1) considerable disorder between the slabs occurs during the acid-exchange leading to the precursor of $Nd_2Ti_3O_9\square$. This disorder is retained in the dehydrated product as indicated by the widening of the diffraction lines, (0.25 to 0.40°) (2) it is likely that the real symmetry is less than I4/mmm, however it is not possible from our experimental data to perform a refinement in a space group with lower symmetry. This point may be a possible explanation for the short value of Ti2-O2 bond. However despite these potential errors, this refinement allows us to conclude that the condensation of the perovskite layers to give a 3D framework does not occur and that part of the neodynium cations are distributed within the inter-slab space and the intra-layer sites. At 900°C, the X-ray powder diffraction pattern of the final product can be indexed in a primitive cell with a=a_p and c ≈ a_p, suggesting cationic disorder inside the perovskite like cavity.

CONCLUSION

The layered perovskite $Na_2Nd_2Ti_3O_{10}$ exhibits an ordered distribution of the Nd^{3+} and Na^+ cations. The Nd^{3+} cations are located in the intra-slab perovskite-like cavity whereas the Na^+ cations occupy the site in the interlayer space. In the protonated form the perovskite layers are linked by hydrogen bonds. The most basic oxygen of the perovskite layer is situated at the periphery of the slab and forms the OH groups in the anhydrous acid form $H_2Nd_2Ti_3O_{10}$; this oxygen O4, which is bonded to only one titanium Ti2, is therefore strongly protonated. During the dehydroxylation process between 300 and 650°C oxygen vacancies are formed and one third of the Nd^{3+} cations migrate from the perovskite site to the interlayer space without any other structural changes aside from a continuing contraction of c. Finally the condensation of the layers occurs at 900°C leading to a 3D tetragonal perovskite in which the Nd^{3+} cations are randomly distributed in the perovskite sites.

Acknowledgment : we thank Pr. P. Davies for help in the writing up of the paper.

[1] R. Marchand, L. Brohan and M.Tournoux, Mater. Res. Bull., **15**, 201, (1980)
[2] M. Tournoux, R. Marchand and L. Brohan, Prog. Sol. St. Chem., **17**, 33, (1986)
[3] T.P. Feist and P.Davies, J. Solid State Chem., **101**, 275, (1992)
[4] T.P. Feist, S.J. Mockarski, P.K. Davies, A.J. Jacobson and J.T. Lewandowski, Solid State Ionics, **28-30**, 1338, (1988)
[5] H. Izawa, S. Kikkawa and H. Koizumi, J. Phys. Chem. Solids, **86**, 5023, (1982)
[6] J. Gopalakrishnan, Inorg.Chem., **26**, 4301-4303, (1987)
[7] M. Gondrand, J.C. Joubert, Rev. de Chim. min., **24**, 33-41, (1987)
[8] M. Dion, M. Ganne and M. Tournoux, Mater. Res. Bull., **16**, 1429, (1981)
[9] S.N. Ruddlesden and P. Popper, Acta Cryst., 11, 54-55, (1957)
[10] A.D. Murray and A.N. Fitch, "A multipattern Rietveld refinement Program for Neutron X ray and Synchroton Radiations", (1989)
[11] R. Deblieck, J. van Landuyt and S. Amelinckx, J. Solid State Chem., **59**, 379-387, (1985)
[12] S. Anderson and A.D. Wadsley, Acta. Chem. Scand., **15**, 663, (1961)
[13] M. Guillen and E.F. Bertaut, C. R. Acad. Sc. Paris, **262**, B 962-965, (1966)
[14] W.G. Mumme and A.D. Wadsley, Acta Crystallogr., **B24**, 1327-1333, (1968)
[15] A. Verbaere, M. Dion and M. Tournoux, J. of Solid State Chem., **11**, 60-66, (1974)
[16] P.B. Moore and J. Louisnathan, Science, **156**, 1361, (1967)
[17] Yu K. Egorov-Tiamenko, M.A. Simonov and N.V. Belov, Sov. Phys. Dokl., **23**, 289, (1978)
[18] P. A. Thomas, "Materials for non linear and electrooptics.", Inst. Phys. Conf. Ser., **103**, Part. 1,53, (1989)

Materials Science Forum Vols. 152 - 153 (1994) pp. 251-254

A PROTONATED FORM OF THE TITANATE WITH THE LEPIDOCROCITE-RELATED LAYER STRUCTURE

T. Sasaki, M. Watanabe, Y. Fujiki and S. Takenouchi

National Institute for Research in Inorganic Materials,
1-1 Namiki, Tsukuba, Ibaraki 305, Japan

Keywords: Layered Titanate, Protonic Oxide, Solid-Acid, Ion-Exchange, Intercalation

ABSTRACT

The Cs-titanate, $Cs_xTi_{2-x/4}\square_{x/4}O_4$ ($\square$: vacancy, $x \sim 0.7$), with the lepidocrocite-related layer structure is converted into a protonated form by treating it with acid solutions. The resulting protonic titanate $H_xTi_{2-x/4}\square_{x/4}O_4 \cdot H_2O$ is a unifunctional solid acid with interlayer oxonium ions as an exchangeable site. The alkali metal ion-exchange proceeded evolving two defined hydrate structures; monolayer hydrates (<70 % consumption of the exchange capacity) and bilayer ones (>70% conversion). The acidity of the material was strong enough to interact with the weak base such as pyridine.

INTRODUCTION

A variety of layered alkali titanates, e. g., $Na_2Ti_3O_7$, $K_2Ti_4O_9$ and $Cs_2Ti_5O_{11}$, have been synthesized[1-4] and their soft-chemical aspects (protonation, ion-exchange, intercalation, slow-thermolysis etc.) have been extensively investigated[5-15]. Recently Grey et al.[4,16] have prepared a Cs-titanate, $Cs_xTi_{2-x/4}\square_{x/4}O_4$ ($\square$: vacancy), a new member of the layered titanates. The material is characterized by a structural relative of lepidocrocite $FeO(OH)$ and charge-compensation through vacancies at Ti sites. Similar compounds have been synthesized by incorporating various di- or trivalent cations in the host framework[17,18], e. g., $Cs_xTi_{2-x/2}Mg_{x/2}O_4$. However, there have been few reports concentrating on their ion-exchange and intercalation properties except one by England et al. [19]. The present study is undertaken, by transforming $Cs_xTi_{2-x/4}\square_{x/4}O_4$ into a protonated form, to examine its acid-base properties and compare them with those of the other types of protonic titanates.

EXPERIMENTAL

Polycrystalline samples of the Cs-titanate were synthesized by calcining a mixture of Cs_2CO_3 and TiO_2 at 800°C (20 hr × 2). The single phase was obtained in the molar range of 1/5.0 - 1/5.5 for Cs_2CO_3/TiO_2. The nonstoichiometric parameter, x, in the formula was 0.67 - 0.73.

The interlayer Cs ions were extracted by treating the titanate with aqueous HCl at room temperature. The solution to solid ratio was 100 cm^3 g^{-1} and the solution was replaced to a fresh one every 24 hours.

The titration was carried out batchwise by equilibrating a weighed amount (0.2 g) of the protonic titanate with a 20 cm^3 of (MCl-MOH) mixed solution (M: alkali metal) at 25 ± 0.5°C. The titanate with a stoichiometry of $x = 0.7$ was used and the ionic strength of the titrant was adjusted to 0.1. After 7 days, supernatant solutions were analyzed for their residual cation contents and pH values. The solids, washed and conditioned at a relative humidity of 70 %, were characterized by powder XRD, IR, TGA and chemical analysis.

RESULTS AND DISCUSSION

Preparation of the Protonic Titanate

Approximately 98% of the interlayer Cs ions was eluted after 3 cycles of the HCl treatment. The body-centered orthorhombic layer structure was preserved upon the Cs ion removal, which expanded the interlayer separation. A change in unit cell dimensions for a typical stoichiometry of x = 0.7 was as follows: a, 3.837(1) Å → 3.783(2) Å; b, 17.198(3) Å → 18.735(8) Å; c, 2.960(1) Å → 2.978(2) Å.

A few more cycles of the treatment achieved a further removal of the Cs ions to a negligible level, which, however, tended to introduce a stacking disorder, an irregular displacement of the host layers parallel to the a axis. The reflections with indices of $hk0$ and hkl became broad or diffuse while the other diffraction lines remained substantially unchanged. The layer-to-layer registry was restored when pinning guests were incorporated again in the interlayer space.

To sum up, the protonic titanate with the lepidocrocite-related layer structure is obtained as formulated below.

$$Cs_xTi_{2-x/4}\square_{x/4}O_4(s) + xH^+(aq) + H_2O(aq) \rightarrow H_xTi_{2-x/4}\square_{x/4}O_4 \cdot H_2O(s) + xCs^+(aq) \qquad (1)$$
$$[8.6 \text{ Å}] \qquad\qquad\qquad\qquad\qquad [9.4 \text{ Å}]$$

where numerals in square parentheses denote the interlayer spacing, d_{020}.

As illustrated in Fig. 1, the material accommodates one water molecule in each pseudo-cubic

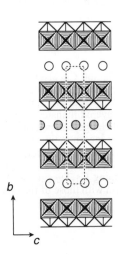

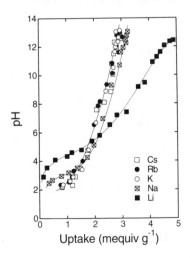

Figure 1. Ideal representation of the crystal structure for $H_xTi_{2-x/4}\square_{x/4}O_4 \cdot H_2O$ projected onto [100]. Open and stippled circles represent H_2O (or H_3O^+) at different levels along the a axis.

Figure 2. Alkali metal ion uptake as a function of pH. □: Cs, ●: Rb, ○: K, ⊠ : Na, ■ : Li.

cavity in the interlayer space, 70 % of which is protonated to the oxonium ion. Unlike the other types of protonic titanates, $H_2Ti_3O_7$, $H_2Ti_4O_9 \cdot 1.2H_2O$ and $H_2Ti_5O_{11} \cdot 3H_2O$ [6,8,11,15], the oxonium ion is only protonic species present in the material (no hydroxyls), which is supported by IR data.

Ion-Exchange Behavior

The protonic titanate $H_xTi_{2-x/4}\square_{x/4}O_4 \cdot H_2O$ readily undergoes ion-exchange and intercalation reactions at ambient temperature. The titration with alkali metal ions (Fig. 2) suggests that the material is a unifunctional solid acid. Multifunctional nature has been established for the other protonic titanates, $H_2Ti_4O_9 \cdot 1.2H_2O$ and $H_2Ti_5O_{11} \cdot 3H_2O$[8,9,11]. This difference may be attributable to the geometry of the interlayer space.

The protonic titanate preferred the heavier members of alkali metal ions to the lighter ones, i. e., Cs ~ Rb ~ K > Na > Li in a weakly acidic media (pH < 5), while the order was inverted in a higher pH region. Similar selectivity series have been observed in $H_2Ti_4O_9 \cdot 1.2H_2O$ and $H_2Ti_5O_{11} \cdot 3H_2O$ [8,9,11] and this is elucidable in terms of hydration energy of the cations and steric constraints between them.

Up to a 70 % consumption of the theoretical ion-exchange capacity (4.12 mequiv g^{-1}), the interlayer separation contracted continuously from 9.4 to 9.3 Å for Li, 8.9 Å for Na, 9.0 Å for K, 8.6 Å for Rb and 8.8 Å for Cs. A further exchange beyond 70 % conversion took place only for Li and Na ions and evolved highly hydrated phases with an expanded interlayer spacing, 11.3 Å and 11.5 Å, respectively. These phenomena indicate that the ion-exchange processes can be divided into two regions at the threshold conversion of 70 % where two changes took place. One is the cessation of the exchange reaction for K, Rb and Cs ions and the other is the discontinuous expansion of the lattice for Li and Na ions.

The lattice parameter refinements for the ion-exchanged phases give an evidence that the host layer of lepidocrite-type was maintained throughout. The cations except Li were incorporated keeping the body-centered symmetry. In contrast, the Li ion-exchange brought about a change in lattice type from *I* to *P* in the conversion range 0 - 70 %, and to *C* for further uptakes. This means lateral displacements of the adjacent host layers with respect to one another by $(a+c)/2$ and then by $a/2$.

Hydrate Structures

The modestly hydrated phases with the interlayer distance of ~9 Å are due to monolayer arrangements of cations and water molecules (see Fig. 3a). This configuration is substantiated by chemical compositions and, more straightforwardly, by Rietveld refinements, the results of which will be described elsewhere. On the other hand, the highly swollen phases formed at 70 % loading and above may be ascribed to bilayer hydrates (Fig. 3b). The difference in interlayer separation (~2.5 Å) for these two hydrate stages is reasonable, being consistent with the size of water molecules.

The threshold conversion (= 70 %) involving the two defined hydrate structures is associated with the charge density of the host layer. The threshold uptake corresponds to the occupation of half

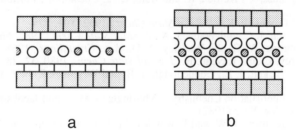

a   b

Figure 3. Possible arrangements of interlayer cations and water molecules. (a) monolayer hydrate (<70 % conversion), (b) bilayer hydrate (>70 %). Open and shaded circles represent water molecules and alkali metal ions, respectively.

of the interlayer units (cavities encircled by eight oxygen atoms on neighboring host layers).

$$0.7 \ (= \text{threshold}) \times 0.7 \ (= \text{layer charge per formula weight}) \approx 0.5 \qquad (2)$$

The incorporated cations below the threshold are likely to be shielded effectively with dielectric spacers of H_2O (or H_3O^+) on both sides (see Fig. 3a). The uptake beyond the threshold inevitably gives rise to a close contact of cations which are situated in one cavity and in another next to it. The exchange reaction for K, Rb and Cs ions does not proceed beyond this point to avoid the unfavorable proximity. The cations (Li, Na) with higher hydration energy produce bilayer arrangements of cation/water clusters where severe repulsion may be tolerated.

Schöllhorn et al. have reported similar hydrate series for layered dichalcogenides of transition metals[20]. The formation of the two types of hydrate structures is dominated principally by charge/radius ratio of interlayer cations and, being contrasted with the case found here, not by a population of cations.

Pyridine Intercalation

The action of liquid pyridine yielded a complex $(C_5H_5N)_{0.2}H_xTi_{2-x/4}\square_{x/4}O_4 \cdot H_2O$ (d_{020} = 11.4 Å), the composition of which was determined from the TGA-IR data. It is to be pointed out that $H_xTi_{2-x/4}\square_{x/4}O_4 \cdot H_2O$ is a stronger Brønsted acid than $H_2Ti_3O_7$ and $H_2Ti_4O_9 \cdot 1.2H_2O$, which, according to the literatures[7,21], do not take up pyridine. The Rietveld refinement demonstrates a unique gliding of the host layers along the a axis by ~1 Å as well as a swelling of the interlayer separation by ~2 Å. This structural modification makes the interlayer space suitable for pyridine to be packed efficiently and to interact with surface oxygen atoms of the host layers.

REFERENCES

1) Andersson, S. and Wadsley, A. D.: *Acta Crystallogr.*, 1961, **14**, 1245.
2) Verbaere, A. and Tournoux, M.: *Bull. Soc. Chim. Fr.*, 1973, **4**, 1237.
3) Dion, M., Piffard, Y. and Tournoux, M.: *J. Inorg. Nucl. Chem.*, 1978, **40**, 917.
4) Grey, I. E., Madsen, I. C., Watts, J. A., Bursill, L. A. and Kwiatkowska, J.: *J. Solid State Chem.*, 1985, **58**, 350.
5) Marchand, R., Brohan, L. and Tournoux, M.: *Mater. Res. Bull.*, 1980, **15**, 1129.
6) Tournoux, M., Marchand, R. and Brohan, L.: *Prog. Solid St. Chem.*, 1986, **17**, 33.
7) Clément, P. and Marchand, R.: *C. R. Acad. Sci. Paris Ser. II*, 1983, **296**, 1161.
8) Sasaki, T., Watanabe, M., Komatsu, Y. and Fujiki, Y.: *Inorg. Chem.*, 1985, **24**, 2265.
9) Sasaki, T., Komatsu, Y. and Fujiki, Y.: *Inorg. Chem.*, 1989, **28**, 2776.
10) Sasaki, T. and Fujiki, Y.: *J. Solid State Chem.*, 1989, **83**, 45.
11) Sasaki, T., Komatsu, Y. and Fujiki, Y.: *Chem. Mater.*, 1992, **4**, 894.
12) Izawa, H., Kikkawa, S. and Koizumi, M.: *J. Phys. Chem.*, 1982, **86**, 5023.
13) Izawa, H., Kikkawa, S. and Koizumi, M.: *Polyhedron*, 1983, **2**, 741.
14) Miyata, H., Sugahara, Y., Kuroda, K. and Kato, C.: *J. Chem. Soc., Faraday Trans. 1*, 1988, **84**, 2677.
15) Feist, T. P. and Davies, P. K.: *J. Solid State Chem.*, 1992, **101**, 275.
16) Grey, I. E., Madsen, I. C. and Watts, J. A.: *J. Solid State Chem.*, 1987, **66**, 7.
17) Reid, A. F., Mumme, W. G. and Wadsley, A. D.: *Acta Crystallogr.*, 1968, **B24**, 1228.
18) Groult, D., Mercy, C. and Raveau, B.: *J. Solid State Chem.*, 1980, **32**, 289.
19) England, W. A., Birkett, J. E., Goodenough, J. B. and Wiseman, P. J.: *J. Solid State Chem.*, 1983, **49**, 300.
20) Schöllhorn, R.: "Intercalation Chemistry", Whittingham, M. S. and Jacobson, A. J. (Eds.) Academic Press, New York (1982).
21) Jacobson, A. J., Johnson, J. W. and Lewandowski, J. T.: *Mater. Res. Bull.*, 1987, **22**, 45.

Materials Science Forum Vols. 152 - 153 (1994) pp. 255-258
© 1994 Trans Tech Publications, Switzerland

THE LAYERED PHOSPHATONIOBIC ACID HNbP$_2$O$_7$·xH$_2$O: SYNTHESIS, STRUCTURE, THERMAL BEHAVIOR AND ION EXCHANGE PROPERTIES

J.J. Zah Letho [1,2], P. Houenou [1], A. Verbaere [2], Y. Piffard [2] and M. Tournoux [2]

[1] Laboratoire de Chimie Minérale, Faculté des Sciences et des Techniques, 22 BP 582, Abidjan 22, République de Côte d'Ivoire

[2] Institut des Matériaux de Nantes, UMR 110 CNRS, Université de Nantes, 2 rue de la Houssinière, F-44072 Nantes Cédex 03, France

Keywords: Solid Acid, Niobium Diphosphate, Ion Exchange

ABSTRACT. - The layered phosphatoniobic acid HNbOP$_2$O$_7$·xH$_2$O has been prepared by an ion-exchange process in acidic medium, starting from CsNbOP$_2$O$_7$. Its structure is similar to that of the Cs precursor, with corrugated covalent layers [NbOP$_2$O$_7^-$]$_n$. X-ray and electron diffraction studies and thermal analyses provide evidence for the existence of three phases (x = 2, 1 or 0) between 20 and 490°C, with different interlayer distances. In this temperature range, the dehydration process is reversible, but at higher temperatures, the title acid decomposes. Titration of HNbOP$_2$O$_7$·xH$_2$O with alkali hydroxide solutions as well as ion-exchanges in alkali nitrate solutions have been performed. They show an interesting ion-exchange behavior for the acid.

I - INTRODUCTION

Some acidic niobium phosphates have been recently reported [1]. Their structure contains [NbOPO$_4$]$_n$ layers of the α-NbOPO$_4$ type, and the acidic character comes from molecules, such as H$_3$PO$_4$, intercalated into the interlayer space. As part of a search for compounds exhibiting a more acidic character and ion-exchange properties, we have already prepared the layered acids HM(PO$_4$)$_2$·xH$_2$O, M = Sb, Ta, via an ion-exchange process in acidic medium, starting from KM(PO$_4$)$_2$ [2-4], whose structures are closely related to that of α-Zr(HPO$_4$)$_2$·H$_2$O [5]. In order to obtain similar niobium V compounds, the same strategy was applied starting from CsNb(PO$_4$)$_2$ or CsNbOP$_2$O$_7$, both with a layered structure [4,6]. Attempts to ion-exchange CsNb(PO$_4$)$_2$ did not succeed, under various conditions such as Cs$^+$-K$^+$ exchange in fused KNO$_3$, before acidic treatment, or Cs$^+$-H$^+$ exchange in fused benzoïc acid.

We report here on the acid obtained from CsNbOP$_2$O$_7$.

II - PREPARATION

The precursor $CsNbOP_2O_7$ was first prepared from a solid state reaction at 800°C for 24 hours in air, starting from a stoichiometric mixture of $CsNO_3$, Nb_2O_5 and $NH_4H_2PO_4$ in a platinum crucible. The solid obtained corresponds to the compound previously described [6]. A suspension of 1 g of the precursor was stirred for 2 hours in 100 ml of a 9N HNO_3 solution at ambient temperature. The solid was separated from the solution, and two further treatments were performed to complete the Cs^+-H^+ exchange. The solid was washed with a pH 3-4 nitric acid solution in order to prevent hydrolysis from occuring, and then dried under vacuum to afford $HNbOP_2O_7 \cdot xH_2O$, $1 \leq x \leq 2$. X-ray and electron diffraction show that washing with water, or failing to first dry under vacuum, or both, lead to the formation of layered compounds exhibiting the same layer periodicity as in a $NbOPO_4 \cdot xH_2O$. After $HNbOP_2O_7 \cdot xH_2O$ was dried in an oven at 80°C, $HNbOP_2O_7 \cdot H_2O$ is obtained ; it must be kept in a dry atmosphere in order to prevent the slow formation of $HNbOP_2O_7 \cdot 2H_2O$, which slowly hydrolyzes, leading to the layered compounds mentioned above. Spectrophotometric analyses of Cs^+ in the filtrates, and analyses of Cs^+ in the final solid by energy dispersive spectrometry, using a microprobe, indicate a complete Cs^+-H^+ exchange.

III - THERMAL BEHAVIOR AND STRUCTURE

Both the TG curve (figure 1) and the X-ray powder diffraction (XRPD) studies in the temperature range 20-800°C show the thermal evolution :

$$HNbOP_2O_7 \cdot 2H_2O \xrightarrow{50°C} HNbOP_2O_7 \cdot H_2O \xrightarrow{100°C} HNbOP_2O_7$$

in the reversible dehydration domain (20 - 490°C). Upon heating at temperatures above 490°C, the acid decomposes, and mixtures of $NbP_2O_{7.5}$ [7] and $Nb_2P_3O_{12.5}$ [8] are obtained. The DSC curve for $HNbOP_2O_7 \cdot H_2O$ (figure 2) only shows the thermal phenomena associated with dehydration and decomposition.

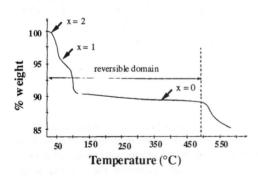

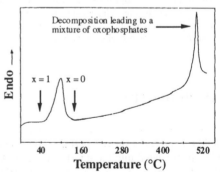

Figure 1. - TG curve for $HNbOP_2O_7 \cdot 2H_2O$.   Figure 2. - DSC curve for $HNbOP_2O_7 \cdot H_2O$.

The two reversible water loss steps of the acid involve two modifications of the interlayer distance, which decreases from 9.4 Å for x = 2, to 7.9 Å for x = 1, and then increases to 8.5 Å for x = 0. A similar increase, associated with the loss of the remaining water molecules, has already been observed for $HTa(PO_4)_2 \cdot xH_2O$ [3], and it can be related to a diminution of hydrogen bonding between the layers.

The title acid contains $[NbOP_2O_7]_n$ layers similar to those in the Cs precursor (figure 3), with P_2O_7 groups and NbO_6 octahedra sharing all their vertices except one which points into the interlayer space ; this space can accomodate one or two H_2O molecules. These structural conclusions can be inferred from the following observations :

- The acid is prepared by a Cs^+-H^+ exchange, and, when reacted with CsOH (see below), it leads to a compound whose XRPD pattern is very similar to that of the Cs precursor.

- Electron diffraction studies show that the periodicity within a layer in the protonic phase is very close to that in the parent Cs compound. (Because of the presence of diffuse streaks, perpendicular to the layers, and probably arising from stacking faults, the whole lattice cannot be unambiguously obtained).

- The interlayer distance reversibly varies with the water content x.

For $HNbOP_2O_7 \cdot H_2O$, refinement of the monoclinic unit-cell parameters leads to:

$$a = 4.891 \, (1) \, \text{Å} \quad , \quad b = 8.815 \, (6) \, \text{Å} \quad , \quad c = 15.54 \, (1) \, \text{Å} \quad , \quad \beta = 90.59 \, (6)°,$$

in fair agreement with the above conclusions.

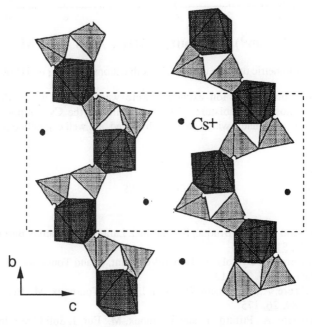

Figure 3 - $CsNbOP_2O_7$: [100]* view of the structure.

IV - ION EXCHANGE BEHAVIOR

In order to assess initially ion-exchange behavior of the acid, titrations were performed, using a Metrohm 702 SM automatic system, at ambient temperature. 110 mg of the solid were added to 50 ml of H_2O, before titration (constant rate of 2.5 ml/hour) with an equimolar AOH and ACl 0.09 M solution (A = alkali or NH_4). Figure 4 shows the titrations curves, for which the theoretical endpoints corresponding to the full H^+-A^+ exchange are at 4.05 ml of titrant. The curves show that the extent of exchange is related to the size of the alkali ion : 85 % for Cs^+, nearly 100 % for Li^+, and 90-92 % for the ions with intermediate size.

Exchange reactions were also carried out by adding 150 mg of the acid to 50 ml of a ANO_3 0.1 M solution, A = alkali, and stirring for 24 hours at room temperature. Chemical analyses of the filtrates and of the solids show an extent of exchange of 80 % for A = Cs, and of at least 95 % in the other cases.

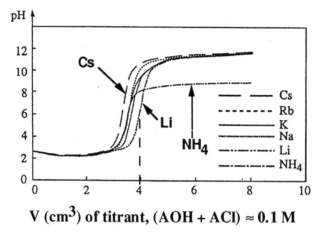

V (cm^3) of titrant, (AOH + ACl) $\approx$ 0.1 M

Figure 4 - Potentiometric titration curves for alkali cations and NH$_4$ on HNbOP$_2$O$_7$·H$_2$O.

The compounds obtained by ion exchange were washed with water, dried at 50°C and equilibrated under atmospheric moisture. The crystallinity of the Cs compound is rather poor, whereas the Li and Na compounds, which are more hydrated, are well crystalline products.

REFERENCES

1) Cantero, M., Moreno Real, L., Bruque, S., Martinez Lara M. and Ramos Barrado, J.R.: Solid State Ionics, 1992, 51, 273
2) Piffard, Y., Verbaere, A., Oyetola, S., Deniard-Courant, S. and Tournoux, M.: Eur. J. Solid State Inorg. Chem., 1989, 26, 113
3) Oyetola, S., Verbaere, A., Guyomard, D., Piffard, Y. and Tournoux, M.: Eur. J. Solid State Inorg. Chem., 1989, 26, 175
4) Oyetola, S., Verbaere, A., Piffard, Y. and Tournoux, M.: Eur. J. Solid State Inorg. Chem., 1988, 25, 259
5) Clearfield, A. and Smith, G. D.: Inorg. Chem., 1969, 8, 431
6) Nikolaev, V.P., Sadikov, G.G., Lavrov, A.V. and Porai-Koshits, M.A.: Dokl. Akad. Nauk SSSR, 1982, 264, 859
7) Levin, E.M. and Roth, R.S.: J. Solid State Chem., 1970, 2, 250
8) Zah-Letho, J.J.: Thesis, Abidjan, 1993

Materials Science Forum Vols. 152 - 153 (1994) pp. 259-262

NEW PHASES OBTAINED BY ACID DELITHIATION OF LAYERED LiMO$_2$ (M = Co,Ni)

E. Zhecheva and R. Stoyanova

Institute of General and Inorganic Chemistry, Bulgarian Academy of Sciences,
BG-1113 Sofia, Bulgaria

Keywords: LiCoO$_2$, Li$_x$Ni$_{2-x}$O$_2$, Acid Delithiation, Lithium Exchange with Protons

ABSTRACT

Metastable layered phases Li$_{1-x-y}$H$_y$MO$_2$ (0≤y<x<1) were obtained by acid digestion of LiMO$_2$ (M=Co, Ni) at room temperature. It has been shown that the layered MO$_2$ framework of the parent LiMO$_2$ is retained during acid treatment. For LiCoO$_2$, lithium extraction and exchange of lithium ions with protons proceed concomitantly and as a result lithium containing cobalt oxyhydroxides Li$_{1-x-y}$H$_y$CoO$_2$ (x≤0.55, x+y<1) were obtained. For Li$_x$Ni$_{2-x}$O$_2$ (0.6<x<1, where a long-range cation order is developed), acid delithiation proceeds only in the LiO$_2$-layers and is accompanied by removal of the statistically distributed impurity nickel ions in the same layer. The acid delithiation of Li$_x$Ni$_{2-x}$O$_2$ (x>0.9) with a long-range cation ordered structure occurs as in the case of LiCoO$_2$, i.e. with partial exchange between the lithium ions from the depleted LiO$_2$- layers and protons from the acid solution.

INTRODUCTION

Acid delithiation of alkali-transition metal oxides offers a new route for materials design. The effect of acid treatment may be generalized as alkali ion extraction or proton exchange. Usually, the parent crystal structure is preserved during alkali ion extraction [1], while alkali ion exchange with protons produces structural varieties of the initial crystal structure [2]. In the present paper, we have studied the acid delithiation of layered oxides LiMO$_2$ (M=Co, Ni), which are of potential interest as cathode materials in lithium batteries.

EXPERIMENTAL

The acid digestion of LiMO$_2$ was achieved by treating 2 g of the samples at room temperature with 100 ml of 0.1, 1 and 8N HCl or H$_2$SO$_4$. The duration of treatment was experimentally determined with a view to attaining at least 50-60% dissolution of the samples. The solid residues thus obtained were rinsed with water and dried at room temperature.

RESULTS AND DISCUSSION

LiCoO$_2$ has a hexagonal α-NaFeO$_2$-type structure in which the lithium and cobalt ions are located in alternating layers of octahedral sites. The X-ray diffraction patterns of acid treated LiCoO$_2$ are similar (in terms of line positions) to that of initial LiCoO$_2$ (s. g. R$\overline{3}$m) which indicates clearly that the structural framework of parent $LiCoO_2$ is retained during acid digestion (Fig.1). The extra peak in the XRD pattern of $LiCoO_2$ treated with 1N H_2SO_4 was also observed in the XRD patterns of completely delithiated to CoO_2 samples in a non-aqueous medium [3]. The chemical compositions of the acid treated samples are summarized in Table 1. These results reveal that two competitive reactions, depending on the acid concentration, proceed within the initial layered framework: lithium extraction and exchange of lithium ions with protons, as a result of which new metastable layered phases $Li_{1-x-y}H_yCoO_2$ ($x\leq0.55$, $x+y<1$) are obtained. Both reactions are concomitant with Co^{4+}ions clustering as demonstrated by EPR and magnetic susceptibility measurements [4] and provoke a strong expansion of the interlayer spacing in the parent hexagonal type structure. The simultaneous distribution of Li^+ and H^+ in the CoO_2 matrix is also proved by the thermochemical properties of $Li_{1-x-y}H_yCoO_2$ (Fig.2): on heating, the enhanced lithium and/or proton mobility in the lithium depleted layers induce some structural variation in the crystal lattice of $Li_{1-x-y}H_yCoO_2$ (exo-effect at 180°C) before its dehydration to a lithium-rich cobalt spinel oxide (endo-effect at 280°C).

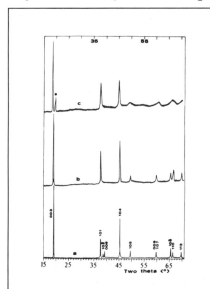

Fig.1 XRD patterns of $LiCoO_2$ (a) and acid treated $LiCoO_2$ with 0.1N H_2SO_4 (b) and 1 N H_2SO_4 (c).

Table 1
Lithium content (wt.%), mean oxidation state (OS) of cobalt ions, d_{003}-spacing and XRD intensity ratio I_{003}/I_{104} of the acid treated $LiCoO_2$ samples till a dissolution degree of 50%. The chemical compositions are calculated according to the formula: $Li^+_{1-x-y}H^+_yCo^{4+}_xCo^{3+}_{1-x}O_2$.

H_2SO_4	Li, wt %	OS	Li^+	H^+	Co^{3+}	Co^{4+}	d_{003}	$\dfrac{I_{003}}{I_{104}}$
			1-x-y	y	1-x	x	Å	
0.1 N	5.14	3.15	0.71	0.14	0.85	0.15	4.659	1.32
1 N	2.65	3.42	0.36	0.22	0.58	0.42	4.665	1.12
8 N	0.58	3.16	0.08	0.76	0.84	0.16	4.758	1.05

Stoichiometric $LiNiO_2$ is isostructural with $LiCoO_2$ but, in contrast to $LiCoO_2$, some cation disorder exists in the layers, i.e. "not-perfect" LiO_2- and NiO_2-layers are formed. Moreover, stoichiometric $LiNiO_2$ is difficult to prepare and, usually, the layered lithium-nickel oxides contain Ni^{2+} ions: $Li_xNi_{2-x}O_2$ ($0.6<x<1$). The impurity Ni^{2+} ions frustrate the layered crystal structure and, in the limiting case, when $x_c=0.62$, short range cation order only is preserved in the cubic structural framework. According to EPR studies of ours [5,6], short-range cation order in $Li_xNi_{2-x}O_2$, when $x=0.62$, consists in partial Li^+, Ni^{3+} and Ni^{2+} segregation in three consecutive cubic planes. Using XRD-measurements and EPR of low-spin Ni^{3+}, we have established that short-range and long-range cation order in $Li_xNi_{2-x}O_2$ specify their behaviours towards acids. The X-ray diffraction patterns of the acid treated oxides are similar to those of the initial samples, thus indicating preservation of the structural framework during acid treatment (Fig.3). For $Li_xNi_{2-x}O_2$ ($0.6<x<1$), lithium extraction proceeds only in the LiO_2-layers and is accompanied by removal of the statistically distributed

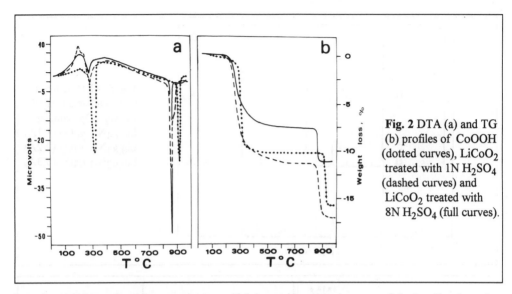

Fig. 2 DTA (a) and TG (b) profiles of CoOOH (dotted curves), LiCoO$_2$ treated with 1N H$_2$SO$_4$ (dashed curves) and LiCoO$_2$ treated with 8N H$_2$SO$_4$ (full curves).

impurity nickel ions from the same layers, which is manifested by increase of the I_{003}/I_{104} intensity ratio (Table 2). Only for layered Li$_{0.92}$Ni$_{1.08}$O$_2$, DTA and TG measurements [7] show some proton insertion in the layered structure of this sample during acid treatment.

Table 2
Final chemical composition Li$_{x-z-y}$H$_y$Ni$_{2-xa}$O$_2$ (as calculated from the chemical analysis and the weight loss curves) and relative XRD intensity ratio I_{003}/I_{104} for acid-treated Li$_{0.64}$Ni$_{1.36}$O$_2$ (S 1), Li$_{0.83}$Ni$_{1.17}$O$_2$ (S 2) and Li$_{0.92}$Ni$_{1.08}$O$_2$ (S 3).

HCl	S 1			$\dfrac{I_{003}}{I_{104}}$	S 2			$\dfrac{I_{003}}{I_{104}}$	S 3			$\dfrac{I_{003}}{I_{104}}$
	z	y	2-xa		z	y	2-xa		z	y	2-xa	
Fresh	0	0	1.36	0.17	0	0	1.17	0.73	0	0	1.08	1.02
0.01 N					0.10	0	1.17	0.87				
1.0 N					0.14	0	1.16	0.96	0.10	0.10	1.0	1.34
8.0 N	0.02	0	1.36	0.22	0.17	0	1.11	1.00				

EPR studies were fulfilled in order to elucidate the behaviour of the NiO$_2$ layers during the acid treatment (Fig.4). For the samples with a partial ordered structure ($0.6<x\leq0.9$), the EPR spectra originate from Ni^{3+} ions (S=1/2) which interact antiferromagnetically with Ni^{2+} from a neighbouring layer [5] but for the samples with x>0.92 the EPR signal is attributed to the ferromagnetically interacting Ni^{3+} ions from the NiO$_2$ layer [6]. After the acid treatment, the EPR spectra show sharply outlined magnetic transition, this suggesting a composition homogeneity. The acid treatment of the partially ordered does not change their ferrimagnetic transition temperatures which explicitly shows that the Ni^{2+}-Ni^{3+} clusters are not attacked by the acids. The slight broadening of the EPR signal of these samples reflects lattice strains induced by lithium extraction. In contrast to the above samples, acid delithiation of layered Li$_{0.92}$Ni$_{1.08}$O$_2$ causes a significant broadening of the EPR signal of the Ni^{3+} ions in the NiO$_2$-layers, but the temperature dependence of the EPR line width remains the same. These results can be interpreted by stabilization of low-spin Ni^{4+} ions in the NiO$_2$ layers.

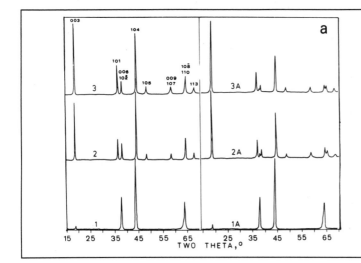

Fig. 3. XRD patterns of initial and acid treated $Li_xNi_{2-x}O_2$ samples: $Li_{0.64}Ni_{1.36}O_2$ (1,1A) $Li_{0.83}Ni_{1.17}O_2$ (2,2A), $Li_{0.92}Ni_{1.08}O_2$ (3,3A).

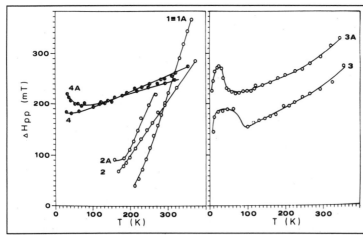

Fig. 4 Temperature dependence of the EPR peak-to-peak line width (ΔH_{pp}) of low-spin Ni^{3+}. Sample notation: 1, 1A; 2, 2A; 3, 3A- as in Fig. 3; 4, 4A- initial and acid treated $Li_{0.92}(Co_{0.2}Ni_{0.8})O_2$.

In conclusion, $LiCoO_2$ and $Li_xNi_{2-x}O_2$ behaviours in acids depend on both the potential ratios of O_2/H_2O and M^{3+}/M^{4+} redox pairs and the transport properties of the crystal structures with a long-range or short-range cation order.

ACKNOWLEDGEMENT. Financial support from the National Research Foundation of Bulgaria is gratefully acknowledged.

REFERENCES
1) Hunter, J.C.: J. Solid State Chem., (1981), <u>39</u>, 142.
2) Delmas, C., Borthomieu, Y., Faure, C., Delahaye, A., and Figlarz, M: Solid State Ionics, (1989), <u>32-33</u>, 104 .
3) Wizansky, A.R., Rauch, P.E., and Disalvo, F.J.: J. Solid State Chem., 1989, <u>81</u>, 203.
4) Zhecheva, E., and Stoyanova, R.: J Solid State Chem., in press.
5) Stoyanova, R., Zhecheva, E., and Angelov, S.: Solid State Ionics, (1993), <u>59</u>, 17.
6) Stoyanova, R., Zhecheva, E., and Friebel, C.: J. Phys. Chem. Solids, (1993), <u>54</u>, 9.
7) Stoyanova, R., and Zhecheva, E.: J. Solid State Chem., in press.

Materials Science Forum Vols. 152 - 153 (1994) pp. 263-266
© 1994 Trans Tech Publications, Switzerland

TITANIUM-CARBON COATINGS PREPARED BY CHEMICAL METHOD AT MILD CONDITIONS

M. Wysiecki [1], A. Biedunkiewicz [1], W. Jasinski [1], S. Lenart [1] and A.W. Morawski [2]

[1] Institute of Material Engineering, Technical University of Szczecin, Al. Piastów 19, PL-70-310 Szczecin, Poland

[2] Institute of Inorganic Chemical Engineering, Technical University of Szczecin, ul. Pulaskiego 10, PL-70-322 Szczecin, Poland

Keywords: Titanium-Carbon Coatings, Chemical Deposition

ABSTRACT

A method of producing carbon-titanium coatings at temperatures below 500°C has been presented. The first step was to prepare an intermediate compound of titanium chloride encapsulated in carbon network on activated metallic base. Then reduction with hydrogen was carried out. The resultant layers were examined by SEM, TEM, XRD, EDX and XRFS methods.

INTRODUCTION

Low-energy technology [1–4] has entered the production of new materials including the carbon-titanium layers [5–7]. Investigations of low-temperature methods based mainly on CVD (Chemical Vapour Deposition) process [8–12] are increasing.

CVD methods consists in depositing titanium carbide layers with the use of gaseous mixture: titanium tetrachloride, hydrocarbon and hydrogen in the temp. range 900 – 1200 °C. The high temperature accompanying this process makes it impossible to use the method for coating alloy steel undergone neat treatment, also glass, diamond and boron nitride as high temperature affects the structure of the substrate.

Titanium tetrachloride reduction with hydrogen and thermal decomposition of hydrocarbon have an effect on temperature in the process of CVD. Both reactions have high activation energy.

In the course of thermodynamic analysis of the CVD method the optimum conditions for gaining titanium carbide were determined. They are: dilution with hydrogen , lowering pressure, lowering stoichiometric Cl : Ti rate . These conditions were used in the CVD process modifications [8-12] such as: moderate temperature CVD, low pressure CVD, photon CVD, plasma assisted CVD and hybrid methods.

The authors present a way of producing carbon-titanium deposits in temperature below 500°C. What is new in the proposed methods is that contrary to CVD technology the reactions are carried out in liquid state. The main purpose of the proposed method is to lower the activation energy when forming the layer by a synthesis of transition compound. The free enthalpy then (-ΔG) should be lower than the corresponding change during direct synthesis of the final product.

EXPERIMENTAL

The tests were carried out in a glass reactor with a reflux condenser. Hydrogen was bubbled through the titanium tetrachloride solution in hydrocarbon in which samples of oxidized Armco iron were immersed. Chemically pure reagents were used. Liquid chain and ring hydrocarbons were introduced. The reactions were carried out in the following conditions: temperature from 20°C to boiling temperature of the solution, atmospheric pressure, flow rate H_2 5–15 l/h, reaction time 0,5–7 h.

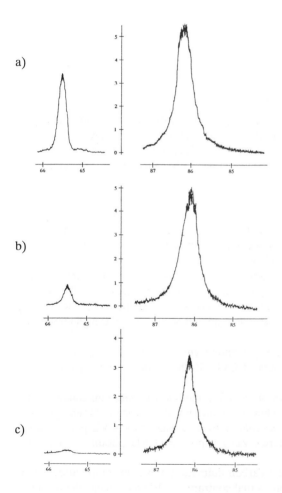

a)

b)

c)

Figure 1. Comparison of intensity of fluorescence lines Kα1 Cl and Ti

The test involved two hydrocarbons: n-hexane and benzene. Next, the resulting products underwent hydrogen reduction in a flow reactor in the temperature range 300 – 500 $^{\circ}$C under atmospheric pressure. The total reduction time was 20 h.

Phase composition was analysed using X-ray diffractometer and TEM while chemical composition — X-ray fluorescence spectroscopy; surface morphology was analysed with a scanning microscope and the X-ray microanalysis — with the EDX method.

RESULTS

Figure 1 illustrates changes in intensity of chlorine and titanium Kα1fluorescence lines on each stage of material preparation. The results of chemical composition analysis point out at:

- presence of chlorine and titanium after the first stage of synthesis (figure 1a),
- distinct fall of chlorine contents while that of titanium constantly grew after reduction in temp. 300°C and 400°C (figure 1 b),
- decrease of chlorine contents to trace amounts with slight change in that of titanium after reduction in temp. 500°C (figure 1 c).

X-ray and electron diffraction indicated that the $Fe_{2-x}Fe_{2x}Ti_{1-x}O_4$ phase and new, unknown phase (fig. 2) were present in the resulting products. Additionally, no free carbon, titanium or titanium carbide of structure corresponding to TiC were observed.

The surface layer analysis (fig. 3) carried out on a scanner equipped with an X-ray microanalyser proved the presence of titanium and carbon. Surface distribution of the elements is illustrated in photos (fig. 4). The results of the tests indicate a correlation in the distribution of titanium and carbon on the surface of the metallic base. Additionally, density of points on the "surface map" in the same places is observed, which suggests a constant relation Ti : C.

Figure 2. Scanning electron micrographs of a sample surface after 30-min process.

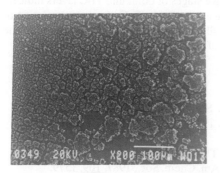

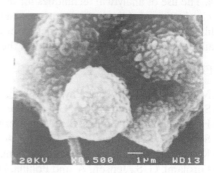

Figure 3. The surface of obtained layer shown on the SEM.

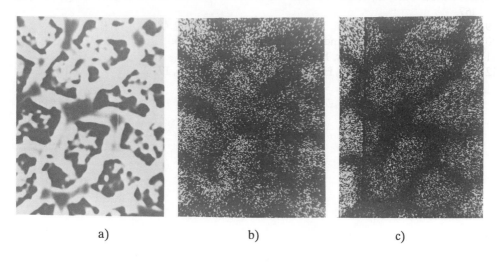

a)   b)   c)

Figure 4. Chemical composition of the surface: a) photography of analysed picture (AEI), b) surface carbon distribution, c) surface titanium distribution.

The obtained layers Ti-C require further separate studies to determine structure and character of atom connections within the layer.

CONCLUSIONS

1. The results presented above prove that an entirely chemical method of coating the metallic base with carbon-titanium layers is possible. The chemical reactions during the process occur in the conditions milder than those during CVD.

2. The method consists of two main stages:
 – heating of the metallic base in the solution containing liquid hydrocarbon and titanium tetrachloride in hydrogen for forming transition compound.
 – reduction of the transition compound in hydrogen in temperature up to $500^{o}C$.

3. It is suggested that transition layer of $Fe_{2-x}Fe_{2x}Ti_{1-x}O_4$ connecting metallic base with titanium-carbon layer is formed.

4. The use of analytical techniques for the successive stages of obtaining Ti-C layers indicate the elimination of chlorine from surfaces and creating a connection Ti-C of constant relation titanium and carbon. The catalytic action of bed-metallic material can also be concluded. The detailed structure of Ti-C layer should be the subject of separate studies.

REFERENCES

1) Matsumura, H.: J. Appl. Phys., 1989, 66, 3612.
2) Siefering, K. L. and Griffin, G. L.: J. Electrochem. Soc., 1990, 137, 1206.
3) Honghua, D. and Gallois, B.: J. Am. Ceram. Soc., 1990, 73, 764.
4) Setton, R.: Synthetic Metals, 1988, 23, 467.
5) Kaloyeros, A., Hoffman, M. and Williams, W. S.: Thin Solid Films, 1986, 141, 237.
6) Girolami, G. S., Jensen, A. and Pollina, D. M.: J. Am. Chem. Soc., 1987, 109, 1579.
7) Wang, M., Schmidt, K., Reicheld, K., Dimigen, H. and Hübsch, H.: Surface and Coatings Technology, 1991, 47, 691.
8) Drouin-Ladouce, B., Piton, J. P. and Vandenbulcke, L.: Journal of Phys., Collog., 1989, 50, 367.
9) Michalski, J. and Wierzchoń, T.: J. Mater. Sci. Lett., 1990, 9, 480.
10) Bunshah, R. F. and Deshpandey, C.: Vacuum, 1989, 39, 955.
11) Wahl, G.: Vacuum-Technik, 1989, 38, 195.
12) Motojima, S., and Mizatani, H.: Thin Solid Films, 1990, 186, 17.

Materials Science Forum Vols. 152 - 153 (1994) pp. 267-270

SYNTHESIS AND CHARACTERIZATION OF SILICA GELS OBTAINED IN LAMELLAR MEDIA

T. Dabadie [1,2], A. Ayral [1], C. Guizard [1], L. Cot [1], C. Lurin [2], W. Nie [2] and D. Rioult [2]

[1] Laboratoire de Physicochimie des Matériaux, CNRS URA 1312, ENSCM, 8 rue de l'Ecole Normale, F-34053 Montpellier Cédex 1, France

[2] Kodak European Research, Centre de Recherches et Technologies, F-71104 Chalon-sur-Saône, France

Keywords: Lyotropic Liquid Crystal, Silica Gel, Rheology, Low Angle X Ray Diffraction

ABSTRACT

This study deals with the synthesis of inorganic materials with an anisotropic texture by the sol-gel process. The formation of polymeric silica gels by the hydrolysis and condensation of an alkoxide precursor, the tetramethylorthosilicate (TMOS), is achieved in the presence of a lyotropic lamellar phase made with a non-ionic surfactant, an octylphenyl polyether alcohol.

The gelation of the lamellar sol is displayed by rheological measurements. Low angle X ray diffraction allows to follow the structural evolution of the lamellar phases in the sol.

The study of the sol-gel transition and of the gels allows to propose a schematic model of gelation in these lamellar phases.The formation of an anisotropic material requires the orientation of the lamellar microdomains in the sol. The "Rocking Curve" X ray diffraction method illustrates the possibility to evaluate the orientation in lamellar films .

I INTRODUCTION

In the sol-gel process, the inorganic gels produced from alkoxide precursors or colloidal solutions are generally amorphous and present an isotropic porous texture. The use of amphiphilic molecules such as non-ionic surfactants can induce a structuration in the gelation medium. In the case of the formation of a lyotropic liquid crystal, hexagonal or lamellar phases can create an anisotropic medium acting as a template [1].

The final aim of this study is the synthesis of oxide gels with an anisotropic texture using a lyotropic lamellar phase as gelation medium. Films from these materials present potential applications for optical devices or separative membranes.

Here, we consider the formation of polymeric silica gels by the hydrolysis and the condensation of tetramethylorthosilicate. The lyotropic lamellar phase is obtained by mixing a non-ionic surfactant (an octylphenyl polyether alcohol), water and decanol.

The evolution of the lamellar sol during the sol to gel transition is studied by rheological and low angle X ray diffraction measurements. Moreover, ^{29}Si NMR and nitrogen adsorption-desorption on lamellar gels contribute to purpose a gelation model of the lamellar gel. The textural orientation of lamellar films is then estimated by X ray diffraction "Rocking Curve" method.

II EXPERIMENTAL

Materials

The selected surfactant is an octylphenyl polyether alcohol $C_8\Phi E_{10}$ (Triton X100 from Rohm and Haas). We have chosen the $C_8\Phi E_{10}$/Decanol/Water system due to its large liquid crystal area in the phase diagram [2]. Tetramethylorthosilicate (TMOS) is used as silica precursor. The formation of a lamellar gel depends on the composition of the starting sol [3]. The initial composition of the lamellar gel L considered in this paper is : 41.8 wt % of $C_8\Phi E_{10}$, 7.7 wt % of decanol, 39.5 wt % of water and 11 wt % of alkoxide. It corresponds to a molar ratio of water to alkoxide h equal to 30 and a potential volume fraction of SiO_2 equal to 2.6 %.

We have also synthesized an isotropic sol in methanol, I, having the same alkoxide concentration and the same hydrolysis ratio than L.

For the textural study, the gels were thermally treated during two hours at 450°C, the treated gel consisting in the silica network.

Thick films of lamellar material were achieved on glass substrates using a tape-casting device (doctor blade method) at different thicknesses and shear rates. The lamellar sols were spread few minutes after their preparation.

- Characterization

The visco-elastic measurements were done with a Couette type viscosimeter in the oscillation mode and a parallel plate cell. The frequency and the amplitude were 0.1 Hz and 1° respectively.

^{29}Si NMR spectra were recorded using a spectrometer working at 59.62 MHz in the Cross-Polarization and Magic Angle Spinning configurations (CP-MAS). Chemical shifts were considered in comparison with TMS. The peaks are referred to the different condensed species Q_i which could exist in a siloxane network (i is the number of bridging oxygen atoms of the SiO_4 group) [4].

The porous texture has been studied by adsorption-desorption of nitrogen. The mesoporous volume V_{meso} was determined by the BJH method and the microporous volume V_{micro} by difference between the total pore volume V_T and the mesoporous volume.

Low angle X ray difffraction allows to study the structure of the lamellar phases. Diffraction patterns of lamellar phases are effectively characterized by a serie of diffraction lines with the following Bragg spacings : d, d/2, d/3, d/4 where d is the interlamellar spacing [5]. The "Rocking Curve" method has been used to determine in a semi quantitative way the orientation of polycrystalline systems by measuring the Full Width at Half Maximum (FWHM) for a given diffraction peak [6].

III RESULTS AND DISCUSSION

1. Study of the sol-gel transition

1.1. Rheological evolution of the lamellar sols

The high viscosity of the lamellar medium hinders the macroscopic observation of the gelation process from the flow behavior of the sol. The evolution versus time of the storage modulus G', of the loss modulus G" and tan δ = G"/G' are reported in Figure 1 for the lamellar sol L. The storage modulus progressively increases with time and after about 125 min, the tan δ

curve presents a maximum which characterizes the sol-gel transition [7,8] and allows to determine a gelation time. We can note that for I, the gelation time macroscopically determined is clearly longer (about 7 days).

1.2. Structural evolution of the lamellar sols

The sol to gel transition of L has been followed with low angle X ray diffraction (Figure 2). We can note on the X ray diffraction pattern a splitting for the first and second order peaks as well as a strong decrease in intensity for the second order peak. This splitting phenomenon has to be related to the existence of two lamellar structures with slightly different periodicities.

There is a decrease of the value of d versus time with Δd equal from -1.4 Å to -3.5 Å. A gelation mechanism is proposed in order to explain these significant variations of the parameter d. One can assume that the alkoxide is located outside lamellar microdomains and, during sol to gel transition, a part of the water contained in the liquid crystal phase is sucked up in order to promote the hydrolysis and condensation reactions. The extraction of water from the lamellar phases for a stoichiometric hydrolysis of the silicon alkoxide (h = 4) corresponds to a calculated value Δd = -3.2 Å which has the same order of magnitude than the experimental variations.

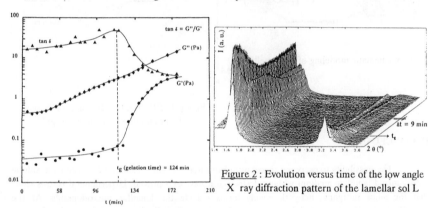

Figure 2 : Evolution versus time of the low angle X ray diffraction pattern of the lamellar sol L

Figure 1 : Rheological evolution versus time of the lamellar sol L

2. Characterization of the lamellar gels

2.1. Texture of the lamellar gels

Let us first consider the ^{29}Si NMR results from spectra recorded after gelation of L and I. The ratios Q_n/Q_{n+1} decrease when we pass from the isotropic system to the lamellar one. This behavior seems to be inconsistent with that we can imagine considering a rather bidimensionnal polymerization in the lamellar structures with poisoning of superficial alkoxide groups. These results are better related to a confinement of the silica network in a interphase around the lamellar microdomains.

Concerning the textural properties of the thermally trated gels, L exhibits a total pore volume which is higher than that of I ($0.632 \text{ cm}^3/\text{g}$ compared to $0.334 \text{ cm}^3/\text{g}$). Moreover, its microporous volume is lower ($0.05 \text{ cm}^3/\text{g}$ compared to $0.15 \text{ cm}^3/\text{g}$). These results can be interpreted on the assumption of a gelation outside the lamellar microdomains (Figure 3.a). The gel formed in this confined medium must effectively be more condensed and less microporous than the isotropic one. The departure of the surfactant must also induce a more important mesoporosity in L.

2.2. Orientation of the lamellar coatings

The orientation of the microdomains is necessary in order to obtain a material with an anisotropic texture (Figure 3.b). Experimentally, the orientation effect of the lamellar coatings has been evaluated from the Full Width at Half Maximum (FWHM) of the first order peak of the lamellar phase. The FWHM value decreases and so the orientation increases as a function of the shear rate applied during the coating process (Figure 4). These results can be correlated with the thixotropic behavior of the pure lamellar phases [3].

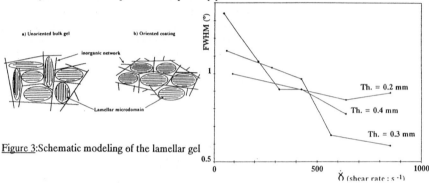

Figure 3:Schematic modeling of the lamellar gel

Figure 4 : Effect of the shearing conditions
on the orientation of the lamellar gel L

IV CONCLUSION

The possibility to carry out reactions of hydrolysis and condensation of a silicon alkoxide in an organized lamellar medium without destroying it has been proved. Experimental results show that the silica inorganic network mainly forms outside the lamellar microdomains. At the microdomain scale, the material is anisotropic but the random distribution of these microdomains involves that the created inorganic network has an isotropic texture. The possibility to orientate the microdomains under shear stress, for instance during coating on a substrate therefore allows to obtain anisotropic films.

REFERENCES

[1] J.S. Beck, J.C. Vartuli, W.J. Roth, M.E. Leonowicz, C.T. Kresge, K.D. Schmitt, C.T-W. Chu, D.H. Olson, E.W. Sheppard, , S.B. McCullen, J.B. Higgins, J.L. Schlenker :
J. Am. Chem. Soc., 1992, 114, 10834.
[2] A. Jurgens : Tenside Surf. Det., 1989, 26 n°3, 222.
[3] T. Dabadie, A. Ayral, C. Guizard, L. Cot, C. Lurin, W. Nie, D. Rioult : to be published .
[4] I. Artaki, M. Bradley, T.W. Zerda, J. Jonas : J. Phys. Chem., 1989, 89, 4399.
[5] V. Luzzati, H. Mustacchi, A. Skoulios, F. Husson : Acta Cryst., 1960, 13, 660.
[6] B. K. Tanner in "X Ray Diffraction Topography"
(Pergamon International Library, New-York, 1976).
[7] C.J. Brinker, G.W. Scherer in "Sol-Gel Science, the Physics and Chemistry of Sol-Gel Processing" (Academic Press, New-York, 1990).
[8] C. Guizard, J.C. Achddou, A. Larbot, L. Cot : J. of Non-Cryst. Solids, 1992, 147 & 148, 681.

Materials Science Forum Vols. 152 - 153 (1994) pp. 271-276
© *1994 Trans Tech Publications, Switzerland*

IS SOFT CHEMISTRY ALWAYS SO SOFT ?

N. Allali[1], J.F. Favard[1], M. Rambaud[2], A. Goloub[1,3] and M. Danot[1]

[1] Laboratoire de Chimie des Solides, IMN, Unité Mixte CNRS, Université de Nantes, UMR 110, 2 rue de la Houssinière, F-44072 Nantes Cédex 03, France

[2] IUT de Lannion, Université de Rennes I, rue Edouard Branly, BP 150, F-22302 Lannion Cédex, France

[3] Permanent address: Institute of Organo-Element Compounds, 117813, Vavilov st. 28, Moscow, Russia

Keywords: Potassium Iron Disulfide, Ionic Exchange, Deintercalation

ABSTRACT : The chain structure of $KFeS_2$ allows exchange reactions to be performed. If potassium is replaced by calcium, the $[FeS_2]$ structural framework is retained so that this reaction can be considered as a Soft Chemistry process. It is not the case of the silver exchange which induces a change to a chalcopyrite structure. Potassium can be extracted from $KFeS_2$ but the FeS_2 obtained is pyrite. This drastic structural change precludes this reaction to be considered as a Soft Chemistry process.

According to the definition given by A.R. West [1], Soft Chemistry, also known as "Chimie Douce", consists in low-temperature processes which allow new compounds to be obtained and retain the structural framework of the precursor. Such a phenomenon can be illustrated by alkali-metal intercalation into titanium disulfide [2]: A lot of similar examples could be found in intercalation, deintercalation, and ionic-exchange reactions. However, the structural framework of the precursor can sometimes be drastically changed through low-temperature processes such as deintercalation and cationic exchange, as it will here be shown on the example of potassium iron sulfide $KFeS_2$.

I THE HISTORICAL BACKGROUND AND THE PRESENT STUDY

1°) $KFeS_2$

Potassium iron sulfide was first prepared in the late sixties by R. Schneider [3]. Its structure was established in 1942 by J.W. Boon and C.H. Mac Gillavry [4]. Iron is tetrahedrally surrounded by sulfur, and every tetrahedron shares two opposite edges with its two neighbors so that $[FeS_2]_\infty$ chains are formed which run along the c-axis (fig. 1), according to a SiS_2 structural model [5]. Potassium ions are located in large spaces between these chains. $KFeS_2$ can so be considered as a potassium-intercalated SiS_2-like structure and then, according to this picture, it appears to offer a good opportunity for deintercalation and ionic-exchange properties to be observed.

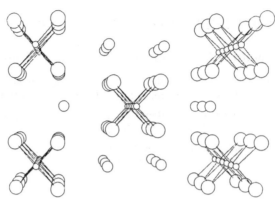

FIG. 1 : *Perspective view of the structure of $KFeS_2$ along the c-axis. Small, medium-sized, and large circles respectively represent iron, potassium, and sulfur.*

2°) Exchange reactions using $KFeS_2$ as a precursor

R. Schneider performed some exchange reactions on $KFeS_2$. For instance, with silver, he obtained a compound with the formula $AgFeS_2$ [6]. From X-ray diffraction data, J.W. Boon [7] deduced that this compound likely belongs to the chalcopyrite [8] structural type, which means that an important structural change has occured, especially concerning the iron positions [4, 7]. In such a case this exchange reaction would not obey the usual definition of Soft Chemistry.

H. Boller [9] exchanged potassium with alkaline earth cations. For example, with calcium, he obtained a phase with the formula $(Ca_{0.50},xH_2O)FeS_2$. From the conservation of the c periodicity, he deduced that the $[FeS_2]$ framework is retained. In that case the exchange reaction could be considered as a Soft Chemistry process.

3°) The present work

Due to the one-dimensional character of $KFeS_2$, its X-Ray diffraction-diagrams are far from being excellent. Moreover, the crystallites are severely damaged during the exchange reactions. The structural conclusions of J.W. Boon [7] and H. Boller [9] were thus drawn from poor-quality spectra. Mössbauer spectroscopy appears as a convenient technique for additional structural information to be obtained concerning exchanged compounds prepared from $KFeS_2$. Here is the reason for which the present Mössbauer study was undertaken, on $AgFeS_2$ and $(Ca_{0.50},xH_2O)FeS_2$.

To our knowledge, nothing has been reported about the deintercalation of potassium from $KFeS_2$. We attempted to do it with the hope of a new FeS_2 to be obtained, with the SiS_2 structure i.e. without the $KFeS_2$ chain arrangement to be broken.

II EXPERIMENTAL

$KFeS_2$ was prepared according to ref.[10]. The exchange reactions were carried out as previously reported [6, 7, 9].

Attempts to deintercalate potassium were made using classical oxidizing treatments, with iodine or ferric chloride solutions in acetonitrile.

For reasons which will be indicated below, we tried to perform exchange reactions with the ammonium ion. For that purpose, we first used ammonium chloride solutions, and then we heated $KFeS_2 + NH_4Cl$ mixtures in evacuated pyrex tubes, at temperatures ranging from 100°C to 350°C.

^{57}Fe Mössbauer measurements were performed at room and liquid nitrogen temperatures. The isomer shift values will refer to metallic iron at 300 K.

The Mössbauer parameters of our $KFeS_2$ starting compound are in good agreement with what previously reported (isomer shift $\delta = 0.19$ mm/s, quadrupole splitting $\Delta = 0.49$ mm/s [11]). Effectively our refined values are as follows :

$\delta = 0.18 \pm 0.02$ mm/s, and $\Delta = 0.50 \pm 0.02$ mm/s.

III THE SILVER AND CALCIUM EXCHANGES

1°) The silver exchange

As expected, the obtained silver-exchanged compound allows only poor-quality X-ray diffraction diagrams to be obtained. Only some weak and broad lines can be observed, which are consistent with the cell-parameters (a = 5.66 Å and c = 10.30 Å) reported by J.W. Boon [7] for a chalcopyrite-like structure.

The Mössbauer spectrum (fig. 2) exhibits two components : a quadrupolar doublet, and a magnetic sextuplet for which the refined hyperfine parameters are :

$\delta = 0.21 \pm 0.02$ mm/s, $\Delta = 0.00 \pm 0.02$ mm/s, H = 37.2 $\pm$ 0.5 T.

As for the central doublet, it can be due to the smallest particles resulting from the splitting of the crystallites during reaction. Some oxidized small particles [12] could also contribute to this part of the spectrum.

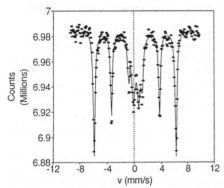

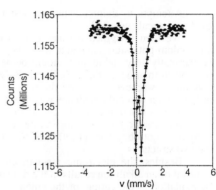

FIG. 2 : *Mössbauer spectrum (RT) of AgFeS₂.*

FIG. 3 : *Mössbauer spectrum (RT) of (Ca₀.₅₀.xH₂O)FeS₂.*

The experimental and calculated spectra are respectively represented by the crosses and the line.

2°) The calcium exchange

As in the case of the silver compound, the diffraction diagram is of poor quality. The (few) observed lines are in agreement with the tetragonal unit-cell (a = 11.31 Å, c = 5.46 Å) reported by H. Boller [9].

The Mössbauer spectrum (fig. 3) is a paramagnetic doublet with $\delta = 0.16 \pm 0.02$ mm/s and $\Delta = 0.43 \pm 0.02$ mm/s. No magnetic component can be detected.

3°) Discussion

- In $KFeS_2$ the Fe-Fe magnetic interactions are essentially one-dimensional [13, 14], due to the chain structure. On the contrary, strong three-dimensional magnetic interactions exist in the 3-D chalcopyrite structure. The Mössbáuer characteristics of $KFeS_2$ and $CuFeS_2$ clearly illustrate this difference : for $KFeS_2$ (non-magnetic RT spectrum [15]), the ordering-temperature is lower than for $CuFeS_2$ (magnetic RT spectrum [16, 17]), and the saturation hyperfine field weaker (23.7 T [18] for $KFeS_2$, ~ 37 T for $CuFeS_2$ [16, 17]).

- For the calcium-exchanged compound, the RT Mössbauer spectrum (Fig. 3) is non-magnetic (as for $KFeS_2$) in agreement with a 1-D iron arrangement : the calcium exchange-reaction retains the structural framework of the initial compound and thus obeys the definition of Soft Chemistry.

- For the silver-exchanged compound, the hyperfine field is 37.0 T at room temperature and 38.6 T at 78 K. The closeness of these two values indicates that the ordering-temperature is high. Besides, these fields are close by that of $CuFeS_2$. For that two reasons it clearly appears that $AgFeS_2$ effectively belongs to the chalcopyrite structural type.

- The Ag/K exchange induces an important change of the pristine structural framework since the $KFeS_2$ --> chalcopyrite transformation requires the migration of every second iron atom [7], which means that the chain arrangement of iron is not retained. Despite the fact that this exchange is performed at room-temperature it cannot be considered as a Soft Chemistry process.

- The so-obtained $AgFeS_2$ decomposes at only 150°C which precludes the magnetic-ordering temperature to be measured.

IV POTASSIUM DEINTERCALATION

Attempts to deintercalate potasssium from $KFeS_2$ were unsuccessful, whatever the oxidizing agents we used.

We then imagined another route for our aim to be reached. The first step was an ammonium-potassium exchange for $(NH_4)FeS_2$ to be obtained. Such a compound should not be thermally very stable, and should be decomposed (second step) by a gentle heating with possibly formation of the SiS_2-like FeS_2 we wished to prepare. Whatever the experimental conditions, we never observed the formation of the $(NH_4)FeS_2$ exchanged compound. The only reaction we could evidence was direct formation of FeS_2, but with the pyrite structure.

Such a reaction could be considered as a deintercalation because potasium is removed. However, despite the low-temperature reaction (150°C), the transition from $KFeS_2$ to pyrite FeS_2 does not obey the definition of Soft Chemistry since no simple relation exists between the two structures.

Besides, the electronic mechanism of this "deintercalation" is rather unusual. The removal of an alkali ion results in the oxidation of the host-structure, which is usually realized by oxidation of the cation, or the anion, or both. In the present case, the oxidation of iron to the +IV state does not occur, which is not surprizing in a sulfide, and as could be expected, the anion is oxidized to the -I state, as sulfur pairs $(S_2)^{-II}$. This oxidation could have concerned every second sulfur atom, so that the +III state of iron in $KFeS_2$ could have been retained in the obtained FeS_2. However, a better stability is reached with complete oxidation of sulfur to the -I state and concomitant reduction of iron to the ferrous state. Per formula unit, one sulfur is oxidized due to potassium removal and the other due to iron reduction. What is unusual in this mechanism is that the global oxidation of the matrix involves the reduction of the cation, which results from an internal electron transfer.

V CONCLUSION

- For the calcium exchange, contrary to the potassium removal, the structural framework can be retained because the sulfur-sulfur inter-chain repulsions remain weak due to the presence of large $[Ca^{2+}, xH_2O]$ species between the chains.

- The silver exchange is not an "*ionic* exchange" since the potassium ion is replaced by silver which establishes essentially covalent bonds with its tetrahedral sulfur surrounding. In that case the iron sublattice cannot be preserved, because of the structural strains due to the size difference between the two metal atoms. The iron migration occurs in order to reduce these structural strains.

REFERENCES

1) West, A.R.: in Solid State Chemistry and its Applications, John Wiley and Sons Ed., 1986, p.30
2) Bichon, J., Danot, M., et Rouxel, J.: C. R. Acad. Sc., 1973, 276, 1283
3) Schneider, R.: Ann. Physik, 1869, 136, 460
4) Boon, J.W. and Mac Gillavry, C.H.: Rec. Trav. Chim., 1942, 61, 910
5) Büssem, W., Fischer, H., and Grüner, E.: Naturwissenschaften, 1935, 23, 740
6) Schneider, R.: J. Prakt. Chem., 1888, 38 (2), 569
7) Boon, J.W.: Rec. Trav. Chim.,1944, 63, 69
8) Pauling, L. and Brockway, L.O.: Z. Krist., 1932, 82, 188
9) Boller, H.: Monatsch. Chem., 1978, 109, 975
10) Brauer, G.: in Handbook of Preparative Inorganic Chemistry, N.Y., 1965, p. 1507
11) Raj, D. and Puri, S.P.: J. Chem. Phys., 1969, 50 (8), 3184
12) Kündig, W., Bömmel, H., Constabaris, G., and Lindquist, R.H.: Phys. Rev., 1966, 142 (2), 327
13) Nishi, M. and Ito, Y.: Solid State Commun., 1979, 30, 571
14) Mauger, A., Escorne, M., Taft, C.A., Furtado, N.C., Arguello, Z.P., Arsenio, T.P.: Phys. Rev. B, 1984, 30 (9), 5300
15) Kerler, W. Neuwirth, W., Fluck, E., Kuhn, P., and Zimmermann, B.: Z. Physik, 1963, 173, 321
16) Ok, H.N. and Kim, C.S.: Il Nuovo Cimento, 1975, 28, (1), 138
17) Goodenough, J.B. and Fatseas, G.A.: J. Solid State Chem., 1982, 41, 1
18) Zink, J. and Nagorny, K.: J. Phys. Chem. Solids, 1988, 49 (12), 1429

Materials Science Forum Vols. 152 - 153 (1994) pp. 277-280
© 1994 Trans Tech Publications, Switzerland

SYNTHESIS, STRUCTURE AND REACTIONS OF OXIDES WITH THE HEXAGONAL MoO_3 STRUCTURE

Y. Hu and P.K. Davies

Department of Materials Science and Engineering, University of Pennsylvania,
3231 Walnut St., Philadelphia, PA 19104-6272, USA

Keywords: Brannerite, Hexagonal MoO_3, Solvolysis, Cation-Exchange, Ion-Insertion

ABSTRACT

In dilute HCl, the layered brannerites $A(VMo)O_6$ (A=Li, Na) show no evidence for ion-exchange, but instead participate in partial solvolysis reactions, leading to the formation of new compounds with the stoichiometry $A_{0.13}(V_{0.13}Mo_{0.87})O_3 \cdot nH_2O$ (A = Li_xH_{1-x}, Na; n≈0.26) and the structure of "hexagonal MoO_3". Complete dehydration of the hydronium form, $H_{0.13}(V_{0.13}Mo_{0.87})O_3 \cdot nH_2O$, yields $(V_{0.13}Mo_{0.87})O_{2.935}$ which retains the structure of hexagonal MoO_3 up to 455°C. Both H and Li are readily inserted into this "open form" of hexagonal MoO_3; $H_x(V_{0.13}Mo_{0.87})O_{2.935}$ was prepared for x<1.13 and $Li_x(V_{0.13}Mo_{0.87})O_{2.935}$ with 0≤x≤1.65. Although the solvolysis reactions of Li brannerite proceed via a simple dissolution-precipitation mechanism, the partial solvolysis of Na brannerite appears to be facilitated by a structural relationship between the brannerite "solute" and the hexagonal "precipitate".

INTRODUCTION: The proton exchange and low temperature dehydration reactions of layered inorganic compounds have led to the preparation of a variety of new metastable oxide structures by chimie douce. In an attempt to apply these methods to the formation of new V-Mo oxides, we have examined the reactions of layered oxides $A(VMo)O_6$ [A = Li, Na] with the "brannerite" structure (see Figure 1a)[1,2,3]. In dilute HCl, the Li and Na brannerites show no evidence for proton exchange, but instead participate in partial solvolysis reactions that lead to the formation of new compounds with the stoichiometry $A_{0.13}(V_{0.13}Mo_{0.87})O_3 \cdot nH_2O$(A=$Li_xH_{1-x}$, Na; n≈0.26) and the structure of "hexagonal-MoO_3"(see Figure 1b). Previously, this structure had only been stabilized by incorporating large cations into the channels[4,5], and attempts to stabilize an open form in which the channels are empty were unsuccessful[6,7]. We have found that the open structure can be prepared by complete dehydration of the hydronium form of h-MoO_3, $H_{0.13}(V_{0.13}Mo_{0.87})O_3 \cdot nH_2O$. In this paper we describe the formation, structure and ion-insertion properties of this new family of hexagonal MoO_3 compounds.

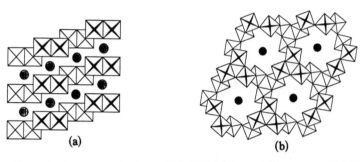

Figure 1. Structure projections of (a) $A(VMo)O_6$; and (b) hexagonal MoO_3

EXPERIMENTAL: The brannerites $A(VMo)O_6$ (A=Li, Na) were prepared by solid-state-reaction of AVO_3 and α-MoO_3 as reported by Galy et al [8]. Li brannerite was stirred in 0.25M HCl at 60°C for 2 days. The solids were collected and re-stirred in 0.5M HCl at room temperature for several hours and then dried in air. The Na brannerite was stirred in 0.25M HCl at 60°C for 16 hours, the process was repeated to ensure the complete conversion of brannerite to the h-MoO_3.

To explore the mechanism of the partial solvolysis reaction of brannerites, we examined the direct precipitation of the Li and Na forms of h-MoO_3 from aqueous A-V-Mo (A=Li, Na) solutions. Mixtures of α-MoO_3 and V_2O_5 were dissolved in 1.0M AOH; the resulting solutions were acidified by the addition of 1.0M HCl until the desired pH was obtained; the acidified solutions were then refluxed at 60-90°C. The precipitates were recovered by vacuum filtration and dried in air.

RESULTS AND DISCUSSION: Unlike many layered transition-metal oxides which undergo ion-exchange reactions in dilute acid, the Li and Na brannerites participate in a partial solvolysis reaction where the more soluble V ions are preferentially leached into the acidic solution, leaving a final product with $A_{0.13}(V_{0.13}Mo_{0.87})O_3 \cdot nH_2O$ (A=H, Na; n≈0.26). Although Li and Na brannerite both react to form an h-MoO_3 product, their reaction paths are quite different. For the Li isomorph, complete dissolution of Li brannerite was observed during the early stages of the reaction followed by

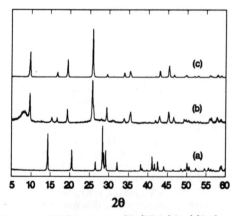

Figure 2. XRD patterns of $Li(VMo)O_6$ (a)before; (b)during; and (c)after acid treatment.

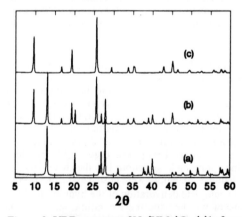

Figure 3. XRD patterns of $Na(VMo)O_6$ (a)before; (b)during; and (c)after acid treatment.

the formation of a two-phase mixture of h-MoO$_3$ with the stoichiometry (Li$_x$H$_{.13-x}$)(V$_{0.13}$Mo$_{0.87}$)O$_3$·nH$_2$O and an amorphous V-rich phase (see Figure 2). When this two-phase mixture was subsequently immersed in 0.5M HCl, the amorphous phase was dissolved and the Li ions in the hexagonal phase were replaced by protons giving a final product with H$_{0.13}$(V$_{0.13}$Mo$_{0.87}$)O$_3$·nH$_2$O. During the partial solvolysis of Na brannerite solid phases are present throughout the entire reaction which proceeds through a two-phase mixture of the starting brannerite and the final product Na$_{0.13}$(V$_{0.13}$Mo$_{0.87}$)O$_3$·nH$_2$O (see Figure 3).

H$_{0.13}$(V$_{0.13}$Mo$_{0.87}$)O$_3$·nH$_2$O readily undergoes ion-exchange reactions with monovalent and some divalent cations. The monovalent exchanges were achieved by stirring in 1.0M ACl at 60°C; complete exchange of H$^+$ and formation of A$_{0.13}$(V$_{0.13}$Mo$_{0.87}$)O$_3$ ·xH$_2$O was observed for A = Li, Na, K, NH$_4$. The exchange between H and divalent cations was more complicated. Upon stirring in 1.0M ACl$_2$ at 60°C, no reaction was observed for small cations such as Mg^{2+} and Cu^{2+}. Exchange of H$^+$ by large cations such as Sr^{2+} and Ba^{2+} was also possible, however these reactions were accompanied by the formation of unidentified by-products. Among the divalent cations studied, only Ca^{2+}, which has an ionic radius comparable to that of Na$^+$, could fully replace H$^+$ to give Ca$_{0.065}$(V$_{0.13}$Mo$_{0.87}$)O$_3$·xH$_2$O. Figure 4 shows TGA scans of A$^{m+}_{0.13/m}$(V$_{0.13}$Mo$_{0.87}$)O$_3$·xH$_2$O obtained from the cation-exchange reactions. As expected the degree of hydration and the strength of the water-framework bond, are highly dependent upon the chemistry of the cations in the channels of the h-MoO$_3$ structure.

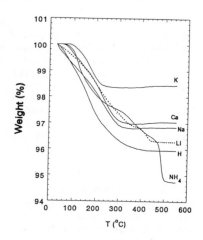

Figure 3.　　　TGA scans of A$^{m+}_{0.13/m}$(V$_{0.13}$Mo$_{0.87}$)O$_3$·xH$_2$O (heating rate: 5°C/min)

The complete dehydration of H$_{0.13}$(V$_{0.13}$Mo$_{0.87}$)$_3$·xH$_2$O results in the formation of (V$_{0.13}$Mo$_{0.87}$)O$_{2.935}$, which retains the h-MoO$_3$ structure up to 455°C. This compound represents the open form of h-MoO$_3$ and contains empty one dimensional channels with a free diameter approaching 3.3Å. The open nature of this structure was confirmed by powder neutron diffraction. Compared to the other polymorphs of MoO$_3$, the 'porosity' (volume per formula unit of MoO$_3$) of (V$_{0.13}$Mo$_{0.87}$)O$_{2.935}$ is significantly larger than the thermodynamically stable α form, and the ReO$_3$-related metastable β and β' forms (see Table 1).

Table 1. Comparison of the 'porosity' of different polymorphs of MoO$_3$

Structure	α-MoO$_3$	β-MoO$_3$	β'-MoO$_3$	(V$_{0.13}$Mo$_{0.87}$)O$_{2.935}$
V$_{per MoO_3}$ (Å^3)	50.8	53.2	53.4	59.8

The ion insertion properties of (V$_{0.13}$Mo$_{0.87}$)O$_{2.935}$ were examined utilizing a variety of chemical methods. As expected the open structure of this phase is suited to the insertion of a range of ions. Using hydrogen spillover methods, H$_x$(V$_{0.13}$Mo$_{0.87}$)O$_{2.935}$ was prepared for x<1.13 though considerable structural degradation was observed for x>0.9. Lithium insertion compounds Li$_x$(V$_{.13}$Mo$_{.87}$)O$_{2.935}$ (0≤x≤1.65) were synthesized using n-butyl lithium, again reduction beyond x=1.0 caused some irreversible changes in the structure.

The structures of $A_{0.13}(V_{0.13}Mo_{0.87})O_3 \cdot nH_2O$ (A=H, Li, and Na) were investigated by powder neutron diffraction. From the results of our Rietveld refinements and the x-ray determinations of Darriet and Galy, a clear correlation was found between the size and the location of the A cations inside the channels. Larger cations such as K, Rb, Cs and NH_4 are located in the center of the channel and are coordinated by 6 terminal oxygen atoms in the (V,Mo) framework; Na and Ca occupy sites away from the channel centers and are octahedrally coordinated by four terminal framework anions and two water molecules in the channels. The smaller Li ions occupy positions closer to the framework and are tetrahedrally coordinated by three terminal oxygens and one water molecule whereas H is attached directly to the terminal oxygen.

To explore whether the leaching reactions of brannerite involve simple dissolution-precipitation, or if they are unique and facilitated in some way by a structural relationship between the brannerite "solute" and hexagonal "precipitate", we explored methods for the direct precipitation of the Li and Na forms of hexagonal MoO_3 from aqueous A-V-Mo (A=Li, Na) solutions. For the Li-V-Mo solutions, it was found that $(Li_xH_{0.13-x})(V_{0.13}Mo_{0.87})O_3 \cdot nH_2O$ could be prepared within a narrow window of pH, solution composition and temperature. The reaction paths of aqueous solutions with the same pH and composition as those formed during the early stage of the $Li(VMo)O_6$ leaching reaction, and the chemistry and properties of the h-MoO_3 precipitates were identical to those observed during the brannerite reaction. This would suggest that the crystallography of the solute is not important and supports a simple dissolution-precipitation reaction mechanism.

However, differences were observed for the sodium-containing isomorphs. Precipitates with the h-MoO_3-structure and a stoichiometry $Na_{0.17}\{V_xMo_{1-x-y}(\square_M)_y\}H_{6y+x-0.17} \cdot nH_2O$ ($0 \leq x \leq 0.14$, $0.01 \leq y \leq 0.09$, $n \approx 0.33$, $\square_M$ = transition-metal vacancy) could be prepared from acidified Na-V-Mo aqueous solutions. However, the concentration of Na in these h-MoO_3 precipitates ($Na_{.17}$) was consistently higher than in those formed from brannerite ($Na_{.13}$) and the thermal stability of the solution derived Na-h-MoO_3 samples were significantly lower (maximum stability limit, $350°C$) than those obtained via the brannerite reaction (stability limit, $400°C$). Furthermore, aqueous solutions prepared with the same pH and composition as those formed during the $Na(VMo)O_6$ leaching reaction did not yield h-MoO_3. These observations would suggest that the crystallography of the brannerite solute may in some way affect the chemistry of the first formed complex ions in solution and stabilize the formation of $Na_{0.13}(V_{0.13}Mo_{0.87})O_3 \cdot nH_2O$.

CONCLUSIONS: In dilute acid, layered brannerites $A(VMo)O_6$ (A=Li, Na) undergo a partial solvolysis reaction, leading to new materials with the stoichiometry $A_{0.13}(V_{0.13}Mo_{0.87})O_3 \cdot nH_2O$ (A = Li_xH_{1-x}, Na; $n \approx 0.26$) and the structure of hexagonal MoO_3. While the reaction of $Li(VMo)O_6$ appears to be a simple dissolution-precipitation reaction, the reaction of $Na(VMo)O_6$ may be facilitated by a unique structural relationship between the brannerite solute and the hexagonal precipitate. Complete dehydration of the hydronium isomorph yields $(V_{0.13}Mo_{0.87})O_{2.935}$ which retains the hexagonal MoO_3 structure and has empty channels. The open structure of this compound permits the formation of a series of reduced forms $A_x(V_{0.13}Mo_{0.87})O_{2.935}$ through the insertion of H (for $x < 1.13$) and Li (for $x < 1.65$).

REFERENCES
1. Feist, T.P. and Davies, P.K. Chemistry of Materials, 1991, 3, 1011
2. Hu, Y. and Davies, P.K.: Solid State Ionics, 1992, 53-65, 325
3. Hu, Y. and Davies, P.K.: J. Solid State Chem., in press
4. Darriet, B. and Galy, J.: J. Solid State Chem., 1973, 8, 189
5. Olenkova, I.P., Plyasova, L.M. and Kirik, S.D.: React. Kinet. Catal. Lett., 1981, 16, 81
6. McCarron, E.M., Thomas, D.M. and Calabrese, J.C.: Inorg. Chem., 1987, 26, 371
7. Craiger, N.A., Crouch-Baker, S., Dickens, P.G. and James, G.S.: J. Solid State Chem., 1987, 67, 369
8. Galy, J., Darriet J. and Darriet B.: C. R. Acad. Sc. Paris, Serie C, 1967, 264, 1477

Materials Science Forum Vols. 152 - 153 (1994) pp. 281-288
© *1994 Trans Tech Publications, Switzerland*

EXFOLIATION OF GRAPHITE INTERCALATION COMPOUNDS: CLASSIFICATION AND DISCUSSION OF THE PROCESSES FROM NEW EXPERIMENTAL DATA RELATIVE TO GRAPHITE-ACID COMPOUNDS

A. Hérold, D. Petitjean, G. Furdin and M. Klatt

Laboratoire de Chimie du Solide Minéral, Université de Nancy I, CNRS,
B.P. 239, F-54506 Vandoeuvre-Les-Nancy, France

Keywords: Graphite Intercalation Compounds, Graphite-Acid Compounds, Expanded Graphite, Irreversible Exfoliation, Endothermic Exfoliation, Exothermic Exfoliation, Adiabatic Exfoliation

ABSTRACT: Expanded graphite (E.G.) produced by exfoliation of graphite intercalation compounds (G.I.C.) is a material of scientific and industrial interest. The usual process through rapid heating of a G.I.C. which induced vaporisation and dissociation of the intercalated species, can be called "endothermic exfoliation in a gaseous medium" and several new processes classified by taking into account the medium (gaseous or liquid) and the thermal effect of the transformation will be described and discussed.

INTRODUCTION

By rapid heating of a graphite intercalation compound (G.I.C.) prepared with natural graphite flakes as pristine material, a voluminous vermicular product is obtained which is called expanded graphite (E.G.) (figure 1.a). It can be characterized by its apparent density, or its expansion coefficient (ratio of the apparent volumes after and before exfoliation), and also by optical and scanning microscopy. The most precise characterization is obtained by adsorption of an inert gas at low temperature. Its surface area is often of the order of 20 to 80 m^2/g corresponding to groups of 80 to 20 graphene sheets on average. The surface is quite regular and allowed Duval and Thomy [1,2] to study the two dimensionnal physical chemistry of the adsorbed gases. Expanded graphite became an industrial pristine material for making graphite foils (grafoil of Union Carbide Corporation, papyex of Le Carbone Lorraine...) which are largely used in several industries. The first patent relative to the exfoliation phenomenon was taken in 1891 [3], and the compression of E.G. into foils in 1915 [4]. More recently, a new kind of powdered graphite, made of monocrystalline particles of diameter 1 to

60 µm and of thickness 0.05 to 0.1 µm was obtained by grinding E.G. under suitable conditions. This material is called flat micronic graphite, in French "graphite micronique plat" (G.M.P.) [5,6,7].

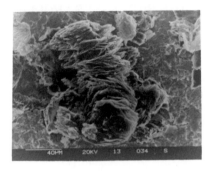

Figure 1.a.

Figure 1.b.

Figure 1 a, b: Scanning electron microscope photographs of exfoliated graphite:
    (a) obtained with the industrial process,
    (b) obtained with the adiabatic process (see § 3).

In this article, the different exfoliation processes will be described. They are characterized by taking into account the nature of the medium (gaseous or liquid) and the thermal effect (endothermal, adiabatic, exothermal exfoliation).

EXPERIMENTAL.

1. Pristine materials.

Almost all experiments have been carried out using Madagascar graphite flakes of 0.5 to 1 mm in diameter.
The Brönsted acids HNO_3, H_2SO_4, $HClO_4$ and the Lewis acid SO_3, all of which are industrial products, have been used as intercalates.

2. Endothermal exfoliation in a gaseous medium.

The industrial process of rapid heating of a G.I.C., generally exfoliates a graphite sulfate under a pressure close to one atmosphere. The vaporisation and dissociation of the intercalate ($H_2SO_4 \longrightarrow H_2O + SO_3$ and $SO_3 \rightleftharpoons SO_2 + 1/2 O_2$) absorbs thermal energy: this exfoliation is endothermal. The vapour and gases produced in the galleries escape at the edges of the particles, and simultaneously exert a pressure, p, perpendicular to the graphene sheets, which can lead to irreversible exfoliation. This latter occurs when p>P+Π, in which Π is the lowest pressure required for separating two adjacent graphene sheets and P is the external pressure, generally atmospheric.

Taking into account theses remarks, it is clear that an efficient exfoliation leading to large specific surface areas, requires rapid heating.

- *The thermal shock* can be obtained by introducing the G.I.C into a preheated furnace, or by heating it in the flame of a blow-pipe, which is often used in industrial plants. According to a recent work the largest exfoliation is obtained by heating with a laser [8].

A second important parameter is the particle diameter, since the intercalate more easily escapes from smaller particles: thus, H_2SO_4-G.I.C or HNO_3-G.I.C. in particles of a diameter lower than $40\ \mu m$ [9] cannot be exfoliated by the endothermal process. Of course, in the polycrystalline carbonaceous materials, the diameter is not that of the particles themselves, but that of the elemental crystallites.

- *The nature of the intercalate* also plays a role: endothermic exfoliation of a graphite sulfate or nitrate [10], is less efficient than that of graphite-ferric chloride [11]: in the first case only vapour and gases evolve (H_2O, NO, O_2), whereas in the second case $FeCl_3$ decomposes into volatile Cl_2 and very less volatile $FeCl_2$, which remains in the galleries and hinders the effluence of Cl_2 from the center to the edges of the particles.

- *The role of the stage* of the pristine G.I.C. was studied in our laboratory [10,12]. Figure 2 shows that the stage of SO_3-G.I.C. and HNO_3-G.I.C. weakly affect the specific surface area of the corresponding E.G..

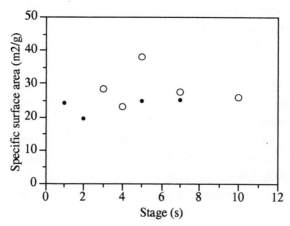

Figure 2: Evolution of the specific area of E.G. versus stage of the HNO_3-G.I.C.(o), and SO_3-G.I.C.(•).

These results can be explained by taking into account the pleated layer model [13], which supposes that a G.I.C. of stage s contains intercalate in each gallery, but only over a fraction 1/s of its surface area. However we do not know the reason why the compounds of odd stages lead to higher surface areas than the even ones (figure 2). Specific surface areas of about 40 m^2/g have also been measured after exfoliation of H2SO4 - GIC's of stage 8. These industrial products are obtained by washing with water and drying of stage 2 compounds synthesized in a large excess of concentrated sulfuric acid. The direct synthesis of high stage compounds - for instance by intercalation of nitric acid in the vapour phase - might lead to a less expansive and cleaner process.

3. Adiabatic exfoliation in a gaseous medium.

In this process the G.I.C. is heated to a temperature T_1 in a small volume so that its dissociation pressure becomes high, whereas the main part of the intercalate remains in the graphite matrix. When T1 is reached the volume is abruptly open, the vapour and gases contained in the volume expand and the G.I.C. particle exfoliates. The thermal energy $\int_{T_0}^{T_1} C\ (T)\ dT$ accumulated by the graphite matrix (C is its heat capacity) during heating from the ambient temperature T_0 to T_1 is used in the exfoliation process: thus, this can be considered as an adiabatic process.

The enthalpy of exfoliation is the sum of the enthalpy of dissociation ΔH_d and the energy ΔH_s required for separating the graphene sheets. ΔH_d is given by the Van t'Hoff relationships and ΔH_s can be neglected because only a small number of adjacent graphene sheets are separated. In our experiments about 300 mg of a graphite nitrate, contained in a tube of pyrex glass are heated in a metallic apparatus of a large volume ($700\ cm^3$) in which a pressure lower than one Torr is etablished. When the temperature T1 is reached, the tube is broken by the fall of an iron cylinder previously retained by a magnet.

For $T_1 = 158°C$, only a very small exfoliation is observed.

For $T_1 = 250°C$, the thermal energy contained in the graphite matrix is enough for complete dissociation of the G.I.C..

Nevertheless, as figure 1.b shows, exfoliation remains incomplete: the expanded product can be exfoliated again by heating with a blow-pipe.

For $T_1 = 384°C$, the graphite particles are dislocated, but not exfoliated. They do not contain intercalate. It is clear, that at this temperature, the graphite nitrate is completely decomposed into graphite and liquid HNO_3, following the general behaviour of the G.I.C. [14].

We do not have any sure explanation of the small efficiency of this process, which certainly deserves being optimisized.

4. Exothermic exfoliation in a gaseous medium.

It is based on the exothermic dissociation of perchloric acid. The pure acid and its blue first stage compound synthesized by spontaneous intercalation easily explode. Thus the reaction was moderated by using 70% commercial perchloric acid. This mixture which contains 29.6 moles of $HClO_4$ for 70.4 moles of H_2O, does not spontaneously intercalate at room temperature. Intercalation was carried out by the addition of 98% HNO_3 [10] or by heating graphite in the 70% $HClO_4$ at a suitable temperature [15,16,17].

The data of table 1 allow concluding that all kinds of carbonaceous materials which are able to intercalate dilute perchloric acid can be exfoliated by this process. However, the isotherms of adsorption of krypton show that the surfaces of the exfoliated materials by the perchloric acid are less regular. Exfoliation by endothermic dissociation of $HClO_4$ might to obtain new kinds of expanded graphites.

Samples	Natural graphites		Pitch cokes	
	Madagascar	Brazillian	graphitized 2600°C	heated at 1000°C
Particle size (μm)	500<ϕ<1000	ϕ<50	ϕ<1	ϕ<1
Specific area of the pristine carbons (m^2/g)	0.3	2.8	1.0	0.5
Specific area of the exfoliated compound (m^2/g)	153	74	26	25.3

Table 1: Exothermal exfoliation process: characteristics and specific surface areas of the pristine and exfoliated carbonaceous materials.

5. Exfoliation in a liquid medium.

Endothermic exfoliation in a liquid medium is theoritically possible, but, as far as we know has never been tested.

Exothermic exfoliation is produced by a chemical reaction which produces a gas in the galleries: hydrogen in the reaction of water with an alkali metal G.I.C. [18], oxygen is the reaction of hydrogen peroxide with a chromiun trioxide G.I.C. [19] and nitrogen by the reaction of hydrazine with a molybdenum chloride G.I.C. [20]. These reactions lead to expanded products containing a large proportion of impurities.

We have studied exfoliation in concentrated sulfuric acid. Its intercalation needs an oxidising agent. By using an excess of ammonium (or potassium) persulfate or an excess of hydrogen peroxide, a part of the $H_2S_2O_8$ acid or of the hydrogen peroxide intercalates with H_2SO_4. The decomposition of these reagents in the galleries with oxygen production leads to a progressive exfoliation of the pristine graphite. After exfoliation the expanded graphite was separated from the liquid by filtration, washed and dried by heating near 110°C, and its apparent density was determined.

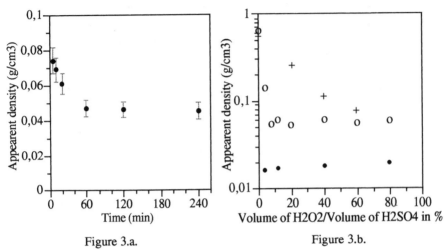

Figure 3.a.   Figure 3.b.

Figure 3.a. Apparent density versus time of Madagascar graphite exfoliated by the mixture of two volumes of 50% H_2O_2 and five volumes of 98% H_2SO_4 (massic percentages).

Figure 3.b. Apparent density of two kinds of graphite exfoliated by a mixture of 50% H_2O_2 and 98% H_2SO_4 (+) Brazil graphite, (o) Madagascar graphite, (•) Madagascar reexfoliated in flame of a torch.

The data of figure 3.a show that exfoliation is achieved in about one hour. Figure 3.b allows comparing two kinds of graphite. Madagascar graphite (0.5<Ø<1mm) exfoliates easier than brazillian one (Ø<50 µm).

Curiously, Madagascar graphite expanded in the liquid phase can be reexfoliated by heating, although after compressing, it gives the [00l] X-ray diagram of pure graphite. The rapid dissociation of a small amount of oxygen and hydroxide groups chemisorbed during graphite oxidation could explain this phenomenon.

Exfoliation in the liquid phase, which operates at room temperature without any evolution of toxic or corrosive gas is a true soft chemistry process, which could be of technical interest.

CONCLUSION

This article shows that much scientific and technical progress is possible in the field of G.I.C. exfoliation. A complete understanding of the exfoliation mechanisms and an efficient industrial use of the scientific data are still long term aims.

Acknowledgements: This work was partially supported by "Carbone Lorraine S.A." (Groupe Péchiney). The authors wish to thank V.Krebs (Laboratoire de Chimie du Solide Minéral-Nancy-) for her participation to the measurements of the apparent density.

REFERENCES.

1) X.Duval and A.Thomy, *C.R.Acad.Sci.*, 1964, 259, 4007.

2) X.Duval and A.Thomy, *Adsorption et Croissance Cristalline* Colloques Internationnaux du C.N.R.S., 1965, 152, 81.

3) P.Willy Luzi, *Patentschrift,* Condensite Compagny of America, 1891, 66804.

4) Jonas W.Aylsworth, *United States Patent Office*, Union Carbide Corporation, 1915, 1,137,373.

5) G.Furdin, J.F.Marêché, A.Hérold, *Int.Pat.* PCT/EP 92/02317.

6) G.Furdin, J.F.Marêché, A.Hérold, *International Symposium on Intercalation Compounds*, 1993, (I.S.I.C.7., Louvain La Neuve, Belgique), to be published in *"molecular crystals and liquids"*, Gordon and Breach.

7) C.Hérold, J.F.Marêché, A.Mabchour, G.Furdin, *This Symposium*, Poster C9, book of abstracts, 1993, 247.

8) A.Thomy, J.C.Ousset, G.Furdin, J.M.Pelletier and A.B.Vannes, *Journal de Physique,* Colloque C7, 1987, 12, 115.

9) J.Maire, Private communication.

10) M.Klatt, *Thesis*, Nancy, 1985.

11) A.Thomy, *Thesis* , Nancy, 1968.

12) M.Klatt, G.Furdin, A.Hérold, N.Dupont-Pavlovsky, *Carbon,*1986, 24, 6, 731.

13) N.Daumas and A.Hérold, *C.R.Acad. Sci.* Paris, C, 1969, 268, 373.

14) A.Hérold, *Synthetic Metals,* 1989, 34, 21.

15) D.Petitjean, *Thesis,* Nancy, 1992.

16) D.Petitjean, M.Klatt, G.Furdin and A.Hérold, *Accepted for publication in "Carbon".*

17) D.Petitjean, G.Furdin, A.Hérold and N.Dupont Pavlovsky, *International Symposium on Intercalation Compounds*, 1993 (I.S.I.C.7.,Louvain La Neuve, Belgique), to be published in *"molecular crystals and liquids"*, Gordon and Breach.

18) R.Schögl and H.P.Boehm, *Carbon,* 1984, 22, 4-5, 351.

19) J.M.Skrowronski, *Journal of Materials Science,*1988, 23, 2243.

20) A.Messaoudi, M.Inagaki and F.Beguin, *Materials Science Forum,* 1992, 91-93, 811.

Materials Science Forum Vols. 152 - 153 (1994) pp. 289-292

NOVEL PROPERTIES OF (LiMo$_3$Se$_3$)$_n$, A POLYMERIC CHEVREL CLUSTER COMPOUND: THIN FILMS OF MACROMOLECULAR WIRES CAST FROM SOLUTION

J.H. Golden, F.J. DiSalvo and J.M.J. Fréchet

Cornell University, Dept. of Chemistry, Baker Labs, Ithaca, N.Y. 14853, USA

Keywords: Inorganic Polymer, Chevrel Phase, (LiMo$_3$Se$_3$)$_n$, Thin Films

ABSTRACT We have found that DMSO solutions of the linear chain compound (LiMo3Se3)$_n$, can be cast into lustrous metallic thin films after heating in vacuo. The morphology of these films can vary from transparent ultra thin films to a conducting free standing foil like material. Herein we describe the preparation and analysis of this new material and the unique physical properties it displays.

INTRODUCTION

Interest in organic and inorganic low dimensional solids has increased steadily over the past decade, as part of a search for new materials displaying novel physical properties including: anisotropic conductivity, superconductivity, non-linear optical phenomena, and piezoelectric behavior [1]. In 1980, Potel et al. [2] described a series of metallic linear chain compounds, (MMo3X3)$_n$(M = a monovalent main group metal, X = Se, S). The structure of MMo3X3 is based on the condensation of octahedral clusters of molybdenum face capped by chalcogen atoms, to form linear (Mo3X3$^-$)$_n$ chains. The structure can also be viewed as antiprismatically stacked triangles of Mo, with edge bridging chalcogen atoms, as shown in figure 1.

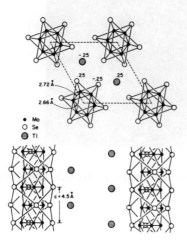

Figure 1. Projection of (TlMo3Se3)$_n$ structure onto (001) and (11$\bar{2}$0)planes. Note that Mo-Mo bondlengths are close to Mo metal lengths of 2.723 Å.

These linear chain compounds, are one member of the family of materials generally known as the Chevrel phases [3]. Depending on the identity of the interstitial cation M sitting between the chains, the $(MMo_3X_3)_n$ class of compounds displays a variety of physical properties [4-7]. One of the most interesting physical properties arises when lithium is the M interstitial cation. When $(LiMo_3Se_3)_n$ is treated with highly polar solvents including propylene carbonate, DMSO, or water, the compound dissolves to form highly absorbing burgundy/red solutions which are indefinitely stable if kept oxygen free [8]. The strong Mo-Mo bonding in the $(Mo_3Se_3^-)_n$ "inner core" imparts considerable stiffness to the chains, producing solutions of rigid rod polyanions and solvated lithium cations [8].

We have discovered that DMSO solutions of the inorganic polymer, $(LiMo_3Se_3)_n$, can be cast or spin coated to form lustrous thin films and peelable free-standing foils after heating in vacuo. As shown in figure 2, an UHV (10^{-12} torr) STEM image shows that these films consist of sheets of locally ordered, flat lying linear chains and chain bundles.

Figure 2. In this 6×10^6 times magnified bright field image, each black line represents an individual 6 Å diameter $(LiMo_3Se_3)_n$ "wire".

Ultra-thin films (< 1μm), prepared by spin coating onto salt discs under argon, are transparent or reflective, depending on the viewing angle. I.R. spectra of these films display a plasma edge at 1195 cm^{-1} [9], while the U.V.-VIS spectra is consistent with spectra of highly dilute solutions [8]. Thick (>1μm) metallic foils of $(LiMo_3Se_3)_n$ are easily prepared by casting 10^{-2} M $(LiMo_3Se_3)_n$ DMSO solutions into a petri dish, followed by heating in vacuo. Free standing lustrous foils of the inorganic polymer can then be peeled off the glass substrate for further analysis.

These cast films display thermally activated conductivity. As shown in figure 3, four-probe temperature vs. resistance studies yielded a resistivity value of 5×10^{-3} $\Omega \cdot$ cm at 23 °C, close to the 10^{-4} value observed for single crystals [7], where M ≠ Li. Semiconducting-like behavior is observed below 100 K, perhaps caused by a one-dimensional Peierls's distortion [7].

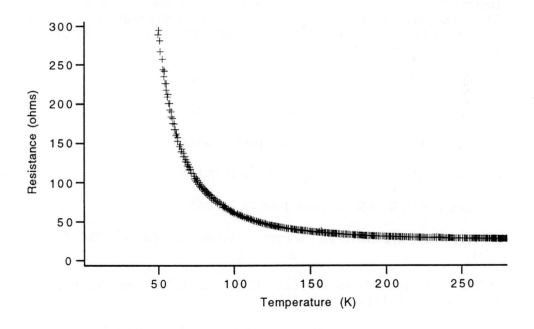

Figure 3. Four probe temperature vs. resistance plot of a 1 μm thick film of $(LiMo_3Se_3)_n$ cast from 10^{-2} M $(LiMo_3Se_3)_n$ in DMSO.

Highly oriented thin films of $(LiMo_3Se_3)_n$ have been prepared by flowing solutions over oriented Teflon substrates. After heating in vacuo, the thin films exhibit order over a length of 160 μm, as confirmed by absorbance polarized light microscopy.

Our current efforts are focused on the preparation of oriented and homogeneous inorganic-organic polymer nanoscale composites. These new materials have been shown to consist of the $(LiMo_3Se_3)_n$ inorganic polymer dispersed in an insulating organic matrix. Possible applications include use as an anti-static coating or as an absorption polarization grating.

REFERENCES

[1] For accounts on the subject of one-dimensional solids, see :
(a) Miller, Joel S. *Extended Linear Chain Compounds* ; Plenum: New York, 1982.
(b) Monceau, Pierre *Electronic Properties of Inorganic Quasi - One - Dimensional Compounds* ; Reidel: Holland, 1982. (c) Rouxel, J.*Crystal Chemistry and Properties of Materials with Quasi - One - Dimensional Structures* ; Reidel: Holland, 1986. (d) Cowan, Dwaine O., Wiygul, Frank M. *The Organic Solid State* ; *Chem. Eng. News*, **1986**,64, 28 - 45. (e) DiSalvo, F.J. *Solid State Chemistry : A Rediscovered Chemical Frontier*; *Science* , **1990**, 247, 649 - 655. (f) Rouxel, J., *Acc. Chem. Res.* **1992**, 25, 328 - 336.

[2] Potel et al. *J. Solid State Chem* . **1980**, 35, 286 - 290.

[3] Chevrel, R. et al. *J. Solid State Chem.* **1971**, 3, 515.

[4] Armici, J.C. et al. *Solid State Comm* . **1980**,33, 607-611.

[5] Gougeon, P. et al. *Ann. Chim. Fr.* **1984**, 9, 1087-1090.

[6] Mori, T. et al. *Solid State Comm* .**1984**,49, 249-252.

[7] Tarascon, J.M.; DiSalvo, F.J.; Waszczak J.V.*Solid State Comm* . **1984** , 52, 227 - 231.

[8] Tarascon, J.M.; DiSalvo, F.J.; Chen, C.H.; Carroll, P.J.; Walsh,M.; Rupp, L. *Solid State Comm.*. **1985**, 58, 227 - 231.

[9] Vassiliou, J.K.; Ziebarth, R.K.; DiSalvo,F.J. *Chem. Mater.* **1990**, 2, 738-741.

Materials Science Forum Vols. 152 - 153 (1994) pp. 293-296
© 1994 Trans Tech Publications, Switzerland

REACTIONS OF SYNTHESIS AND REDUCTION OF MIXED Cu-Ti PEROVSKITES: STRUCTURE AND ORDER-DISORDER CHARACTERISTICS

M.R. Palacín, A. Fuertes, N. Casañ-Pastor and P. Gómez-Romero

Institut de Ciència de Materials de Barcelona (CSIC), Campus U.A.B.,
E-08193 Bellaterra, Barcelona, Spain

Keywords: Copper, Titanium, Perovskites, Synthesis, Ceramic, Oxide Precursors, Sol-Gel, Neutron Diffraction

ABSTRACT

This work summarizes a synthetic study of the mixed perovskite La_2CuTiO_6. The synthesis of this oxide was carried out following three different strategies: i) classical ceramic methods, ii) use of a layered oxide precursor (La_2CuO_4), and iii) sol-gel method. We found an orthorhombically distorted perovskite of the $GdFeO_3$ type, with a distortion proportional to the temperature of synthesis. The different reaction pathways that take place for each method produce a faster reaction for method ii) as compared to i) (at the same temperature), and a substantially reduced temperature of synthesis for iii). Neutron diffraction shows no long-range order of Cu(II) and Ti(IV) in the structure.

INTRODUCTION

The control of the structure or the order-disorder characteristics in solids is a central interest of Solid State Chemistry. In recent years, the discovery of high Tc superconductivity has reinforced the interest for controlling the distribution of metals in mixed perovskites of copper and other metals. But despite the large number of mixed perovskite oxides of formula $A_2BB'O_6$ prepared over the years, the problem of the ordering of ions B and B' is only partly understood. Thus, a novel three-dimensional perovskite of formula La_2CuSnO_6 with alternating layers of CuO_6 and SnO_6 octahedra has been recently characterized,[1] whereas the similar La_2CuIrO_6 (among others) presents a "Na-Cl" type ordered structure.[2]

The oxide La_2CuTiO_6 also presents the perovskite structure and is closely related to the Cu-Sn compound. We have studied this perovskite, first from a synthetic point of view, comparing classical ceramic syntheses and alternative precursor and sol-gel methods, and secondly, to try to determine the ordering scheme of Cu and Ti in the structure by means of x-ray and neutron powder diffractometry.

EXPERIMENTAL RESULTS

We have synthesized this phase following three different approaches:
(1).- Classical ceramic synthesis starting from simple oxides. 1100-1400ºC
(2).- Ceramic synthesis starting from a layered oxide precursor (La_2CuO_4 + TiO_2). 1100-1400ºC
(3).- Sol-gel synthesis from a solution of nitrates and tetrapropylorthotitanate in propanol. 675ºC

The reactions were tried at temperatures lower than the indicated in each case but in order to proceed to completion within 48 hours (with regrindings) the temperatures shown had to be used (see references 3 and 4 for further experimental details).

Whereas the final compound resulting from each of the three methods was the same, there were substantial differences between them concerning the velocity and the temperature of the reactions. We have reported in previous works[3,4] X-ray diffractometry experiments that show how method (2) leads to a faster reaction (at the same temperature) than (1), while method (3) allows for a substantially reduced temperature of synthesis (675°C)

Figure 1 shows Diferential Scanning Calorimetry experiments for each of the three reactions. First we see a very different reaction mechanism in the sol-gel case (Fig. 1a) as compared with the other two, consistent with the amorphous nature of the precursor. Methods (2) and (3) yield more similar results except for the presence, in the classical ceramic synthesis, of peaks assigned to phase transitions of the reacting simple oxides at 400°C and 560°C and the formation of La_2CuO_4 at 1060°C (Figure 1b). The final formation of La_2CuTiO_6 takes place at 1135°C in these two cases. The formation of La_2CuO_4 as an intermediate in the reaction starting from simple oxides has been also detected by X-ray diffraction. The formation of such intermediate explains why the reaction proceeds faster when La_2CuO_4 is directly used as starting reagent.

On the other hand, the formation of the intermediate phase $La_2O_3 \cdot 2TiO_2$ during both methods (1) and (2), also observed by X-ray diffraction, is assigned to the exothermic peak at 1005°C in Figure 1b and c, and indicates the existence of a reaction pathway alternative to the formation of La_2CuO_4.

Figure 2 shows the electron diffraction patterns of La_2CuTiO_6 as obtained by method (b) at 1100°C. The indexes of the zone axes indicated correspond to the simple cubic perovskite.

Figure 1 *DSC curves for the synthe-ses of La_2CuTiO_6: (a) Sol-gel (b) from simple oxides (c) from layered La_2CuO_4*

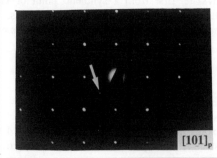

Figure 2. *Electron diffraction patterns of La_2CuTiO_6 along the zone axes indicated (referred to a simple cubic perovskite).*

We can detect in these patterns superstructure spots that force a redefinition of the unit cell from that of a cubic perovskite of dimension a_p to an orthorhombic cell with $a \approx \sqrt{2}a_p$, $c \approx \sqrt{2}a_p$ and $b \approx 2a_p$. This observations agree well with the splitting of peaks in the x-ray diffraction pattern of this oxide[4] (see also Fig. 3a). Thus, in contrast to previous works that reported a cubic phase,[5-7] our X-ray and electron diffraction experiments show that La_2CuTiO_6 presents an orthorhombic distorted perovskite structure (Pnma, a= 5.55, b= 5.61, c=7.83Å) related to the $GdFeO_3$ structure.

La_2CuTiO_6 can be reduced by treatment at 650°C under Ar/H_2(5% v/v) atmosphere for 5 hours. TGA experiments conducted under these conditions indicate a loss of one atom of oxygen per formula unit, whereas the diffraction pattern of the resulting solid shows the preservation of the basic perovskite structure (see Fig. 3). This indicates the obtention of a reduced phase with formula La_2CuTiO_5, where both copper and titanium would be reduced to Cu(I) and Ti(III) respectively. In Fig 3 we can see broader peaks and poorer crystallinity for the reduced derivative, as compared with La_2CuTiO_6. This prevents an unambiguous determination of the exact symmetry of its lattice. The dots on Fig. 3b mark two peaks corresponding to a small amount of metallic copper present as impurity, probably due to the incipient decomposition of the sample under the conditions indicated above. On the other hand, comparison of both patterns on Fig. 3 confirms unambiguously that the perovskite structure is maintained upon reduction.

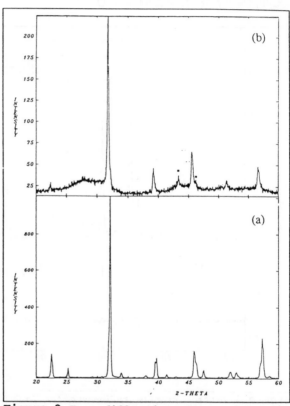

Figure 3 *X-ray diffraction patterns of La_2CuTiO_6 (a) and its reduced derivative La_2CuTiO_5 (b).*

In order to obtain a precise structural description of La_2CuTiO_6 and also to determine unambiguously any possible order of Cu and Ti in this perovskite, a neutron diffraction study was necessary. Although the X-ray diffractograms show symmetry consistent with a disordered distribution of copper and titanium within the octahedral sites of the perovskite structure (space group Pnma), we have shown that copper and titanium are not discerned properly by powder X-ray techniques,[4] and, to avoid any possibility of missing weak x-ray superstructure peaks a neutron diffraction study must be carried out . We have reported recently such a study for the compound obtained by the ceramic precursor method[4] and include here a neutron profile refinement of data collected for the product of the sol-gel reaction.

Neutron powder diffraction on a sample of La_2CuTiO_6 synthesized by the sol-gel method at 675°C was carried out at room temperature on the diffractometer D1A (wavelength 1.9845Å). Profile refinements, using the Rietveld method, were carried out with the help of the program FULLPROF.[8] Data from 10 to 150° (2θ) (step 0.05°) were used for the refinement. The background was estimated by linear interpolation between 32 selected points. Pseudo-Voigt peak shape was selected for the refinement. Scattering lengths used in the calculations were La, 0.8240; Cu, 0.7718; Ti, -0.3300 and O, 0.5805.

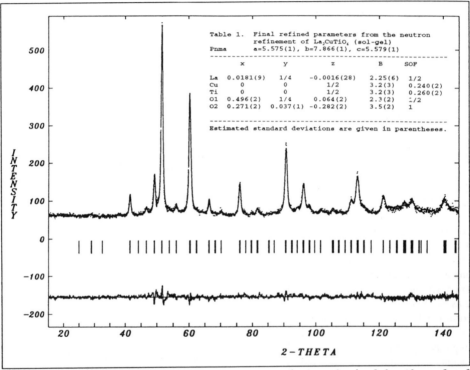

Table 1. Final refined parameters from the neutron refinement of La_2CuTiO_6 (sol-gel)

Pnma   a=5.575(1), b=7.866(1), c=5.579(1)

	x	y	z	B	SOF
La	0.0181(9)	1/4	-0.0016(28)	2.25(6)	1/2
Cu	0	0	1/2	3.2(3)	0.240(2)
Ti	0	0	1/2	3.2(3)	0.260(2)
O1	0.496(2)	1/4	0.064(2)	2.3(2)	1/2
O2	0.271(2)	0.037(1)	-0.282(2)	3.5(2)	1

Estimated standard deviations are given in parentheses.

Figure 4 *Neutron profile refinement for La_2CuTiO_6 synthesized by the sol-gel method. Experimental points (dots); calculated profile (solid line) difference Y(obs)-Y(calc) (bottom line).*

The final cycles converged with 21 refined parameters (11 atom parameters and 10 profile and cell parameters) for 124 reflections. The final reliability factors for background corrected counts, were R_p = 4.20, R_{wp}=5.24, $R_{expected}$=3.67, χ^2=2.04. Figure 4 shows the final refined profile. Table 1 summarizes the structural results obtained, where the disorder of Cu and Ti is definitively confirmed.

REFERENCES

1) Anderson, M.T. and Poeppelmeier, K.R.: *Chem.Mater.*, 1991, **3**, 476

2) Blasse, G.: *J.Inorg.Nucl.Chem.*, 1965, **27**, 993

3) Gómez-Romero, P.; Palacín, M.R.; Casañ, N.; Fuertes, A. and Martínez, B.: *Solid State Ionics* , 1993 in press

4) Palacín, M.R.; Bassas, J.; Rodríguez-Carvajal, J. and Gómez-Romero, P.: *J.Mater.Chem.*, 1993 in press

5) Ramadass, N.; Gopalakrishnan, J. and Sastri, M.V.C.: *J.Inorg.Nucl.Chem.*, 1977, **40**, 1453-4.

6) Parkash, O.M.; Kumar, D.; Gangopadhayay, D K. and Bahadur,D.: *Phys.Stat.Solidi A*, 1986, **96**, K79

7) Jones, R. and McKinnon, W.R.: *Sol.State Com.*, 1990, **76**(3), 397-400

8) Rodríguez-Carvajal, J.: PROGRAM FULLPROF (Version 2.2 -June92- ILL) (unpublished).

Materials Science Forum Vols. 152 - 153 (1994) pp. 297-304

STRUCTURAL STUDY OF A NEW IRON VANADIUM OXIDE
$Fe_{0.12}V_2O_{5.15}$ SYNTHESIZED VIA A SOL-GEL PROCESS

S. Maingot [1], Ph. Deniard [2], N. Baffier [1], J.P. Pereira-Ramos [3],
A. Kahn-Harari [1] and R. Brec [2]

[1] ENSCP, Laboratoire Chimie Appliquée Etat Solide, URA 1466, 11 rue Curie,
F-75005 Paris, France

[2] IMN, Laboratoire de Chimie des Solides, 2 rue de la Houssinière, F-44072 Nantes, France

[3] CNRS, LECSO, UM 28, 2 rue H. Dunant, F-94320 Thiais, France

Keywords: V_2O_5, Intercalation, Rietveld, Iron Vanadium Oxide, $Fe_{0.12}V_2O_{5.15}$

ABSTRACT

A new vanadium and iron $Fe_{0.12}V_2O_{5.15}$ oxide has been synthesised starting from V_2O_5 xerogel through an exchange reaction. Its structure, determined by Rietveld calculation, is built from the well known V_2O_5 ribbon bond to each other by (FeO_6) apex sharing octahedra. The electrochemical cyclability behavior of this new double oxide presents interesting performances that can be explained from the structural features.

INTRODUCTION

Starting from molecular precursors to get an oxide network, the sol-gel process provides new approaches in the preparation of materials and a better control of the syntheses. By this route, new low-dimensional V_2O_5-based compounds can be obtained. Either $M_xV_2O_5$ bronzes with a monoclinic symmetry (M = Na, Li, K, Ag)[1] or $M_xV_2O_5$ oxides with an orthorhombic one (M = Fe, Al, Ni)[2] have been synthesized from vanadium oxide gels[3]. These phases present a layered structure close to that of the orthorhombic V_2O_5 oxide, exhibiting a strong preferred orientation in relation with their layered organization. Such a behavior allows for instance the $Na_{0.33}V_2O_5$ bronze to reversibly intercalate lithium, making this phase a good cathode in non aqueous lithium secondary batteries[4]. The synthesis of $Fe_{0.12}V_2O_{5.15}$ was made by ion exchange reaction from the pentoxide vanadium xerogel phase spread on a glass plate[2] according to the reaction :

V_2O_5 gel   *drying*   **xerogel**

$V_2O_5(0.36OH^-),0.36H_3O^+, nH_2O$ ⟹ $V_2O_5(0.36OH^-),0.36H_3O^+, 1.2H_2O$

                                    20°C

ionic exchange   **iron Xerogel**   *500°C. air*

     ⟹ $V_2O_5(0.36OH^-),0.12Fe^{3+}$,   ⟹ $Fe_{0.12}V_2O_{5.15}$*

FeCl₃, 0.1M   2.2H₂O   *1 hour*

*The detailed formula is: $Fe^{3+}_{0.12}V^{5+}_{1.94}V^{4+}_{0.06}O^{2-}_{5.15}$ in agreement with charge balance achieved by the occurrence of 3% of V^{4+}, already present in the xerogel.

SAMPLE PREPARATION

The as synthesized material led to X-Ray diagrams presented a strong preferential orientation along the c axis, perpendicular to the film surface. Therefore a sample preparation was necessary to obtain data allowing structure refinement calculation.

Although an ultrasonic grinding in ethanol medium was performed., the particles presented a plate-like geometry and random orientation could not be entirely achieved. The Bragg Brentano geometry using flat plate samples was thus forsaken although the used diffractometer allowed for energy discrimination, a very useful item to minimize iron atoms fluorescence. Fortunately, after seaving the oxide powder to 20 μm, the amount of powder was sufficient to fill a 0.1 mm diameter Lindemann capillary, allowing a diffraction study with a curved position sensitive detector (INEL CPS 120) in a Debye-Scherrer geometry.

The structure refinement was carried out using the Rietveld procedure with the GSAS program[5]. This package permits the use of peaks profiles including hkl dependant line broadening due to anisotropic particle size. The observed and calculated diagrams are represented in figure 1.

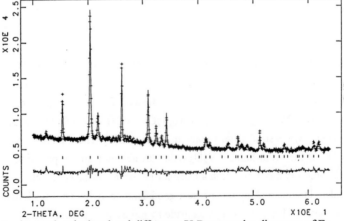

Figure 1: observed, calculated and difference X-Ray powder diagrams of Fe $_{0.12}V_2O_{5.15}$

STRUCTURE REFINEMENT RESULTS

The cell parameters of $Fe_{0.12}V_2O_{5.15}$ are not very different from the V_2O_5 ones: A slight decrease of the c parameter and increase of a parameter are observed as compared with V_2O_5 ones (a = 11.5119(8), b = 3.5646(2) and c = 4.3748(3) Å). For this reason, the calculations were started assuming that both vanadium and oxygen atoms were located in the same positions of the Pmmn space group as in V_2O_5. This led to a successful refinement and iron atoms were located in 2a sites, between four oxygen atoms of the V_2O_5 slab. The calculation used 19 variables and 1799 observations. It gave the following confidence factors: R_{wp} = 0.038, R_p = 0.030 and R_{exp} = 0.013, the light oxygen atoms having their Atomic Displacement Parameter (ADP) fixed for stability reasons. Atomic positions and cell parameters are given in table I. Fe^{3+} ions are not stable in square coordination and since the d_{Fe-O} distances matched that of the Fe^{3+} in oxygen octahedral surrounding, two additional oxygen atoms had to be localized above and below the Fe ions, along the c axis, in order to achieve an octahedral oxygen environment at the Fe-O distance corresponding to the average refined ones (2.180 Å), (due to very weak contribution to the overall diffraction diagram, these two extra atoms positions were not introduced. Indeed they could not be detected on the Fourier map difference.)

Space Group: Pmmn; a = 11.5415(7), b = 3.5642(2), c = 4.3596(3) Å

atom	x	y	z	occ.	100 x U_{iso}
V(1)	0.3994(6)	1/4	0.071(2)	1	4.2(4)
O(2)	0.388(1)	1/4	0.384(4)	1	1
O(3)	0.562(1)	1/4	0.967(5)	1	1
O(4)	1/4	1/4	0.004(7)	1	1
Fe(5)	3/4	1/4	0.03(2)	0.110(8)	1
O*	3/4	1/4	-0.468	/	/

Table I : atomic positions in $Fe_{0.12}V_2O_{5.15}$. APD were fixed for oxygen atoms (O*: additional oxygen not introduced in the calculation completing the Fe^{3+} octahedral environment.)

Perspective structure of $Fe_{0.12}V_2O_{5.15}$ is reported in figure 2. Extra oxygen atoms (O*) complete the iron octahedral environment.

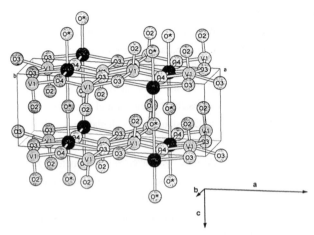

Figure 2: Perspective structure of $Fe_{0.12}V_2O_{5.15}$ plotted with the Molview program[6]. Fe^{3+} (●) is located within the ribbons of V_2O_5. Extra oxygen atoms (O^*) complete the iron octahedral environment. This is an average description: not all Fe sites are occupied.

Distances and angles are gathered in table II and III.

interatomic distances (Å)		angles (degrees)	
V(1)-V(1)	2.991(11) x 2	O(2)-V(1)-O(3)	96.4(8)
V(1)-O(2)	1.369(17)	O(2)-V(1)-O(3)	109.0(13)
V(1)-O(3)	1.844(5) x 2	O(2)-V(1)-O(3)	96.4(8)
V(1)-O(3)	1.932(15)	O(2)-V(1)-O(4)	94.2(14)
V(1)-O(4)	1.749(9)	O(3)-V(1)-O(3)	75.2(6)
V(1)-Fe(5)	2.520(16) x 2	O(3)-V(1)-O(3)	150.3(11)
		O(3)-V(1)-O(4)	102.8(7)
		O(3)-V(1)-O(3)	75.2(6)
		O(3)-V(1)-O(4)	156.8(12)
		O(3)-V(1)-O(4)	102.8(7)

Table II: vanadium environment distances and angles.

interatomic distances (Å)		angles (degrees)	
Fe(5)-V(1)	2.520(16) x 4	O(3)-Fe(5)-O(4)	89.3(7)
Fe(5)-O(3)	2.186(26) x 2	O(3)-Fe(5)-O(3)	165.1(17)
Fe(5)-O(4)	1.789(9) x 2	O(3)-Fe(5)-O*	97.44
Fe(5)-O*	2.180 x 2	O(3)-Fe(5)-O*	82.55
		O(4)-Fe(5)-O*	84.97
		O(4)-Fe(5)-O*	95.03
		O(4)-Fe(5)-O(4)	170.0(7)
		O*-Fe(5)-O*	180.0

Table III: iron environment distances and angles.
(O^*: unrefined additional oxygen completing the Fe^{3+} octahedral environment.)

The extra oxygen atoms increase the oxygen content of the $Fe_{0.12}V_2O_5$ ribbons, in complete agreement with the formulation: $Fe_{0.12}V_2O_{5.15}$ as determined by chemical analysis and with the presence of 3% of V^{4+} (see above charge balance).

DISCUSSION

A possible substitution of vanadium by iron has not been considered because of the short time and moderate temperature of the substitution reaction. Due to charge balance, an infinite cluster of FeO_6 octahedra cannot be considered (in addition, it would lead to only one extra oxygen atom per iron ($Fe_{0.12}V_2O_{5.12}$)). For the same reason, isolated FeO_6 octahedra cannot exist (here, two additional oxygen atoms per iron should gave a $Fe_{0.12}V_2O_{5.24}$ formulation instead of $Fe_{0.12}V_2O_{5.15}$). A small peak around 12.41 ° 2θ is observed in figure 1. This position corresponds exactly to the 010 line with a superstructure doubling the b parameter.

These above observations suggest that FeO_6 octahedra are arranged to form short chains along the c axis whereas an order between these chains is possible along the b axis. A preliminary Mössbauer study [7] seems to confirm that iron atoms are octahedrally coordinated and that their oxidation state is 3+. In addition, differently distorted octahedral environments are in good agreement with some finite chains of octahedra along the c axis where terminal Fe^{3+} ions are different from in-chains ones. These structural features correspond to V_2O_5 layers linked to each other in the c direction, increasing the 3D character of the phase and leading to the peculiar electrochemical behavior upon lithium intercalation as compared with V_2O_5 pristine material.

ELECTROCHEMICAL BEHAVIOR

The above structural results can be correlated with the electrochemical behavior of the compound. The electrochemical measurements of the $Li/Fe_{0.12}V_2O_{5.15}$ system were performed on cells using electrolytes based on twice distilled propylene carbonate (PC) obtained from Fluka and used as received, and anhydrous lithium perchlorate dried under vacuum at 200°C for 12h. The electrolyte was prepared under purified argon atmosphere. The working electrode consisted of a stainless steel grid with a geometric area of $1cm^2$ on which the vanadium oxide mixed with graphite (90 wt%) was pressed. Lithium was used both as the reference and as the auxiliary electrode.

A first galvanostatic electrochemical study of lithium intercalation in $Fe_{0.12}V_2O_{5.15}$ (i=20µA, c/40) has indicated that the phase was quite promising [8]. Four reduction steps appeared in the 3.5-1.8 V potential range , involving the insertion of about 3 lithium ions per mole of oxide (Figure 3).

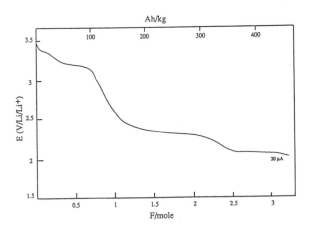

Figure 3: Discharge curve at low constant current density ($20\mu A/cm^2$) for
a $Fe_{0.12}V_2O_{5.15}$ electrode in 1M LiClO$_4$/PC solution at 20°C.

A theoretical specific capacity of 440Ah/kg was then available. Comparison of O.C.V. curves performed for the first two steps of lithium insertion into V_2O_5 and $Fe_{0.12}V_2O_{5.15}$ is shown in figure 4.

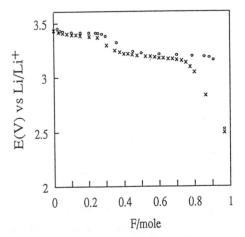

Figure 4: Coulometric titration curves for the first two steps of lithium
insertion in $Fe_{0.12}V_2O_{5.15}$ (×) and in V_2O_5 (O) at 20°C.

Though a similar shape is found for both compounds, a lower lithium uptake is reached for $Fe_{0.12}V_2O_{5.15}$. For a cut off voltage of 3V, about 0.80 lithium ions are accommodated in this compound versus ≈ 0.95 Li in V_2O_5 . This may be correlated to the additional oxygen occupying sites which are normally available for lithium intercalation. However, the presence of ferric ions in the host lattice has been proved to induce interesting features especially in terms of cycle life at a high discharge-charge rate (C/4) [9]. A specific capacity of 200Ah/kg is recovered after 40 cycles performed in the potential window 3.8-2 V while results lower by 25% are obtained for the parent oxide. These first structural results seem to indicate that the FeO_6 octahedra chains linking the V_2O_5 layers are responsible for a better stability of the structural frame on lithium cycling performances. Further Mössbauer investigations should confirm the occurrence of finite octahedral chains. It has still to be proved that a better long range ordering is maintained after many cycles, and additional X-Ray diffraction measurements have to be performed to confirm iron octahedral chains ordering along the b axis.

Acknowledgments. The authors thank Dr. E. Tronc for communicating the preliminary Mössbauer data.

REFERENCES

1)L. Znaidi, N. Baffier and M. Huber, *Mat. Res. Bull*, 1989, 24,1501

2)N. Baffier, L. Znaidi and M. Huber, *Mat. Res. Bull*, 1990, 25,705

3)P. Aldebert, N. Baffier, N. Gharbi and J. Livage, *Mat. Res. Bull*, 1981, 16,669

4)J.P. Pereira-Ramos ,L. Znaidi, N. Baffier and R. Messina, *Sol. St. Ion*, 1988, 28-30,886

5)A.C. Larson and R.B. von Dreele, *LANSCE MS-H805 Los Alamos National Laboratory, Los Alamos,NM 87545*

6)Molview, J.M. Cense, Ecole Nationale Supérieure de Chimie de Paris, France

7)S. Maingot et all, to be published

8)S. Maingot, J.P.Pereira-Ramos, N. Baffier and P. Willmann, J. Electrochem. Soc., (in press)

9)J.P. Pereira-Ramos,R. Baddour, N. Baffier and P. Willmann, French Patent n°9115495

Materials Science Forum Vols. 152 - 153 (1994) pp. 305-308
© *1994 Trans Tech Publications, Switzerland*

SYNTHESIS AND ACID-BASE PROPERTIES OF "ALPON" NITRIDED ALUMINOPHOSPHATES

R. Conanec [1], R. Marchand [1], Y. Laurent [1], Ph. Bastians [2]
and P. Grange [2]

[1] URA 1496 CNRS "Verres et Céramiques", Université de Rennes I,
F-35042 Rennes Cédex, France

[2] Catalyse et Chimie des Matériaux Divisés, Université Catholique de Louvain,
B-1348 Louvain-la-Neuve, Belgium

Keywords: ALPON, Oxynitrides, Aluminophosphates, Acid-Base Properties, Catalysts, Knoevenagel Condensation

ABSTRACT

Aluminophosphates "ALPO" can be prepared at low temperature as reactive powders using the soft chemistry route. Their nitridation under ammonia flow leads to the high surface area "ALPON" oxynitrides. The nitrogen/oxygen substitution deeply modifies the acid-base properties of the solids. It has been evidenced by TPD CO_2 that the ALPON behave as very strong basic catalysts whereas $AlPO_4$ is known as an acidic solid. The characteristics of these new materials have been tested in a Knoevenagel condensation. At 50°C, the reaction of benzaldehyde with ethylcyanoacetate in toluene brings total conversion after 20h.

In heterogeneous catalysis, phosphates, especially aluminophosphates , are well-known through their acid-base properties, either as catalysts or catalyst supports[1]. In this study new oxynitrides, called "ALPON", have been characterized in the Al-P-O-N system and evaluated as catalysts.

Aluminophosphates, in particular $AlPO_4$, can be prepared as amorphous and high dispersed powders, using the soft chemistry route. This step proves to be essential for the following reasons : it is required for the subsequent nitrogen introduction and also for obtaining high surface area powders $(200\text{-}400 \ m^2g^{-1})$.

The synthesis is based on gel method [2]. Suitable amounts of propylene oxide are slowly added to a vigorously stirred $AlCl_3,6H_2O$ / H_3PO_4 M mixture, cooled at 0°C:

$$AlCl_3, 6\,H_2O + H_3PO_4 + 3\,CH_3\text{-}CH\text{-}CH_2 \;\rightarrow\; AlPO_4 + (3\text{-}x)\,CH_3\text{-}CHCl\text{-}CH_2OH$$
$$\underset{O}{\diagdown\diagup}$$
$$+\; x\,CH_3\text{-}CHOH\text{-}CH_2Cl + 6\,H_2O$$

After standing several hours, the transparent gel is thoroughly washed with propanol-2, dried and calcined in the 500-800°C temperature range. Depending on precursors ratio, it is possible to prepare, as single phase powders, a series of non-stoichiometric aluminophosphates with atomic ratios Al/P≥1. All these powders are characterized by very high surface area values.

Nitridation has been carried out under ammonia flow in both low (650-800°C) and high (800-1000°C) temperature ranges. Nitrogen substitutes for oxygen leading to the new "ALPON" oxynitrides. At low temperatures, the Al/P atomic ratio of the starting aluminophosphate is maintained. For example:

$$AlPO_4 + x\,NH_3 \;\rightarrow\; AlPO_{4\text{-}3/2x}N_x + 3/2x\,H_2O\uparrow \quad 0 < x \leq 2$$

Above 800°C, ammonia atmosphere partially reduces phosphorus P^V, with formation of volatile species, leading to phosphorus deficient aluminophosphates of general formula $AlP_{1\text{-}a}O_{4\text{-}3/2x\text{-}5/2a}N_x$.

The white and fluid oxynitride powders are X-ray amorphous. The SEM micrographs display a great similarity with the starting oxide powders. BET analysis confirms that high values of porosity and surface area are kept after nitridation.

In this paper the oxide precursors of the ALPON samples have been preferentially selected with an Al/P atomic ratio equal to 1. The characteristics of both oxide and oxynitride studied powders are collected in Table 1.

Starting oxide	S_{BET} m^{2}g^{-1}	Oxynitride composition	N wt.%	S_{BET} m^{2}g^{-1}
AlPO$_4$-1	270	AlPO$_{1,81}$N$_{1,46}$ (AlPON-1)	19	230
AlPO$_4$-2	390	AlPO$_{1,72}$N$_{1,52}$ (AlPON-2)	20	290

Table1. BET Surface area and nitrogen content of the studied samples.

The acid and basic properties of the solids have been evidenced by TPD of NH_3 and of CO_2 respectively [3]. The total amount of desorbed gas characterizes the number of acid or basic sites. The temperature of the maximum of the TPD curves is significative of the acid or basic strength.

The obtained results point out the different behaviour of ALPON and AlPO$_4$ for their acid and basic properties. The desorption of NH$_3$ versus temperature for both AlPO$_4$-1 and corresponding ALPON-1 is illustrated on Figure 1. The total amount of desorbed NH$_3$ and the temperature at which the desorption curve maximum is observed are quite different for the two solids. While $AlPO_4$ is known as an acidic solid, ALPON presents a relatively low number of acid sites as well as a weak acid character of the sites.

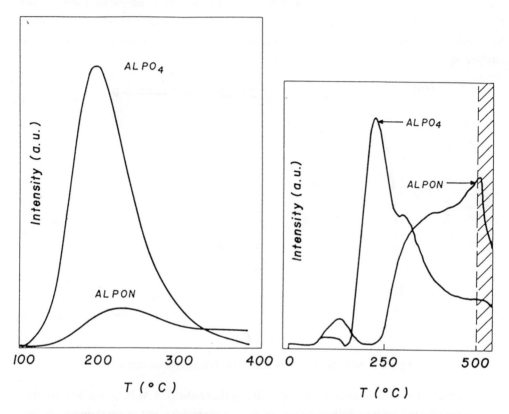

Figure 1.TPD of NH_3, after treatment at 500°C/He and adsorption of NH_3 at 100°C for 15 min.

Figure 2.TPD of CO_2, after treatment at 500°C/He and adsorption of CO_2 at 20°C for 60 min.

The basic character of the AlPO4-2 and ALPON-2 solids is illustrated by the TPD curves of CO_2 on Figure 2. It is shown that the maximum of the desorption is shifted for ALPON to a temperature nearly 300°C higher than that observed for $AlPO_4$. This fact indicates that, in addition to its strongest basic character, ALPON presents strongest basic sites. Even at 500°C, a large amount of CO_2 is not yet desorbed.

In order to evaluate the catalytic properties of these new solids, preliminary experiments of the Knoevenagel condensation of benzaldehyde with ethylcyanoacetate have been performed.This condensation is known as representative of basic catalysts and several examples reported in the literature evidence the need of strong basic sites for the reaction [4-6].

$$C_6H_5CHO + CH_2(CN)CO_2C_2H_5 \rightarrow C_6H_5CH=C(CN)CO_2C_2H_5 + H_2O$$

As an example, the reaction in liquid phase, using 200 mg ALPON-2 at 50°C in toluene as solvent, brings 85% conversion after 2h and 98% conversion after 20h. Figure 3, which gives the conversion of both benzaldehyde and ethylcyanoacetate versus time in stirred reactor, clearly shows that the reaction rate is very high at the beginning of the reaction. This is the first example of catalytic Knoevenagel condensation on nitrided aluminophosphate compounds which show strong basic catalytic sites.

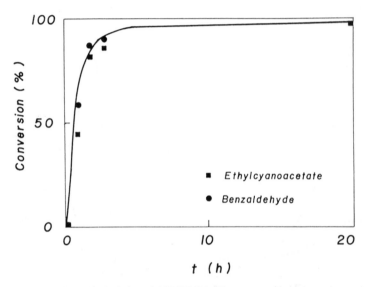

Figure 3. Activity of AlPON-2 in Knoevenagel condensation.

Since it is possible to adjust the N/O ratio during the synthesis of ALPON, it is believed that it is then possible to tune the acid-base properties of these oxynitrides, and to use them as valuable basic catalysts.

REFERENCES

1) Moffat, J.B.: Catal. Rev.-Sci. Eng., 1978, 18, 199.

2) Conanec, R., Marchand, R., Laurent, Y.: High Temp. Chem. Processes ,1992, 1, 157.

3) Malinowski, S., Marczewski, M.: *Acid and base catalysts*, Catalysis, 1989, 8, 107.

4) Tanabe, K., Misono, M., Ono, Y., Hattori, H.: *New solid acids and bases*, Stud. Surf. Sci. Catal., 1989, 51, 211.

5) Lednor, P.W.: Catalysis Today, 1992, 15, 243.

6) Cabello, J.A., Campelo, J.M., Garcia, A., Luna, D., Marinas, J.M.: J. Org. Chem., 1984, 49, 5195.

Materials Science Forum Vols. 152 - 153 (1994) pp. 309-312
© *1994 Trans Tech Publications, Switzerland*

CONTROL OF POROSITY AND SURFACE AREA IN SOL-GEL PREPARED TiO$_2$-Al$_2$O$_3$ MIXED OXIDES BY MEANS OF ORGANIC SOLVENTS

T.E. Klimova and J.R. Solis

Departamento de Ingeniería Química, UNAM, México
Cd. Universitaria, México D.F., México

Keywords: Mixed Metal Oxides, Sol-Gel, Surface Area, Controlled Porosity

ABSTRACT.

The role of organic solvents in controlling the porosity and surface area of a TiO$_2$-Al$_2$O$_3$ mixed oxide (TiO$_2$:Al$_2$O$_3$=1:1), prepared by the sol-gel method, has been examined. Additives used before precipitation only have an effect when their adsorption on the precipitate is strong enough to allow replacement of the adsorbed water. Washing with alcohol after precipitation produced a significant increase of surface area, cumulative pore volume and average pore diameter, apparently as a result of reduced surface tension, growth of the gel particles and the filler effect of the organic chain of additive. The washing time precipitates is an important parameter which affects the quantity of micro- and mesopores in the support. The effect of different functional groups and molecular weights were studied.

INTRODUCTION.

In many catalytic systems, the performance is known to depend not only on the inherent catalytic activity of the active phase, but also on the textural and physicochemical properties of the support. Over the past years, molybdenum catalysts supported on alumina have been used in hydrotreatment catalysts. In this attempt to obtain better catalysts the use of some new supports, such as Ti-Al mixed oxides, has shown promising results since, in this case, greater catalytic activities have been found.

In general, sol-gel methods have been preferred to produce this mixed oxide system. However, the control of the surface area and porosity of the catalysts remains a problem, since small diameter pores of the order of 20-30Å are obtained. In this work, the effectiveness of the use of organic solvents, added at different stages during the preparation, in controlling the porosity and surface area of a TiO$_2$-Al$_2$O$_3$ mixed oxide (TiO$_2$:Al$_2$O$_3$=1:1), prepared by the sol-gel method, using Ti and Al alkoxides as precursors, has been tested.

EXPERIMENTAL.

The samples were prepared using Ti and Al isopropoxides as precursors and n-propyl alcohol as the solvent. Water in an excess of 30 times the stoichiometric amount required, was used to produce the formation of the metallic hydroxides. A serie of additives such as alcohols (C_1-C_7) and acetone was used in order to analyze the effect of the functional group and the molecular weight on surface area and porosity. The solids, formed after drying at 373 K during 24 hr and calcining at 773 K during 12 hr, were characterized by surface area and porosity using a BET nitrogen physisorption commercial apparatus.

RESULTS.

Samples of TiO_2-Al_2O_3 mixed oxides were first prepared in the absence of additives. The reproducibility of preparation was found to be reasonable.

All solvents added before precipitation had little effect on the surface areas and porosities (except for n-heptanol, with which the pore volume was increased to 1.12 cm^3/g, the surface area to 433 m^2/g and average pore diameter to 95Å). Washing the precipitates formed without additives with different solvents before drying and calcination produced a large incease in surface area and pore volume, and significantly changed the pore size distribution (Table 1). The medium pore diameter and cumulative pore volume increased with the molecular weight of the solvent. Surface areas tended to follow the same trend, but no regular behavior could be observed.

Table 1. Samples prepared by washing the precipitate with the organic additives and water on the filter.

Solvent	Surface Tension, Dynes/cm	Surface Area, m^2/g	Pore Volume, cm^3/g	Average Pore Diameter, Å
H_2O	72.8	233	0.230	30.0
CH_3COCH_3	23.7	320	0.286	29.0
CH_3OH	22.5	250	0.391	32.5
C_2H_5OH	21.9	279	0.420	48.5
n-C_3H_7OH	23.6	406	0.896	68.8
n-C_4H_9OH	24.6	385	1.228	118.0
n-$C_7H_{15}OH$	---	393	1.026	125.0

A series of similar experiments was then carried out, where time of washing was increased to 24 hr (Table 2). Results showed a further increase of average pore diameter. n-heptanol was found to reduce slightly the surface area and pore volume.

Table 2. Samples prepared by washing the precipitate with organic additives and water for 24 hours.

Solvent	Surface Area, m^2/g	Pore Volume, cm^3/g	Average Pore Diameter, Å
H_2O	214	0.335	33
CH_3OH	335	0.410	35
C_2H_5OH	299	0.916	59
$n-C_4H_9OH$	371	1.106	125
$n-C_7H_{15}OH$	318	0.909	127

DISCUSSION.

Although the use of alcohols and organic solvents at various stages of the preparation of oxide supports is known to have significant effect on surface area and porosity, the role of the additive is still not clear. The suggestion that replacement of water by the molecules of additive leading to lower surface tension and less collapse of pores during drying and calcination [1,2] did not correlate well with the results (Table 1). A substantial increase in surface area in the case of acetone is accompanied by a small increase of pore volume, indicating an increase in the number of micropores owing to the reduction of surface tension. In this case, a monomodal pore size distribution was obtained. On the other hand, alcohols behave in a more complex manner. The curves dV/dD vs. D show a bimodal pore size distribution (Figure 1). A substantial amount of pores between 100 and 500Å was found for C_3-C_7 alcohols. Other possibilities can be invoked to explain the observed effects. Alcohols have been suggested to interfere with crystallite ordering by replacing the "solvent barrier" at the surface of the precipitate [3]. The solubility of the hydroxide is decreased in the alcohol medium leading to enhanced aggregation-cementation, therefore particle bridging is encouraged. The result of this being a greater amount of mesopores.

For higher molecular weight alcohols, the organic chain itself is believed to act as a filler and, as a result, pore volume at higher pore diameters is increased. Such an increase of mesoporosity would not be expected to be linear, because of coiling of the organic chain can occur, in agreement with our observations.

The increase of the time of washing to 24 hr resulted in the growth of the quantity of mesopores and reduced the amount of micropores, probably by the further increasing of the average size of gel particles.

In the case of additives incorporated before the precipitation step, solids with higher surface areas and micro- and mesoporosity were obtained. This result may be explained by assuming a combined effect. On one hand it may be possible the stabilization of small crystallites by the presence of the additive leads to more micropores and enhanced surface area, on the other hand, a filler effect of the organic chain of the additive leads to mesoporosity. However, this case is more complex and is incompletely understood. Additional work is under way.

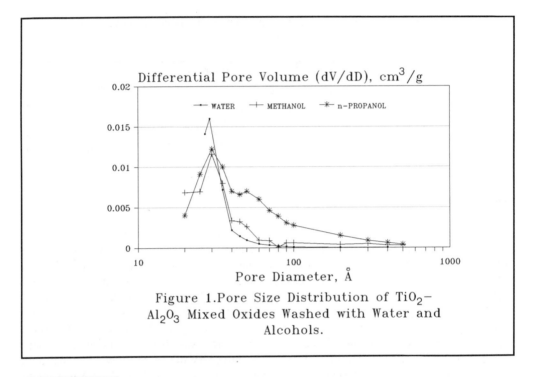

Figure 1.Pore Size Distribution of TiO$_2$–
Al$_2$O$_3$ Mixed Oxides Washed with Water and
Alcohols.

CONCLUSIONS.

Organic solvents, used as additives before the precipitation process only have an effect when their adsorption on the precipitate is strong enough to allow replacement of the adsorbed water. (The case of high molecular weight alcohols.) Washing with the organic solvent after precipitation increases surface area and mesoporosity in the calcined support, apparently as a result of:

1. Replacement of water by additive, leading to lower surface tension and less pore collapse during drying and calcination.

2. Growth of the gel particles, as a result of reduction of solubility of hydroxides in the organic solvent medium.

3. Filling the pores by the organic chain of additive.

The effect of reduced surface tension, of the solvent filling the pores, on the porosity of the solid, was found to be small. The washing time of precipitates is an important parameter which affects the quantity of micro- and mesopores in the support.

REFERENCES.

1) Johnson, M.F.L.; Mooi, J.: J. Catal., 1968, 10, 342.

2) White, A.; Walpole, A.; Huang, Y.; Trimm, D.L.: Appl. Catal., 1989, 56, 187.

3) Cormack, B.; Freeman, J.J.; Sing, K.S.W.: J.Chem. Tech. Biotechnol., 1980, 30, 367.

Materials Science Forum Vols. 152 - 153 (1994) pp. 313-318

CHEMISTRY OF HYBRID ORGANIC-INORGANIC MATERIALS SYNTHESIZED VIA SOL-GEL

C. Sanchez, F. Babonneau, F. Banse, S. Doeuff-Barboux, M. In and F. Ribot

Chimie de la Matière Condensée, URA CNRS 1466, Université Pierre et Marie Curie, 4 place Jussieu, F-75252 Paris, France

Keywords: Sol-Gel, Hybrid Materials, Siloxane, Transition Metal Oxo-Polymers, Tin, Vanadium, Polypyrrole

ABSTRACT - Design and synthesis of hybrid organic-inorganic networks, where both components are chemically bonded, are reported. These nano composites can be obtained by hydrolysis and condensation reactions of organically functionalized alkoxide precursors. Striking examples, taken from silicon, tin and transition metal alkoxide chemistry, are presented.

1. INTRODUCTION

Sol-gel chemistry is based on the polymerization of molecular precursors such as metal alkoxides $M(OR)_n$ [1,2]. Hydrolysis and condensation of these alkoxides lead to the formation of metal oxo-polymers. Such synthesis being performed at room temperature chemically bonded organic groups can be kept inside the inorganic gel matrix, leading to hybrid materials. These new materials offer many opportunities for applications in different fields such as patternable optical devices, photonics, sensors, catalysis [3-6].Siloxane based hybrids [7,8] can be easily synthesized because Si-C bonds are rather covalent and therefore they are not broken upon hydrolysis. Similar chemistry can be developed from tin alkoxides. This is no longer valid with transition metals for which the more ionic M-C bond is easily cleaved by water. Complexing organic ligands have then to be used. Such groups can be functionalized for any kind of organic reactions such as organic polymerization and lead to hybrid organic-inorganic copolymers [9,10]. This paper summarizes previous work on the design and synthesis of hybrid organic-inorganic gels in which organics are chemically bonded to the inorganic gel network.

2. SILOXANE BASED HYBRID MATERIALS

Organic groups can be bonded to an inorganic network for two different goals, as network modifier or network former. Both functions have been achieved in the so-called ORMOSILS [7]. The precursors of these compounds are organo substituted silicic acid esters of general formula $R'_nSi(OR)_{4-n}$, where R' can be any organofunctional group. If R' is a just non hydrolyzable group bonded to silicon through a Si-C bond, it will have a network modifying effect ($Si-CH_3$). On the other hand, if R' can react with itself (R' contains a methacryl group for example) or additional components, it will act as a network former.

Network modifiers and network formers can also introduce other physical properties. The hydrolysis and condensation of $(RO)_3Si-(CH_2)_n-NH_2$ allow to synthesize compounds where anions such as $CF_3SO_3^-$ can be solvated. The associated counter ions (protons) are free and can be used for conduction [11].

Recently, a new class of hybrid materials with electronic properties have been synthesized from precursors such as TMSPP (N-3 trimethoxy-silyl-propyl-pyrrole) [12]. TMSPP, pyrrole and THF or TMSPP, TMOS, pyrrole and THF were hydrolyzed in the presence of nucleophilic catalysts to yield transparent sols. The oxidative polymerisation of grafted and free pyrrole was performed with FeCl$_3$. The black xerogels obtained after drying exhibit electrochemical properties and an electrical conductivity of about 6.10^{-3} S.m^{-1} at 1 GHz [12]. These hybrid organic-inorganic nano-composites made with a siloxane network and polypyrrole oligomers correspond to the schematic structure I in figure 1.

Hybrid materials can also be prepared by cross linking preformed polymeric/oligomeric species (siloxane or organic backbone) with metal alkoxides. Among those preformed species one can mentioned hydroxyl terminated polydialkyl siloxane, hydrolyzed triethoxysilane end-capped polytetramethylene oxide (PTMO) or hydrolyzed (N-triethoxysilylpropyl) o-polyethylene oxide urethane (MPEOU) [13,16].

Aryl bridged polysilsesquioxanes have been recently produced by sol-gel processing of bis(triethoxysilyl)aryl and bis(trichlorosilyl)aryl monomers. Phenyl, biphenyl and terphenyl groups act as rigid spacers, of different length, between the siloxane chains [17,18].

3. TIN OXO SPECIES BASED HYBRID MATERIALS

Tin is a very interesting element because its characteristics make it intermediate between silicon and transition metals. Like the latter, tin exhibits several coordination numbers (generally from 4 to 6) and coordination expansion makes hydrolysis-condensation reactions of tin alkoxides fast. But, as for silicon, the Sn-C bond is stable, especially towards nucleophilic agents such as water. This last characteristic allows to chemically link organic moieties to the tin oxo polymers/oligomers. Hydrolysis of BuSn(OPr^i)$_3$ (H$_2$O/Sn>3) leads to a crystalline compound with the following formula [(BuSn)$_{12}$(μ_3-O)$_{14}$(μ_2-OH)$_6$](OH)$_2$(HOPr^i)$_4$ [19]. This compound is made of a tin oxo-hydroxo cluster with an equal amount of six fold and five fold coordinated tin atoms. This cage-like cluster is surrounded by twelve n-butyl chains which prevent further condensation. The ^{119}Sn-{^{1}H} NMR signature of this cluster, in solution, corresponds to two peaks (-281 and -449 ppm vs TMT) , each with two sets of tin-tin coupling satellites [19].

Hydrolysis of ButenylSn(OAm^t)$_3$ (H$_2$O/Sn>3) yields clear solutions. The ^{119}Sn-{^{1}H} NMR spectra of such solutions are very close to those observed for the previously described cluster (-283 and -468 ppm vs TMT, similar tin-tin coupling values). These results indicates the very likely formation of a closo-type tin oxo-hydroxo cluster which is surrounded by butenyl chains. The organic polymerization of these butenyl groups was performed, in solution, at 80°C with azobisisobutyronitrile (AIBN) as a radical promoter and yielded a turbid solution. A noticeable decrease of the intensity and a broadening of the ^{119}Sn NMR resonances (still located at -283 and -468 ppm) were observed. ^{13}C NMR experiments also revealed a strong decrease of the resonances located at 115 and 140 ppm (unsaturated carbons of the butenyl). Moreover, Quasi Elastic light Scattering experiments evidenced entities with diameters around 200 Å. These experiments indicates that the polymerization of the butenyl groups had occurred and attached the tin oxo hydroxo clusters though polybutane chains (schematic structure II in figure 1). Stable M-C bond towards hydrolysis can also be found for some divalent metals, such as Hg and Pt, and can be used to synthesize hybrid materials [20]. However the introduction of transition metal in covalently bonded hybrid compounds generally need other chemical strategies.

4. HYBRID MATERIALS BASED ON M-O-Si-C BONDS

Copolymerization at a molecular level has been described for the synthesis of hybrid siloxane-oxide materials. These compounds have been synthesized from hydrolysis-condensation of diethoxydimethylsilane (DEDMS) and various metallic alkoxides, M(OR)$_n$ with M=Si, Ti, Zr, Al and V [12,21,22]. These materials can be described as copolymers of D species (-O-Si(CH$_3$)$_2$-O-) and "Q species" (SiO$_4$, O=VO$_3$) which act as cross linking agents (scheme III), or as nano composites made of metal oxo-polymers linked to siloxane chains probably through M-O-Si bonds (M=Ti, Zr, V), as pictured in scheme IV.

The structure of these hybrids can be tailored by a careful adjustment of chemical conditions. For the system obtained from the hydrolysis-condensation of DEDMS and VO(OAm^t)$_3$ different structures and redox behavior were recently reported [12]. For hydrolysis at neutral pH, these hybrid gels show no reduction of V(V) after a few months. With acidic hydrolysis conditions, the reduction of V(V) to V(IV) occurs in a few days. For neutral pH, discrete tetrahedral coordinated V(V) are covalently bonded to polysiloxane moieties (scheme III) and this sequestering of V(V) in tetrahedral

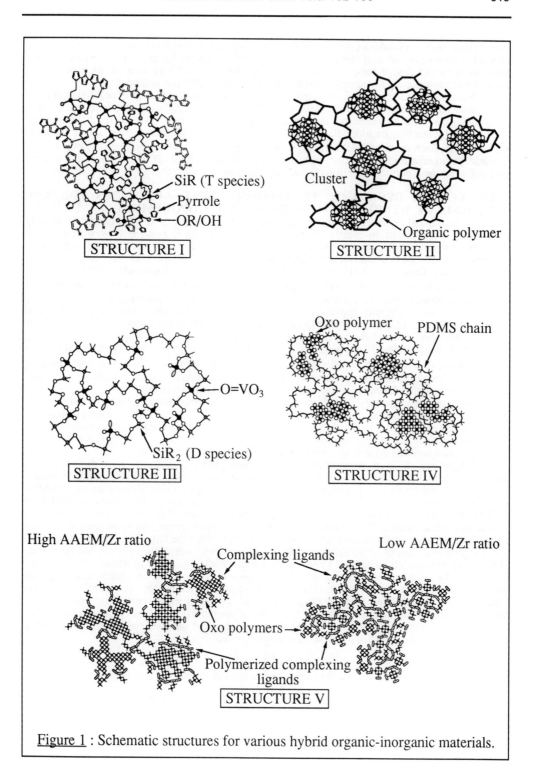

STRUCTURE I

SiR (T species)
Pyrrole
OR/OH

STRUCTURE II

Cluster

Organic polymer

STRUCTURE III

O=VO₃

SiR₂ (D species)

STRUCTURE IV

Oxo polymer   PDMS chain

High AAEM/Zr ratio   Low AAEM/Zr ratio

Complexing ligands

Oxo polymers

Polymerized complexing
ligands

STRUCTURE V

<u>Figure 1</u> : Schematic structures for various hybrid organic-inorganic materials.

coordination probably inhibit its reduction by alcohol and light. Under acidic conditions, the cleavage of Si-O-V bonds is easier and vanadium oxo-polymeric species are formed and cross link siloxane units (scheme IV). These vanadium oxo-polymers allow the V(V) to increase its coordination number and consequently favours the reduction by stabilizing V(IV) oxo compounds.

Recently P. Judeinstein [23] reported the synthesis and characterization of negatively charged macromolecules based on poly-oxo-metallates (POM) which are organically functionalized through W-O-Si-C links and anchored to an organic polymeric backbone. The modified POM were synthesized by reacting trichlorosilane, $RSiCl_3$ (R= vinyl, allyl, methacryl, styril), with lacunary $K_4SiW_{11}O_{39}$. Each modified POM ($[SiW_{11}O_{40}(SiR)_2]^{4-}$) carries two unsaturated organic groups. Organic polymerization performed with a radical initiator yields POM attached together by polymethacrylate or polystyrene chains (schematic structure II).

5. MOLECULAR DESIGN OF TRANSITION METAL ALKOXIDES FOR THE SYNTHESIS OF HYBRID ORGANIC-INORGANIC COPOLYMERS

The chemical tailoring performed with systems containing a Si-C bond cannot be directly extended to pure transition metals because the more ionic M-C bond is broken down upon hydrolysis. Organic modification can however be performed by means of strong complexing ligands. The best are ß-diketones and allied derivatives, polyhydroxylated ligands such as polyols, and α- or ß-hydroxyacids. These ligands (HL) react readily with transition metal alkoxides $M(OR)_4$ (M= Ce, Ti, Zr,...) to yield new precursors $M(OR)_{3-x}(L)_x$ [9]. Upon hydrolysing these new precursors, most of the alkoxy groups are quickly removed while all strong complexing ligands cannot be completely removed. Complexing ligands appear to be quite stable towards hydrolysis because of chelate and steric hindrance effects. Thus, they allow to anchor organic groups to transition metal oxo-polymeric species and synthesize new hybrid organic-inorganic materials.

Organically modified TiO_2 gels, which give photochromic coatings, were synthesized from an allyl acetylacetone modified $Ti(OBu^n)_4$ alkoxide [9]. Double polymerization was performed by partial hydrolysis of the alkoxy groups and radical polymerization of the allyl functions. However polymerization of allyl functions is slow and polymerization degree stays low.

More reactive methacrylic acid can also be used as a polymerizable chelating ligand. The sol-gel synthesis of zirconium oxide based monoliths synthesized by UV copolymerization of zirconium oxide sols and organic monomers was recently reported [24]. However carboxylic functions being weak ligands, they are largely removed upon hydrolysis [9,24] and thus a large amount of the chemical bonds between organic and inorganic networks is lost in the sol state.

Therefore a new approach was chosen with different ligands, such as acetoacetoxyethylmethacrylate (AAEM) and methacrylamidosalicylic acid (MASA), which present both a strong chelating part and a highly reactive methacrylate group [10]. Zirconium-oxo-PAAEM copolymers were synthesized from zirconium propoxide modified at the molecular level with AAEM [10]. These hybrid organic-inorganic copolymers are made of zirconium oxo-polymers and polymethacrylate chains. The zirconium oxo species, in which zirconium is oxygen seven fold coordinated, are chemically bonded to methacrylate chains through the ß-diketo complexing function. The complexation ratio (AAEM/ Zr) is the key parameter which controls the structure and the texture of these hybrid materials (schematic structure V). A careful adjustment of this parameter leads to the tailoring of the ratio between organic and inorganic components and also to zirconium oxo species with more or less open structures. The ratio inorganic/organic increases when the complexation ratio decreases. For a high complexation ratio (0.75) both networks interpenetrate intimately at the nanometer scale, while for a low ratio (0.25) the size of the inorganic domains increases to the sub-micron range. Organic and inorganic growths are not independent and such systems exhibit probably similar behavior to the so-called interpenetrating polymer networks.

REFERENCES

[1]   Brinker C.J. and Scherrer G.: "Sol-Gel Science, the Physics and Chemistry of Sol-gel Processing" (Academic press, San-Diego, 1989).

[2]   Livage J., Henry M. and Sanchez C.: Progress in Solid State Chemistry, 1988, 18, 259.

[3]   Avnir D., Levy D. and Reisfeld R.: J. Phys. Chem., 1984, 88,5956.

[4]   Dunn B. and Zink J.I.: J. Mater. Chem., 1991, 1, 903

[5]   (a) Griesmar P., Sanchez C., Pucetti G., Ledoux I. and Zyss J.: Molecular Engineering, 1991, 1(3), 205.
      (b) E.Toussaere, J.Zyss, P.Griesmar and C.Sanchez: Nonlinear Optics, 1991, 1(3).

[6]   Breistcheidel B., Zieder J. and Schubert U.: Chem. Mater., 1991, 3, 559.
[7]   Schmidt H., Scholze H. and Kaiser A.: J. Non-Cryst. Solids, 1984, 63 ,1.
[8]   Wang B., Wilkes G.L., Smith C.D. and McGrath J.E., Polymer Comm., 1991, 32, 400.
[9]   Sanchez C., Livage J., Henry M. and Babonneau F.: J. Non-Cryst. Solids, 1988, 100, 650.
[10] Sanchez C. and In M.: J. Non-Cryst. Solids, 1992, 147-148, 1.
[11] Charbouillot Y., Ravaine D., Armand M. and Poinsignon C.: J. Non-Cryst. Solids, 1988, 103, 325.
[12] Sanchez C., Alonso B., Chapusot F., Ribot F.and Audebert P.: J. Sol-Gel Science and Technology, in print.
[13] Huang H.H., Orler B.and Wilkes G.L.: Macromolecules, 1987, 20, 1322.
[14] Glaser R.H. and Wilkes G.L.: Polym. Bull., 1988, 19, 51.
[15] Burkhardt E.W., Burford R.R. and Deatcher J.H.: Chem. Mater., 1989, 1, 767.
[16] Boulton J.M., Fox H.H., Neilson G.F. and Uhlmann D.R.: Mat. Res. Soc. Proc. Symp. Vol. 180. 1990 Materials Research Society. p.773.
[17] Shea K.J. and Loy,D.A.: Chem. Mater., 1989, 1, 572.
[18] Corriu R., Moreau J., Thepot P. and Wong Chi Man M.: Chem. Mater., 1992, 4, 1217.
[19] Banse F., Ribot F., Toledano P. and Sanchez C.: Mat. Res. Soc. Proc. Symp. Vol. 271. 1992 Materials Research Society. p.45.
[20] Bonhomme C., Henry M.and Livage J.: J. Non-Cryst. Solids, 1993, 159, 22.
[21] Diré S., Babonneau, F. Sanchez C. and Livage J.: Journal of Mater Chem., 1991, 2, 239.
[22] Babonneau F., Polyhedron, in print.
[23] Judenstein P., Chem. Mater. 1991, 4, 4.
[24] Naß R.and Schmidt H. in "Sol-Gel Optics", Mackenzie J.D. and Ulrich D.R. (Eds.) (Proc. SPIE 1328, Washington, 1990). p.258.

Materials Science Forum Vols. 152 - 153 (1994) pp. 319-322
© *1994 Trans Tech Publications, Switzerland*

A NEW SOFT CHEMISTRY SYNTHESIZED VANADIUM OXYSULFIDE

G. Tchangbédji, E. Prouzet and G. Ouvrard

Institut des Matériaux de Nantes, CNRS UMR 110,
2 rue de la Houssinière, F-44072 Nantes Cédex 03, France

Keywords: Ion Condensation, Vanadium, Oxysulfide

ABSTRACT

In the frame of a quest of new positives for lithium batteries, a new vanadium oxysulfide has been prepared by an ion condensation reaction in water at room temperature, from a mixture of water solutions of a vanadyl salt and sodium sulfide. Chemical and thermal analyses conclude to a formulation $V_2O_{6-x}S,xH_2O$. Proton NMR experiments show that the water molecules are tightly bonded to vanadium. EPR, XAS and magnetic susceptibility measurements allow to determine the oxidation states of vanadium and sulfur, IV and -II, respectively, confirming the general formula, $V_2O_3S,3 H_2O$. Infrared experiments show the existence of a vanadyl group and an EXAFS study at the vanadium K edge allows to characterize the coordination of vanadium. Vanadium atoms are surrounded by a square plan of oxygen atoms, one vanadyl group and a sulfur atom.

INTRODUCTION

In order to compromize between transition metal oxides and sulfides as positives for lithium batteries, preparations of new transition metal oxysulfides were attempted. Very few examples of such compounds are known, especially if we consider phases whith an open structure favorable for lithium insertion. Poorly crystallized molybdenum [1] and titanium [2] oxysulfides have been obtained and tested as positives in lithium batteries. Both of them evidence a good reversibility and a mean voltage intermediate between those generally observed for oxides and sulfides. They have been prepared under mild synthesis conditions which probably allow to stabilize a mixed oxygen and sulfur environment for the transition metal. By using a soft chemistry reaction we were able to prepare a new vanadium oxysulfide by an ion condensation reaction [3].

I. SYNTHESIS AND ANALYSIS

The new phase is prepared by room temperature reaction between water solutions of vanadyl salts ($VOCl_2,2H_2O$ or $VOSO_4,5H_2O$) and hydrated sodium sulfide, $Na_2S,9H_2O$. A gray precipitate is immediately obtained and the reaction is complete in about 24 hours upon mechanical stirring. This phase appears to be amorphous from an X-ray diffraction experiment, even for long time exposures. EDX analysis performed in a SEM indicates a mean V/S atomic ratio of 2 with a very small dispersion for various samples and different parts of them. This is confirmed by a chemical analysis (excluding hydrogen) which concludes to a formula V_2O_6S. By a simple comparison with $VOSO_4,5H_2O$, Infrared spectroscopy experiments attribute the unexpectededly large amount of oxygen to water molecules. At the same time they show that the new phase do not contain sulfate groups. TGA experiments made on the compound obtained at room temperature and before any thermal treatment, evidence two distinct weight losses (Figure 1). A mass spectrometry analysis of the evolved gases proves that the first loss, below 120°C, corresponds mainly to water (attributable to adsorbed molecules) and the second one, between 120°C and 300°C, to remaining water and sulfur. The final plateau corresponds to VO_2. From these various experiments, a $V_2O_{6-x}S,xH_2O$ ($x \approx 3$) formula can been deduced. The water content x, and the nature of these water molecules depend largely of the heat treatment of the rough powder. Adsorbed water can be removed either by heating at 120°C at ambient pressure or by drying at 70°C under dynamic vacuum. Lyophilisation can also be used to dry the compound. In any case, it is impossible to remove completely water without losing sulfur. This indicates that a fraction of water stabilizes the phase. A proton NMR study confirms this fact. Mainly two types of line width may be observed depending on the way water molecules are bonded to vanadium atoms. If water is only adsorbed, the corresponding spin-spin relaxation time T2 is rather long (100 μs), and the line is sharp. On the contrary, if oxygen is bonded directly to vanadium atom, T2 is very short (6μs) and the line is broad. An interesting evolution in the line shape is observed, depending on the procedure used to dry the phase. The general tendency is that the fewer water molecules in the phase, the more tightly they are bonded to vanadium.

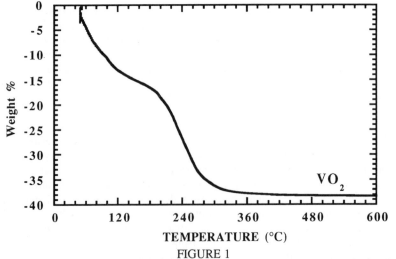

FIGURE 1
TGA analysis of the rough vanadium oxysulfide under argon flow (heating rate:2°C/mn).

PHYSICAL AND STRUCTURAL CHARACTERIZATION.
    Magnetic susceptibility measurements show a temperature dependant paramagnetic behaviour with a curvature of the χ^{-1} versus T variation which can be analyzed as superimposed Temperature Independant Paramagnetism (χ_{TIP}) and Curie-Weiss paramagnetism [3]. The last one correponds to an observed magnetic moment of 1.09 μB per vanadium atom, lower than the expected value for a vanadium IV in a d^1 configuration (1.73 μB). Nevertheless such a discrepancy has already been encountered in vanadium derivatives with a superimposed TIP. The existence of the vanadium IV

oxidation state is confirmed by EPR spectroscopy results which show a large signal at room temperature with g=1.97.

An XAS study was undertaken in order to more precisely define the oxidation states for vanadium and sulfur atoms. It is known that the positions of various singularities at the XAS edges are very sensitive to the oxidation state (pre-peak, main edge, first peak...). This can be seen on figure 2 where the vanadium K edge moves towards lower energies when the vanadium oxidation state decreases. The main edge for the vanadium oxysulfide is found at the same position as for $VOSO_4,5H_2O$, confirming the oxidation state IV for vanadium. A similar study performed at the sulfur K edge with various reference compounds proves the anionic nature of sulfur with the probable oxidation state -II. Taking into account these oxidation states, the formula of the new phase may be written $V^{IV}_2O^{-II}_3S^{-II},3H_2O$.

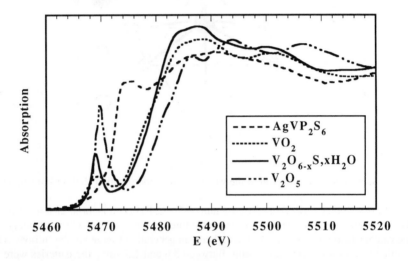

FIGURE 2

Vanadium K edge X Ray absorption spectra for the new vanadium oxysulfide and some reference compounds with various vanadium oxidation states.

Despite the amorphous nature of the new phase, it is important to determine the vanadium environment to better understand its physical properties and electrochemical behaviour. To do so, an EXAFS study was performed at the vanadium K edge [4]. It is not possible to discriminate the presence of sulfur in the vanadium radial distribution fonction (RDF) deduced from the EXAFS signal. The same phenomenon was observed before for V_2O_5, where the large distribution in the V-O distances, especially the vanadyl group, can not be seen on the RDF [5]. Infrared spectroscopy has shown the existence of the vanadyl group in the new compound. Some considerations about the shape of the EXAFS signal corresponding to the first coordination shell, by comparison with vanadium sulfate, leads to conclude to the existence of a square plan of oxygen atoms as encountered in $VOSO_4,5H_2O$. We may then consider that vanadium is surrounded by five oxygen atoms, one of them forming a vanadyl bond (d_{V-O}=1.58 Å) and four others describing the square plan (d_{V-O}=2.02 Å). A refinement of the EXAFS signal has been made in order to explorate the possible existence of a sixth atom which could be either oxygen or sulfur. The results tend to prove unambiguously the presence, in the first vanadium coordination shell, of one sulfur atom at 2.23 Å, in good agreement with the expected vanadium-sulfur distance for the considered oxidation states and environment. Figure 3 schematizes the vanadium first coordination shell. A refinement of the EXAFS signal corresponding to the second vanadium coordination shell, assuming that it is only constituted by vanadium atoms, leads to a coordination number of 2 and a vanadium-vanadium distance of 3.14 Å. This would correspond to chains formed by O-O edge sharing of the above described groups, in such a way as to find vanadium atoms alternatively above and below the chains. Nevertheless, a particle size effect may reduce the apparent coordination number in large proportion and it is impossible to describe precisely the actual arrangement of the VO_5S polyhedra shown on figure 3.

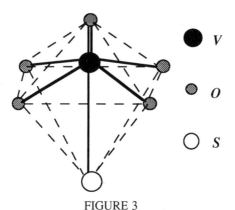

FIGURE 3

Schematic representation of the vanadium first coordination shell in $V_2O_3S,3H_2O$.

The new compound, dried under vacuum at 70°C, has been incorporated as positive in lithium cells [3]. As expected for a partially sulfur substituted oxide, the Open Circuit Voltage is lower than in the parent vanadium oxide V_2O_5, but higher than generally observed in the sulfides. The behaviour in cycling conditions is very promising. Between 3.6 and 1.5 volts, the batteries were cycled more than 100 times under a regime of C/8, with about 0.5 lithium atom per vanadium corresponding to a capacity and an energy of 120 Ah kg^{-1} and 300 Wh kg^{-1} respectively.

REFERENCES

1) Abraham, K.M., Pasquariello, D.M. and Willstaedt, E.B.: J. Electrochem. Soc., 1989, 136, 576
2) Meunier, G., Dormoy, R. and Levasseur A.: Mater. Sc. Eng., 1989, B3, 19
3) Tchangbédji, G., Odink, D. and Ouvrard, G.: J. Power Sources, 1993, 43-44, 577
4) Tchangbédji, G., Thesis, University of Nantes, 1993

ACKNOWLEDGEMENTS : Authors thank DRET for its financial support under the contract number 91-1265 A.

Materials Science Forum Vols. 152 - 153 (1994) pp. 323-326

LITHIUM DOPING OF COBALT-NICKEL SPINEL OXIDES AT LOW TEMPERATURE

E. Zhecheva and R. Stoyanova

Institute of General and Inorganic Chemistry,
Bulgarian Academy of Sciences, BG-1113 Sofia, Bulgaria

Keywords: Lithium-Cobalt-Nickel Spinels, Cobalt-Nickel Hydroxide-Nitrates, Cobalt-Nickel Oxide-Hydroxides, Thermal Hydrolysis, Ion Exchange, Thermal Decomposition

ABSTRACT

Cobalt-nickel spinel oxides containing lithium $Li_y(Ni_xCo_{1-x})_{3-y}O_4$ ($y \approx 0.13$, $0 \leq x \leq 0.4$) are prepared by thermal decomposition of mixed cobalt-nickel hydroxide-nitrates with a double-chain structure, $Ni_xCo_{1-x}(OH)(NO_3).H_2O$, in a lithium nitrate medium at 300°C. It has been shown that in air $Ni_xCo_{1-x}(OH)(NO_3).H_2O$ decompose to nickel-cobalt spinels via layered hydroxide-nitrates $Ni_xCo_{1-x}(OH)_{1+s}(NO_3)_{1-s}$, but in a lithium nitrate medium layered hydroxide-oxides $Ni_xCo_{1-x}OOH$ are obtained as an intermediate product due to complete hydrolysis of the initial hydroxide-nitrates. Above the melting point of $LiNO_3$ ($T_m = 255$°C), $Ni_xCo_{1-x}OOH$ decompose to lithium containing cobalt-nickel spinel oxides, $Li_y(Ni_xCo_{1-x})_{3-y}O_4$ which occurs by a topotactic mechanism preceded by partial ion exchange of protons from $Ni_xCo_{1-x}OOH$ with Li^+ from the melt.

INTRODUCTION

Lithium doping of transition metal oxides is one of the ways of controlling their electrochemical and catalytic properties. Usually, lithium substitution in oxides is achieved by solid state reactions between lithium salts and transition metal oxides at temperatures above 500°C. To the best of our knowledge there are no data on lithium doping of spinel cobalt-nickel oxides owing to their low temperature stability (up to 400°C). Having in mind the soft-chemistry "rules", we have tried a new route for the preparation of lithium-cobalt-nickel spinel oxides, and namely, thermal decomposition of mixed cobalt-nickel precursors in a lithium salt containing medium. In the present study we have investigated the thermal decomposition of low-dimensional cobalt-nickel hydroxide-nitrates and oxide-hydroxides in a lithium nitrate melt.

EXPERIMENTAL

To obtain an intimate mixture of Co and Ni hydroxide-nitrate and $LiNO_3$, solid Li_2CO_3 was added to a 75% solution of $Co(NO_3)_2.6H_2O$ and $Ni(NO_3)_2.6H_2O$ ($0 \leq Ni/(Ni+Co) \leq 0.4$) with intensive stirring, the Li/(Ni+Co) being 1.05/1. This mixture was evaporated to a dry residue at 80°C. The dry residue was heated at 150°C for 4 hours, then at 300°C for 2 hours, after which the sample was quickly cooled. To eliminate the lithium nitrate from the hydroxide nitrates thus obtained, the samples were washed with acetone.

CoOOH was obtained by oxidation with oxygen at room temperature of freshly precipitated β-Co(OH)$_2$.

RESULTS AND DISCUSSION

When solid Li$_2$CO$_3$ is added to a concentrated cobalt and nickel nitrate solution (0≤ Ni/(Ni+Co)≤0.4), mixed Co-Ni hydroxide-nitrates are precipitated which have the composition Ni$_x$Co$_{1-x}$(OH)(NO$_3$).H$_2$O and are isostructural to the "double-chain" Zn(OH)(NO$_3$).H$_2$O [1] and Ni(OH)(NO$_3$).H$_2$O [2] formed during very slow thermal hydrolysis (about 100 days) of the corresponding nitrate salts (Fig.1a). During heating up to 140°C, a gain in crystallinity is achieved as evidenced by the better resolution of the (300), (102), (402) and (202) diffraction lines (Fig.1b). The thermochemical behaviour of Ni$_x$Co$_{1-x}$(OH)(NO$_3$).H$_2$O displays very well the relationships between crystal structure and OH$^-$/NO$_3$ ratios. From 140 to 160°C, Ni$_x$Co$_{1-x}$(OH)(NO$_3$).H$_2$O with a double-chain structure hydrolyses to Ni$_x$Co$_{1-x}$(OH)$_{1+s}$(NO$_3$)$_{1-s}$ with a layered structure evolving gaseous HNO$_3$ and H$_2$O (Fig. 1c). However, in the LiNO$_3$ medium, where the water vapour evolution is hindered, a complete oxidative hydrolysis to an oxide-hydroxide, Ni$_x$Co$_{1-x}$OOH, occurs (Fig.2c). On further heating up to 300°C, both hydrolysing products (hydroxide-nitrate and oxide-hydroxide) decompose to cobalt-nickel spinel oxides (Fig.1d, Fig.2d), but in the LiNO$_3$ melt, the cobalt-nickel spinels thus formed contain lithium: Li$_y$(Ni$_x$Co$_{1-x}$)$_{3-y}$O$_4$, y≈0.13 (Fig. 3).

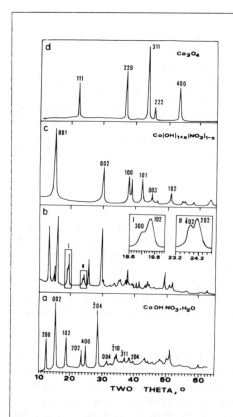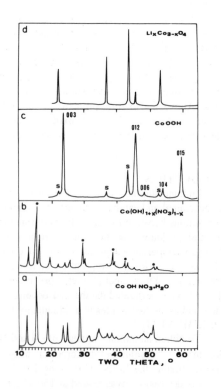

Fig. 1. XRD patterns of Co(OH)(NO$_3$).H$_2$O during heating at 100°C (a), 140°C (b), 155°C (c) and 200°C (d).

Fig. 2. XRD patterns of Co(OH)(NO$_3$).H$_2$O+ LiNO$_3$ during heating at 100°C (a), 140°C (b) 155°C (c) and 250°C (d).

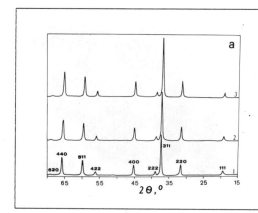

Fig. 3. XRD patterns of the decomposition products of co-precipitated cobalt-nickel hydroxide-nitrates in a $LiNO_3$ melt:
1 - Ni/(Ni+Co)=0; 2 - Ni/(Ni+Co)=0.2 and 3 - Ni/(Ni+Co)=0.4.

In order to elucidate the mechanism of lithium doping of nickel-cobalt spinels at low temperature, we have studied the thermal decomposition of CoOOH (obtained by oxidation of β-$Co(OH)_2$) in a $LiNO_3$ melt. As compared to pure CoOOH, the decomposition temperature of CoOOH in a $LiNO_3$ melt decreases by about 30°C and the shape of the DTG-curve changes (Fig.4), which gives evidence of interactions between the undecomposed CoOOH and the $LiNO_3$ melt. During the thermal decomposition, platelike particles of $Li_yCo_{3-y}O_4$ with a diameter of about 100 nm are formed and the shape of the spinel particles coincide with that of the mother-phase particles (Fig.5). Selected area electron diffraction patterns of an isolated grain of the spinel containing lithium show a single crystal lying on the (111) plane (Fig.5). The morphology of the spinel crystallites show that CoOOH transforms topotactically to a lithium-cobalt spinel in the $LiNO_3$ melt. In addition, the unit cell dimensions and intensity ratios of diffraction lines reveal that the lithium ions are statistically distributed over 8a and 16d spinel sites. Hence, it can be concluded that the formation of a mixed lithium-cobalt spinel by thermal decomposition of CoOOH in a $LiNO_3$ melt is determined by preceding ion-exchange reactions between Li^+ from the melt and H^+ from CoOOH. A proof for the formation of an intermediate "$H_{1-z}Li_zCoO_2$" phase before the

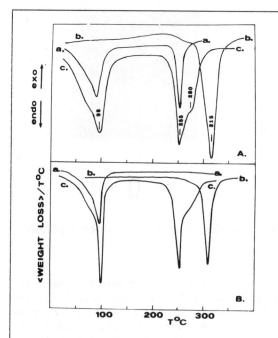

Fig. 4. DTA- and DTG-curves (A and B, respectively) for (a) $LiNO_3.3H_2O$, (b) CoOOH, and (c) a CoOOH-$LiNO_3.3H_2O$ mixture.

CoOOH decomposition is the fact that we have succeeded to prepare a single $Li_{1-x}H_zCoO_2$ phase (x<0.55, x+z<1) by acid delithiation of $LiCoO_2$ [4]. As the thermal decomposition of CoOOH also

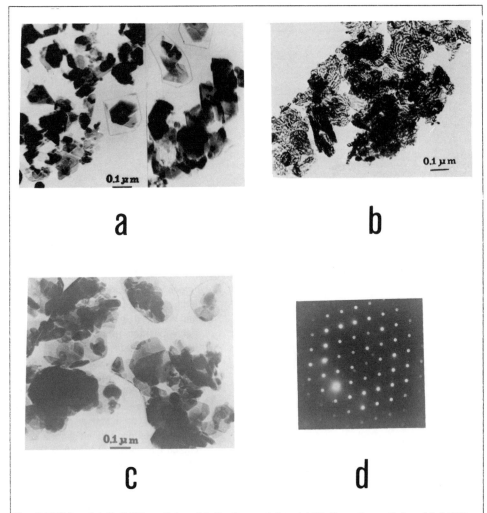

Fig. 5. TEM on (a) CoOOH particles; (b) Co_3O_4 particles; (c) $Li_yCo_{3-y}O_4$ particles; (e) SAED on a $Li_yCo_{3-y}O_4$ particles.

proceeds topotactically, it is noteworthy that the formation of the intermediate phase "H_1-$_zLi_zCoO_2$" does not perturb the transformation mechanism of CoOOH to a spinel in the $LiNO_3$ melt. A distinction between both processes can only be made according to the texture of the reaction products: the smaller degree of the lattice shrinkage during the transformation of CoOOH to $Li_yCo_{3-y}O_4$ than is the case of CoOOH to Co_3O_4 results in the formation of larger particles with a not well-pronounced porous texture.

REFERENCES

1) L. Eriksson, D. Louer and P.E. Werner, J. Solid State Chem., **81**, 9 (1989).

2) M. Louer, D. Louer, D. Grandjean, Acta Crystallogr. **B29**, 1707 (1973).

3) E. Zhecheva and R. Stoyanova, Mater. Res. Bull. **26**, 1315 (1991).

4) E. Zhecheva and R. Stoyanova, J. Solid State Chem., in press.

Materials Science Forum Vols. 152 - 153 (1994) pp. 327-330
© *1994 Trans Tech Publications, Switzerland*

PREPARATION OF SOLIDS CONSTITUTED OF NANOCRYSTALLITES BY REACTIONS IN MOLTEN SALTS

B. Durand [1,2], J.-P. Deloume [2] and M. Vrinat [3]

[1] Laboratoire de Matériaux Minéraux, URA 428, ENSCMu, 3 rue A. Werner,
F-68093 Mulhouse Cédex, France

[2] Laboratoire de Chimie Minérale 3, URA 116, Université Claude Bernard Lyon 1,
43 Boulevard du 11 Novembre, F-69622 Villeurbanne Cédex, France

[3] Institut de Recherches sur la Catalyse, 2 Avenue Einstein,
F-69626 Villeurbanne Cédex, France

Keywords: Nano-Sized Crystallites, Reactions in Molten Salts

ABSTRACT

Oxide powders, precipitated either by reactions of alkaline mixed oxides with molten divalent metal chlorides or by reactions of transition metals salts with molten alkali nitrates, are characterized by a satisfying degree of purity and a good chemical homogeneity. The latter reactions produce powders with nano-sized crystallites and large specific surface areas allowing uses in heterogenous catalysis and sintering.

INTRODUCTION

The use of molten salts as reaction medium promotes the formation of simple or mixed oxides, likely to find industrial applications owing to their characteristics.

It is well known, that molten salts may be used as flux. For instance, magnesium iron mixed oxide $MgFe_2O_4$ can be obtained at temperatures as low as 900°C by reaction of the oxides MgO and Fe_2O_3 in presence of molten Li_2SO_4-K_2SO_4 eutectic according to equation 1, whereas the same transformation in solid state requires a temperature overcoming 1200°C.

$$\lozenge\ MgO(s) + Fe_2O_3(s) \xrightarrow{Li_2SO_4\text{-}K_2SO_4(l)} MgFe_2O_4(s) \tag{1}$$

Yet, the versatility of the process is considerably improved when molten salts are directly involved in the chemical reaction. Two types of transformations are considered in the present communication :

* reactions between a mixed alkaline oxide and a molten divalent metal chloride as illustrated by equation 2 for the formation of the same mixed oxide $MgFe_2O_4$ which can be prepared at temperatures as low as 600°C.

$$\lozenge\ 2\ LiFeO_2(s) + K_2MgCl_4(l) \longrightarrow MgFe_2O_4(s) + 4\ (Li/K)Cl(l) \tag{2}$$

$\lozenge$ in the equations, s = solid, l = liquid, g = gas.

* reactions between a transition metal salt and a molten alkali nitrate as illustrated by equation 3 for the preparation of zirconia which can be obtained at temperature as low as 450°C

◊ $ZrOCl_2(s) + (K, Na)NO_3(l) \rightarrow ZrO_2(s) + NO_2(g) + 0.5\ Cl_2(g) + (K, Na)Cl(l)$ (3)

REACTION MECHANISM

Two mechanisms have to be considered, an exchange process or a double decomposition process.

* The exchange mechanism is restricted to :
– some reactions between a mixed alkaline oxide and a molten chloride performed at temperatures lower than 400°C. The substitution of the alkaline ion in solid state by a divalent metal ion leads to a metastable solid with vacancies, exhibiting a structure very close to the one of the starting material [1-3].
– reactions of zirconium oxichloride with nitrates carried out at temperatures lower than 400°C and leading to amorphous solids containing significant amounts in nitrates and chlorides [4].

* The double decomposition process concerns all the transformations performed above 400°C. It involves a partial dissolution of the starting solid in the molten salt according to equations 4 and 6 and the precipitation of the final product via equations 5 and 7 which shifts the dissolution equilibriums until complete transformation

$Li\ FeO_2 \rightarrow Li^+ + FeO_2^-$ (4)

$Mg^{2+} + 2\ FeO_2^- \rightarrow MgFe_2O_4$ (5)

$ZrOCl_2 \rightarrow ZrO^{2+} + 2\ Cl^-$ (6)

$ZrO^{2+} + O^{2-} \rightarrow ZrO_2$ (7)

For the precipitation of magnesium iron oxide, the anions FeO_2^- are generated by the dissociation of the molten salt via equation 8. For the precipitation of zirconia, the oxide ions O^{2-} are formed by oxido-reduction in the molten salt via equation 9.

$K_2MgCl_4 \rightarrow 2\ K^+ + 4\ Cl^- + Mg^{2+}$ (8)

$NO_3^- + Cl^- \rightarrow NO_2 + O^{2-} + 0.5\ Cl_2$ (9)

ARGUMENTS IN FAVOUR OF THE DISSOLUTION - PRECIPITATION PROCESS

* Analogy with solubility equilibriums of salts in water
The reactions of lithium iron oxide $LiFeO_2$ with a molten chloride containing two divalent metals $K_2M(II)_{1-y}\ M'(II)_y\ Cl_4$ lead to homogeneous solid solutions $M(II)_{1-x}\ M'(II)_x\ Fe_2O_4$. The subtitution rate in the solid is not necessary the same that the one in the molten salt [5].

* Formation of homogeneous solid solutions
The simultaneous reactivity of zirconium oxychloride and yttrium chloride towards molten sodium-potassium nitrates leads to tetragonal or cubic homogeneous solid solutions x mol % Y_2O_3-ZrO_2 (x in the range 0-10) [6].

* Influence of the stability of species able to dissolve in the molten medium.

The investigation of the reactivity of lithium aluminium oxide $LiAlO_2$ towards chlorides $K_2M(II)Cl_4$ [M(II) = Zn, Co, Mg, Ni] shows that with cobalt and zinc salts the aluminates $M(II)Al_2O_4$ are obtained whereas with magnesium salt, the reaction leads to the aluminate with a small proportion of magnesium oxide, and with nickel salt to a mixture of alumina and nickel oxide [7]. On the contrary, the reactions of lithium iron oxide and chromium iron oxide with the same chlorides always only lead to the corresponding ferrites $M(II)Fe_2O_4$ and chromites $M(II) Cr_2O_4$. It has to be concluded that the stability of FeO_2^- and CrO_2^- anions in molten chlorides is higher than the one of AlO_2^- anions. It is noticeable that in the reactions of $LiAlO_2$, the formation of the oxide MO is all the more significant as its solubility constant is lower.

* First results of infrared spectroscopy

In the transformations between alkali nitrates and transition metal sulphates, infrared spectroscopy performed on quenched baths shows a three step reaction. The second one is characterized by a lowering of the symmetry of sulphate ions indicating the formation of complex species which decomposition leads in the third step to the formation of the final oxide [8]. The dissolution of the starting material in the molten salt is then corroborated.

CHARACTERISTICS OF OXIDE POWDERS PRERARED IN MOLTEN SALTS.

Powders precipitated according to the double decomposition process from molten media are characterized by :
– a degree of purity satisfying for various applications. As example, such prepared manganese pyrophosphate can be used for magnetic calibration [9].
– a high chemical homogeneity
– a nano crystallite size with a narrow distribution, especially for powders elaborated in nitrate medium at 450°C
– a specific surface area exceeding 100 m^2g^{-1} linked to the crystallite size.
– an agglomeration state dependent upon the extraction conditions.

In the double decomposition process the nucleation and the growth of particles can be influenced by numerous factors :
– reaction parameters such as temperature, duration, heating profil, viscosity.
– precursor characteristics such as solubility or anion reactivity.
– molten salt characteristics such as melting point, acido-basic properties, complexing efficiency.
– external factors such as ultrasonication, stirring or bubbling.

CONCLUSION

A wide variety of divalent metals mixed oxides has been prepared successfully by reaction of mixed alkaline oxides towards divalent metals molten chlorides [10]. More recently, the reactivity of transition metals salts, chlorides or sulphates, has been used to prepare zirconia containing compounds, pure [4] and yttrium stabilized [6] zirconia, zirconia-alumina dispersions [11] and titanium dioxide [12]. On account of their morphological and textural characteristics, the latter oxides are attractive in heterogeneous catalysis for the preparation of supports [13-14]. Zirconia powders also lead to high density ceramic bodies by natural sintering [15].

Reactions with molten salts appear as a simple and easy reproducible preparation method of numerous simple or mixed oxides. The properties of the obtained solids can be modified acting on different parameters. The method can be extended to the preparation of other solids such as sulphides [16].

REFERENCES

1/   Durand, B., Pâris, J. and Pâris , R. : C.R. Acad. Sci. Paris, 1973, 276, 1557.
2/   Durand, B. and Pâris, J. : C.R. Acad. Sci. Paris, 1976, 282, 591.
3/   Durand, B. and Loiseleur, H. : C.R. Acad. Sci. Paris, 1979, 288, 197.
4/   Jebrouni, M., Durand, B. and Roubin, M.: Ann. Chim. Science des Matériaux, 1991, 16, 5569.
5/   Durand, B., Pâris, J. and Pâris, R. : C.R. Acad. Sci. Paris, 1975, 280, 1309.
6/   Jebrouni, M., Durand, B. and Roubin, M. : Ann. Chim. Science des Matériaux, 1992, 17, 143.
7/   Durand, B.and Pâris, J. : Ann. Chim. Fr., 1979, 4, 123.
8/   Deloume, J.P. : (unpublished results)
9/   Durand, B., Pâris, J. and Poix, P. : C.R. Acad. Sci. Paris, 1975, 281, 1007.
10/   Durand, B., Ceramic Powders : Preparation, Consolidation and sintering, Ed. P. Vincenzini, Elsevier Sci. Publ. Co., Amsterdam 1983, p. 413.
11/   Jebrouni,M., Durand, B. and Saïkali, Y. : An. Chim. Fr., (to be published)
12/   Harle, V., Deloume, J.P., Mosoni, L., Durand, B., Vrinat, M.and Breysse, M. : Europ. J. Solid State and Inorg. Chem. (to be published).
13/   Hamon, D., Vrinat, M., Breysse, M., Jebrouni, M., Durand, B., Roubin, M.and Magnoux, P. : Catalysis Today, 1991, 10, 613.
14/   Durand, B., de Mareuil, D., Vrinat, M. and des Courières, T.: Brevet Elf Aquitaine n°90 05 049.
15/   Descemond, M., Jebrouni, M., Durand, B., Roubin, M., Brodhag, C. and Thevenot, F. : J.mater., Sci., 1993, 28, 2283.
16/   Geantet, C., Kerridge, D.H., Descamp, T., Durand, B., Breysse, M. : Mater. Sci. Forum, 1991, 73-75, 693.

Materials Science Forum Vols. 152 - 153 (1994) pp. 331-334
© 1994 Trans Tech Publications, Switzerland

SOLUTION SYNTHESIS OF VANADIUM AND TITANIUM PHOSPHATES

A. Ennaciri[1] and P. Barboux[2]

[1] Département de Chimie, Université Cadi Ayyad-Semlalia, Marrakech, Morocco

[2] Chimie de la Matière Condensée, Université Pierre et Marie Curie,
4 Place Jussieu, F-75252 Paris , France

Keywords: Sol-Gels, Titanium Phosphates, Vanadium Phosphates, Solvent Effects

ABSTRACT:
      The reaction of vanadium oxo-alkoxides ($VO(OR)_3$) with phosphoric acid (H_3PO_4) yields $VOPO_4 \cdot nH_2O$ precipitates. The presence of water and the nature of the precursor as well as the solvent have a drastic effect on the crystallinity and the morphology of the resulting vanadium phosphate. In alcohol, a reversible slow condensation yields crystalline platelets whereas in tetrahydrofuran, submicronic grains are obtained in an irreversible fast reaction.
      Such aprotic solvents allow to adjust the precipitation kinetics of vanadium phosphates to those of titanium phosphates. Thus, homogeneous precipitation of mixed titanium-vanadium phosphates can be achieved in the $(TiOH)_x(VO)_{1-x}PO_4$ system. These amorphous metastable phases demix upon heating above 700°C into crystalline titanium phosphate and amorphous vanadium oxide.

I INTRODUCTION

      Vanadium phosphates have been used in recent years as intercalation materials [1] and catalysts for selective mild oxidation [2]. Vanadyl phosphate dihydrate is readily obtained by refluxing vanadium oxide in phosphoric acid for several days [3]. In previous works, we have demonstrated that the use of different precursors such as alkoxides allows the synthesis of the same material but having different morphologies depending on the nature of the precursor and the hydrolysis [4].
      Indeed, the reaction of phosphoric acid with vanadium alkoxides (in the parent alcohol) corresponds to a nucleophilic substitution at the V site:
$$-P-OH + RO-V -----> -P-O-V + ROH. \quad (1)$$
      The formation of solid vanadium phosphate must be considered as a polycondensation of V-O-P-O-V bonds through reaction (1).The reversibility of reaction (1) has been oberved by ^{51}V liquid NMR of dilute vanadium phosphate solution in alcohol. Chemical exchange broadening of the single peak was observed. The precipitation kinetics depend on the solvent used and on the starting alkoxide. The reaction of vanadium alkoxide with phosphoric acid in the parent alcohol is slow In the case of primary and secundary alcohols (R=propyl), water ($0 \leq H_2O/V \leq 10$) acts as an inhibitor of the reaction. There is an equilibrium between condensed species oxo (V-O-V) and phosphate (-V-O-P-O-V). The slow condensation process always leads to the crystallized single phase $VOPO_4, nH_2O$ ($0 \leq n \leq 2$), independant of the starting P/V ratio. However, the morphology of the crystals depends on the reaction mechanism (with or without hydrolysis) and on the reaction kinetics. Extended layers (width 50 μm) are obtained in anhydrous conditions whereas aggregates of small platelets showing a rosette morphology are obtained in the presence of water (diameter 2 μm).

Vanadium phosphates are slightly soluble in protic solvents such as water and alcohol in a process similar to the inverse of reaction (1). In the following, we will show that using an aprotic anhydrous solvent such as tetrahydrofuran (THF) allows a rapid precipitation of vanadium phosphate. Such rapid kinetics could be used to obtain metastable mixtures of titanium-vanadium phosphates.

II EFFECT OF SOLVENT ON THE PRECIPITATION OF VANADIUM PHOSPHATE

Reactions have been performed by mixing a solution of freshly distilled vanadyl isopropoxide ($VO(OR)_3$ with R=isopropyl) with a 1M solution of anhydrous H_3PO_4 (Fluka) in the appropriate solvent. Resulting precipitates have been filtered and dried in air in order to equilibrate with the ambient moisture and remove the residual alcohol groups adsorbed at the surface.

When isopropanol is used as the solvent, a long initiation time (hours) is observed before precipitation occurs and yields the well crystallized $VOPO_4.2H_2O$ phase after equilibration in air [4]. Upon heating at 700°C, the $\alpha_I VOPO_4$ phase is obtained. When using THF as the solvent, immediate precipitation occurs and a less crystalline phase is obtained. Upon heating at 700°C, the powder transforms to the β-$VOPO_4$ phase (with some α_{II}-$VOPO_4$ as minor phase [5]). As observed in Scanning Electron Microscopy, samples obtained from the reaction in alcohol have a platelet-like shape (Fig. 1a) whereas fine compact particles are obtained in THF (Fig. 1b). The specific area of these powders is 8 m^2/g for the reaction in isopropanol, to be compared to 40 m^2/g for the THF solvent and 1.8 m^2/g for the classical synthesis method of Ladwig [3]. It is easily understood that the use of THF better than the parent alcohol displaces reaction (1) towards the right. This has a drastic effect on the nucleation rate, yielding a larger number of particles, then smaller grains. The grain size and the amount of reduced vanadium may affect the crystal structure formed at 700°C.

Fig. 1: SEM of a $VOPO_4$ heat-treated at 700°C, a) isopropanol b) tetrahydrofuran

III MIXED VANADIUM-TITANIUM PHOSPHATES

In a similar way, we have tried to perform the coprecipitation of titanium and vanadium phosphates by adding phosphoric acid solutions (1M) to mixtures of vanadium isopropoxide and titanium n-propoxide solutions. Reactions were studied on the line $Ti_xV_{1-x}P_1$ of the phase diagram. We first recall that immediate precipitation is always observed when reacting titanium alkoxide with phosphoric acid. Amorphous materials are always obtained. The composition $Ti(HPO_4)_2.xH_2O$ is the most stable and the fastest to form since it yields a stable six-fold coordination of Ti by 6 phosphate groups[6]. In alcohols, the kinetics of the analogue of reaction (1) are slow. Coprecipitation experiments show the precipitation of titanium phosphate in the P/Ti=2 ratio leaving unreacted vanadyl alkoxide in the solution.

However, chemical analysis of the precipitates made from THF shows that total precipitation of V, Ti and P occured for all x. These precipitates, show, upon substitution of V by Ti, a progressive broadening of the X-ray diffraction peaks characteristic of $VOPO_4.2H_2O$ (Fig. 2). The mixture is amorphous for the ratios Ti/V above 1.

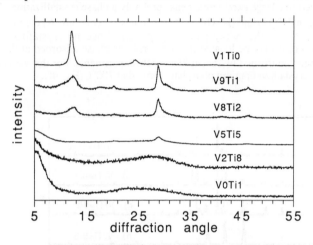

Figure 2: X-ray diffraction on V/Ti phosphates just precipitated (CuKα).

A similar trend is observed in the ^{31}P MAS-NMR spectra (Fig. 3). For x=0, a peak, characteristic of the $VOPO_4.2H_2O$ phase is observed at +5.8 ppm. Some other peaks around +1 ppm may correspond to crystalline defects in this phase observed when organics have not been completly removed during the condensation of the vanadium phosphate layers [4]. Upon substitution of V for Ti, a broad peak grows around -8 ppm. It corresponds to an amorphous phase. At the ratio V/Ti=5/5, the peak characteristic of the $VOPO_4.2$ H_2O phase is still present but, as shown by its weak intensity, it only corresponds to a minor phase. The origin of the broad peak at -8 ppm is still not well understood. Amorphous titanium phosphates exhibit a broad peak at -10 ppm (for the ratio P/Ti=1) and a sharp peak at -21 ppm (for the ratio P/Ti=2) [6]. If phosphate groups were bound to titanium only in the $V_{0.5}Ti_{0.5}PO_4$ mixture they would be found in the ratio P/Ti=2 and correspond to a different chemical shift. We then conclude that the distribution of vanadium and titanium around the phosphate groups must be homogeneous.

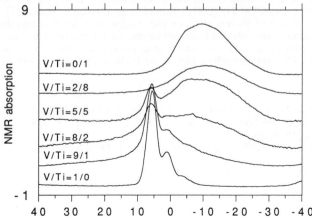

Fig. 3: ^{31}P MAS-NMR spectra of mixed $V_{10-x}Ti_x$ phosphates just precipitated (ref. 0 ppm= H_3PO_4)

After heating at 700°C, the titanium-free sample yields the β-VOPO$_4$ phase as shown in Fig. 4a. For the titanium-vanadium mixtures, amorphous materials are obtained after heating at 300°C, the end temperature for calcination of the residual organics and dehydration (around 20% weight loss). Progressive phase partitioning occurs between 500°C and 700°C with an acceleration between 650°C and 700°C associated to a large exothermic peak (probably a phase crystallization). This phase seggregation occurs between TiP$_2$O$_7$ and β-VOPO$_4$ for the lowest Ti contents. At the ratio V/Ti=5/5, a pure TiP$_2$O$_7$ phase appears in the X-ray pattern (Fig. 4d). Since no crystallized vanadium containing phase is observed, it may be located in an amorphous phase adsorbed at the surface of TiP$_2$O$_7$ grains. This has been confirmed by recent ^{51}V NMR experiments. For the V=0 content, we have not yet identified the crystalline titanium phosphate formed at 700°C(Fig. 4f).

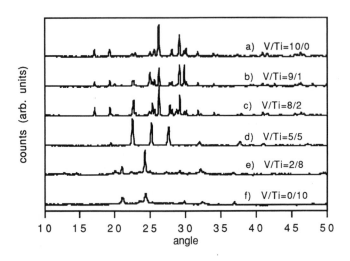

Fig. 4: X-ray diffraction on P/Ti phosphates heat-treated at 700°C under oxygen (CuKα)

CONCLUSION

As a conclusion, we have shown that powders with different morphologies can be obtained by changing the nature of the solvent and of the metal precursor. Changing the functionality of phosphoric acid by using esters (PO(OH)$_x$(OR)$_{3-x}$)would be an alternative method [7]

The appropriate choice of solvent allows rapid condensation kinetics and the synthesis of homogeneous multi-element materials that could not be obtained from classical reactions of salts in aqueous solutions or from solid state reactions. These amorphous mixtures are metastable and seggregate above 500°C. They are currentlty under testing for their catalytic properties.

REFERENCES

1) J.W Johnson, A.J. Jacobson, J.F. Brody, S.M. Rich, Inorg. Chem., 1982, 21, 3820.
2) N. Wustneck, H. wolf, H. Seeboth, React. Kinet. Catal. Lett., 1982, 21, 479.
3) G. Ladwig, Z. Chem, 1979, 19 , 368
4) S.A. Ennaciri, C. R'Kha, P. Barboux and J. Livage, Eur. J. Solid State Inorg. Chem., 1993, 30, 227.
5) B. Jordan and C. Calvo, Can. J. Chem., 1973, 51, 2621.
6) J. Livage, P. Barboux, M.T. Vandenborre, C. Schmutz and F. Taulelle, J. Non Cryst. Solids, 1992, 147&148, 18.
7) C. Schmutz, E. Basset and P. Barboux, J. Mat. Chem. , 1993, 3, 757.

Materials Science Forum Vols. 152 - 153 (1994) pp. 335-338
© 1994 Trans Tech Publications, Switzerland

SYNTHESIS AND STRUCTURE OF CALCIUM ALUMINATE HYDRATES INTERCALATED BY AROMATIC SULFONATES

V. Fernon[1,2], A. Vichot[2], P. Colombet[2], H. Van Damme[1] and F. Béguin[1]

[1] CRMD, 1B rue de la Férollerie, F-45071 Orléans Cédex, France

[2] TECHNODES SA, Groupe Ciments Français, BPO1, F-78931 Guerville Cédex, France

Keywords: Arenesulfonates, Poly(alkyl-aryl)sulfonates, Calcium Aluminate Hydrate, Lamellar Double Hydroxides, Coprecipitation, Organoceramics

ABSTRACT :

Calcium-aluminate hydrates intercalated by aromatic sulfonates are new phases resulting from the precipitation of calcium aluminate hydrates in the presence of arenesulfonate or poly(alkyl-aryl)sulfonate anions. These new lamellar compounds are structurally similar to the layered double hydroxides (LDH's) obtained from anionic exchange. The organization of the lamellar structure appears as being strongly influenced by the molecular weight of the intercalated aromatic sulfonates.

INTRODUCTION :

Tetracalcium aluminate hydrate C_4AH_x (C=CaO, A=Al$_2$O$_3$ and H=H$_2$O) is a layered double hydroxide (LDH) [1], i.e an antitype of clay minerals. It consists of positively charged hexagonal layers of composition $[Ca_2Al(OH)_6]^+$ separated by OH$^-$ anions and water molecules (figure 1a). The general composition of C_4AH_x may be formulated as : $[Ca_2Al(OH)_6]^+$, OH$^-$, nH_2O where n=6, 3 or 2 depending on the relative humidity [2]. As commonly observed for other LDH's, C_4AH_x readily undergoes inorganic anion-exchange, such as CO_3^{2-} and SO_4^{2-} for OH$^-$. It also reacts with various organic molecules and anions such as amines, alkyl-chain alcohols, carboxylic acids [1,3], arene or secondary alkane-sulfonates [1], forming pillared-LDHs. In the case of (poly(vinylalcohol)) as the guest species, intercalation compounds have recently been synthesized by precipitation from ionic solutions containing the polymer as co-solute [4].

In this paper, we report the synthesis and the structure analysis of new calcium-aluminate hydrate LDHs containing arenesulfonates or poly(alkyl-aryl)sulfonates.

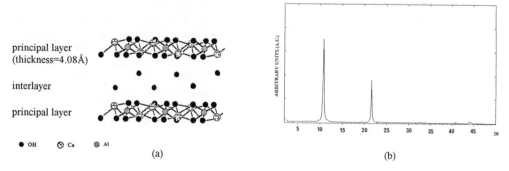

principal layer
(thickness=4.08Å)

interlayer

principal layer

● OH   ⊘ Ca   ⊙ Al

(a)

ARBITRARY UNITS (A.U.)

(b)

Figure 1 : (a) view along (110) of the C_4AH_x structure where the water molecules are not shown and (b) X-Ray diffractogram of as-prepared $C_4A\overline{C}_{0.5}H_x$

EXPERIMENTAL :
     Pristine calcium-aluminate hydrate was prepared by reacting a small excess of a saturated $Ca(OH)_2$ solution with an aqueous solution of $NaAlO_2$. Although much care was paid to avoid CO_2 contamination, carbonatation could not be avoided, yielding $C_4A\overline{C}_{0.5}H_x$ (figure 1b, $\overline{C}=CO_2$) with carbonate instead of hydroxyl groups as interlamellar anions [2]. The poly(alkyl-aryl) or arene-sulfonate C_4AH_x derivatives were synthesized following the same procedure. Sodium arenesulfonates (β-naphtalenesulfonate (2NSA) (Fluka); 1,5-naphtalenedisulfonate (1,5NDS) (Aldrich) and 1,3,6,8-pyrenetetrasulfonate (TSP) (Kodak)), and sodium poly(alkyl-aryl)sulfonates (polystyrenesulfonate with various molecular weights : 5400 Daltons, 35000 Daltons (PSS5400 and PSS35000) (Interchim) and polydispersed product (PSSpoly) (Aldrich)) were first dissolved in the $Ca(OH)_2$ saturated solution. The synthesis was performed using a continuous batch reactor with a sufficient residence time (one hour) to allow the equilibrium to be reached. After precipitation, the suspension was filtered. The solid phase was dried under vacuum at 80-100°C and stored in a CO_2, H_2O-free atmosphere.

     The analytical techniques employed for the characterization of the organo-mineral compounds were Elemental Analysis, X-Ray Diffraction (λ=1.5405 Å, Siemens D-500 goniometer), Infra-Red Spectrometry (1% dispersion in KBr pellets, Nicolet 710), Scanning Electron Microscopy (SEM) and Transmission Electron Microscopy (TEM).

RESULTS AND DISCUSSION :
     The elemental compositions of the organo-mineral compounds are reported in Table 1.

sulfonate	% C	% S	% H	%Ca	%Al	Ca/Al molar ratio	Ca/-SO_3^- molar ratio
2NSA	25.64	6.64	3.86	16.42	5.73	1.93	1.97
1,5NDS	12.51	6.01	2.39	21.18	8.10	1.76	2.81
TSP	14.49	8.93	3.08	18.94	6.14	2.08	1.69
PSS5400	22.48	3.59	4.67	14.86	4.87	2.06	3.30
PSS35000	26.44	7.33	4.29	14.71	4.55	2.18	1.60
PSSpoly	24.06	7.53	4.46	15.04	4.81	2.11	1.60

Table 1 : Elemental composition of the different organomineral compounds synthesized in this work.

     In all cases the Ca/Al ratio is equal to about 2 suggesting the existence of $Ca_2Al(OH)_6^+$ units. The Ca/interlayer anions molar ratios are sometimes higher and sometimes lower than for C_4AH_x. Ratios lower than 2 could be due to cointercalated OH^- or CO_3^{2-}.
     The SEM observations of the powders revealed a plate-like morphology for the arenesulfonate compounds as for $C_4A\overline{C}_{0.5}H_n$, with an arrangement in spherical or pseudo-spherical particles (Ø =20 to 50 μm) giving a "rose-des sables" aspect. These individual spherical particles appeared to be composed of lamellae emanating from a central core (figure 2a) as observed previously for poly(vinylalcohol) materials [4]. The lamellae are not apparent with the polystyrenesulfonate adducts (figure 2b,2c). The layered structure of the materials was confirmed by the XRD patterns showing a set of 001 lines (figure 3). The interlayer spacing is shifted from 8.22 Å in $C_4A\overline{C}_{0.5}H_x$ to 13.57-19.55 Å in the organic sulfonate containing compounds (table 2).

     hk0 lines at 4.9 Å, 2.87 Å and 2.49 Å already existing for C_4AH_x, were also detected. They are characteristic of the normal separation between Ca and Al atoms in positive $[Ca_2Al(OH)_6]^+$ layers showing that the layer parameters are not affected by the nature of the intercalated anions.

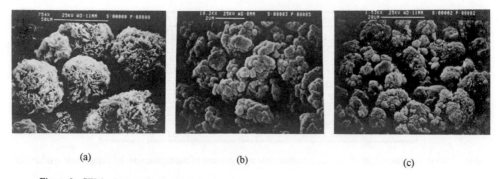

(a)   (b)   (c)

Figure 2 : SEM micrographs on the C_4AH_x with   (a) simple arenesulfonate , (b) low molecular weight polystyrenesulfonate and (c) polydispersed polystyrenesulfonate

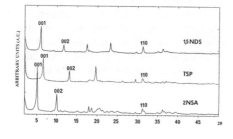

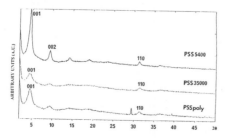

Figure 3 : X-Ray diffractograms of the C_4AH_x with (I) arenesulfonates, (II) polystyrenesulfonates.

The line width which may be related to the degree of crystallinity of the samples was observed to be dependent strongly on the molecular weight : a decrease of the degree of the crystallinity was observed in the PSS derivatives (figure 3 II) as the molecular weight increased. With small size anions the interlayer spacing was found to be close to the theoretical value calculated assuming a $[Ca_2Al(OH)_6]^+$ layer and an intercalated sulfonate anion with its mean plane parallel to the c axis (figure 4). In the case of 2NSA, a double anionic layer may be assumed as to allow the compensation of the positive charges of the C_4AH_x as suggested by Meyn [1]. The lamellar aspect of the organomineral compounds was also confirmed by Transmission Electron Microscopy showing well defined 002 lattice fringes.

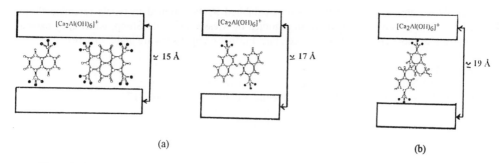

(a)   (b)

Figure 4 : Structural model proposed for the C_4AH_x with (a) arenesulfonates and (b) polystyrenesulfonate.

compounds	d in Å
C_4AH_{13}	7.94 [2]
$C_4A\overline{C}_{0.5}H_{12}$	8.19 [2]
$C_4A\overline{C}_{0.5}H_x$	8.22
2NSA	17.66
1,5NDS	14.97
TSP	13.57
PSS5400	19.29
PSS35000	19.55
PSSpolydispersed	19.31

Table 2 : Interlayer spacing (d) of the calcium aluminate hydrates and of the organomineral compounds synthesized.

wavenumber in cm-1	interpretation
300-420	Ca-O vibration (a)
500-680	Al-O vibration in AlO_6^- (a)
780	Al-O vibration in AlO_6^- (a)
1042 and 1184	S-O stretching in SO_3^- (b)
3200-3600	OH stretching of interlayer water (a) and (b)
>3615	OH stretching (a)

Typical FT-IR spectra are shown in figure 5 and are compared to that corresponding to $C_4A\overline{C}_{0.5}H_x$. Lines at 1415-1430 cm^{-1} clearly correspond to carbonate groups. In the case of the sulfonate adducts, lines characteristic of the $Ca_2Al(OH)_6^+$ layers [5] along with lines characteristic of the sulfonate anions may be observed (Table 3).

Table 3 : Main bands observed in the product obtained by coprecipitation with sulfonate anions. (a) and (b) refer to $Ca_2Al(OH)_6^+$ layers and to sulfonate anion respectively.

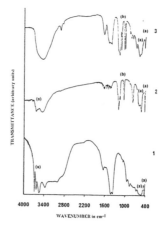

Figure 5 : IRTF spectra of (1) carbonated C_4AH_x, (2) compounds with arenesulfonate and (3) compounds with polystyrenesulfonate composites. (a) and (b) refer to $Ca_2Al(OH)_6^+$ layers and to sulfonate anion respectively.

As shown above, many of the features exhibited by the studied materials resemble those exhibited by the poly(vinylalcohol) derivative studied by Messersmith and Stupp, which they named nanocomposite or more specifically organoceramic. Further comparisons are under progress.

REFERENCES :
1) MEYN,M.; BENEKE, K.; LAGALY, G.: Inorg. Chem.,1990, 29,5201.
2) TAYLOR,H.F.W. : Cement chemistry, Academic: London, 1990, pp167-173.
3) DOSCH,W.:Neues Jahrb. Mineral. Abhandl.,1967,106,200.
4) MESSERSMITH,P.B.;STUPP,S.I.: J. Mater. Res.,1992,7,2599.
5) HOUTEPEN,C.M.J.;STEIN,H.N.:Spectrochim. Acta,1976,32A,1409.

Materials Science Forum Vols. 152 - 153 (1994) pp. 339-342
© 1994 Trans Tech Publications, Switzerland

PREPARATION AND MORPHOLOGICAL CHARACTERIZATION OF FINE, SPHERICAL, MONODISPERSE PARTICLES OF ZnO

D. Jezequel, J. Guenot, N. Jouini and F. Fievet

Laboratoire de Chimie des Matériaux Divisés et Catalyse, Université Paris 7 - Denis Diderot, 2 place Jussieu, F-75251 Paris Cédex 05, France

Keywords: Zinc Oxide, Varistors, Monodisperse Particles, Spherical Particles, Preparation Method, Polyols

ABSTRACT

A very simple method is proposed to prepare powdery ZnO with monodisperse, spherical, non porous particles in the submicrometre range which are suitable for varistors fabrication. Zinc acetate dihydrate is dissolved in diethyleneglycol. After heating up to 180°C ZnO precipitation occurs. The ZnO particles obtained are spherical, monodisperse and in the submicrometre range but porous. After heating up to 500°C they remain non agglomerated and become almost non porous, so they are suitable for varistors fabrication.

INTRODUCTION

One important application of zinc oxide is the fabrication of varistors [1]. In order to obtain the highly non-linear voltage-current varistor characteristics powdered ZnO is doped with a number of additives such as Bi_2O_3, Sb_2O_3, CoO, Cr_2O_3 and MnO before sintering or hot pressing. Classically, bulk ZnO is powdered by ball milling but in order to obtain an uniform sintered microstructure and, consequently, reliable and stable characteristics it would be better to use monodisperse, equiaxed, non porous particles [2]. So, a lot of methods have been proposed for ZnO powder synthesis with particles of well-defined shape : alkoxide hydrolysis [2], spray pyrolysis [3], evaporative decomposition of solutions [4], bis(acetylacetonato)zinc(II) hydrolysis [5], metal vapour oxidation [6], controlled precipitation by heating a zinc hydroxysalt solution with an ammonia-alcohol-water mixture as solvent [7], nitrate mixture heating [8], pyrolytic process [9], ageing of dilute solutions in presence of urea [10,11]. All these methods are rather complex to achieve. In the present paper we present a very simple method to obtain monodisperse, spherical, almost non porous ZnO particles suitable for varistors fabrication.

1. METHOD PRINCIPLE

Zinc oxide films have been prepared by spray pyrolysis method (12) and atomic layer epitaxy method (13) by allowing zinc acetate to react with water vapour. So, if Zn acetate and water are dissolved in a solvent with a sufficiently high boiling point it should be hydrolysed with ZnO production. Moreover, in the solvent the particles are generally formed by nucleation, growth and aggregation and it should be possible, by varying the reaction conditions and, consequently, the relative importance of these three factors, to act on particle morphology.

2. PRELIMINARY EXPERIMENTS

Polyols are good candidates as solvents : because of their fairly high dielectric constant they can easily dissolve inorganic compounds and, because of their relatively highly boiling point, reactions at temperature higher than 100°C are made possible. Furthermore, different polyols, pure or mixed, can be used to obtain a continuous variation of solvent characteristics.

By dissolving zinc acetate dihydrate in ethyleneglycol (EG) (which introduces both salt and water) and heating up to 180°C pure ZnO is not obtained whereas it is in diethyleneglycol (DEG) so the latter solvent was used to investigate the influence of other parameters : maximum temperature, heating rate, salt and water concentration.

Typically, *ca.* 0,1 mole of zinc acetate dihydrate is dissolved in one litre of DEG and after heating up to 180°C under stirring, in a pyrex flask fitted with a reflux column, ZnO is obtained. After cooling at room temperature, the solid phase is separated by centrifugation and washed with alcohol. After examination, the ZnO powder appears as formed of spherical, monodisperse, porous particles in the range size 0.2-0.4 µm depending on heating rate.

3. CHARACTERIZATION TECHNIQUES

Crystalline phases were determined by powder X-ray diffraction. Crystallite sizes were calculated from X-ray line broadening using the Scherrer formula. Particle size and morphology was obtained by scanning electron microscopy (SEM) and transmission electron microscopy (TEM). Particle mean sizes and distribution standard deviations were deduced from measuring 200 particles on a TEM micrograph. Powder specific surface area was measured by nitrogen adsorption (BET method).

4. REACTION PARAMETERS

4.1 WATER CONCENTRATION

Starting from a solution of dehydrated zinc acetate in dehydrated DEG, ZnO is not formed. So the hydrolysis mechanism is confirmed. This was not explicitly noted by Collins and Taylor who made some investigations of oxides precipitation in different alcohols and EG [14].

In order to examine the effect of water amount, the hydrolysis molar ratio defined as $h = H_2O/Zn$ was varied in the 0-40 range. The best results were obtained with $h = 2$ which corresponds to the commercial salt $Zn(CH_3COO)_2, 2H_2O$. For $h<2$ or $h>2$, yield is lower and the particles are smaller and polydisperse with irregular shape. Precipitation of ZnO also occurs if acid acetic is added instead of water because water is produced *in situ* by esterification between polyol and acetic acid. However, in this case particle size increases with AcH/Zn ratio.

4.2 NATURE OF THE SOLVENT

Besides DEG and EG, other polyols have been tested under otherwise identical conditions, particularly tetraethyleneglycol (TTEG), polyethyleneglycol (M = 300) and glycerol. Precipitation of ZnO does not occur at all in glycerol. The other polyols lead to ZnO formation but with worse morphology (irregular and polydisperse particles).

4.3 SALT CONCENTRATION

When the salt concentration is lower than 0.18 mol.L^{-1} spherical monodisperse particles of submicrometre size are obtained. In the range 0.18 - 0.25 mol.L^{-1} approximately, the particles become irregular and very large. So, the final product is not satisfactory.

Above the latter concentration no aggregation occurs and tiny non spherical particles are obtained.

4.4 MAXIMUM REACTION TEMPERATURE AND HEATING RATE

The maximum reaction temperature was varied from 100°C to 220°C (at this temperature DEG thermal degradation begins to be important). In this temperature range particle shape and size remain almost unchanged but to obtain a yield of the order of 90 % a more or less long time is required after maximum temperature is attained (12 h at 100°C).

On the contrary, particle size greatly depends on the heating rate : the size is reduced by a rate increasing. For example, for a 6°C.min^{-1} heating rate the mean particle size is 0.35 μm and for a 14°C.min^{-1} it decreases to 0.20 μm - figure 1. So, heating rate is an interesting parameter for particle tailoring.

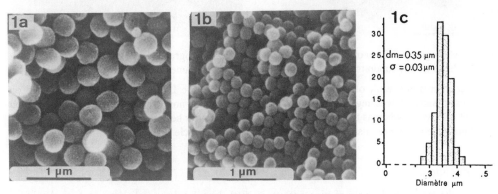

Fig.1 - Micrographs (SEM) of ZnO particles obtained from Zn(CH$_3$CO$_2$)$_2$,2H$_2$O in DEG at 180°C. a) Heating rate : 6°C.min^{-1} b) Heating rate : 14°C.min^{-1} c) Size distribution of sample a.

5. FINAL PRODUCT CHARACTERISTICS

The characteristics of the particles obtained in the best conditions are described hereafter.

SEM shows the ZnO particles to be non agglomerated and nearly spherical with a mean size lying between 0.20 and 0.35 μm and a narrow size distribution - figure 1. By X-ray diffraction they appear to be well crystallised. Particle porosity is evidenced by TEM imaging - figure 2. It can also be seen that the particles are made of very small subunits (mean size ≅ 8 nm) which indicates the importance of aggregation phenomenon (besides nucleation and growth) in the formation of the final particles. This phenomenon is clearly differentiated of the others because, as the reaction proceeds, one can observe that the solution becomes opalescent before precipitation. This aggregation no longer takes place beyond a salt concentration of 0.25 mol.L^{-1}.

The porous nature of the particles is reflected in the relatively high specific surface area as determined by BET method (60 m^2.g^{-1}). X-ray diffraction broadening gives a crystallite size of 10 nm near the subunit size.

6. ZnO THERMAL TREATMENT

Porous particles are not ideally suitable to obtain dense ceramics by sintering. Consequently, with the aim of reducing particle porosity, powdery ZnO samples were heated under air flow at 500 and 730°C during a few hours. It is noteworthy to notice that the particles remain non agglomerated whereas their porosity is drastically reduced : crystallite size increases to reach 20 nm and specific surface area is reduced to 5 m^2.g^{-1} - figure 3. Thus, after heating we have succeeded in obtaining almost non porous, monodisperse, non agglomerated particles.

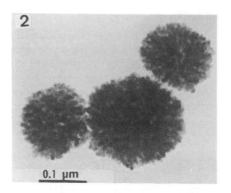

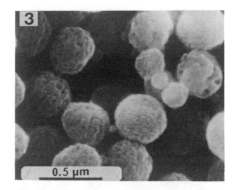

Fig.2 - Micrograph (TEM) of porous ZnO parti-   Fig.3 - Micrograph (SEM) of non porous ZnO
cles made up of tiny subunits.   particles obtained after heating (500°C - 21.30 h).

Another interesting result obtained by heating in air is that the DEG which remains adsorbed on particles in spite of washing is eliminated. Carbon analysis of non heated samples gives a carbon amount of 5 wt. % whereas this amount becomes negligible after heating.

Finally, temperature raising up to 730°C results in interparticle sintering.

CONCLUSION.

By the simple hydrolysis of zinc acetate in diethyleneglycol monodisperse, spherical, almost porous ZnO particles lying in the submicrometre range are obtained. By heating, the particles become almost non porous and are suitable for varistors fabrication.

REFERENCES

1) Matsuoka, M. : Ceram. Trans., 1989, 3 (Adv. Varistor Technol.), 3.
2) Heistand, R.H., II; Chia Y.-H. : Mater. Res. Soc. Symp. Proc., 1986, 73 (Better Ceram. Chem. 2), 93.
3) Liu, T.-Q.; Sakurai, O.; Mizutani, N.; Kato M. : J. Mater. Sci., 1986, 21, 3698.
4) Sproson, D.W.; Messing, G.L.; Gardner, T.J. : Ceram. Int., 1986, 12, 3.
5) Kamata, K.; Hosono, H.; Maeda, Y.; Miyokawa, K. : Chem. Lett., 1984, 2021.
6) Suyama, Y. : Mater. Sci. Monogr., 1987, 38B (High Tech Ceramics), 1175.
7) Haile, S.M.; Johnson, D.W., Jr.; Wiseman, G.H.; Bowen, H.K. : J. Am. Ceram. Soc., 1989, 72, 2004.
8) Louer, D. : Patent FR 2584389, 9 Jan 1989.
9) Kuntz, M.; Bauer G.; Grobelsez, I. : Patent PCT Int. Appl. WO 90 14307, 29 Nov 1990.
10) Castellano, M.; Matijevic, E. : Chem. Mater., 1989, 1, 78.
11) Tschuchida, T.; Kitajima, S. : J. Mater. Sci., 1992, 27, 2713.
12) Eberspacher, C.; Fahrenbruch, A.L.; Bube, R.H. : Thin Solid Films, 1986, 136, 1.
13) Tammenmaa, M.; Koskinen, T.; Hiltunen, L.; Niinistö, L.; Leskelä, M. : Thin Solid Films, 1985, 124, 125.
14) Collins, I.R.; Taylor, S.E. : J. Mater. Chem., 1992, 2, 1277.

Materials Science Forum Vols. 152 - 153 (1994) pp. 343-346

INTERCALATION OF ORGANIC PILLARS IN [Zn-Al] AND [Zn-Cr] LAYERED DOUBLE HYDROXIDES

M. Guenane, C. Forano and J.P. Besse

Laboratoire de Physico-Chimie des Matériaux, CNRS URA 444, Université Blaise Pascal, F-63177 Aubière Cédex, France

Keywords: Pillared Layered Structures, Layered Double Hydroxides, Organic Anions Intercalation

ABSTRACT.

The coprecipitation method is used to prepare a serie of LDH containing organic anions. The importance of the synthesis parameters on the nature of the obtained phases are presented. Large expansions of the LDH interlayer is observed depending on the charge and the stereochemistry of the organic molecules. Structural evolution of some of these hybrid organic-inorganic sandwiches under thermal treatment reveal an unusual stability.

INTRODUCTION

The Layered Double Hydroxides (LDH), also named Synthetic Anionic Clays are lamellar compounds (figure 1) under intense investigations, since few years. Their easily tunable chemical composition, based on the following formula:
$[M^{2+}_{1-x}M^{3+}_x(OH)_2]^{x+}[A^{m-}_{x/m},nH_2O]^{x-}$ also abbreviated as $_x[M^{2+}-M^{3+}-A]$, their wide exchange and intercalation properties make them interesting materials for catalysis, adsorption process, electrochemistry or conductivity applications.

The intercalation of organic anions in LDH is a very promising field. The obtention of new microporous materials might be obtained by the preparation of LDH with higly charged, long and stable organic pillars. Moreover, such materials should constitute nanoreactors for supported catalysis reactions. We present here an overview study of the preparation of LDH containing organic anions, some of them presenting an interesting pillaring effect which stabilize the hybrid organic-inorganic lamellar structure.

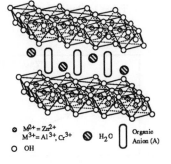

$M^{2+} = Zn^{2+}$
$M^{3+} = Al^{3+}, Cr^{3+}$ H₂O   Organic Anion (A)
OH

Figure 1: Structural presentation of Layered Double Hydroxides.

ELABORATION OF LDH WITH ORGANIC PILLARS

Organic anions-containing LDH have been mainly prepared by anionic exchange reactions [1,2]. We have developped the coprecipitation method which allows to intercalate a wider variety of organic molecules. The accurate control of the synthesis parameters allows to obtain pure and well crystallized pillared organic LDH. A systematic screening of the synthesis conditions must be developped for each particular anion, in order to determine the corresponding optimal pH of coprecipitation. This optimal pH of coprecipitation is defined as the pH at which the LDH phase is supposed to be obtained pure and best crystallized.

The nature of the (M^{2+}, M^{3+}) system has also a great effect on the conditions of the optimized preparation. For the (Zn-Al) system, the LDH phases are usually obtained for a pH range comprised between 7.0 and 11.0, while pure [Zn-Cr] LDH must be prepared at a pH=5.0-6.0. But this is not always the case and it seems overall that no simple relation between all theses parameters can be drawn.

The figure 2 and the following table present our results for either [Zn-Al] or [Zn-Cr] LDH containing different organic anions.

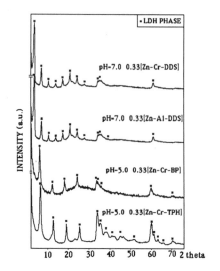

Figure 2: PXRD spectra of four coprecipitated organic anion - LDH.

pH opt. cop.	BENZ	TPH	OPH	BP	DDS
0.33[Zn-Cr]	----	5.0	5.0	5.0	7.0
0.33[Zn-Al]	10.0	8.0	8.0	7.0	7.0

BENZ: Benzoate; TPH: Terephtalate; OPH: Orthophtalate; BP: Butyl -1,4 diphosphonate; DDS: Dodecylsulfate.

Table 1: Optimal coprecipitation pH for a series of organic anions containing LDH.

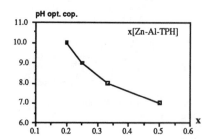

Figure 3: Optimal pH of coprecipitation versus x for x[Zn-Al-TPH] LDH.

The optimal pH of coprecipitation depends also on the trivalent to divalent metals ratio, x. In the case of [Zn-Al-TPH], the increase of Al content (x) leads to a decrease of this pH value (figure 3).

This route of preparation appears a convenient alternative when direct anionic exchange reactions are unsuccessful. Both synthesis methods allow to intercalate negatively charged organic molecules of very different nature.

STRUCTURE OF EXPANDED ORGANIC PILLARED LDH

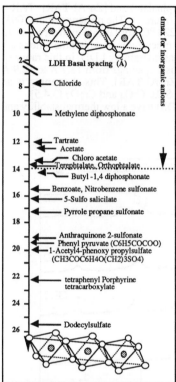

Figure 4: Basal spacings scale for a series of expanded LDH.

Long functionalized hydrocarbon chains with anionic groups can be easily intercalated in LDH and can led to basal spacings, obtained from the d_{001} distances on the diffraction patterns, higher than those encountered for inorganic anions containing LDH (figure 4). They lie with the anionic functions hydrogen bonded to the hydroxilated sheets.

The intercalation of (ω,ω') alkyl dicarboxylates, $COO(CH_2)_nCOO$, leads to a perpendicular intercalation of the anions according a simple relation, d (Å) = 10.10 + 1.02n (for n≥2 and even) (figure 5). The oxalate anion (n=0), with a shorter basal spacing, is oriented slantwise .

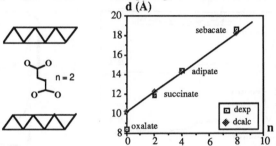

Figure 5: Structural model and basal spacing variation of [Zn-Cr] LDH containing (ω,ω') alkyl dicarboxylates.

[Zn-Al] LDH containing Porphyrin tetraphenyl tetracarboxylate has successfully been prepared by the coprecipitation method. Interestingly, the orientation of the macrocyclic polyanion (figure 6), strongly bonded onto the hydroxylated layers, leaves the complexation site free for metallation. Provided, the remaining LDH interlayer space is accessible for external molecules, reactions such as stereospecific oxydation should take place.

Monoanionic molecules, as dodecylsulfate, do not pillar the LDH as strongly as polyanions. The interlayer ordering corresponds, usely, to an overlapping of two layers of organic anions, giving rise to variable d values (figure 6). In the case of alkyl chains bearing anionic functions, the interlamellar domain is constituted of an hydrophobic region, with possible expansion, bordered by two hydrophilic interfaced regions.

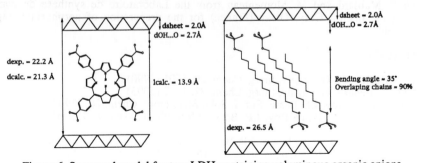

Figure 6: Structural model for two LDH containing voluminous organic anions.

PILLARING UNDER CALCINATION

The thermal evolutions for the calcined [Zn-Al-TPH] (TPH = terephtalate) and [Zn-Al-TAR] (TAR = tartrate) phases are shown on the powder X rays diffractograms (figure 7). The presence of the (003) diffraction line up to 350 °C for the terephtalate compound and 220 °C for tartrate evidence that an hybrid organic-inorganic lamellar structure is retained at these temperatures of which the anions are evolved. This behavior has not been observed for inorganic anions-containing LDH, in this range of temperature. The calcination also involves a contraction of the basal spacings from 13.7 Å to 12.9 Å and from 12.1 Å to 9.1 Å for respectively [Zn-Al-TPH] and [Zn-Al-TAR]. This shortening is an evidence of a pillaring effect, already shown for LDH containing SO_4, CrO_4 and Cr_2O_7 [3,4,5].
On the contrary, LDH with monoanion such as dodecylsulfate (DDS) have a low thermal stability and display a structure collapse near 100°C.

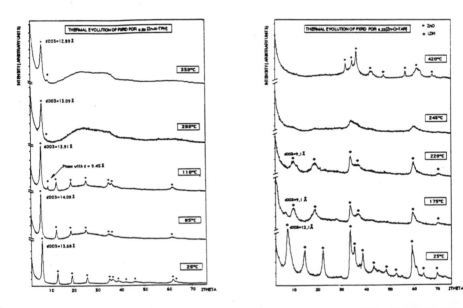

Figure 7: Powder X rays diffractograms of calcined [Zn-Al-TPH] and [Zn-Al-TAR] LDH's.

AKNOWLEDGEMENT

We thank P. Maillard and M. Momenteau from the Laboratoire de synthèse de composés hétérocycliques, Institut Cury, Orsay (France) for the preparation of porphyrin tetraphenyl tetracarboxylate.

REFERENCES

1) M. Meyn, K. Beneke, G. Lagaly, Inorg. Chem. 29, 5201 (1990)
2) M. Meyn, K. Beneke, G. Lagaly, Inorg. Chem. 32, 1209 (1993)
3) K. El Malki, A. de Roy, J.P. Besse, Eur. J. Solid State Inorg. Chem. 26, 339 (1989)
4) C. Depege, C. Forano, A. de Roy, J.P. Besse, Molecular Crystals and Crystal Liquids, to be published.
5) A. de Roy, C. Forano, K. El Malki, J.P. Besse, Anionic clays: Trends in pillaring Chemistry, Synthesis of Microporous Materials, Vol. 2, 108-169,Ed. M.L. Occelli, H. Robson, Van Nostrand Reinhold, New York, 1992.

Materials Science Forum Vols. 152 - 153 (1994) pp. 347-350
© *1994 Trans Tech Publications, Switzerland*

A NEW VARIETY OF MICRONIC GRAPHITE AND THE REDUCTION OF ITS INTERCALATION COMPOUNDS

C. Hérold, J.F. Marêché, A. Mabchour and G. Furdin

Laboratoire de Chimie du Solide Minéral (URA 158 - CNRS), Université de Nancy I, B.P. 239, F-54506 Vandoeuvre-les-Nancy Cédex, France

Keywords: Graphite, Micronic, Flat, Intercalation, Reduction, Percolation

ABSTRACT

By grinding mixtures of exfoliated natural graphite and organic solvents small particles are obtained. These particles are mostly single crystals and have average dimensions of 10 microns wide and 0.1 micron thick. They are used in composite materials with a low percolation threshold for electrical conductivity. This micronic graphite can be intercalated with transition metal halides and further reduction leads to included metallic clusters or, under mild conditions, to intercalated transition metals.

ELABORATION

Natural single crystal graphite flakes are intercalated with acids. Rapid heating provokes the expansion of the initial flakes in the c-axis direction. The resulting material is very light (average density 3-12 g/l). Scanning electron microscopy shows that the particles are not completely separated. The expanded graphite is thoroughly mixed with an organic solvent (cyclohexan) and submitted to cycles of tangential shearing with an Ultra-Turrax ™ mill and high power ultrasonic disintegrator. The resulting mixture is then freeze dried to avoid compaction of the particles [1].

PROPERTIES

The resulting particles are mostly single crystals and have average dimensions of 10 microns wide and 0.1 micron thick. Scanning electron microscopy shows them as flat sheets laying on their support (Figure 1). This view, tilted 60° allows appreciating the thickness of the particle. Electron microdiffraction of the whole area displays the diagram of single crystal graphite (Figure 2). The specific area of the product is 20 m^2/g. The morphology of this micronic graphite favors a good orientation when deposited on a support (mosaic spread of ± 15°) (Figure 3) and a good coverage of any surface.

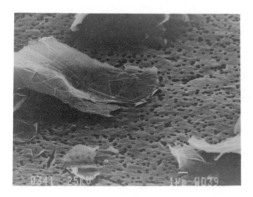

Figure 1 : Scanning electron micrograph
of a graphite particle (tilt 60°).

Figure 2 : Electron microdiffraction diagran
of a graphite particle.

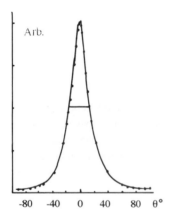

Figure 3 : Mosaicity (FWHM of the scan
around the 002 reflexion position).

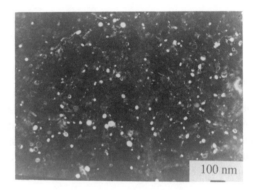

Figure 4 : Clusters in a graphite $FeCl_3$ compound
reduced with potassium at 300°C.

This material has been mixed with polyurethane paints to make conducting composites with a low electrical conductivity threshold, around 2% in volume. Films have an anisotropic conductivity, the higher value in the deposition plane [2].

INTERCALATION

This graphite retains roughly the same reactivity as that of natural graphite and thus can be intercalated with various reactants, such as transition metal halides. Ferric trichloride was intercalated to first stage in the classical two bulb tube under a chlorine atmosphere, with a small temperature gradient to avoid the condensation of the excess halide on the compound. The particles are pressed on a beryllium foil under inert atmosphere, placed in a sample holder closed by a Mylar film for X-Ray examination. The transmission diffractogram shows the 00l reflexions. They are somewhat broadened by the misorientation of the individual particles with respect to that obtained from a highly oriented pyrographite based compound, but the stage purity and fidelity remain easy to check.

REDUCTION

The intercalated graphite is transferred under inert atmosphere in a two bulb tube and the (excess) alkali metal is introduced the same way, and then distilled beside the compound. After evacuation the tube is sealed and placed in a furnace with a slight temperature gradient to prevent condensation of the metal and ensure gas phase reaction. Alkali metal vapour pressure and activity depend strongly on the temperature. So we varied the reduction temperature from 400°C to 100°C, the lower corresponding to quite mild conditions. Heavy alkali metals are able to intercalate and to reduce in situ the intercalated halide and to form alkali metal halide as a by product. As the resulting halide and iron phases occupy less gallery space than the starting materials, the alkali metal fills all the empty space to give a MC_8 phase. This pyrophoric phase is removed by an overnight slow oxidization of the product in argon containing 10 - 100 ppm oxygen.

Full reduction is confirmed by Mössbauer spectroscopy, showing the disappearence of the Fe^{3+} and Fe^{2+} lines [3]. Scanning electron microscopy shows that the particles are not very distorted. For transmission electron microscopy, the samples are dispersed in alcohol, filtered on a Nuclepore TM filter covered with carbon. A new layer of carbon is deposited on the particles and the polymer is dissolved with chloroform on the microscope grid. Products reduced at 400, 300 and 200°C for three days contain several phases [4] resembling that obtained by other authors on bigger graphite particles [5] : potassium chloride microcrystals, included or placed on the surface of the graphite particles, and α iron clusters (Figure 4). The average cluster diameter was determined using the Scherrer formula applied to the 110 X-Ray diffraction line and depends on the reduction temperature : 28 nm for 200°C, 35 nm for 300 and 400°C. Electron microscopy shows a great spread in the cluster dimensions. The quantity of small (10 nm) clusters is strongly enhanced when the reduction temperature is lowered. Some iron clusters are on the surface of the particles, and can be at least partially oxidized by the atmosphere. Electron Energy Loss Spectroscopy shows that the small clusters are quite fully oxidized, and the bigger ones only partially. These clusters can be removed by treatment with a hydrochloric acid solution. They represent no more than 25% of the total iron. The other clusters are included, this situation can be viewed by electron microscopy examination of the side, showing the graphitic layers bent around the iron clusters.

Reduction at 100°C is much longer (1 - 2 weeks) and adds new phases to the previous ones. Included KCl crystallites have a tendency to have their basal planes parallel to the graphite layers. Some of the iron is intercalated between the graphene sheets with an hexagonal two dimensional lattice the parameter of which is 353 pm. Mössbauer spectroscopy shows that this iron phase is not ferromagnetic due to the long distance between atoms.

Reduction was carried out using other alkali metal vapours : rubidium and cesium. The same kind of phases are present, but when the reduction temperature is raised, the quantity of two dimensional metal decreases and the number and size of α iron clusters are greater. At a reduction temperature of 100°C, the heavier the alkali metal, the higher the yield in intercalated iron. The reduction with cesium gives 80% of intercalated iron, measured using Mössbauer spectroscopy [3]. The two dimensional iron lattices depend strongly on the reducing agent [6]. The reduction by rubidium vapour gives two lattices oriented with respect to the in-plane graphite axes : a square lattice a = 410 pm and an hexagonal lattice a = 252 pm both a-axes parallel to a-axis of graphite. The reduction by cesium gives a square lattice a = 348 pm which is oriented ± 15° with respect to the **a** direction of graphite. Sometimes another square lattice a = 492 pm is present which is a lacunar version of the previous. Included RbCl and CsCl microcrystals are strongly oriented with respect to the three graphite axes : the square basal planes of the cubic lattice are parallel to the graphene sheets and the other faces oriented with respect to the in-plane graphite axes by 0° (CsCl) or ± 15° (RbCl).

The products were heated up to 730°C. The general result was a loss of the orientation of all the above mentioned lattices with respect to the graphite lattice. The square two dimensional lattice a = 348 pm (reduction with Cs) remains but the other two dimensional lattices transform partly to this one, partly to α iron. This square lattice is the most stable of the intercalated iron phases, possibly due to the commensurability of this lattice and that of the graphite along one of their axes.

Similar reductions were carried out on cobalt and copper chloride intercalated graphite [7,8] with similar results : higher yields in intercalated transition metals with heavier alkali metals, two dimensional arrangements of the atoms strongly dependent on the reducing agent and oriented crystals of the alkali metal chlorides.

CONCLUSIONS

This new kind of micronic graphite retains, on the micron scale, the morphology of the starting flakes; i.e. dimensions in the plane much larger than the thickness. This geometrical consideration is the base of the material's ability to deposit on a surface with a quite small mosaicity, covering ability and specific area. As the particles can be oriented parallel to the deposition plane of a composite material, we can obtain an anisotropic conductivity. In the case of scanning electron microscopy, the particles are thin and quite transparent to the electrons and the support is visible under the particles. This is of particular interest for transmission electron microscopy, and the fact that most of the particles remain single crystals helps to have simpler diffraction diagrams, where the relative orientations are clearly visible. These qualities were very useful when we decided to shed new light on the quite old subject of the reduction of transition metal chloride intercalated graphite. The small size and good quality of the particles helped to obtain complete reduction by a good diffusion of the reducing agent. The small thickness permitted the accommodation of the defects created by the inclusion of iron or alkali metal halide clusters by bending of the graphene layers. It was also necessary for the microanalysis to avoid superposition of clusters. The cristallinity of the particles was fundamental for the obtention of interpretable diffraction diagrams. These studies have given for the first time the planar arrangement of the transition metal atoms. The differences observed for the different alkali metals as reducing agents employed in identical experimental conditions help to understand the spread of the results obtained by previous authors [9-11].

REFERENCES
1) Furdin, G., Hérold, A., Marêché, J.F.: Int Pat , PCT / EP92 / 02317
2) Furdin, G., Marêché, J.F., Mabchour, A., Hérold, A. : Mol. Cryst. Liq. Cryst. (in press)
3) Hérold, C., Marêché, J.F., Gérardin, R., Mabchour, A., Furdin, G : Mat. Res. Bull., 1992, 27, 185
4) Mabchour, A., Furdin, G., Marêché, J.F. : C. R. Acad. Sci. Paris Série II, 1991, 312, 1293
5) Messaoudi, A., Erre, R., Béguin, F. : Carbon, 1991, 29, 515
6) Hérold, C., Marêché, J.F., Furdin, G. : Microsc. Microanal. Microstruct., 1991, 2, 589
7) Hérold, C., Marêché, J.F., Furdin, G., Hubert, N. : Proc. 5th Int. Carbon Conf. "Carbon'92" (22-26 June, Essen, Germany), 1992, 606
8) Marêché, J.F., Hérold, C., Furdin, G., Hubert, N. : Mol. Cryst. Liq. Cryst. (in press)
9) Shuvayev, A.T., Helmer, B.Yu., Lyubeznova, T.A., Kraizman, V.L., Mirmilstein, A.S., Kvacheva, L.D., Novikov, Yu.N., Vol'pin, M.E. : J. Phys. France, 1989, 50, 1145
10) Meyer, C., Yazami, R., Chouteau, G. : J. Phys. France, 1990, 51, 1239
11) Touzain, Ph., Chamberod, A., Briggs, A. : Mat. Sci. Eng., 1977, 31, 77.

Materials Science Forum Vols. 152 - 153 (1994) pp. 351-354
© 1994 Trans Tech Publications, Switzerland

NOVEL PREPARATION OF CdS NANOCRYSTALS IN A SODIUM BOROSILICATE GLASSY MATRIX

W. Granier [1], L. Boudes [1], A. Pradel [1], M. Ribes [1], J. Allegre [2], G. Arnaud [2], P. Lefebvre [2] and H. Mathieu [2]

[1] Laboratoire de Physicochimie des Matériaux Solides URA D0407 CNRS, CC003, Université de Montpellier II, F-34095 Montpellier Cédex 5, France

[2] Groupe d'Etudes des Semi-conducteurs URA 357 CNRS, Université de Montpellier II, F-34095 Montpellier Cédex 5, France

Keywords: Sol-Gel Process, CdS Nanocrystallite, Nonlinear Optics, Sodium Borosilcate Glass

ABSTRACT

CdS nanocrystallite doped glasses were prepared using the sol-gel process. The gel was formed in an aqueous solution where a preliminary complexation of cadmium with 5-sulphosalicylic acid was made. An "in situ" precipitation of CdS nanocrystallites was further obtained during densification of gel. Optical absorption measurements carried out on these glasses revealed a quantum confinement effect of carriers and a shifting of optical absorption edge to the blue wavelength with decrease of crystallite size. Raman scattering experiments were performed and the first results are presented. The influence of quantum confinement on the scattering of light by Longitudinal Optical (LO) phonons is discussed.

INTRODUCTION

Quantum confinement effects in two - or one - dimensional systems e.g. quantum well layers, superlattices or quantum filaments are well known to enhance the semiconductor electronic and optical properties [1].

In the past years, in the context of search of reduced dimensional semiconductor structures, a new type of materials i.e., semiconductor crystallite doped glasses, were considered. In these materials, the quantum confinement of carriers occurs in the all three space dimensions when the crystallite size is close to the bulk Bohr radius. Beyong their fundamental interest for the physics of low dimensionality, these glasses, showing third order optical nonlinearity likely to be improved by quantum size effects could have important applications in high technologies such as all-optical communications and computers (optical bistables and switches).

Current experimental studies are carried out using commercial CdS_xSe_{1-x} doped glasses. These materials do not exhibit any evident quantum size effect because of the large distribution of crystallite sizes due to a bad solubility of the starting components and to the high melting temperatures of the glass matrix. Recently, to prevent these problems, the sol-gel process was proposed as an alternative route to the preparation of crystallite doped glasses [2]. The main results to date concern xerogels or porous glasses in which the semiconductor nanocrystallites are unstable and sensitive to oxydation [3].

The present study examines a method for preparing nanocrystalline CdS particle doped Na_2O B_2O_3 SiO_2 glasses from a gel formed in an aqueous solution. This new process is attractive since it allows the complete dissolution of precursors, the complexation of cadmium in the sol and the fully densification of vitreous matrix. Optical absorption and Raman experiments were carried out in order to confirm the occurence of quantum size effects in these materials.

EXPERIMENTAL PROCEDURE

Glass preparation

Firstly, complexation of cadmium was obtained by dissolving adequate quantities of cadmium precursor, i.e. 3 $CdSO_4$, 8 H_2O and $HOSO_2$ C_6H_3 (CO_2H) OH (5-sulphosalicylic acid) in a 1:1 molar ratio into 50 cc of distilled water. 1.24 g of H_3BO_3 boric acid was then dropped into this solution. This mixture was stirred and heated at 60°C until a clear solution was obtained. 2.84 g of Na_2SiO_3, 9 H_2O sodium metasilicate was further introduced into the homogeneous solution to initiate the gelation. The stiff gel was dried at 100°C for 12 hours and melted at 750°C. Quenching of this melt gave a transparent glass of $1Na_2O$ $1B_2O_3$ $1SiO_2$ composition doped with CdS nanocrystallites which concentration ranged from 0.003 mol% to 0.3 mol% for the various prepared samples.

Thermal analysis of complex degradation

Sulphosalicylate with Cd^{2+} cadmium leads to a very stable complex with pK = 16.7. Thermal degradation of this complex in the absence of glass matrix precursors was followed by thermal analysis in order to understand the "in situ" formation of CdS during conversion of gel into glass. Differential scanning calorimetry (DSC) and thermal gravimetric analysis (TGA) data were respectively obtained on a SETARAM 121 analyzer with a heating rate of 10°C/min and a V5.1A DUPONT 2000 equipment with a heating rate of 5°C/min. The X-ray diffraction measurements were carried out using a SEIFERT diffractometer.

Measurements of properties of the CdS nanocrystallite doped glasses.

Optical absorption spectra of glasses were recorded on a JOBIN YVON spectrometer using a tungsten light. Raman scattering spectra were recorded with a DILOR OMARS 89 spectrometer using the three wavelengths of an Ar^+ laser i.e., 514 , 488 and 458 nm .

RESULTS AND DISCUSSION

Evolution of the Cadmium complex with temperature

DSC curve of the Cd complex formed between sulphosalicylic acid and Cd sulphate dissolved in water is shown in figure 1a. Endothermic peaks below 400°C correspond certainly to the departure of strongly bound molecules of water. Two exothermic peaks at 530°C and 564°C are further observed. The first one can be related to the pronounced peak around 500°C observed on the derived TGA curve (Figure 1b). This peak corresponds to the loss of organic matters due to the combustion of inorganic component. The DSC peak at 564°C which is not related to any weight loss is attributed to crystallisation of cadmium sulphide. This hypothesis is confirmed by X ray diffraction. An X ray diffraction pattern of the powder heated at 560°C and cooled down from this temperature is shown in figure 1a. It corresponds to CdS hexagonal structure.

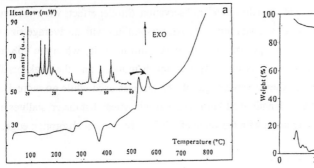
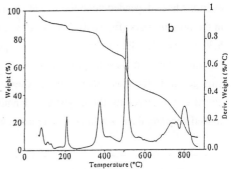

Fig. 1 : DSC and TGA curves of Cd complex and X-ray diffraction pattern of the powder cooled down from 560°C

Optical absorption spectra

Typical optical absorption spectra are shown in figure 2. They reveal blue shifted edges with decrease of nanocrystallite size which is a result of decrease of CdS concentration. Crystallite sizes were calculated as indicated in reference [4]: 13.5 nm particles correspond to 0.3 %, 12.0 nm to 0.03 %, 5.3 nm to 0.01 % and 3.8 nm to 0.003 %.These blue shifts are attributed to quantum size effects of carrier confinement. Usual spectrum of interband absorption localized near the bulk semiconductor energy gap is observed for the biggest particles.When crystallite size decreases, the excitonic structure disappears to give a broadened edge, indicating the independent quantum confinement of electrons and holes.

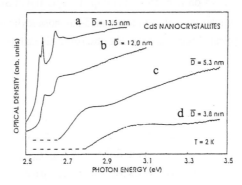

Fig.2 : Optical absorption spectra of prepared glasses (a: 0.3 %, b: 0.03 %, c: 0.01 %, d: 0.003 % in CdS mol)

Resonant Raman spectra

Provided that the energy corresponding to the excitating radiation wavelength be larger than the optical gap, Raman spectra of all glasses prepared as indicated above showed two peaks due to the Longitudinal Optical (LO) phonons at 306 cm^{-1} (1LO) and at 605 cm^{-1} (2LO) in agreement with those generally obtained for pure CdS crystals, which confirmed the formation of CdS crystallites during synthesis. In view of their Raman experiments, Alivisatos and al. [5] reported that the exciting radiation wavelength of a resonant Raman spectrum and the relative intensities of the 1LO and 2LO overtone lines depended upon the size of crystallites under observation. Similar effects were observed in the present work (figure 3). For the glass containing CdS nanocrystallites with an average size of 5.3 nm, the 1LO line in resonant Raman spectrum had its maximum intensity when the 458 nm exciting radiation wavelength was used. This intensity was much reduced and the peak hardly emerged from the glass matrix signal when a wavelength of 488 nm was used while no more 1LO signal was observed for the largest wavelength of 514 nm. These results attest that the crystallites are indeed very small and are in good agreement with the value of ~ 2.7 eV for optical absorption edge.

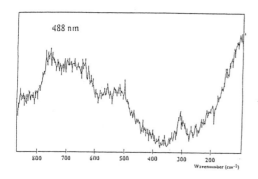 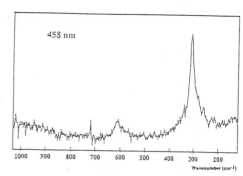

Fig.3 : Raman spectra of 5.3 nm nanocrystallite doped glass under 488 and 458 nm excitation wavelengths

CONCLUSION

Quantum size effects in CdS nanocrystallite doped Na_2O B_2O_3 SiO_2 glasses prepared by a sol-gel process including complexation of cadmium with 5-sulphosalicylic acid were demonstrated by measuring blue shifts of optical absorption edge and observing his manifestation in resonant Raman scattering spectra.

REFERENCES

[1]   Rosencher E., Bois P., Nagle J., Delaitre S. : Electronics Let. 1989, 25, 1063.

[2]   Nogami M., Nagasaka K., Takata M. : J. Non-Cryst. Solids 1990, 122, 101.

[3]   Nogami M., Nagasaka K. : J. Non-Cryst. Solids 1992, 147 & 148, 331.

[4]   Granier W., Boudes L., Pradel A., Ribes M., Allègre J., Arnaud G.,Lefebvre P., Mathieu H.: J. Sol-Gel Sci. Tech.(in press)

[5]   Shiang J.J.,Goldstein A.N.,Alivisatos A.P. :.J. Chem. Phys. 1990, 92,3232

Materials Science Forum Vols. 152 - 153 (1994) pp. 355-358
© *1994 Trans Tech Publications, Switzerland*

PARTIAL CHARGES DISTRIBUTIONS IN CRYSTALLINE MATERIALS THROUGH ELECTRONEGATIVITY EQUALIZATION

M. Henry

Laboratoire de Chimie de la Matière Condensée, Université Pierre et Marie Curie T54-55 E5,
4 place Jussieu, F-75252 Paris Cédex 05, France

Keywords: Electronegativity, Partial Charge Model, Madelung Potentials, Oxides, Solid-State Acidity

ABSTRACT : A structure-dependent version of Sanderson's electronegativity equalization principle allowing a rapid computation of partial charges distributions in crystalline solids is described. The method is first applied to simple oxides ranging from highly ionic 3D-networks (Li_2O, CaO) to highly covalent molecular crystals (P_4O_{10}, OsO_4), then to more complex structures based on polymerized anionic ($K_2Ti_4O_9$) or cationic (zunyite $Al_{13}Si_5O_{20}(OH)_{18}Cl$) frameworks. From these partial charges distributions , it becomes possible to quantify the solid-state acid-base chemistry as a function of any crystalline or molecular environment.

1. INTRODUCTION

The accurate knowledge of partial charges distributions in solid materials is of the utmost importance to get a better understanding of their solid-state chemical reactivity. Unfortunately, *ab initio* computations are not well suited to a solid-state approach, requiring huge computers. The aim of this contribution is to show that a structure-dependent version of the electronegativity equalization principle of R.T. Sanderson [1] allows to compute quite detailed partial charges distributions in crystalline solids. The algorithm is based on a formula derived from Density Functional Theory by W.J. Mortier et al [2,3] that we have matched to the Sanderson electronegativity scale :

$$\begin{cases} \chi_i = <\chi> = \chi_i^0 + k_1\sqrt{\chi_i^0}\, q_i + k_2\sum_{j=1}^{n} M_{ij}q_j \quad \forall i = 1,...n \\ \qquad\qquad\qquad\qquad\qquad\qquad\quad k_1 = 2.415 \text{ and } k_2 = 2.415 \text{ Å} \\ \sum_{i=1}^{n} q_i = 0 \end{cases}$$

$$\tag{1}$$

$$M_{ij} = \frac{14400\pi R^2}{V} \sum_{\substack{\alpha \le 2\pi R|\vec{h}| \\ \vec{h}=h\vec{a}^*+k\vec{b}^*+l\vec{c}^* \\ \vec{x}=x\vec{a}+y\vec{b}+z\vec{c}}} \exp\left[2\pi i\vec{h}.(\vec{x}_j - \vec{x}_i)\right]\frac{\left[\alpha\cos\alpha - 3\sin\alpha + 2\alpha\right]^2}{\alpha^{12}} - \frac{25}{14R}\delta_{ij}$$

In these relationships, χ_i° and q_i are the dimensionless Sanderson electronegativity and partial charge of the ith atom in the unit cell which contains n non-equivalent atoms, $<\chi>$ the mean electronegativity of the network to which all individual electronegativities χ_i

have to equal and k_1, k_2 empirical calibration constants [4]. M_{ij} is the Madelung contribution at site i coming from all sites j in the network. It is computed in the reciprocal space following the method of F. Bertaut [5] and R.E. Jones & D.H. Templeton [6], knowing the unit cell parameters, the atomic coordinates $\bar{x}_i$, $\bar{x}_j$ of sites i or j and R the minimum interatomic distance in the structure. The dimensionless summation index α is defined as $\alpha = 2\pi|\bar{h}|R$, where $|\bar{h}|$ stands for the modulus of any reciprocal space vector and δ_{ij} is the Kroenecker symbol. Knowing the electronegativities and each M_{ij} allows to compute the (n+1) unknowns (n partial charges and the mean electronegativity $\langle\chi\rangle$) by solving the linear system (1). This method have been used to calculate partial charges distributions in oxide-based materials. Knowing partial charges on oxygen atoms q(O), pK_a values are then calculated using equation 2 [4] :

$$pK_a(O) = -42.2q(O) - 19.8 \qquad (2)$$

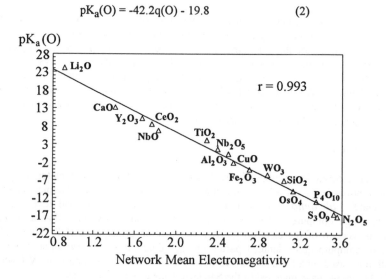

Figure 1 : Theoretical acidity constants (pK_a) of oxide networks against their calculated mean electronegativity. pK_a values for non-equivalent oxygen atoms within the same structure have been properly averaged.

2. RESULTS

Various oxides ranging from highly ionic 3D-networks to highly covalent molecular crystals have first been investigated (figure 1). When oxygen atoms are in eightfold coordination (Li_2O, Na_2O, K_2O) they display strong basic properties ($pK_a > 14$). Conversely, when they are found in terminal position (OsO_4, N_2O_5) they display strong acid behavior ($pK_a < -10$). Generally speaking, oxygen atoms become more and more acidic as their coordination number decreases. For example, with MO oxides the pK_a values as a function of the metal atom are found to be : 13.5 (Sr), 13.2 (Ca), 12.7 (Ba), 9.3 (Mg), 6.8 (Nb), 6.0 (Be), 1.6 (Mn), -2.0 (Cu), -5.4 (Zn), -6.7 (Hg). For a fixed coordination number, pK_a values decreases as the electronegativity of the metal atom increases. Thus for M_2O_3 corundum-type structure (R-3c) we found : 6.1 (Ti), 0.3 (Al), 0.5 (V), -1.0 (Cr) and -4.0 (Fe). However, if we change the structural type, inversions may occur : χ(Al) > χ(Mn) but $pK_a(Al_2O_3)$ = 0.3 (R-3c) > $pK_a(Mn_2O_3)$ =-0.1 (Ia3).

When both the metal atom and the stoichiometry are kept constant, no simple rules can be found. Taking the TiO_2 polymorphs as a test we found pK_a : 4.5 (TiO_2-B), 4.1 (anatase), 4.0 (TiO_2-II), 3.95 (rutile) and 3.9 (brookite). Finally, if within the same structure, several atoms are found in different crystalline environments, the lower the coordination, the higher the acidity. Typical examples are found for molecular crystal : P_4O_{10} [-12.1 (μ_2), -13.6 (t)], Mn_2O_7 [-5.0 (μ_2), -8.2 (t)]; chain-like structures : β–SO_3 [-15.9 (μ_2), -16.6 (t)], CrO_3 [-5.2 (μ_2), -7.9 (t)]; layered compounds : MoO_3 [-7.2 (μ_3), -8.2 (μ_2), -9.9 (t)], V_2O_5 [-2.5 (μ_3), -3.8 (μ_2), -5.3 (t)]; and 3D networks : m-ZrO_2[9.3 (μ_4), 8.7 (μ_3)], B-Nb_2O_5 [2.5 (μ_3), 1.1 (μ_2)].

One of the big advantage of the model is that it applies without any modification to very complex mixed oxides structures. Let us consider a typical example provided by the layered compound $K_2Ti_4O_9$ which can be considered as a $[Ti_4O_9]^{2-}$ 3D-polyanion with readily exchangeable K^+ counter-ions [7]. It has been shown [8] that this compound can be hydrolyzed according to :

$$K_2Ti_4O_9 + 2\ HNO_3 + H_2O\ \rightarrow\ Ti_4O_7(OH)_2.H_2O + 2\ KNO_3$$

leading finally to the TiO_2(B) structure after thermolysis around 500°C. In order to check that potassium atoms were readily exchangeable in this structure and that some oxygen atoms were basic enough to be hydroxylated, we have computed the partial charge distribution on the 60 atoms found in the unit cell of $K_2Ti_4O_9$ (table1).

4.O6 = -0.766	4.O8 = -0.708	4.O5 = -0.703	4.O7 = -0.699	4.O9 = -0.641
4.O3 = -0.629	4.O4 = -0.627	4.O2 = -0.625	4.O1 = -0.620	4.Ti3 = +0.948
4.K1 = +0.983	4.Ti4 = +0.992	4.Ti2 = +0.995	4.K2 = +1.047	4.Ti1 = +1.053

Table 1 : Partial charges distribution in $K_2Ti_4O_9$ computed with electronegativities 3.654 for O, 1.295 for Ti and 0.74 for K, leading to a framework mean electronegativity of 1.726 (Sanderson scale).

As expected it is found that potassium atoms behave as mere counter-ions (q(K) $\approx$ +1) facing a highly covalent $[Ti_4O_9]^{2-}$ framework (q(Ti) $\approx$ +1 << +4, <q(O)> $\approx$ -0.67 >> -2). Such atoms are thus readily exchangeable with $[H_3O]^+$ ions. More interesting is the clear distinction found between the nine oxygen atoms present within the framework. If the charges are converted to pK_a values according to (2) we obtain, in contrast with simple oxides, a totally inverse order as a function of the oxygen coordination number :

$$O6(t) = 12.5 > [O8, O5, O7](\mu_2) = 9.9 > O9(\mu_2) = 7.3 > [O3, O4, O2, O1](\mu_3, \mu_4) = 6.6$$

The rather high pK_a value found for the terminal O6 atom suggests that it must be protonated in contact with water, releasing OH^- ions in solution. Moreover, outer μ_2-oxo groups (O5, O5 and O8) although less basic, can also be hydroxylated allowing the complete transformation into the TiO_2(B) structure at high temperature. Finally, the other inner μ_2-oxo (O9), μ_3-oxo (O3) and μ_4-oxo (O4, O2 and O1) groups are quasi-neutral and should not react with water.

To conclude, we would like to investigate another kind of complex oxide network found in the mineral zunyite and having an ideal formula $Al_{13}Si_5O_{20}(OH)_{18}Cl$ [9].

Formally this structure can be derived from a 3D-polycationic octahedral network $[Al_{13}O_4(OH)_{18}]^{13+}$ having a building unit closely related to the well-known tridecameric polycation $[Al_{13}O_4(OH)_{24}(OH_2)_{12}]^{7+}$ found in hydrolyzed Al^{3+} solutions [10]. Within the voids generated by the tetrahedral packing of these polymerized tridecameric units, discrete pentameric polyanions $[Si_5O_{16}]^{12-}$ and monomeric chloride ions insure charge compensation. Table 2 gives the total partial charge distribution computed on the 300 atoms found in the unit cell of zunyite.

24.O3 = -0.642	48.O4 = -0.489	16.O1 = -0.471	48.O5 = -0.404
16.O2 = -0.376	4.Cl = -0.335	48.H1 = +0.245	24.H2 = +0.426
4.Si1 = +0.514	16.Si2 = +0.520	4.Al1 = +0.747	48.Al2 = +0.788

Table 2 : Partial charges distribution in zunyite computed with electronegativities 3.654 for O, 3.475 for Cl, 2.592 for H, 2.138 for Si and 1.714 for Al, leading to a framework mean electronegativity of 2.660 (Sanderson scale).

The highly covalent nature of this compound is evidenced by its total charge distribution : $[(Al1)(Al2)_{12}(O1)_4(O3H2)_6(O4H1)_{12}]^{+4.1}.[(Si1)(O2)_4(Si2)_4(O5)_{12}]^{-3.8}.Cl^{-0.3}$. Two kind of μ_2-hydroxo groups bridging two Al atoms are found in this structure : a neutral one $pK_a(O3) = 7.3$ and a rather acidic one $pK_a(O4) = 0.8$. Other oxygen atoms in the structure have rather low pK_a values : O1 = 0.1 (OAl$_4$), O5 = -2.7 (OSiAl$_2$) and O2 = -3.5 (OSi$_2$) and are thus hardly protonable.

3. CONCLUSION

This work has tried to show that the electronegativity equalization principle can be a very valuable tool in the field of solid-state acid-base chemistry. It is based on a very simple algorithm (relation 1) which can be implemented on any personal computer. Input data are limited to a crystalline structure (space group, asymmetric unit cell parameters and fractional atomic coordinates) and an electronegativity scale (Sanderson, Pauling, Allred-Rochow, Mulliken ...). With this extremely limited data set, quite detailed partial charges distribution can be computed which throw new light on very complex domains such as aqueous [6] or solid-state inorganic chemistry (*vide supra*).

4. REFERENCES

[1] Sanderson R.T., Science, 1951, 114, 670.
[2] Mortier W.J., Ghosh S.K. and Shankar S., J. Am. Chem. Soc., 1985, 107, 829.
[3] Van Genechten K. and Mortier W.J., Zeolites, 1988, 8, 273.
[4] Henry M. and Taulelle F. in "Application of NMR Spectroscopy to Cement Science", Colombet P. and Grimmer A.R. Eds, Harwood, London (1993).
[5] Bertaut F., J. Phys. Rad., 1952, 13, 499.
[6] Jones R.E. and Templeton D.H., J. Chem. Phys., 1956, 25, 1062.
[7] Verbaere A. and Tournoux M., Bull. Soc. Chim. Fr., 1973, 1237.
[8] Marchand R., Brohan L. and Tournoux M., Mater. Res. Bull., 1980, 15, 1129.
[9] Baur W.H. and Ohta T., Acta Cryst., 1982, B36, 390.
[10] Johansson G., Arkiv Kemi, 1962, 20, 321.

Materials Science Forum Vols. 152 - 153 (1994) pp. 359-364
© 1994 Trans Tech Publications, Switzerland

RUTHENIUM SPECIES IN LAYERED COMPOUNDS
PART 1. THE DIRECT SELF-CATALYSED INTERCALATION OF CATIONIC RUTHENIUM(II) AMMINE COMPLEXES FROM AQUEOUS MEDIA INTO ALPHA-TIN(IV) HYDROGEN PHOSPHATE

M.J. Hudson [1] and A.D. Workman [2]

[1] Department of Chemistry, University of Reading, Box 224, Whiteknights, Reading, Berkshire RG6 2AD, UK

[2] Department of Chemistry, University of Leicester, University Road, Leicester, Leicestershire LE2 7RH, UK

Keywords: Ruthenium Extraction, Tin(IV) Hydrogen Phosphate, Ion Exchange, Self-Catalysed Intercalation

The ruthenium(II)-containing cation $[Ru(NH_3)_6]^{2+}$ has been directly intercalated into α-tin(IV) hydrogen phosphate (SnP). Since the corresponding ruthenium(III) complex cation was not so intercalated, a self-catalysed intercalation mechanism involving labile ammonia ligands from the ruthenium(II) has been proposed. At high loadings, guest ruthenium(II) species were oxidised to ruthenium(III).

Ruthenium compounds have interesting properties with respect to their oxidative and catalytic activities. Consequently, there is a need to prepare PLS with ruthenium-containing pillars but it has proved difficult to intercalate ruthenium containing species into the layered hosts. The inorganic ion exchanger alpha-tin(IV) hydrogen phosphate monohydrate $(Sn(HPO_4)_2.H_2O)$ [SnP] is a layered (hydrogen) phosphate with a basal spacing of 0.76 nm; an area of 21.4 $Å^2$ per phosphate group, and a maximum cation exchange capacity (CEC) of 6.08 mmol g^{-1} for a monovalent ion.[1] Despite this high CEC, SnP is regarded as a poor ion exchanger because the free diffusion of counter ions is limited by the strong layer-layer interactions and the small passageways (ca. 2.6 Å) that connect the interlayer cavities. In this case only surface sorption will occur.[2] Metal complexes are not known to intercalate directly into SnP and consequently amine intercalation compounds have been prepared in order to separate the layers and to reduce the interlayer charge density.[3,4] Intercalation of cations into SnP may be catalysed using hydrated sodium ions which enhances the separation of the layers.[5] In this study ruthenium(II) complex cations have been intercalated directly from aqueous solution.

Experimental

SnP was synthesised according to the published procedure.[6] Its X-ray powder diffraction pattern (XRD) (Spectrolab series 3000 CPS-120 instrument, Ni-filtered Cu-Kalpha =0.154051 nm) indicated a basal spacing of 0.78 nm, and the absence of any impurity peaks. Hexaamminoruthenium(II) chloride was prepared by the method of Fergusson and Love.[7] All extractions were done in duplicate as batch experiments using degassed, doubly deionised water. The cation exchange capacity (CEC) was taken as 6.08 mmol H^+ g^{-1} of dry ion exchanger. The solutions were diluted for atomic absorption analysis using hydrochloric acid (10%, v/v) and lanthanum(III) chloride pentahydrate (0.5%, w/v) so that the range of ruthenium concentration was in the range of 0 to 100 mg dm^{-3}.

Results and Discussion
Sorption Studies

Sorption of the hexaamminoruthenium(II) cation

The extraction curve for hexaamminoruthenium(II) chloride is shown in Fig. 1. The rate of extraction appears to be quite rapid with $t_{\frac{1}{2}}$ (the time for half of the ruthenium to be extracted) of under five min to a capacity of 1.66 mmol g^{-1}, which is 55% of the theoretical exchange capacity of 3.04 mmol g^{-1} for a divalent ion. For corresponding ruthenium(III) complex cation, there was only surface adsorption.

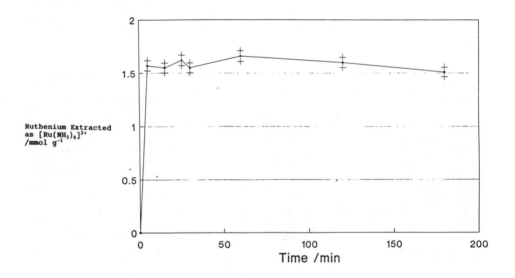

Ruthenium Extracted as $[Ru(NH_3)_6]^{2+}$ /mmol g^{-1}

Time /min

Figure 1: Extraction of the Hexaamminoruthenium(II) $[Ru(NH_3)_6]^{2+}$ cation (A) by SnP as a function of time.

Mechanism of Extraction

Clearly, there is an important difference between the ruthenium(II) and ruthenium(III) species. It is considered that the relative labilities of the ammonia ligands are the key factors concerning the surface sorption and intercalation of ruthenium(II) but surface sorption only of ruthenium(III). Ammonia ligands are labile for the ruthenium(II) species but are unreactive for the ruthenium(III) species and effectively remain bound to the ruthenium atom. The calculated half-life of the ligand exchange process for the ruthenium(III) complex is 3 yr.[8] The rate constant for the exchange of ammonia with water has been measured[9] as 1.24×10^{-3} dm mol^{-1} s^{-1} for the ruthenium(II) complex.[9] Moreover, the pH of the solution in contact with the ruthenium(II) cation is 11.2 which is consistent with free ammonium groups. Thus the exchange of ammonia by another ligand such as water in the hexamminoruthenium(II) complex cation introduces ammonia locally into the edges of the layers of the host. The ammonia molecules are able to react with the proton of the phosphate group in the host SnP. The effective acidity of this phosphate group is increased, and exchange of the cation occurs. In effect this is a self-catalysed intercalation reaction, which does not occur with the ruthenium(III) compound because there are few free ammonia molecules. The intercalation of ruthenium(III) was not catalysed by separate additions of ammonium ions because there are few free ammonia molecules generated locally. It has been suggested previously that trace amounts of sodium can act as catalysts for the intercalation of molecules.[10] However, the two catalysed mechanisms are different because, in the second case, the sodium increases the interlayer distance only and not the effective acidity of the host.

Nature of the Intercalated Species.

It was of interest to establish the nature of the guest species and it was noted that the powder diffraction data changed with extent of the loading of the sample. When there is a restricted amount of exchange and the effective pH within SnP is low, ruthenium(II) appears to dominate but at higher extents of exchange ruthenium(III) is dominant. (Electron spin resonance data supports the conversion from ruthenium(II) to ruthenium(III)). The X-ray powder diffraction patterns of separate samples of the initially ruthenium(II) intercalation compounds are shown in Fig. 2. Up to sample 5, the materials were biphasic with d_{002} at 1.06 nm for the intercalated phase and 0.78 for the host material. It is interesting to note that the peak at 1.06 nm diminished sharply when the concentration of the initial solution increased above 50% CEC and another peak formed indicating a monophasic material with an interlayer spacing of 0.99 nm. The reason why not all of the CEC is used is that the ruthenium species cover sites which are not used for ion exchange. The cross sectional area for the ruthenium(II) ion is approximately 0.24 nm^2 whereas the free area per phosphate is 0.21(4) nm^2.

Conclusions

The ruthenium(II)-containing cation $[Ru(NH_3)_6]^{2+}$ has been directly intercalated into α-tin(IV) hydrogen phosphate. (SnP).

Acknowledgements

This work has been supported by the Department of the Environment as part of their Radioactive Waste Management Programme. The results may be used in the formulation of Government policy but at this stage do not necessarily represent Government policy.

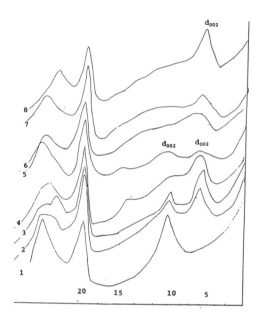

Angle /2θ

Figure 2: The X-ray Powder Diffraction Data for the Host(1) and the Intercalation Compounds. 1. SnP 2. 5% CEC of A in original solution 3. 10% 4. 25% 5. 50% 6. 75% 7. 100% 8. 300%.

References

1. M.J. Hudson, E. Rodriguez-Castellon, P. Sylvester, A. Jimenez-Lopez A. and P. Olivera-Pastor, *Hydrometallurgy*, (1990), **24**, 77-88.

2. G. Alberti, U. Costantino, S. Alluli and M.A. Massucci, *J. Inorg. Nucl., Chem.* (1975), **37**, 1779.

3. M.J. Hudson, E. Rodriguez-Castellon, P. Olivera-Pastor, A. Jimenez-Lopez, P. Maireles-Torres and P. Sylvester P., *J. Canadian Chemistry*, (1989), **67**, 2095-2101.

4. M.J. Hudson, E. Rodriguez-Castellon, P. Olivera-Pastor, A. Jimenez-Lopez, P. Maireles-Torres and P. Sylvester P., *J. Canadian Chemistry*, (1989), **67**, 2095-2101.

5. G. Alberti, U. Costantino and J.P. Gupta, *J. Inorg. Nucl. Chem.*, (1974), **36**, 2109.

6. U. Costantino and A. Gasparoni, *J. Chromatogr.*, (1970), **51**, 289.

7. J.E. Fergusson and J.L. Love, *Inorg. Synth.*, (1972), **13**, 208.

8. J.N. Armor and H. Taube, *Inorg. Chem.*, (1971), **10**, 1570.

9. P.C. Ford, J.R. Kuempel and H. Taube, *Inorg. Chem.*, (1968), **7**, 1976.

10. G. Alberti, U. Costantino, G.P. Gupta, *J. Inorg. Nucl. Chèm.*, (1978), **40**, 87.

Materials Science Forum Vols. 152 - 153 (1994) pp. 365-370

SYNTHESIS, STRUCTURE AND MAGNETIC PROPERTIES OF SOME NEW METAL(II) PHOSPHONATES: LAYERED $Fe(C_2H_5PO_3) \cdot H_2O$ AND $\alpha\text{-}Cu(C_2H_5PO_3)$, TUBULAR $\beta\text{-}Cu(CH_3PO_3)$

B. Bujoli [1], J. Le Bideau [2], C. Payen [2], P. Palvadeau [2] and J. Rouxel [2]

[1] Laboratoire de Synthèse Organique, CNRS-URA 475, 2 rue de la Houssinière,
F-44072 Nantes Cédex 03, France

[2] Institut des Matériaux, CNRS-UMR 110, 2 rue de la Houssinière,
F-44072 Nantes Cédex 03, France

<u>Keywords</u>: Phosphonates, Iron(II), Copper(II)

ABSTRACT. - $Fe(II)(C_2H_5PO_3)$. H_2O has been prepared via prolonged reaction between the lamellar iron oxychloride FeOCl and ethylphosphonic acid in acetone. The lamellar structure is very similar to that of previously reported divalent metal phosphonates $M(II)(RPO_3)$. H_2O (M= Mg, Mn, Co, Ni, Zn). The compound shows sign of 2D antiferromagnetic correlations above the Néel temperature $T_N=24K$ whereas a weak ferromagnetic behavior is observed below T_N. We also report on the preparation methods and the crystal structures of two new anhydrous copper phosphonates $\alpha\text{-}Cu(C_2H_5PO_3)$ and $\beta\text{-}Cu(CH_3PO_3)$ which exhibits an original tubular three-dimensional structure.

1. INTRODUCTION

Metal phosphonates have received considerable attention since the 1970s because of their fundamental interest as well as their various potential practical uses. Our work in this field was originally focused on layered iron(III) phosphonates and we have demonstrated that two types of compounds could be prepared : I $HFe(III)(RPO_3)_4$ and II $HFe(III)(RPO_3)_2$. xH_2O (R = phenyl, methyl, ethyl) [1,2]. In the case of the ethyl group, compound II slowly leads to an iron(II) lamellar derivative $Fe(II)(C_2H_5PO_3)$. H_2O in which the layer arrangement appears clearly to be the same as in the structure of the $M(II)(RPO_3)$. H_2O series (M=Mg, Mn, Co, Ni, Zn ; R= alkyl, phenyl) previously described by Cao et al. [3] and Martin et al. [4]. The first section of this paper will shortly report on the direct synthesis method, single-crystal structure analysis and some magnetic properties of $Fe(II)(C_2H_5PO_3)$. H_2O. Our work has been also extended to the copper(II) phosphonates especially since Zhang and Clearfield [5] showed that the layered structure of $Cu(II)(RPO_3)$. H_2O (R=CH_3, C_6H_5) was different from that of the other divalent metal phosphonates. Section 2 will present the preparation methods and the crystal structures of two new anhydrous copper phosphonates : $\alpha\text{-}Cu(C_2H_5PO_3)$ and $\beta\text{-}Cu(CH_3PO_3)$. The former is layered whereas the β compound has an original tubular three-dimensional structure.

2. $Fe(II)(C_2H_5PO_3)$. H_2O

The layered phosphonates of divalent metals $M(II)(RPO_3)$. H_2O - M=Mg, Mn, Co, Ni, Zn are simply prepared by combining aqueous solutions of a phosphonic acid and a soluble metal salt

[3,4,6]. Surprisingly, this procedure is unsuccessful in the iron(II) case and the compound $Fe(II)(C_2H_5PO_3)$. H_2O has been prepared by prolonged reaction between the lamellar iron oxychloride FeOCl and ethylphosphonic acid in acetone [7]. Nevertheless, the monoclinic (space group P1n1) crystal structure is very similar to that of previously reported $Mn(II)(C_6H_5PO_3)$. H_2O and $Zn(II)(C_6H_5PO_3)$. H_2O compounds [3,4]. Polar groups (phosphonates, oxygen, and lattice water) form two-dimensional (2D) sheets in which the metal ions are nearly coplanar (fig. 1) [7]. The P-C bonds are approximately perpendicular to these metal planes, and the ethyl groups make van der Waals contacts between layers. The environment about the iron atom is a distorted octahedron of oxygen atoms; five coordination sites are occupied by phosphonate oxygens whereas the sixth is occupied by the oxygen from a water molecule.

The quasi 2D square lattice formed by iron(II) ions within a layer gives rise to 2D antiferromagnetic properties above the Néel temperature $T_N = 24$ K [7]. A weak ferromagnetic behavior, which is probably due to a spin canting, is observed below T_N [7]. Taking into account the magnetic properties of this iron(II) phosphonate, we have extended our magnetic study to the other members of the $M(II)(RPO_3)$. H_2O series (M=Mn, Co, Ni ; R=CH_3, C_2H_5, C_4H_9, C_6H_5) [8]. All the studied compounds are 2D antiferromagnets with T_N ranging from 24 K to 3 K [8]. When possible, values for the intralamellar exchange coupling constants have been estimated using the high temperature series predictions. Quite constant interactions are obtained for each metal showing that the atomic arrangement within a layer is not significantly affected by the nature of the organic group. A more surprising result is that the Néel temperature does not depend of the interlamellar distance.

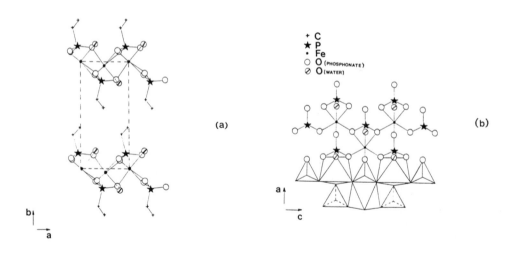

Fig. 1 - Structure of $Fe(C_2H_5PO_3)$. H_2O viewed down the c axis (a) and the b axis (b). In (b) carbon atoms have been omitted for clarity. Figures are taken from reference 7.

3. COPPER(II) PHOSPHONATES

3.1 Synthesis work

The anhydrous $Cu(RPO_3)$ compounds ($R=CH_3$, C_2H_5, C_6H_5), noted α, can be prepared by thermal treatment of the corresponding hydrates $Cu(RPO_3) \cdot H_2O$, but the resulting products are always poorly crystalline. However, the direct synthesis of crystalline α-$Cu(C_2H_5PO_3)$ in an autoclave was successful (Scheme 1) [9]. In the case of the methyl or the phenyl group, this direct method of preparation of the α-$Cu(RPO_3)$ analogs failed. When $R= C_6H_5$, the hydrated form was produced whereas two materials were obtained when $R = CH_3$. The major portion was an anhydrous form of copper methylphosphonate noted β-$Cu(CH_3PO_3)$ because its X-ray and infrared spectra were different from the one of the α form. Traces of a minor product : $Cu_3O(CH_3PO_3)_2 \cdot 2H_2O$ were also present in the reaction medium (Scheme 1). These results are likely a consequence of the stability of the α-$Cu(RPO_3)$ form in the experimental conditions. When α-$Cu(RPO_3)$ samples obtained after thermal treatment of the corresponding hydrates were placed with water in an autoclave at 180 °C for 2 weeks, no modification was observed in the case of the ethyl group while the hydrated form was obtained again in the case of the phenyl group. When $R = CH_3$, β-$Cu(CH_3PO_3)$ and $Cu_3O(CH_3PO_3)_2 \cdot 2H_2O$ were isolated again, in the same respective amounts.

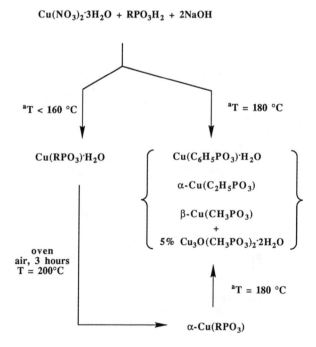

[a]*15 days reaction / H_2O / autoclave*

Scheme 1

3.2 Crystal structures

The structure of α-Cu(C$_2$H$_5$PO$_3$) obtained via the direct preparation method has been solved ab initio from X-ray powder diffraction data using the Rietveld method [9]. The monoclinic (space group P2$_1$/c) structure is layered as formed by 5-coordinate distorted trigonal bipyramidal copper atoms. The three oxygens of the phosphonate groups are all bonded to copper atoms. The ethyl groups are pointing toward the interlamellar space above the CuO$_3$P inorganic layers, making van der Waals contact between these layers. We believe that the structures of the dehydrated materials obtained by thermal treatment of Cu(RPO$_3$). H$_2$O in air are the same as described above, since the IR spectra the X-Ray powder diagrams and the magnetic susceptibility curves of α-Cu(RPO$_3$) prepared by both methods are very similar.

A drastic change of the magnetic susceptibility curves is observed after dehydration of the Cu(RPO$_3$). H$_2$O materials, with stronger intralayer magnetic interactions after the water removal [10]. This is consistent with the structural rearrangement that occurs during the dehydration process (fig. 2) [9].By comparing the structural work made on Cu(RPO$_3$). H$_2$O and on our own study of α-Cu(C$_2$H$_5$PO$_3$), it appears that one of the phosphonate oxygens shifts to occupy the vacant coordination site after the water removal. This reorganization induces a decrease of one of the parameters in the plane of the layer from 7.3 to 5.7 Å. Thus, in the anhydrous form of copper phosphonates, there is a bidimensional network of Cu-O-Cu paths, with intermetallic distances ranging between 3.1 and 3.2 Å, that likely explain the higher magnetic interactions observed for the anhydrous form of copper phosphonates compared to their hydrated form.

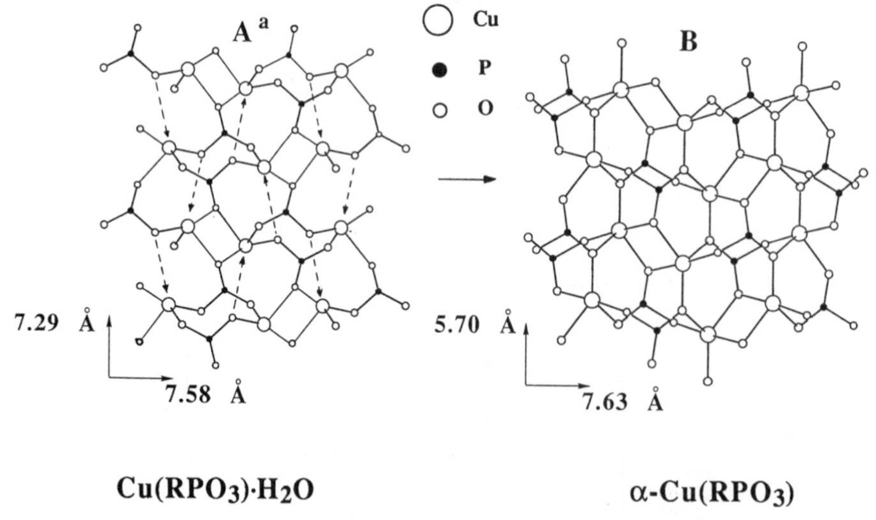

Fig. 2 : Schematic illustration of the structure rearrangement occuring during the dehydration of Cu(II)(RPO$_3$). H$_2$O [9]. The water molecule in A and the organic group R in A and B have been omitted for clarity. [a] according to reference 5.

The structure of β-Cu(CH$_3$PO$_3$) was determined on a single crystal [10]. Figure 3 shows that this compound is three-dimensional and consists of infinite zigzag chains of copper running parallel to the a-axis. Each copper atom is connected to two adjacent copper atoms, with for each one, two bridging oxygens. These chains are linked together in the b,c-plane by O-P-O bridges, forming hexagonal tunnels parallel to the a-axis and the methyl groups are extending in the space of these tunnels. It is worth noting the originality of the β-Cu(CH3PO3) compound. To our knowledge, this is the first example of tridimensional phosphonate, that are usually layered, except of course, when the phosphonic acid used bears a functional group (such as PO3H2, COOH or NH2) that is able to coordinate to metal centers. The three-dimensional character of the structure can be viewed as a consequence of both the small size of the methyl group and the richness ot the copper(II) stereochemistry.

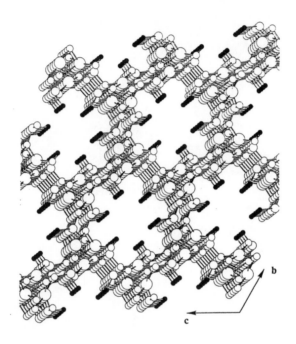

Fig. 3 : Perspective representation of β-Cu(CH3PO3) viewed down the *a*-axis,
illustrating the tubular structure. Small, medium and large circles are for oxygen, phosphorous and
copper atoms respectively. The carbon atoms are figured in black.

REFERENCES

1) Bujoli, B., Palvadeau, P., Rouxel, J.: Chem. Mater., 1990, 2, 582
2) Bujoli, B., Palvadeau, P., Rouxel, J.: C. R. Acad. Sc. Paris, Série II 1990, 310, 1213
3) Cao, G., Lee, H., Lynch, V.M., Mallouk, T.E.: Inorg. Chem., 1988, 27, 2781
4) Martin, K.J., Squattrito, P.J., Clearfield, A.: Inorg. Chim. Acta, 1989, 155, 7
5) Zhang, Y., Clearfield, A.: Inorg. Chem., 1992, 31, 2821
6) Cunningham, D. , Hennelly, P.J.D. , Deeney, T. : Inorg. Chim. Acta, 1979, 37, 95
7) Bujoli, B., Pena, O., Palvadeau, P., Le Bideau, J., Payen, C., Rouxel, J.: Chem. Mater., 1993, 5, 2781
8) Le Bideau, J., Payen, C., Palvadeau, P., Bujoli, B.: to be submitted
9) Le Bideau, J., Bujoli, B., Jouanneaux, A., Payen, C., Palvadeau, P., Rouxel, J. : Inorg. Chem., 1993, in press
10) Le Bideau, J., Payen, C., Palvadeau, P., Bujoli, B.: Inorg. Chem., submitted for publication

Materials Science Forum Vols. 152 - 153 (1994) pp. 371-374

HARD-SOFT CHEMISTRY IN NEW MATERIALS SYNTHESIS

P.F. McMillan [1,4], W.T. Petuskey [1,4], C.A. Angell [1,4], J.R. Holloway [1,2,4], G.H. Wolf [1,4], M. O'Keeffe [1,4] and O. Sankey [3,4]

[1] Department of Chemistry, Arizona State University, Tempe, AZ 85287, USA

[2] Department of Geology, Arizona State University, Tempe, AZ 85287, USA

[3] Department of Physics, Arizona State University, Tempe, AZ 85287, USA

[4] Materials Research Group in High Pressure Materials Synthesis

Keywords: Hard Chemistry, High Pressure Synthesis, Sulphides, Semiconductors, GeSiC Alloys

ABSTRACT

Our major research goals include synthesis and study of several classes of materials under high pressure conditions, utilizing techniques which might be termed "hard chemistry" (to distinguish these from the methods of "soft chemistry", and those of traditional synthetic chemistry). However, several of our syntheses require preparation of suitable precursors for subsequent high pressure treatment. The techniques of "soft chemistry" are often ideally suited to these precursor syntheses, providing an interesting marriage between the two synthesis philosophies. We present examples of this from our recent work on GeSiC tetrahedral alloys, and synthesis of ABS_3 sulphide phases. In addition, we are beginning to explore some potentially exciting directions for future research, in which "hard" chemistry treatments should result in precursor materials for subsequent soft chemistry investigations.

INTRODUCTION

We have recently established a Materials Research Group in High Pressure Materials Synthesis at Arizona State University. The principal goals of this group are (a) to synthesize new solids under high pressure conditions, (b) to investigate the effect of high pressure on materials properties, and (c) to explore the use of the pressure variable in designing novel synthesis routes, and in "tuning" desireable material properties. Whereas much of this chemistry and physics is "hard" in nature, in that it is carried out under extreme conditions of pressure and/or temperature, the techniques of "soft" chemistry are often essential to a successful synthesis, and a marriage of the two philosophies gives rise to interesting new possibilities in materials chemistry.

Some of the materials of interest to our group include group IV semiconductors within the Ge-Si-C system, and chalcogenides (X=S,Se,Te) with the ABX_3 stoichiometry. The former will give rise to a new family of wide band gap semiconductors, and the latter are likely to form new high dielectric or electrically conducting materials, based on the perovskite or hexagonal polytype structures. These provide useful examples of the interplay between soft and hard chemistry techniques in use at ASU.

GeSiC ALLOYS

Within the Ge-Si-C phase diagram, our initial interest focused on GeC, which is not yet known as a compound, but which might be expected to crystallize at high pressures. Before embarking on a "trial and error" program of exploratory syntheses, we carried out a series of ab initio pseudopotential calculations within the local density approximation, to obtain information on the thermodynamic stability of this hypothetical phase, both intrinsic, and relative to its consituent elements [1]. The electronic band structure for Si, C, Ge, SiC and the hypothetical GeC were obtained within the cubic (diamond or zincblende) structure, and the total energy computed as a function of volume. The calculations showed that cubic GeC indeed has a minimum in its potential energy vs volume surface, so that it could exist as a (meta)stable phase. Below approximately 15 GPa, this phase would be unstable to decomposition into Ge + C (diamond), which might indicate likely success of a synthesis at pressures above this value. However, Ge undergoes a transition to a β-Sn structure at 8.9 GPa, which will further stabilize the element at high pressures, so that the reaction Ge+C = GeC is always unfavourable. From these results, no direct syntheses of GeC from its elements were attempted. However, the calculations suggested that SiC and GeC should have similar lattice parameter and electronic structure, so that it should be possible to prepare a GeC-SiC solid solution, most likely with Si and Ge disordered on one sublattice [1]. Such a solid solution might well be stabilized by an entropic ($-T\Delta S$) term, so that a high temperature, high pressure synthesis might succeed. This has so far proved unsuccessful: the run products to date have contained only GeSi alloys (with a few per cent dissolved C) and graphite.

The result from the ab initio calculations, that GeC (and presumably GeC-SiC solid solutions) are intrinsically stable, suggests that it might be possible to design a synthetic scheme via a metastable route. One which we are in the process of exploring is the preparation from amorphous hydrogenated $(Si,Ge,C)H_x$ precursors, prepared via chemical vapour deposition. Pyrolysis of organmetallic (silyl and germyl) precursors has been shown to give amorphous SiGe:H and CSiGe:H materials [2,3]. A common technique in high pressure experimentation is the use of the noble metal container (capsule) as a semi-permeable membrane, allowing the flow of gas (especially hydrogen) across the capsule walls. This is usually used to fix the hydrogen (and hence oxygen) activity within the sample [4]. This property can be exploited in the GeSiC syntheses: amorphous hydrogenated alloy will be loaded into the capsule, and the external atmosphere buffered so that hydrogen diffuses out during the high pressure run, leaving a phase which might well crystallize metastably as a GeSiC tetrahedral alloy.

ABS$_3$ SULPHIDES

The second class of projects described here includes the synthesis of ABS$_3$ ternary sulphides. Oxides with the ABO$_3$ stoichiometry commonly crystallize with the perovskite structure, and give rise to a wide range of commercially important dielectric and non-linear optic materials. The corresponding sulphides crystallize mainly with the hexagonal BaNiO$_3$ structure (2H polytype), or with an orthorhombic structure based on that of NH$_4$CdCl$_3$. These latter structures have potentially interesting low dimensional electrical conductivity and magnetic properties. Based on packing arguments, the perovskite structure, or 3C polytype, is theoretically the most dense of the AX$_3$ layer stacking polytypes, so that transitions to this phase are expected at high pressure.

One of the principal targets of this research has been BaTiS$_3$, to form the sulphide analogue of barium titanate. Direct synthesis of this phase at high pressure from the constituent sulphides (BaS+TiS$_2$) is complicated by reaction with the capsule material (Au,Pt capsules), or loss of S from the charge (ceramic containers). (Reaction of the binary sulphides with the gold capsule at 1-2 GPa appears to yield a new phase in the Ba-Au-S system, which we are in the process of characterizing). Instead, a low pressure form of BaTiS$_3$ is prepared (2H form) as a precursor, and its phase transitions explored at high pressure and moderate temperature. This ternary sulphide shows little or no reaction with Pt capsules. So far, only broadening of the X-ray lines has been observed in experiments up to 9 GPa, indicating some polytype formation. The methods of soft chemistry are essential for the preparation of stoichiometric precursor material. Conventional syntheses at temperatures above 600°C lead to non-stoichiometric BaTiS$_{3-x}$, which enhances the stability of the hexagonal phase [5]. We have successfully prepared stoichiometric BaTiS$_3$ via CS$_2$ treatment of the oxide at 300-400°C and 2-3 atm pressure [6,7]. A second route to synthesis of metastable BaTiS$_3$ phases which we are currently exploring involves pressure treatment of Ba-intercalated TiS$_3$.

We are investigating other families of ABS$_3$ sulphides, with A=Ba,Sr,Pb,Sn and B=Ge,Si,Sn. As for the titanium sulphides, direct reaction of the binary sulphides or elements at high pressure results in complex reactions with all convenient container materials, so that low pressure preparation of precursors is required. In the Ba-Ge-S system, conventional precursor synthesis results in a mixture of barium germanium sulphides: however, reaction of BaGeO$_3$ with CS$_2$ at low temperatures gives directly stoichiometric BaGeS$_3$. This route also works smoothly for the other sulphides, except for

BaSiS$_3$, which does not react to completion. High pressure investigation of these phases, and their solid solutions, is under way.

HARD CHEMISTRY ROUTES TO SOFT CHEMISTRY PRECURSORS?

One other intriguing research direction which we are currently pursuing involves the exploration of high pressure, low temperature, routes to new phases. Considerable interest has been sparked by the phenomenon of "pressure-induced amorphization", in which a crystalline material is made to vitrify by compression at low (often room) temperature [8]. The result of this amorphization is a family of highly metastable materials, which may well exhibit a range of interesting and useful properties. In addition, this class of materials recovered from pressure-induced amorphization experiments is well suited to subsequent low pressure, low temperature reaction chemistry utilizing the free energy stored in these highly metastable materials as a driving force; i.e., to exploration by the methods of soft chemistry.

Acknowledgement

This research is supported by the National Science Foundation Materials Research Group grant DMR-9121570, and by Arizona State University. The high pressure experimental studies would not have been possible without the participation of Dr. Alison Pawley, Facility Manager, and Eileen Dunn, Instrument Maker. Several graduate students, undergraduates, and post-doctoral associates are involved in on-going research of the group, including the studies described here. These include Richard Brooker, Tina Clough, Phil Coffman, Chris Cahill, Alexander Demkov, Jason Dieffenbacher, Mark Drake, Shirley Ekbundit, Diana Fisler, Sharon Furcone, Tor Grande, Andrzej Grzechnik, Melonie Hall, Kris Halvorson, Hervé Hubert, Sarah Jacobs, Pankaj Joshi, Chris Pettinato, Scott Robinson, Tori Swarmer, Yanzhen Xu, and Jian Zhi Zheng.

References

1) Sankey, O.F., Demkov, A.A., Petuskey, W.T. and McMillan, P.F.: Modelling and Simulation in Mater. Sci. Eng., 1993, 1, 1
2) Meyerson, B.S.: IBM J. Res. Develop., 1990, 34, 806
3) Mazerolles, P., Reynes, A., Sefiani, S. and Morancho, R.: J. Anal. Appl. Pyrolysis, 1991, 22, 95
4) Holloway, J.R. and Wood, B.J.: *Simulating the Earth: Experimental Geochemistry*. Unwin Hyman, Boston. 1988
5) Clearfield, A.: Acta Cryst., 1963, 16, 134
6) Colombet, P., Molinie, P. and Spiesser, M.:"Procédé d'élaboration de composés binaires de soufre", 1991, European patent n° 440516S2
7) LeRolland, B., McMillan, P. and Colombet, P.: C.R. Acad.Sci Paris, sér. II, 1991, 312, 217
8) Wolf, G.H., Wang, S., Herbst, C.A., Durben, D.J., Oliver, W.F., Kang, Z.C. and Halvorson, K.: In *High-Pressure Research: Application to Earth and Planetary Sciences*, ed. Y. Syono and M.H. Manghnani, pp. 503-517. Terra Scientific Pub. Co. (Tokyo)/ Am. Geophys. Union, (Washington D.C.), 1992

Materials Science Forum Vols. 152 - 153 (1994) pp. 375-378
© *1994 Trans Tech Publications, Switzerland*

VANADATE-PILLARED HYDROTALCITE CONTAINING TRANSITION METAL CATIONS

F. Kooli [1,2], V. Rives [1] and M.A. Ulibarri [2]

[1] Departamento de Química Inorgánica, Universidad de Salamanca, Facultad de Farmacia, E-37007 Salamanca, Spain

[2] Departamento de Química Inorgánica e Ingeniería Química, Universidad de Córdoba, Facultad de Ciencias, F-14004 Córdoba, Spain

Keywords: Hydrotalcite, Pillaring, Decavanadate, Reconstruction

ABSTRACT

Synthesis of new pillared materials with the hydrotalcite structure, obtained from Ni-Al, Ni-Cr and Mg-Cr Layered Double Hydroxides in the carbonate form, has been attained by reconstruction of the structure with polyoxovanadate aqueous solutions. Such a synthesis was possible at a pH value of 4.5 for the Ni-Al sample, but not for the Ni-Cr one. At pH values higher than 4.5, pillared materials containing chain-like polyvanadates are obtained for both Ni-Al and Mg-Cr samples; however, no pillared materials were obtained from the Ni-Cr solid solution.

INTRODUCTION

A large number of pillared materials can be developed through pillaring of Layered Double Hydroxides (LDHs). The layered materials, the pillared compounds (Pillared Layered Double Hydroxides, PLDHs) and their decomposition products are highly interesting because of their activity in several catalytic processes [1]. Their properties have been recently reviewed [1,2]. Previously procedures to pillar LDHs have been reported; in some cases, swelling with organic compounds has been used to easy the access of polyoxometalates (POMs) [3,4], but this does not seem to be essential [5]. Calcination of LDHs at temperatures around 450°C leads to a mixture of mostly amorphous oxides; reconstruction of the LDH structure with intercalated POMs has been attained by impregnation of these materials with aqueous solution of POMs [6].

In the present study, we report the synthesis and characterisation of new pillared materials, obtained from Ni-Al, Mg-Cr and Ni-Cr LDHs having the hydrotalcite (HT) structure, following the second procedure described above, and using polyoxovanadate (POV) as a pillaring agent. Identification of decavanadate and the other species in the interlayer space was carried out by X-ray diffraction and FT-IR spectroscopy.

EXPERIMENTAL

The Ni-Al, Mg-Cr and Ni-Cr LDHs containing carbonate anion in the interlayer space were prepared at 70°C, by the coprecipitation method [7]. The precipitate was washed to eliminate the excess of sodium carbonate, and was submitted to hydrothermal treatment at 120°C for 48 h, then dried at 80°C overnight. The LDH material was calcined in air at 300, 400, 500°C for 3 h. The product thus obtained was added to a solution of sodium metavanadate at 70°C maintained at a pH value ranging from 4.5 to 8.5 in different preparations, stirred for 1 h and submitted again to hydrothermal treatment at 120°C overnight, being finally washed and dried at 65°C.

Powder X-ray diffraction (PXRD) diagrams were recorded in a Siemens D-500 instrument, using CuKα radiation; the FT-IR spectra were recorded using the KBr pellet technique in a Perkin

Elmer FT-1730 instrument, with a nominal resolution of 2 cm[-1] and averaging 50 scans to improve the signal-to-noise ratio.

RESULTS AND DISCUSSION

Prior to test reconstruction of the LDH structure using decavanadate solutions, it was essayed with water under hydrothermal conditions. PXRD diagrams showed that when the Ni-Al and Mg-Cr samples pre-calcined at 300°C were hydrothermally treated at 120°C overnight they recovered the LDH structure, but the Ni-Cr sample did not. When the pre-calcination temperature is increased up to 500°C, reconstruction is difficult to be achieved, whichever the nature of the divalent and trivalent cations in the oxide solid solution.

Ni-Al Precursor

The PXRD patterns obtained for the Ni-Al precursor after reaction with the vanadate solution are included in figure 1. They show that formation of the PLDH materials is achieved, together with an unreactive phase of NiO. The proportion of the last phase in the final product depends on the pre-calcination temperature, increasing as this does. The presence of an excess of NiO phase can be due to an easy dissolution of aluminum [8].

Assuming that the cationic layers have a thickness similar to that of brucite (4.8 Å), the gallery height is 6.9 Å, corresponding to the presence of $V_{10}O_{28}^{6-}$ anions with an orientation in which the C_{2v} axis is parallel to the host layers [3,9]. The broad peak around 9.8 Å can be due to the presence of $V_4O_{12}^{4-}$ species as a coproduct [9]. Formation of this species can be attributed to the fact that polyvanadate species are stable only over a narrow pH range, in which case the synthesis of a single PLDH phase is difficult to be achieved, even with a careful control of pH.

The influence of the pH on the synthesis of the PLDH materials was also examined. Figure 1 shows that the pillared product prepared at pH=5.5 is formed by a single phase containing a polyvanadate anion. The basal spacing 7.3 Å (d_{003} = 7.3 Å) corresponded to a gallery height of 2.5 Å, due to the formation of VO_3^- chain-like polymeric metavanadate species [10]. Contrary to the results reported by Dutta et al. [11] for Li,Al-LDHs, our results indicated that the nature of the PLDH seems to be independent of the pH during reconstruction, if above 5.5.

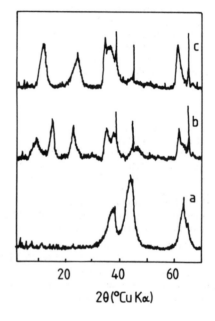

Fig. 1. XRD diagrams of: (a) Ni-Al hydrotalcite calcined at 300°C; (b) sample (a) reconstructed with metavanadate at pH=4.5; (c) ibid. pH=5.5.

Increasing in the temperature during hydrothermal treatment did not affect seriously the chemical nature of the species formed, but increased the cristallinity of the PLDHs materials. Even after hydrothermal treatment at 150°C and pH= 4.5, the first order (003) peak (d=11.69 Å) can be clearly detected.

The FT-IR spectra of the Ni-Al compounds before and after reconstruction with decavanadate are shown in figure 2. Incorporation of the decavanadate anion yield significant changes in the spectrum. New bands appear in the 1000-500 cm[-1] range. The band at 1375 cm[-1] due to carbonate is completely eliminated, indicating that only decavanadate anions exist in the interlayers, and the new bands indicate its presence [12]. The band at 960 cm^{-1} can be ascribed to the V=O terminal stretching mode; polyoxovanadates give rise to a different number of bands in this region, depending on the polymerization degree, and decavanadate gives rise to a single band only [13]. Other bands at 806, 748 and 586 cm^{-1} coincide with those reported by Lopez-Salinas and Ono for $V_{10}O_{28}^{6-}$ -pillared Mg,Al LDHs [12]. Figure 2c shows the FT-IR spectrum of LDH materials pillared at pH 5.5. The spectrum is different from that recorded for the material pillared at pH 4.5, indicating the formation of a different vanadate species and confirming the PXRD results. The band close to 902 cm^{-1} is assigned to $v(VO_2)$ (asy) and the band at 683 cm^{-1} can be attributed to $v(VOV)$, which are the typical bands of $(VO_3)_n^{-n}$ chain [10]. There is, however, a weak band close to 1370 cm^{-1} that can be due to carbonate, indicating the presence of traces of carbonate in this sample, probably as the result of interference during the synthesis by carbon dioxide from the atmosphere.

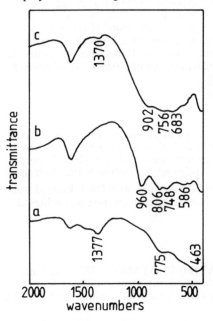

Fig. 2. FT-IR spectra of: (a) Ni-Al hydrotalcite calcined at 300°C; (b) sample (a) reconstructed with metavanadate at pH=4.5; (c) ibid. pH=5.5.

Mg-Cr Precursor

Hydrothermal treatment of the Mg-Cr oxide phase at pH 4.5 caused its dissolution. Then, reconstruction was attempted at pH values higher than 4.5. The pillared materials obtained from Mg-Cr LDH, calcined at 300°C and then reacted with vanadate at pH=5.5, did not show a good cristallinity; the peaks recorded in the PXRD pattern were weak, and characteristic of VO_3^- polyvanadate chain-like anions. Similar weak diffraction peaks were observed in the case of Mg-Al oxide after reconstruction with vanadate anions at pH=4.5 [6]. In our case, a further of pH during reconstruction did not affect the nature of the vanadate species existing in the interlayer space, this being the same polyvanadate species obtained with the Ni-Al precursor. However, an improvement in the cristallinity degree is observed in both systems (Ni-Al, Mg-Cr) if higher pHs are used. This finding can be related to a lower dissolution of the oxides at relatively high pH values.

The nature of the pillared materials obtained depends on the pre-calcination temperature of the Mg-Cr precursors: if precalcination is carried out at 300°C, obtention of pillared materials is attained, but from materials pre-calcined at 500°C, the product formed after attempting reconstruction consists of spinel and oxide phases, without any reconstruction to the LDH structure. This different behaviour can be originated by the high stability of the Cr-containing spinel phase. Attempting reconstruction with materials pre-calcined at intermediate temperatures (e.g., 400°C) gave rise to an amorphous phase, that can represent the transition from a Mg-Cr-O oxide solid solution to spinel and MgO phases.

The FT-IR spectrum of the solids reconstructed at pH≥5.5 confirms the presence of the polyvanadate chain-like species, with characteristic bands 940, 760, 605 and 544 cm^{-1} as in the case of the Ni-Al-O precursor. Minor shifts towards higher wavenumbers for some bands, if compared to those for the Ni-Al material, can be due to the influence of the chromium cation on the configuration of the $(VO_3)_n^{-n}$ groups in the interlamellar space [10].

Ni-Cr precursor

Similar attempts have been carried out for the Ni-Cr solid solution, but no reconstruction was achieved. The crystallographic phases did not change after exposure to the metavanadate solution, whichever the experimental conditions. Only an improvement in the cristallinity of the oxides (sharper PXRD peaks) was observed after the hydrothermal process. The spinel phase ($NiCr_2O_4$) is already detected in the material pre-calcined at 500°C, and the enhanced stability of this precursor may hinder reconstruction. This is in agreement with previous studies [14] that have shown that the ability of a calcined precursor to reconstruct is independent on the type of product formed upon calcination, but is influenced by the presence of metal cations.

CONCLUSION

This study demonstrates that the ability for reconstruction of the oxides obtained by calcination of LDHs is mainly controlled by the properties of divalent cations [15] but also by that of the trivalent ones. Therefore, obtention of the pillared materials with decavanadate at pH=4.5 was possible starting from the Ni-Al precursor, but not from the Ni-Cr one. The nature of the intercalated polyvanadate in the pillared material, $(VO_3)_n^{-n}$ chain-like, did not change at pH$\geq$4.5. Pillared materials are hardly obtained by reconstruction from Ni-Cr precursors. Reconstruction was achieved in all cases by anion exchange [16].

ACKNOWLEDGMENT

Authors thank finantial support from CICYT (MAT91-767 and MAT93-787) and Junta de Castilla y León (Consejería de Turismo y Cultura). FK acknowledges a grant from Junta de Andalucia and Ministerio de Educación y Ciencia (SB92-AE0474745). This work is within the CEA-PLS scheme.

REFERENCES
1 Cavani, F., Trifiró, F. and Vaccari, A.: Catal. Today, 1991, 11, 173.
2 de Roy, A., Forano, C., El Malki, K. and Besse, J.-P.: in "Expanded Clays and Other Microporous Solids" (M. L. Occelli & H. E. Robson, Eds.), Van Nostrand Reinhold, New York, 1992, p. 108.
3 Drezdzon, M.A.: Inorg. Chem., 1988, 27, 4628.
4 Dimotakis, E.D. and Pinnavaia, T.J.: Inorg. Chem., 1990, 29, 2393.
5 Wang, J., Tiang, Y., Wang, R.C. and Clearfield, A.: Chem. Mater, 1992, 4, 1276.
6 Chibwe, K. and Jones, W.: Chem. Mater, 1989, 1, 489.
7 Reichle, W.T.: Solid States Ionics, 1986, 22, 135.
8 Clause, O., Rebours, B., Merlen, E., Trifió, F. and Vaccari, A.: J. Cat,1992, 133, 231.
9 Kwon, T., Tsigdinos, G.A., and Pinnavaia T.J.: J. Am. Chem.Soc. 1988, 110, 3653.
10 Onodera, S. and Ikegami, Y.: Inorg. Chem. 1980, 19, 615.
11 Dutta, P.K. and Twu, J.J.: J. Phys. Chem., 1989, 93, 7863.
12 Lopez-Salinas, E. and Ono, Y.: Bull. Chem. Soc. Jpn., 1992, 65, 2465.
13 Griffith, W.P. and Wickins, T.D.: J. Chem. Soc. (A)., 1966, 1087.
14 Kagunya, W. Ph. D. Thesis, Darwin College, Cambridge. U.K., 1992.
15 Sato, T., Fujita, T., Shimada, M. and Tsunashima, A.: Reactivity of Solids, 1988, 5, 219.
16 Kooli, F., Rives, V. and Ulibarri, M. A.: to be published.

Materials Science Forum Vols. 152 - 153 (1994) pp. 379-382
© *1994 Trans Tech Publications, Switzerland*

INTERCALATION OF THIONINE IN COLLOIDAL
α-ZIRCONIUM PHOSPHATE

E. Rodríguez-Castellón[1], A. Jiménez-López[1], P. Olivera-Pastor[1],
J.M. Mérida-Robles[1], F.J. Pérez-Reina[1], M. Alcántara-Rodríguez[1],
F.A. Souto-Bachiller[2], L. de los A. Rodríguez-Rodríguez[2] and
G.G. Siegel[2]

[1] Departamento de Química Inorgánica, Facultad de Ciencias, Universidad de Málaga,
E-29071 Málaga, Spain

[2] Departamento de Química, Recinto Universitario de Mayagüez, Universidad de Puerto Rico,
Mayagüez, Puerto Rico 00681-5000

Keywords: Intercalation Complexes, Thionine, Layered Zirconium Phosphates, Optical
Properties

ABSTRACT

The reaction of thionine acetate with colloidal n-propylammonium α-zirconium phosphate has been investigated. The dye intercalates by ionic exchange with n-propylammonium up to completion. A complex of formula $Zr(C_{12}N_3H_{11}S)_{0.91}H_{1.09}(PO_4) \cdot 2.21H_2O$ having an interlayer distance of 17.73 Å is obtained at the maximum thionine loading. Monolayer thionine arrays, slanted 56° with respect to the phosphate layers, form as a result of this process. This orientation gives rise to a pronounced blue shift and splitting of the absorption spectrum of thionine measured by diffuse reflectance. The absorption band of the thionine monomer (λ_{max} = 597 nm, H_2O, pH 1) splits upon intercalation showing maxima at 493 nm and 620 nm. This band splitting and the shifts are indicative of the formation of an array of straight chain polymers confined in the interlayer region of the phosphate.

INTRODUCTION

α- Zirconium phosphate (α-$Zr(HPO_4)_2 \cdot H_2O$, α-ZrP) is a swellable layered solid with an interlayer region made up of acid phosphate groups [1]. α-ZrP is not readily swellable with large, voluminous amines. We and others have shown recently that if the phosphate is swelled previously with n-propylamine, then the insertion of large basic organic molecules is possible [2]. Thionine (TH^+) is a well known thiazine dye with a strong tendency to aggregate in aqueous solution, forming dimers, trimers, and higher polymers as the stoichiometric dye concentration increases [3]. The sorption properties and the metachromasia of TH^+ have found interesting applications ever since its discovery by Lauth in 1876 [4]. Staining techniques developed originally for biological materials have been applied also to mineralogy [5]. The sorption of organic dyes such as TH^+ or methylene blue in natural or modified clay minerals can produce colour changes in the clay. Frequently such changes vary depending on the identity of the clay mineral and its composition. Dye metachromasy therefore provides a possible basis to characterise clay-mineral components or clay-type materials. Recent experimentation here and elsewhere confirms this view. Porous solids such as zeolite L can include TH^+ molecules forming head-to-tail linear chains with in-plane transition moments. No metachromasia is observed in this case even at the highest TH^+ concentrations studied: 0.2 M relative to the zeolite L volume [6]. Clay minerals, such as montmorillonite and vermiculite, intercalate TH^+ in the interlayer space producing metachromasia [7]. The relatively low charge density of these silicates causes TH^+ to adopt a flat orientation. Layered phosphates of tetravalent metals (M(IV) = Zr, Sn, Ti) are alternative host matrices to smectite clays and zeolites. They are swellable like smectites but have a higher layer charge density. The orientation of intercalated organic molecules in layered phosphates will depend both, on the geometry, acidity and degree of charge localization of the complexed organic cation, and on the charge density of the phosphate layers [8]. TH^+ has two amino groups located accross the phenothiazine nucleus and is capable to interact simultaneously with acid groups in two opposite phosphate layers. Hence, formation of monomolecular arrays of TH^+ molecules normal to the planes of the phosphates can be anticipated. In this paper we report the preparation and characterisation of well ordered complexes of α-zirconium phosphate with intercalated TH^+ and their striking optical properties.

EXPERIMENTAL

α-ZrP was prepared by well established methods [9]. Colloidal suspensions of n-propylamine α-zirconium phosphate (*n*-PrAH-ZrP) were prepared as described elsewhere [2]. Thionine acetate was obtained from Aldrich and used without further purification. All other chemicals used were the best available analytical grade reagents.

Ion exchange isotherms.— Colloidal *n*-PrAH-ZrP was placed in contact with increasing quantities of thionine acetate dissolved in 100 cm^3 of water (from 0.053 to 10.22 mmol TH^+ added per gram of ZrP_2O_7, *i.e.* [TH^+] varying from $9.36 \cdot 10^{-5}$ to $1.80 \cdot 10^{-2}$ mol dm^{-3}), at 25°C for 1 day. The suspensions were partitioned by centrifugation at 11.000 r.p.m.. The deeply coloured solids were washed well with water until the washings were colourless. Then they were air dried and analysed by CHN and TG/DTA. The concentration of TH^+ in the uppernatant solutions was determined spectrophotometrically. The TH^+ content of the solids was estimated as the difference between the initial and the equilibrium dye concentration in the uppernatant solutions. In all cases, the elemental and thermogravimetric analyses and the spectrophotometric determinations were in good agreement. Table 1 shows observed and calculated compositions for the α-zirconium complexes with TH^+ and *n*-propilammonium.

The intercalation compounds in powder and cast films were characterised by X-ray diffraction (XRD) on a Siemens D-500 diffractometer. DTA-TG analyses were carried out on a Rigaku Thermoflex TG8110 using calcined Al_2O_3 as the internal standard reference and 10 $Kmin^{-1}$ as a heating rate. Thionine was colorimetrically determined using a Kontron-Uvikon spectrophotometer. Diffuse reflectance spectra were recorded on a Shimadzu MPC 3100 spectrophotometer using $BaSO_4$ as reference.

Table 1. Composition of thionine-n-PrAH-ZrP intercalation compounds.

Sample	Observed			Calculated			Formula
	C	H	N	C	H	N	
A	24.56	2.57	7.12	24.70	2.50	7.20	$Zr (C_{12}N_3H_{11}S)_{0.91} H_{1.09} (PO_4)_2 \cdot 2.21 H_2O$
B	21.56	2.87	6.56	20.83	2.54	6.16	$Zr (C_{12}N_3H_{11}S)_{0.61}(C_3H_{10}N)_{0.11} H_{1.26} (PO_4)_2 \cdot 1.13 H_2O$
C	21.22	3.10	6.49	21.39	3.14	6.83	$Zr (C_{12}N_3H_{11}S)_{0.56} (C_3H_{10}N)_{0.51} H_{0.93} (PO_4)_2 \cdot 1.15 H_2O$
D	18.79	2.95	6.10	19.04	3.11	5.97	$Zr (C_{12}N_3H_{11}S)_{0.45} (C_3H_{10}N)_{0.53} H_{1.02} (PO_4)_2 \cdot 1.23 H_2O$
E	14.53	3.22	4.74	15.93	3.78	5.44	$Zr (C_{12}N_3H_{11}S)_{0.22}(C_3H_{10}N)_{0.95} H_{0.82} (PO_4)_2 \cdot 1.51 H_2O$
F	10.54	3.06	4.01	10.80	3.75	3.81	$Zr (C_{12}N_3H_{11}S)_{0.11} (C_3H_{10}N)_{0.75} H_{1.14} (PO_4)_2 \cdot 1.51 H_2O$
G	10.53	3.38	3.76	10.55	3.56	3.89	$Zr (C_{12}N_3H_{11}S)_{0.06} (C_3H_{10}N)_{0.87} H_{1.07} (PO_4)_2 \cdot 1.45 H_2O$
H	11.96	3.71	4.54	12.12	4.08	4.61	$Zr (C_{12}N_3H_{11}S)_{0.03} (C_3H_{10}N)_{1.18} H_{0.79} (PO_4)_2 \cdot 1.39 H_2O$
I	11.38	3.80	4.31	11.38	3.80	2.64	$Zr (C_{12}N_3H_{11}S)_{0.01} (C_3H_{10}N)_{1.12} H_{0.87} (PO_4)_2 \cdot 1.25 H_2O$

RESULTS AND DISCUSSION

The uptake curve for TH^+ on n-PrAH-ZrP is shown in figure 1. At low TH^+ concentrations, the phosphate showed a strong affinity for thionine, the slope being practically 1. The plateau of the curve is reached at relatively low dye additions. The isotherm shows that maximum uptake corresponds to a formula $Zr(C_{12}N_3H_{11}S)_{0.91}H_{1.09}(PO_4)_2 \cdot 2.21H_2O$. The analysis of the materials separated at the points of the isotherm is reported in Table1. In contrast to some layered compounds such as layered dichalcogenides and graphite, where the intercalation takes place by stages, in the case of n-PrAH-ZrP, the diffusion begins at the edges of the crystallites and proceeds, with an advancing phase boundary, to the bulk of the crystallites. The XRD patterns of the intercalation compounds show that with low TH^+ loadings, two phases coexist (figure 2), the original of the precursor n-PrAH-ZrP with a reflection line at 14.67 Å and the new intercalated phase at 17.73 Å. The intercalation of TH^+ results in a progressive ion exchange process where TH^+ is exchanged for n-propylammonium ion. The new reflection at 17.73 Å increased while that of the original interlayer spacing for n-PrAH-ZrP (14.67 Å) decreased proportionately with TH^+ loading. After TH^+ additions of 0.5 mol per formula, the only phase present is that at 17.73 Å. A plot of the phase percentage versus the moles TH^+ added per g α-ZrP (figure 3) is useful to follow the intercalation process.

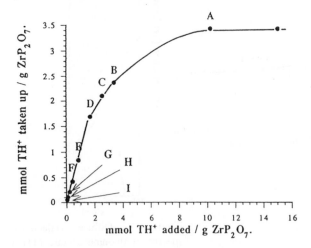

Figure.1 Uptake of thionine by n-PrAH-ZrP at 25ºC.

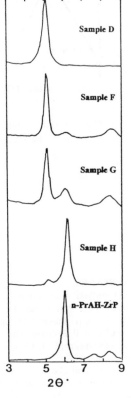

Figure 2. XRD patterns of intercalation compounds at different thionine loadings.

The composition of the intercalate at the highest TH^+ loading (0.91 mol of TH^+ per formula), and the corresponding interlayer distance of 17.73 Å, agree well with the formation of a monolayer of slanted TH^+ molecules inclined ~56° with respect to the phosphate plane. The inclination angle was calculated using a Van der Waals diameter of 13.9 Å along the longitudinal axis of TH^+ and a free height for the phosphate interlayer of 11.53 Å. The free height was estimated as the difference between the basal spacing, 17.73 Å, and the phosphate layer thickness, 6.2 Å.

The diffuse reflectance spectra of all samples (figure 4) show two maxima for TH^+, one blue shifted with respect to the monomer, between 516 and 493 nm, and the other shifted to the red, between

609 and 626 nm. The observed metachromasia and the band splitting are consistent with the formation of parallel, card pack aggregates even at the lowest TH^+ loadings. This is attributed to a very efficient *in situ* aggregation of intercalated TH^+ monomers caused by the disposition of the P-O-H groups toward the interlayer space of the phosphate. The high degree of molecular organization imparted by the phosphate sheets of layered tetravalent metal phosphates and the optical manifestations in the absorption spectra will be dealt with in a forthcoming paper [10].

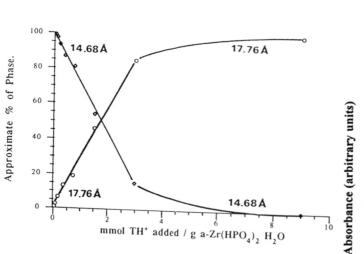

Figure 3. Approximate percentage of the phases present in the intercalacion of TH^+ into n-PrAH-ZrP.

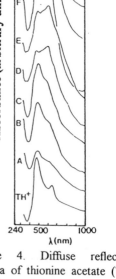

Figure 4. Diffuse reflectance spectra of thionine acetate (TH+) and intercalation compounds at different thionine loadings

REFERENCES
1) (a) Clearfield, A. and Smith, G.D.: Inorg. Chem., 1969, 8, 431. (b) Albertsson, J.; Oskarsson, A; Tellgren, R. and Thomas, J.O.: J. Phys. Chem., 1977, 16, 1574.
2) (a) Alberti, G.; Casciola, M. and Costantino, U.: J. Colloid Interface Sci., 1985, 107, 256. (b) Maireles-Torres, P.; Olivera-Pastor, P; Rodríguez-Castellón, E.: Jiménez-López, A. and Tomlinson, A.A.G.: J. Mat. Chem., 1991, 1, 739.
3) (a) Rabonowitch, E. and Eptein, L.: J. Am. Chem. Soc., 1941, 63, 69. (b) Haugen, G.R.; Hardwick. E.R., J. Phys. Chem., 1963, 67, 725.
4) Lauth. F.: Bull. Soc. Chim. Fr., 1876, 26, 1422.
5) Grim, R.E.: Clay Mineralogy, McGraw-Hill, New York, N.Y.. 1968, p. 407.
6) Calzaferri, G. and Gfeller. N.: J. Phys. Chem., 1992, 96, 3428.
7) Sunwar, C.B. and Bose H.: J. Colloid Interface Sci., 1990, 136, 54.
8) Jiménez-López, A; Maireles-Torres, P.; Olivera-Pastor, P.; Rodríguez-Castellón, E.; Hudson. M and Sylvester, P.: J. Incl. Phenomena, 1990, 9, 207.
9) Alberti, G. and Torraca, E.: J. Inorg. Nucl. Chem., 1968, 30, 317.
10) Rodríguez-Castellón, E.; Jiménez-López, A.; Olivera-Pastor, P.; Souto-Bachiller, F.; Rodríguez-Rodríguez, L. and Siegel, G.G.: J. Phys. Chem. submitted.

Materials Science Forum Vols. 152 - 153 (1994) pp. 383-386
© *1994 Trans Tech Publications, Switzerland*

TaNi$_{2.05}$Te$_3$ AND Ta$_2$Ni$_3$Se$_5$, NEW METAL-RICH TERNARY TANTALUM CHALCOGENIDES

J. Neuhausen [1], W. Tremel [1] and R.K. Kremer [2]

[1] Institut für Anorganische Chemie und Analytische Chemie der Johannes Gutenberg-Universität, J.J. Becherweg 24, D-55128 Mainz, Germany

[2] Max Planck-Institut für Festkörperforschung, Heisenbergstrasse 1, D-70506 Stuttgart, Germany

Keywords: Transition Metal Chalcogenides, Metal Rich Layer Compounds, Electronic Structure

ABSTRACT

TaNi$_{2.06}$Te$_3$ and Ta$_2$Ni$_3$Se$_5$ have been prepared from the elements at high temperature. Both compounds crystallize as layer structures. A characteristic structural feature are infinite chains of rhomb-like edge-sharing Ta$_2$Ni$_2$ clusters sandwiched between chalcogen sheets. Tight-binding band structure calculations indicate that bonding interactions between early and late transition metals are responsible for the electronic stability of these compounds.

INTRODUCTION

Transition metal chalcogenides have attracted wide interest during the past decades primarily because of their interesting structural chemistry, unusual electronic properties and their rich intercalation chemistry. Traditionally, sulfides and selenides received more attention than tellurides. This may have historical reasons - most of the early investigated compounds are minerals - but the main reason was based on the assumption that no significant differences should be found between sulfides, selenides and tellurides. This picture has changed dramatically. The increased covalency of tellurium allows for a large variability in chalcogen-chalcogen bonding. In view of these differences the synthetic efforts in telluride chemistry were intensified within the past few years and many compounds with unusual structures and properties have been synthesized. Meanwhile the chemical "individualism" of the tellurides seems well accepted. Many novel materials, e.g. Ta$_6$Te$_5$ [1] or TaM'$_2$Te$_2$ (M' = Co, Ni) [2], were reported recently. None of them has a counterpart among the sulfides and selenides. The reason for the chemical differences among the chalcogenides are poorly understood. Neither space filling nor electronegativity arguments can be used in a consistent manner to explain their different behaviour. If stabilizing chalcogen-chalcogen interactions were a dominant factor, those structures types with Q-Q separations intermediate between nonbonding and single bond distances should be restricted to the tellurides. However, few examples such as NbSe$_3$ [3], where short Se-Se separations of approximately 2.7 Å are observed, show these factors may be effective for the selenides as well. In order to elucidate the differences between the transition metal chalcogenides we have begun an investigation of the syntheses, structures, and properties of metal-rich early transition metal sulfides and selenides. We report here the synthesis and structure of Ta$_2$Ni$_3$Se$_5$, which is one of the rare examples for selenides with a layer structure.

EXPERIMENTAL

$Ta_2Ni_3Se_5$ was prepared by heating the elements in the desired formula ratio in sealed evacuated quartz tubes at 800°C for 10 days. Single crystals can be obtained by adding a small amount of iodine as a mineralizer. The bulk material obtained in almost quantitative yield consists of needles up to 1 cm long with a metallic luster.

STRUCTURE

The crystal structure of $Ta_2Ni_3Se_5$ has been determined by single crystal X-ray methods. A summary concerning the structure determination is given in Table 1.

Table1: Crystal data for $Ta_2Ni_3Se_5$

Crystal System	orthorhombic	$\mu(MoK_\alpha)$	56.49 mm^{-1}
Space Group	Pnma	$\lambda(MoK_\alpha)$	0.71073 Å
a	12.711(2) Å	min., max. 2θ	4°, 65°
b	3.508(1) Å	Observed Reflections	780 (I>2σ(I))
c	17.252(2) Å	Absorption Correction	empirical (Ψ-Scan)
V	769.27 Å^3	Structure Solution	Direct Methods (SHELXS)
D(calc.)	8.053 g cm^{-3}	Stucture Refinement	Full Matrix Least Squares
T	298 K	No. of Parameters	62
Z	2	$R(R_w)$	0.033(0.028)

$Ta_2Ni_3Se_5$ crystallizes as a layer structure. The unit cell contains two layers, every second one being displaced by $a/2$ with respect to the neighboring ones. Only van der Waals interactions have to be assumed between the layers, the shortest interlayer Se-Se distance being approx. 3.58 Å. Fig. 1 shows an idealized polyhedral representation of the Ta_2Se_5 portion of the structure. The structure contains double chains of edge-sharing octahedra, Ta_2Se_6, which are linked by sharing common vertices.

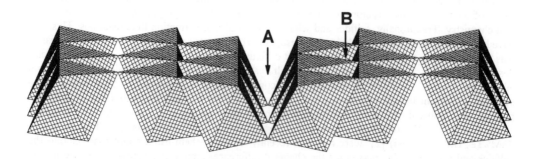

Figure 1: Polyhedral representation of a layer of the $Ta_2Ni_3Se_5$ structure.
           Only $TaSe_6$ octahedra are shown.

Each Ta atom is in an effective five coordination, the remaining Ta-Se distances being 3.678(1) Å and 3.694(1) Å. The distortion from the ideal geometry is not the result of chalcogen-chalcogen bond formation. All Se-Se distances (larger than 3.5 Å) are formally consistent with the presence of Se^{2-} anions. In addition, all metal-chalcogen distances are in reasonable agreement with the values computed from tabulated ionic radii. Therefore, it can be assumed that the Ta coordination sphere is completed by three additional Ni atoms at (average) distances of 2.614 Å and 2.689 (2x) Å, respectively. Considering the Ta five coordination the polyhedral representation in Fig. 1 may seem inappropriate, but - as it will turn out - it facilitates the structure description and the comparison with related compounds. The Ta_2Se_5 octahedral double chain contains two types of tetrahedral voids. One of them (labelled A in Fig. 1) is formed by four octahedra, each two of them sharing common edges, and each two others sharing common vertices. The second one (labelled B in Fig. 1) is formed by three edge-sharing octahedra, and there are two type B sites per formula unit. Filling all tetrahedral voids of the layer with Ni atoms leads to a composition $Ta_2Ni_3Se_5$. Filling the type A sites only leads to a compound with composition Ta_2NiSe_5, while filling the type B sites results in a composition $Ta_2Ni_2Se_5$. Interestingly, the former compound has been reported several years ago, and $Ta_2Ni_3Se_5$ may be considered a stuffed variant of the Ta_2NiSe_5 structure type [4]. There are, however, some significant differences between the structures of $Ta_2Ni_3Se_5$ and Ta_2NiSe_5. (1) Whereas Ta is octahedrally Se-coordinated in the Ta_2NiSe_5 structure (d_{Ta-Se} = 2.523 - 2.678 Å), we observe five coordination of Ta by Se atoms in the $Ta_2Ni_3Se_5$ structure. (2) Metal-metal bonding is more important in the metal-rich structure type. This is immediately clear from a comparison of the short metal-metal contacts in both structure types. Whereas there are only few short metal-metal separations (d_{Ta-Ni} = 2.804 Å (1x) and 2.813 Å (1x), respectively) in the Ta_2NiSe_5 structure, there are numerous short Ta-Ni contacts ($\bar{d}_{Ta-Ni}$ = 2.781 Å (2x), $\bar{d}_{Ta-Ni}$ = 2.689 Å (2x), $\bar{d}_{Ta-Ni}$ = 2.614 Å (1x)) in the $Ta_2Ni_3Se_5$ structure type. From another point of view, the metal-sublattice of the title compound contains rhomb-like Ta_2Ni_2 metal clusters, which are a characteristic structural feature for many of the recently reported metal-rich tellurides. These clusters may share common edges as in the $Ta_2Ni_3Se_5$ structure, common corners or common MM' axes as in the TaM'_2Te_2 structure (M' = Co, Ni) [2]. A major portion of the metal-metal bonding in these structures is carried through Ta-Ni bonds. This result is consistent with Brewer's rules, which state that a strong affinity exists between d-electron-poor ("Lewis acids") and d-electron-rich metals ("Lewis bases"). Interestingly, a similar relationship exists between the recently synthesized tellurides $TaNi_{2.05}Te_3$ and $TaFe_{1.14}Te_3$. Here, the structure of the former compound can be derived by insertion of additional 3d-metal atoms in appropriate tetrahedral voids of the $TaFe_{1.14}Te_3$ structure as shown in Figure 2. The importance of interactions between d-electron poor and rich metals is confirmed by the results of band structure calculations [5].

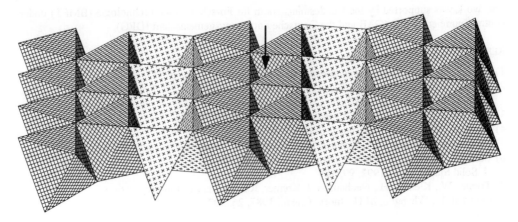

Figure 2: Polyhedral representation of the $TaMTe_3$-portion of the $TaFe_{1.14}Te_3$ and $TaNi_{2.05}Te_3$ structures. Additional tetrahedral sites occupied in $TaNi_{2.05}Te_3$ are indicated by an arrow.

The results presented here indicate that the occurrence of low-dimensional metal-rich chalcogenides is not restricted to tellurides only. New and unexpected possibilities emerge from the synthesis of related selenides and sulfides.

ELECTRONIC PROPERTIES

The electronic stability and the occurrence of other more subtle structural features such as super-structure formation originating from metal-metal bonding have been attributed to the presence of weak *interlayer* Te-Te interactions for the tellurides [6]. While the presence of weak chalcogen-chalcogen interactions may be intuitively clear for the tellurium-rich tellurides, where Te-Te distances can vary over a wide range between 2.7 Å and 3.5 Å, it seems surprising for metal-rich compounds. Band structure calculations (extended Hückel approximation) for $Ta_2Ni_3Q_5$ (Q = Se, Te) [7] and Ta_2NiQ_5 (Q = S, Se) [8] show that weakly bonding intralayer and interlayer interactions exist. The largest Q-Q overlap populations (0.008 and 0.025, respectively) are associated with the intralayer Q-Q contacts (3.48 Å for $Ta_2Ni_3Se_5$ and 3.53 Å for $Ta_2Ni_3Te_5$), the remaining intralayer Q-Q contacts being weakly antibonding. All interlayer contacts are weakly bonding; the largest computed overlap populations being 0.007 for $Ta_2Ni_3Se_5$ and 0.006 for $Ta_2Ni_3Te_5$. On the other hand, small (but positive) interlayer Se-Se overlap populations (approx. 0.002) are computed for Ta_2NiSe_5, whereas all intralayer overlap populations are negative (approx. -0.01). Therefore, weak chalcogen-chalcogen bonding interactions may be of general importance in low dimensional chalcogenide structures, although their influence has been mostly neglected. In agreement with qualitative bonding arguments their strength increases in the order S-S < Se-Se < Te-Te.

PHYSICAL PROPERTIES

$Ta_2Ni_3Se_5$ is diamagnetic and shows semiconducting behaviour. This result is surprising for a metal rich compound, where metallic properties should be expected. However, the observed semiconductivity can be rationalized with the aid of band structure calculations [7].

ACKNOWLEDGEMENTS

This work was supported by the Bundesministerium für Forschung und Technologie (BMFT) under contract number 05 5UMGAB and the Deutsche Forschungsgemeinschaft (DFG).

REFERENCES

1) Conrad, M., Harbrecht, B.: Chemiedozententagung, 1992, Abstr. p. 35
2) Tremel, W.: Angew. Chem., 1992, 104, 230; Angew. Chem., Int. Ed. Engl., 1992, 31, 217
   Tremel, W.: J. Chem. Soc., Chem. Commun., 1991, 1405
3) Meerschaut, A., Rouxel, J.: J. Less-Common. Met., 1975, 39, 498
4) Sunshine, S.A., Ibers, J.A.: Inorg. Chem., 1985, 24, 3611
5) Neuhausen, J., Tremel, W., Kremer, R.K.: Chem. Ber., submitted for publication
6) Canadell, E., Jobic, S., Brec, R., Rouxel, J., Whangbo, M.H.:
   J. Solid State Chem., 1992, 99, 189.
7) Tremel, W., Kleinke, H., Neuhausen, J. Kremer, R.K.: manuscript in preparation
8) Canadell, E., Whangbo, M.H.: Inorg. Chem., 1987, 26, 3974

Materials Science Forum Vols. 152 - 153 (1994) pp. 387-390
© 1994 Trans Tech Publications, Switzerland

NON-STOICHIOMETRY AND SOLID STATE IN SPINEL-TYPE CATALYSTS

F. Trifirò and A. Vaccari

Dipartimento di Chimica Industriale e dei Materiali, Viale del Risorgimento 4,
I-40136 Bologna, Italy

Keywords: Non-Stoichiometry, Spinel-Type Catalysts, Low-Temperature Preparation, Methanol, Selective Hydrogenation

ABSTRACT

The effects induced by non-stoichiometry in various mixed oxides are examined. Non-stoichiometric spinel-type phases may be obtained by low-temperature methods, and formed by different mechanisms as a function of the nature of the elements present and of the heating atmosphere. All these phases are metastable and evolve towards stoichiometric spinels as temperature and time-on-stream increase. Multicomponent catalysts for the hydrogenation of carbon monoxide or organic molecules can thus be obtained, the properties of which can be regulated by properly selecting the elements and appropriately adjusting their composition.

INTRODUCTION

Mixed oxides are widely employed in the chemical industry both as heterogeneous catalysts and as materials having specific properties [1,2]. In the last few years, there has been an increasing interest in preparation methods employing low temperatures, by means of which materials with properties very different from those of the same solids synthesized using ceramic methods [3]may be obtained. Non-stoichiometric spinel-type (NSS) compounds are interesting examples of these unusual solids, which are employed both as solid state gas sensors [4] and as catalysts for the hydrogenation of CO as well as of many organic molecules [5]. A large number of different cations may enter the spinel structure [6], while the presence of an excess of divalent cations inside the structure considerably affects both the texture and the catalytic properties of these phases.

The aim of this work was to study the formation of NSS phases at low temperature as well as the changes in their structure and reactivity in relation to the composition and degree of non-stoichiometry. Starting from the Zn/Cr system, the modifications induced by the partial substitution of zinc ions were investigated in order to illustrate the ease with which the properties of NSS compounds can be manipulated to meet the demands of specific applications.

EXPERIMENTAL

All samples were prepared by coprecipitation at pH 8.0 by adding a solution of metal nitrates to a solution containing a slight excess of $NaHCO_3$ at 333K. The resulting precipitate was washed to minimum Na-content, dried at 363K and calcined at 653K for 24h. Different temperatures and heating atmospheres were also investigated. XRD analyses were carried out by means of a Philips PW 1050/81 diffractometer and Ni-filtered CuK_α radiation. The lattice constants were determined by least square refinements, and the quantitative analysis of oxides was performed according to the method developed by Klug and Alexander [7]. Cation distribution in the cubic spinel-phases was evaluated on the basis of the I400/I440 ratio by an extension of the Bertaut method [8]. XPS experiments were carried out using a Perkin-Elmer PHI 5400 ESCA system. A C.Erba Sorptomatic 1826 apparatus with N_2 adsorption was used to measure the surface area, whereas copper surface area was determined by reaction with N_2O. IR spectra were recorded using a Nicolet MXI FT spectrometer.

The CO hydrogenation catalytic tests were carried out by means of a Cu-lined fixed-bed tubular reactor operating at 6.0MPa and 500-700K, with GHSV= $16,000h^{-1}$ and a gas mixture, $H_2/CO/CO_2$, equal to 65:32:3 (v/v). Reaction products were analyzed by gas-chromatography. The catalysts had been previously activated in-situ by hydrogen diluted in nitrogen. Oxoaldehyde mixture hydrogenation catalytic tests were performed in an autoclave at 7.0MPa and at a temperature range of 400 to 500K, after having previously activated the catalysts [9]. The reaction was controlled by periodically drawing   small samples, which were then analyzed by gas-chromatography. After reaction, all catalysts were cooled at r.t. under a nitrogen flow.

RESULTS AND DISCUSSION

The calcined samples with a Zn/Cr ratio ranging from 33:67 to 50:50 are characterized by the presence of only one microcrystalline spinel-type phase, while crystalline ZnO is also present in those with higher ratios. However, at Zn/Cr ratios >50:50, quantitative XRD determinations show that the amount of crystalline ZnO is always smaller than that expected for a simple phase composition $ZnO + ZnCr_2O_4$ (Fig. 1). Given the substantial analogies with the XRD spectra of some non-stoichiometric   mixed oxides [10], and assuming that the undetected ZnO is inside the

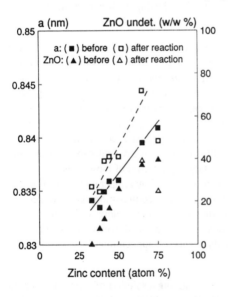

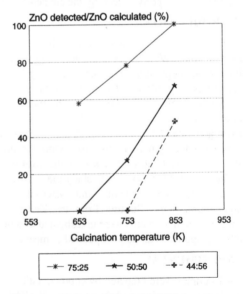

Fig. 1. Amount of undetected ZnO and lattice parameter *a* of the NSS phases (before and after reaction).

Fig. 2. Percentage of ZnO detected as a function of calcination temperature for samples with different atomic ratios.

spinel-type phase, the general formula $Zn_{1+y}Cr_{2-2y/3}O_4$ may be adopted, where y depends on both the Zn/Cr ratio and the calcination temperature [11]. On the basis of an X-ray full profile-fitting, it has been suggested that excess Zn^{2+} ions are located in the octahedral sites of the lattice, with the nearest tetrahedral sites being left vacant. This substitution implies a progressive structural change from the normal spinel structure to a rock-salt type structure and a corresponding increase in the metal/oxygen ratio from 3/4 to 1. Bulk data indicate a progressive variation away from the properties of the stoichiometric spinel, while surface properties show a remarkable variation as soon as departure from stoichiometry occurs [13]. For example, the lattice parameter a shows a regular trend indicating an expansion of the NSS cell (Fig. 1).

As a function of the heating conditions, Zn/Cr NSS phases may come to be formed by amorphous chromate intermediate decomposition or by direct reaction of the oxides [12]. However, the NSS solids are non-equilibrium phases and evolve, with increasing temperature or time-on-stream, towards stoichiometric $ZnCr_2O_4$, with a parallel segregation of ZnO (Fig. 2). The catalytic data show that NSS is the active phase for methanol synthesis, with a decrease in activity for the samples in which a side phase ZnO was also observed; furthermore, if methanol productivity is plotted as a function of zinc surface content (as determined by XPS analysis) a linear correlation is obtained up to Zn/Cr= 41:59 (Fig. 3). In our reaction conditions, no zinc surface enrichment was detected by XPS [14], in agreement with the regular trend of the lattice parameter after reaction (Fig. 1), with the exception of the catalyst richest in zinc, for which considerable ZnO segregation was reported.

The partial substitution of the zinc ions with cobalt or copper ions strongly modifies the catalytic behaviour of the NSS catalysts (Fig. 4). The presence of 2% cobalt causes a considerable decrease in methanol productivity (50% ca.), without any change in selectivity. It is worth noting that no traces of methane and other hydrocarbons were detected for this sample, while for a cobalt content of 8% only the methanation reaction was observed. On the basis of the FTIR spectra of the samples after activation and contact with H_2 or CO, the significant decrease in catalytic activity of the 2% cobalt sample with respect to that of the Zn/Cr sample may be related to a decrease in the amount of

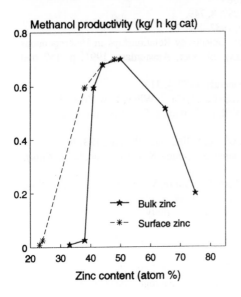

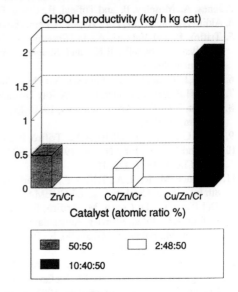

Fig. 3. Methanol productivity for the Zn/Cr catalysts, as a function of the bulk and surface compositions (React. temp.= 583K).

Fig. 4. Methanol productivity for Zn/Cr, Co/Zn/Cr and Cu/Zn/Cr catalysts (React. temp.= 573K).

active adsorbed hydrogen. This effect cannot be easily related to a "geometric effect" and seems to be reasonably ascribable to a strong perturbation of the electronic structure of the bulk catalyst. The tendency of the 8% cobalt sample to form hydrocarbons may instead be related to the presence of metallic particles of cobalt and consequently to its ability to dissociate CO.

The progressive substitution of zinc with copper ions gives rise to considerable differences in catalytic activity as a function of copper content. However, two general behaviours may be observed: 1) at Cu/Cu+Zn ratios of up to 0.5, the presence of copper considerably increases the activity in methanol synthesis; 2) at higher ratios, a dramatic deactivation is reported. Zinc-rich catalysts exhibit methanol productivities comparable to the highest values reported in the literature, although maximum activity is observed for temperatures intermediate between those of the low- and those of the high-pressure methanol catalysts. This increase in activity may be ascribed to the presence of copper ions inside the non-stoichiometric spinel-type structure, which favours the development of Cu-Zn synergetic effects.

Copper-rich catalysts exhibit lower values for both catalytic activity and CO chemisorption capacity, while showing, however, a surprisingly high activity in the hydrogenation of crude oxoaldehyde mixtures [9] or of maleic anhydride [15,16]. Again, the increase in activity and/or selectivity cannot be ascribed to differences in physical properties, but must be associated to the simultaneous presence of both copper and other divalent ions inside the spinel-type structure [15]. It is noteworthy that these catalysts exhibit better performances than the commercial catalysts tested under the same conditions [9,16].

Financial support from the Italian National Research Council (CNR, Rome) is gratefully acknowledged.

REFERENCES

1. Burton, J.J. and Garten, R.L. (Ed.s): "Advanced Materials in Catalysis", Academic Press, New York, 1977.
2. Yanagida, H.: Angew. Chem. (Engl. Ed.), 1988, 27, 1389.
3. Jones, A., Moseley, P. and Tofield, B.: Chem. Brit., 1987, 8, 749.
4. Aldinger, I. and Kalz, H.J., Angew. Chem. (Engl. Ed.), 1987, 26, 371.
5. Trifirò, F. and Vaccari, A.: in "Structure-Activity and Selectivity Relationships in Heterogeneous Catalysis" (Grasselli, R.K., and Sleight, A.W., Ed.s), Elsevier, Amsterdam, 1991, p. 157 and references therein.
6. Hill, R.J., Craig, J.R. and Gibbs, G.V.: Phys. Chem. Minerals, 1979, 4, 317.
7. Klug, H.P. and Alexander, L.E.: "X-Ray Diffraction Procedures", Ch. 7, Wiley, New York, 1974.
8. Bertaut, E.F.: C.R. Acad. Sciences, 1950, 231, 213; Weil, L., Bertaut, E.F. and Bochirol, J.: J. Phys. Radium, 1950, 11, 208.
9. Braca, G., Raspolli Galletti, A.M., Trifirò, F. and Vaccari: A., Italian Pat., 1989, n. 21831A.
10. Hojlund Nielsen, P.E.: Nature, 1977, 267, 822; Von Laqua, N., Dudda, S. and Revtlec, B.: Z. Anorg. Allg. Chem., 1977, 428, 151.
11. Di Conca, M., Riva, A., Trifirò, F., Vaccari, A., Del Piero, G., Fattore, V. and Pincolini, F., in "Proc. 8th Int. Congr. Catalysis", 2, DECHEMA, Frankfurt an Main, 1984, p. 173.
12. Del Piero, G., Di Conca, M., Trifirò, F. and Vaccari, A.: in "Reactivity of Solids" (Barret, P. and Dufour, L.C., Ed.s), Elsevier, Amsterdam, 1985, p. 1025.
13. Bertoldi, M., Fubini, B., Giamello, E., Busca, G., Trifirò, F. and Vaccari, A.: J. Chem. Soc., Faraday Trans. 1, 1989, 84, 1405.
14. Errani, E., Trifirò, F., Vaccari, A., Richter, M. and Del Piero, G.: Catal. Lett., 1989, 3, 65.
15. Castiglioni, G.L., Gazzano, M., Stefani, G. and Vaccari, A.: in " Heterogeneous Catalysis and Fine Chemicals", Elsevier, Amsterdam, 1993, in press.
16. Castiglioni, G.L., Fumagalli, C., Lancia, R., Messori, M. and Vaccari, A.: Chem. Ind. (London), 1993, 510.

Materials Science Forum Vols. 152 - 153 (1994) pp. 391-394
© *1994 Trans Tech Publications, Switzerland*

SYNTHESIS AND PROPERTIES OF HIGH SURFACE AREA Ni/Mg/Al MIXED OXIDES VIA ANIONIC CLAY PRECURSORS

D. Matteuzzi [1], F. Trifirò [1], A. Vaccari [1], M. Gazzano [2] and O. Clause [3]

[1] Dipartimento di Chimica Industriale e dei Materiali, Viale del Risorgimento 4,
I-40136 Bologna, Italy

[2] CSFM (CNR), Via Selmi 2, I-40136 Bologna, Italy

[3] Institut Français du Pétrole, 1 Avenue de Bois-Préau, BP 311,
F-92506 Rueil-Malmaison, France

Keywords: Ni/Mg/Al Mixed Oxides, Anionic Clays, Hydrotalcite, Spinels, Nickel Reducibility

ABSTRACT

The flexibility of the hydrotalcite-type (HT) structure is clearly demonstrated by the fact that pure Ni/Mg/Al HT anionic clays may be obtained by coprecipitation regardless of the Ni/Mg ratio. The thermal decomposition of these phases involves an initial loss of interlayer water followed by the elimination of hydroxide and carbonate , with some differences in this pattern being related to sample composition. The samples obtained by calcination at values up to 1073K exhibit a high surface area and stability, and are characterized by the low reducibility of the NiO particles. At higher temperatures, the formation of spinel phases has negative effects, so that surface area decreases and NiO reducibility increases. For Mg-containing samples, the latter effect is not always observed, on account of the formation of NiO/MgO solid solutions.

INTRODUCTION

High surface area homogeneous mixed oxides are employed in a number of industrial applications, such as adsorbents, catalysts, pigments, sensors, a.s.o. [1-3]. To obtain these materials, preparation techniques ensuring an intimate mixture of the components without high-temperature treatments must be employed. The synthesis of hydrotalcite-type (HT) anionic clays by coprecipitation, followed by thermal decomposition at moderate temperature, fulfils these requirements [4].

High surface area Ni/Al mixed oxides resulting from the thermal decomposition of HT coprecipitates are among the most widely investigated catalyst precursors on account of the remarkable properties of the final catalysts, such as high metallic dispersion and particle stability against sintering, even under extreme conditions [4]. However, the reasons for this stability are still being debated and, in particular, the phases formed as a function of calcination temperature are not fully understood. Furthermore, while several papers have been published describing the preparation and reducibility of high surface area NiO/MgO mixed oxides [5,6], much less is known about ternary NiO/MgO/Al$_2$O$_3$ mixed oxides and, in particular, about the modifications of the properties induced by increasing amounts of MgO. This topic is of great interest, if one considers that MgO is often to be found as a promoter in commercial nickel catalysts and that the progressive dilution of the transition metal may also shed light on the properties of Ni/Al systems.

For all these reasons, the reactions which take place during the preparation of Ni/Mg/Al catalysts deserve a more thorough scrutiny. This paper deals with the preparation and the properties of high surface area Ni/Mg/Al mixed oxides featuring different Ni/Mg ratios obtained from HT anionic clays. As in HT precipitates all cations are to be found inside the brucite-type layers, the specific properties of each element may be studied without any interference in relation to phase segregation and/or physical dishomogeneity.

EXPERIMENTAL

All samples were prepared by coprecipitation at pH 8.0 by adding a solution of the metal nitrates to a solution containing a slight excess of NaHCO3 at 333K. The resulting precipitate was washed to minimum Na-content, dried at 363K and calcined at different temperatures for 14h. XRD analyses were carried out using a Philips PW1050/81 diffractometer and Ni-filtered CuK_a radiation. The lattice constants were determined by least square refinements. Both oxide and spinel crystal sizes were determined by the Scherrer equation using the average values of (200), (220) and (220), (400) line widths, respectively. The Warren correction was used for instrumental line broadening, while the possible contribution of disorder effects and/or lattice strains was not taken into account. The quantitative composition of NiO/MgO solid solutions was determined on the basis of the lattice parameter a using Vegard's law [7], with a linear interpolation between the values reported in the ICDD files.

IR spectra were recorded according to the KBr disk technique using a Perkin-Elmer 1750 FTIR spectrometer. Surface area and pore volume were determined on a C. Erba Sorptomatic model 1700 apparatus, by means of N_2 adsorption. Thermogravimetric (TG) analyses were performed using a Perkin-Elmer TGS-2 thermobalance with a He-flow of $3dm^3/h$ and a heating rate of 600K/h. After various thermal treatments temperature-programmed reductions (TPR) of the samples were obtained by means of the previously described TG apparatus, employing a H_2-He mixture (37:63 v/v) at a flow rate of $3dm^3/h$ and a heating rate of 600K/h [8,9]. Apparent activation energy was calculated by analysing the reduction curves according to the empirical rate constant method [10]; so that the reduction had to be carried out isothermally within the various temperature ranges.

RESULTS AND DISCUSSION

The XRD spectra of the precipitates which were dried at 363K exhibit the presence, for all samples, of only a well crystallized hydrotalcite-type phase, in agreement with the M(II)/M(III) ratio and nature of the ions [4]. Notwithstanding, the crystallographic parameters a and c (Fig. 1) calculated on the basis of a rhombohedral symmetry show that the structure of Mg/Al HT anionic clay exhibits the best packing and that the partial substitution with Ni^{2+} ions increases the disorder more than the partial substitution of Mg^{2+} ions inside the Ni/Al structure. It may therefore be deduced that HT anionic clays feature an elastic structure, a fact which is compatible with the slight alterations encountered for different compositions.

TG analyses confirm the absence of side phases and show the presence of only two weight losses: the first, at about 503K, is attributable to the elimination of the water molecules from the interlayers, while the second, at about 673K, is due to the dehydroxylation of the brucite-type layers and to the elimination of the carbonate anions from the interlayers [4,9]. As magnesium content increases, this latter loss is displaced towards the higher temperatures and is only a partial one, on

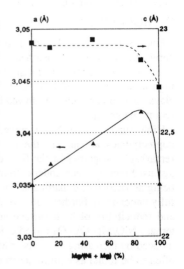

Fig. 1. Crystallographic parameters a and c of the HT precipitates dried at 363K

account of the higher affinity of the Mg^{2+} ions for the CO_2 [11,12]. This phenomenon is also observed when the samples are calcined at 923K for 14h; the FTIR spectra of these samples confirm the presence of residual carbonates.

Up to 523K, XRD analysis of the calcined samples (Fig. 2) reveals the presence of only a HT phase, with changes in the pattern being ascribable to the loss of water molecules from the interlayer. When temperature is increased up to 1023K, the only pattern to be observed is that of poorly crystallized oxide phase, which cannot, however, be identified with absolute certainty on account of the similarity of MgO and NiO patterns (ICDD 4-829 and 4-835, respectively). These data are in good agreement with the formation of rock-salt-type mixed oxides previously reported for Mg/Al clays calcined at 773-1073K [13]. A further increase in calcination temperature leads to a structure rearrangement, accompanied by the segregation of oxide and stoichiometric spinel phases.

The values of lattice parameter a for both the oxide and spinel phases as a function of calcination temperature are reported in Table 1. Parameter a values for all samples calcined at 723K were seen to be smaller than those reported for pure oxides, a fact which may be attributed to the presence of Al(III) ions inside the oxide lattice [14]. Lattice distorsion for Ni-rich samples becomes negligible at the higher calcination temperatures, while, by increasing Mg-content these distorsions are always to be found also in

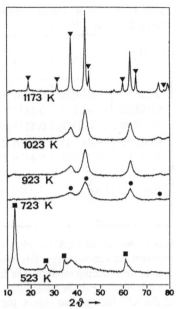

Fig. 2. XRD powder patterns of the sample Ni/Mg/Al = 34:37:29 (at. ratio %) calcined at different temperatures. (■) HT phase; (●) oxide phase; (▼) spinel phase.

the samples calcined at 1023K, which demonstrates that these samples are characterized by a higher tendency to retain the Al(III) ions inside the oxide lattice. $NiAl_2O_4$ and NiO were seen to be present in the Ni/Al sample calcined at T≥ 1173K, while the a values of the oxide phases in the Ni/Mg/Al samples clearly reveal the formation of NiO/MgO solid solutions, the quantitative composition of which may be determined by means of Vegard's law, using a linear interpolation between the a values reported in the ICDD data file [16].

Calcination temperature affects surface area to a greater extent than composition (Fig. 3). A two- to three-fold increase of the surface area is generally observed for the samples obtained by calcination at about 723-823K, while a dramatic decrease is associated with spinel formation at higher temperatures. The extent of this increase seems to suggest that exiting steam and CO_2 escape through holes in crystal surface without any extensive change in crystal morphology [17]. This mechanism may also explain the pore size distribution pattern observed in these samples, with a narrow peak centred around the most frequently occurring pore radius.

TPR analyses show that by increasing calcination

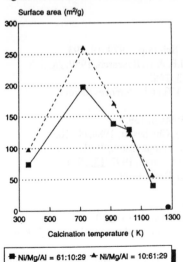

Fig. 3. Surface area as a function of calcination temperature for a Ni- and a Mg-rich HT precursor.

Table 1. Values of lattice parameter a for the oxide and spinel phases as a function of sample composition and calcination temperature

Ni/Mg/Al	Oxide phase					Spinel phase	
	723	923	1073	1173	1273	1173	1273
71:0:29	0.414(2)	0.4167(4)	0.4172(1)	0.4177(2)	n.d.	0.8051(3)	n.d.
61:10:29	0.415(1)	0.4173(6)	0.4173(3)	0.4178(4)	n.d.	0.8048(3)	n.d.
34:37:29	0.4165(7)	0.4172(6)	0.4173(2)	0.4184(4)	0.4195(3)	0.8056(5)	0.8076(6)
10:61:29	0.417(1)	0.417(1)	0.4184(9)	0.4207(4)	0.4205(2)	0.8080(5)	0.8079(7)
0:71:29	0.418(1)	0.4179(8)	0.418(1)	0.4216(4)	n.d.	0.8086(7)	n.d.

$NiO = 0.4177nm$ (ICDD 4-835); $MgO = 0.4213nm$ (ICDD 4-829); $NiAl_2O_4 = 0.8048nm$ (ICDD 10-339); $MgAl_2O_4 = 0.8083nm$ (ICDD 21-1152).

temperature of the Ni/Al samples, a decrease in reducibility occurs. On the basis of the values of the apparent activation energy, this trend can be attributed to a decrease in the accessibility of the Ni^{2+} ions to the reducing mixture. The formation of a spinel phase has a negative effect, giving rise to a remarkable increase in reducibility. With regards the Mg-containing samples, however, the effect of the Al^{3+} ions present within the oxide lattice has to be taken into consideration. The increase in reducibility associated with spinel formation in these samples is not necessarily to be observed as a function of the formation of NiO/MgO solid solutions. In previous papers [9,18], a model has been proposed for Ni/Al mixed oxides obtained from HT anionic clays involving the formation of NiO and Ni-doped alumina phases, which strongly interact with a spinel-type phase present at their interface. The validity of this model may also be extended to Ni/Mg/Al mixed oxides, with some adaptations being required depending on composition and on the phases which are formed at different calcination temperatures.

REFERENCES

1. Dadyburior, D.D., Jewur, S.S. and Ruckenstein, E.: Catal. Rev. Sci. Eng., 1979, 19, 293.
2. Lew, S., Jothimurugesan, K. and Flytzani-Stephanopulos, M.: I.E. & C. Research, 1989, 28, 535.
3. Suresh, K., Kumar, N.R.S. and Patil, K.C.: Adv. Mater., 1991, 3, 148.
4. Cavani, F., Trifirò, F. and Vaccari, A.: Catal. Today, 1991, 11, 173 and references therein.
5. Bond, G.C. and Sarsam, S.P.: Appl. Catal, 1988, 38, 365.
6. Arena, F., Licciardello, A. and Parmaliana, A.: Catal. Letters, 1990, 6, 139.
7. West, A.R.: "Solid State Chemistry and its Applications", Wiley, Chichester, 1984, ch. 10.
8. Clause, O., Gazzano, M., Trifirò, F., Vaccari, A. and Zatorski, L.: Appl. Catal., 1991, 73, 217.
9. Clause, O., Rebours, B., Merlen, E., Trifirò, F. and Vaccari, A.: J. Catal., 1992, 133, 231.
10. Delmon, B.: "Introduction à la Cinétique Hétérogène", Technip, Paris, 1969.
11. Ross, G.J. and Kodama, H.: Amer. Miner., 1967, 52, 1036.
12. Brindley, G.W. and Kikkawa, S.: Amer. Miner., 1979, 64, 836.
13. Sato, T., Wakabayashi, T. and Shimada, M.: I.E.&C. Prod. Res. Dev., 1986, 25, 1
14. Ross, J.R.H.: in "Catalysis, Specialist Periodical Reports" (Bond, G.C. and Webb, G., Ed.s), 7, Royal Society of Chemistry, London, 1985, p. 1.
16. Clause, O., Goncalves Coelho, M., Gazzano, M., Matteuzzi, D., Trifirò, F. and Vaccari, A.: Appl. Clay Sci., 1993, 8, 1.
17. Reichle, W.T., Kang, S.Y. and Everhardt, D.S.: J. Catal., 1986, 101, 352.
18. Beccat, P., Roussel, J.C., Clause, O., Vaccari, A. and Trifirò, F.: in "Catalysis and Surface Characterisation" (Dines, T.J., Rochester, C.H. and Thomson, J., Ed.s), The Royal Society of Chemistry, Cambridge, 1992, p. 32.

AUTHOR INDEX

Achddou, J.C. 149
Alberti, G. 87
Alcantara-Rodríguez, M. 379
Allali, N. 271
Allegre, J. 351
Angell, C.A. 371
Arnaud, G. 351
Ayral, A. 267

Babonneau, F. 313
Baffier, N. 297
Banse, F. 313
Barboux, P. 331
Bassas, J. 197
Bastians, Ph. 305
Béguin, F. 335
Bem, D.S. 183
Besenhard, J.O. 13
Besse, J.P. 343
Bhardwaj, C. 115
Bhuvanesh, N.S.P. 175
Biedunkiewicz, A. 263
Borthomieu, Y. 201
Boudes, L. 351
Bouquet, V. 205
Brec, R. 143, 297
Breysse, M. 221
Brohan, L. 245
Bujoli, B. 365
Burel, L. 205

Cahen, D. 187
Campet, G. 217
Casañ-Pastor, N. 193, 197, 293
Chernyak, L. 187
Chevrel, R. 205
Chien, S. 35

Clause, O. 391
Clearfield, A. 115
Cody, J.A. 35
Colombet, P. 335
Conanec, R. 305
Coowar, F. 213
Cors, J. 205
Cot, L. 149, 267

Dabadie, T. 267
Dance, I. 137
Danot, M. 271
Davies, P.K. 277
de los A. Rodríguez-Rodríguez, L.
.... 379
Decroux, M. 205
Delmas, C. 131, 201, 217
Deloume, J.-P. 327
Deloume, J.P. 221
Demourgues-Guerlou, L. 201
Deniard, Ph. 297
DiSalvo, F.J. 209, 289
Doeuff-Barboux, S. 313
Durand, B. 221, 327
Durand, S. 149

Ennaciri, A. 331
Even-Boudjada, S. 205

Favard, J.F. 271
Férey, G. 125
Fernon, V. 335
Fievet, F. 339
Figlarz, M. 55
Fisher, K. 137
Forano, C. 343
Fréchet, J.M.J. 209, 289

Fuertes, A. 193, 197, 293
Fujiki, Y. 251
Furdin, G. 281, 347

Gartsman, K. 187
Gazzano, M. 391
Golden, J.H. 209, 289
Goloub, A. 271
Gomez-Romero, P. ... 193, 197, 293
Gopalakrishnan, J. 175
Grange, P. 305
Granier, W. 351
Guan, J. 109
Guenane, M. 343
Guenot, J. 339
Guizard, C. 149, 267
Guo, J.D. 99
Guyomard, D. 213

Han, S.D. 217
Harlé, V. 221
Harris, K.D.M. 155
Henry, M. 355
Hérold, A. 281
Hérold, C. 347
Holloway, J.R. 371
Houenou, P. 255
Houmes, J.D. 183
Hu, Y. 277
Hudson, M.J. 359

Ibers, J.A. 35
In, M. 313

Jacobson, A.J. 1
Jasinski, W. 263
Jegaden, J.C. 205
Jezequel, D. 339
Jiménez-López, A. 379
Jouini, N. 339

Kahn-Harari, A. 297
Kasthuri Rangan, K. 175
Klatt, M. 281
Klimova, T.E. 309
Kooli, F. 375
Kremer, R.K. 383

Lassègues, J.C. 217
Laurent, Y. 305
Le Bideau, J. 365
Lee, Y.S. 237
Lefebvre, P. 351
Lenart, S. 263
Li, J. 99
Livage, J. 43
Loiseau, T. 125
Lurin, C. 267
Lyubomirsky, I. 187

Mabchour, A. 347
Maingot, S. 297
Mansuetto, M.F. 35
Marchand, R. 305
Marêché, J.F. 347
Mathieu, H. 351
Matteuzzi, D. 391
McKinnon, W.R. 213
McMillan, P.F. 371
Mérida-Robles, J.M. 379
Morawski, A.W. 263
Mosoni, L. 221
Mouchet, C. 149
Murcia Mascarós, S. 87

Neuhausen, J. 383
Nie, W. 267

O'Connor, C.J. 237
O'Keeffe, M. 371
Olivera-Pastor, P. 379

Ondoño-Castillo, S. 193, 197
Ortiz-Avila, C.Y. 115
Ouvrard, G. 143, 319

Pagnoux, C. 241

Palacín, M.R. 293
Palvadeau, P. 365
Payen, C. 365
Pereira-Ramos, J.P. 297
Pérez-Reina, F.J. 379
Petitjean, D. 281
Petuskey, W.T. 371
Pierre, A. 217
Piffard, Y. 241, 255
Pinnavaia, T.J. 109
Poeppelmeier, K.R. 163
Portier, J. 217
Pradel, A. 351
Prouzet, E. 143, 319

Rambaud, M. 271
Ribes, M. 351
Ribot, F. 313
Richard, M. 245
Riou, D. 125
Rioult, D. 267
Rives, V. 375
Rodríguez-Castellón, E. 379
Rouvière, J. 149
Rouxel, J. 143, 365

Sanchez, C. 313
Sankey, O. 371
Sasaki, T. 251
Scolnik, Y. 187
Sergent, M. 205
Siegel, G.G. 379
Solis, J.R. 309
Souto-Bachiller, F.A. 379
Stafsudd, O. 187

Stein Sr., E.W. 115
Stoyanova, R. 259, 323
Subramanian, M.A. 115
Suchaud, M. 241

Takenouchi, S. 251
Tarascon, J.M. 213
Taulelle, F. 241
Tchangbédji, G. 319
Tomczak, D.C. 163
Tournoux, M. ... 169, 241, 245, 255
Tremel, W. 383
Treuil, N. 217
Triboulet, R. 187
Trifirò, F. 387, 391

Ulibarri, M.A. 375
Uma, S. 175

Vaccari, A. 387, 391
Van Damme, H. 335
Verbaere, A. 241, 255
Vichot, A. 335
Vivani, R. 87
Vrinat, M. 221, 327

Watanabe, M. 81, 251

Whittingham, M.S. 99
Wolf, G.H. 371
Workman, A.D. 359
Wu, B. 237
Wysiecki, M. 263

Yamanaka, S. 69
Zah Letho, J.J. 255

Zavalij, P. 99
Zhecheva, E. 259, 323
zur Loye, H.-C. 183

KEYWORD INDEX

Acid
  Delithiation 259
  -Base 169
    Catalysis 163
    Properties 305
  -Exchange 245
Acidity, Bronsted ~ 175
Adiabatic Exfoliation 281
Akoxides 43
Alkali Polychalcogenide 35
ALPON 305
Aluminophosphates 305
Ammonolysis of Metallates 183
Amorphous Semiconductor 237
Anionic Clays 391
Arenesulfonates 335

Batteries 13
  Lithium.......................... 131, 213
  Ni-Cd................................... 131
Bonded and Encapsulated Fluorine ......
....125
Botallackite 69
Brannerite 277
Bronsted Acidity 175

Calcium Aluminate Hydrate 335
Catalysis, Acid-Base 163
Catalysts 305
Cation Mixing 213
Cation-Exchange 277
CdS Nanocrystallite 351
Ceramic(s) 293
  Nanophase ~ 149
Chalcogenide(s) 35, 143, 187
Chalcopyrites 187
Charge Reorganization 193
Chemical Deposition 263
Chevrel Phase 205, 209, 289
Chimie Douce 131
Clay 69
  Anionic ~ 391
Cluster 137

Co 137
Coatings, Titanium-Carbon 263
Cobalt and Iron Substituted Nickel
  Hydroxides 201
Cobalt-Nickel
  Hydroxide-Nitrates 323
  Oxide-Hydroxides 323
Colloidal Dispersions 1
Controlled Porosity 309
Copper 137, 293
  Oxides 193
Copper(II) 365
Coprecipitation 335
Crystal Structures 125
$Cs_{0.68}CuTiTe_4$ 35
$CsTiUTe_5$ 35
$Cs_4Zr_3Te_{16}$ 35
Cu 137, 293
  Oxides 193
Cu(II) 365
Cu_3NbSe_4 35
Cuprates 187

Dangling Bond 217
Decavanadate 375
Decomposition, Thermal 323
Deintercalation 143, 271
Deposits 197
Device 187
Diffusion 187
Double Hydroxides, Layered ~ 343

Electrochemical
  Insertion Reactions 13
  Oxidation 193
Electrodeposition 197
Electrodes 13
Electromigration 187
Electronegativity 355
Electronic Structure 383
Electronics 187
Electrophoresis 197
Exfoliation 1

Exfoliation *(continued)*
   Endothermic 281
   Exothermic 281
Expanded Graphite 281

Fe 137
$Fe_{0.12}V_2O_{5.15}$ 297
$FeWN_2$ 183
Flat 347
Fluorine, Bonded and Encapsulated .....
  125

Geometry of Templates 125
GeSiC Alloys 371
Grafting 69
Graphite 347
   -Acid Compounds 281
   Intercalation Compounds 281
   Thin Films 81

H-V-W-O Oxides 175
Hard Chemistry 371
Hexagonal MoO_3 277
High Pressure Synthesis 371
High Surface Area Oxides 163
Hybrid Materials 313
Hydrogenation, Selective 387
Hydrolysis, Thermal...................... 323
Hydrotalcite 375, 391
Hydroxides 131

$(InMo_3Se_3)_n$ 209
Incommensurate Solids 155
Infrared 201
Inorganic Polymer 209, 289
Intercalation 1, 209, 251, 297
   Complexes 379
   Compounds, Graphite 281
   ~/Deintercalation 175
   Lithium ~ 213
   Organic Anions ~ 343
   Reduction Percolation 347
   Self-Catalysed ~ 359
Intermetallics 237
Interstratification 201

Ion Condensation 143, 319
Ion Exchange 241, 251, 255,
   271, 323, 323, 359
   of Metal Phosphonates 115
Ion-Insertion 277
Ionic Exchange 271
Iron Vanadium Oxide 297
Iron(II) 365
Irreversible Exfoliation 281

Junction 187

$KCuZrQ_3$ 35
KCu_2NbSe_4 35
$K_2CuNbSe_4$ 35
K_3NbSe_4 35
$K_4Hf_3Te_{17}$ 35
$K_4Ti_3S_{14}$ 35
Knoevenagel Condensation 305
Lamellar
   Perovskites 169
   Double Hydroxides 335

Laser Ablation 137
Layer(ed)
   and Chain Structures 1
   Compounds, Metal Rich 383
   Double Hydroxides 343
   Perovskite Oxides 175
   Perovskites 245
   Structure(s) 69
      Pillared 343
   Titanate 251
   Zirconium Phosphates 379
$(LiMo_3Se_3)_n$ 209, 289
Li-Selectivity 81
$LiCoO_2$ 259
$LiNiO_2$ 131
$Li_xNi_{2-x}O_2$ 259
Liquid Crystal, Lyotropic 267
Lithium
   Aluminum Oxides 163
   Batteries 131, 213
   -Cobalt-Nickel Spinels 323
   Exchange with Protons 259

Insertion 217
Intercalation 213
Low Angle X-Ray Diffraction 267
Low-Temperature Preparation 387
Lyotropic Liquid Crystal 267

Madelung Potentials 355
Manganese Oxide 213
Mass Spectrometry 137
Mesoporous Materials 81
Metal 137
  Phosphonates
    Synthesis 115
    Self Assembly of 115
  Rich Layer Compounds 383
Metallates, Ammonolysis of 183
Methanol 387
Micronic 347
Microporous Compounds 125
Mixed Conductor Compounds 13
Mixed Metal Oxides 309
Mn 137
$MnWN_2$, α- ~ 183
$MnWN_2$, β- ~ 183
Molecular Dynamics 155
Molecular Precursors 43
Molten Salts, Reactions in 327
Monodisperse Particles 339

$Na_2Ti_2Se_8$ 35
$NaCuZrQ_3$ 35
$NaNiO_2$ 131
Nano-Sized Crystallites 327
Nanocrystallite 217
  CdS.. 351
Nanophase Ceramics 149
NASICON and $KTiOPO_4$-Type
  Phosphates 175
Neutron Diffraction 293
Nickel 137
  -Cd Batteries 131
  Hydroxides, Cobalt and Iron
    Substituted 201
  /Mg/Al Mixed Oxides 391
  Reducibility 391

Niobium Diphosphate 255
Non-Stoichiometry 387
Nonlinear Optics 351

Optical Properties 379
Organic Anions Intercalation 343
Organoceramics 335
Oxide Precursors 183, 293
Oxides 355
Oxides Fluorides 125
Oxygen Deficiency 213
Oxygen Doping 193
Oxyhydroxides 131
Oxynitrides 305
Oxysulfide 319

Partial Charge Model 355
$PbMo_6S_8$ 205
Perovskite Oxides, Layered 175
Perovskites 293
  Lamellar...................................... 169
  Layered...................................... 245
pH Dependant Reactions 125
Phase Equilibria 193
Phosphatoantimonic Acids 169
Phosphonates 365
Pillared Layered Structures 343
Pillaring 69, 375
Poly(alkyl-aryl)sulfonates 335
Polychalcogenide 35
Polymer, Inorganic.............. 209, 289
Polyols 339
Polypyrrole 313
Potassium Iron Disulfide 271
Precursors
  Molecular.................................... 43
  Oxide 183, 293
Preparation Method 339
Proton Conduction of
  Sulfophosphonates 115
Protonic Conductivity 241
Protonic Oxide 251

Quaternary 35

Reactions in Molten Salts 327
Reactive Flux Method 35
Reconstruction 375
Relations Synthesis-Structure 125
Rheology 267
Rietveld 297
Ruthenium Extraction 359

Secondary Structure-Control 81
Selective Hydrogenation 387
Self Assembly of Metal Phosphonates
....115
Self-Catalysed Intercalation 359
Semiconductor(s) 187, 371
     Amorphous................................. 237
Shape-Control 81
Silica Gel 267
Silicoantimonic Acids 241
Siloxane 313
Sodium Borosilcate Glass 351
Soft Chemistry 55, 169, 209
     Structural Aspects of 55
     Thermodynamic Aspects of 55
Sol-Gel(s) 43, 293, 309, 313, 331
Sol-Gel Process 149, 351
Solid Acid 255
Solid Inclusion Compounds 155
Solid-Acid 251
Solid-State Acidity 355
Solvent Effects 331
Solvolysis 277
Spherical Particles 339
Spin Glass 237
Spinel-Type Catalysts 387
Spinels 391
Structural Aspects of Soft Chemistry ...
....55
Structure 137
     Determination 245
Sulfide 137
Sulphides 371
Superconductivity 193, 197
Superconductor 187, 205
Surface Area 309
Synthesis 293

Synthesis-Structure, Relations 125

TEM Diffraction 245
Ternary 35
     Nitrides 183
tert-Butyllithium 209
Theory 137
Thermal Decomposition 323
Thermal Hydrolysis 323
Thermodynamic Aspects of Soft
     Chemistry 55
Thin Films 289
     Graphite ~ 81
Thionine 379
Tin 313
Tin(IV) Hydrogen Phosphate 359
$TiO_2(B)$ 169
Titanate
     Fibers 81
     Layered....................................... 251
Titania 149
Titanium 293
     -Carbon Coatings 263
     Phosphates 331
Transition Metal
     Chalcogenides 1, 383
     Oxo-Polymers 313

V_2O_5 297
Vanadium 313, 319
Vanadium Phosphates 331
Varistors 339

WO_3 Polymorphs 55

X-Ray Diffraction, Low Angle 267
XAS 143

Zinc Oxide 339
Zirconia 149
Zirconium Phosphate(s) 379
     Layered ~ 379